ADVANCES
IN
NUCLEAR PHYSICS

Proceedings of the International Symposium on

ADVANCES IN NUCLEAR PHYSICS

(Fifty Years of Institutional Physics Research in Romania)

Bucharest, Romania   9 – 10 December 1999

Organized by

Horia Hulubei National Institute of Nuclear Physics
and Engineering (IFIN-HH)

Sponsored by

UNESCO Venice Office (UVO-ROSTE)
JINR Dubna
National Agency for Science, Technology and Innovation
Romanian Commercial Bank, via PROFIZICA NGO

Editors

Dorin Poenaru and **Sabin Stoica**

Horia Hulubei National Institute of Nuclear Physics
and Engineering, Bucharest, Romania

World Scientific
Singapore • New Jersey • London • Hong Kong

Published by

World Scientific Publishing Co. Pte. Ltd.
P O Box 128, Farrer Road, Singapore 912805
USA office: Suite 1B, 1060 Main Street, River Edge, NJ 07661
UK office: 57 Shelton Street, Covent Garden, London WC2H 9HE

British Library Cataloguing-in-Publication Data
A catalogue record for this book is available from the British Library.

ADVANCES IN NUCLEAR PHYSICS
Proceedings of the International Symposium

ISBN 981-02-4276-X

Printed in Singapore by Uto-Print

PREFACE

The interest in understanding the physical world that we live in, the origin of its formation and evolution, is reflected by the enormous amount of activities that are carried out in Europe, USA and Japan to develop further advanced theoretical studies and set up powerful research facilities providing beams of radioactive nuclei of various kinds, as well as beams of extremely large energies. Complex and large detector arrays with improved technical capabilities are built around these facilities.

On the other hand, the energy spectrum of cosmic rays exceeds the upper limits provided by artificial accelerators. The installation of an extremely large area (3000 km^2) detector array AUGER, named after the famous French cosmic ray physicist Pierre Auger, will be used in the future by an international collaboration to study the highest particle energies in the Universe.

The results of this continuous effort can be seen every year. Recently, dramatic progress has been reported in the field of superheavy nuclei, cold binary and ternary fission, nuclear structure, relativistic heavy ion physics, nuclear astrophysics, cosmic rays physics, high energy and particle physics, atomic physics, etc. Some of these scientific achievements are described in the present Proceedings of the Symposium held in Bucharest.

In 1949 the Romanian Physicist Horia Hulubei, Correspondent Member of the Academy of Sciences in Paris, Directeur de Recherches at CNRS, France, and a former J. Perrin's Ph. D. student, inaugurated the Institute of Physics of the Romanian Academy which shortly became the INSTITUTE OF ATOMIC PHYSICS (IFA).

The Institute of Physics and Nuclear Engineering (IFIN), was created as the largest fraction of IFA during a reorganization of research activities in 1976. Since 1996 it has continued to exist under its current name, Horia Hulubei National Institute of Physics and Nuclear Engineering (IFIN-HH).

Amid political transformations which started in 1990, a number of wrong decisions concerning Science and Technology in Romania have been taken. Two Institutions (the State Committee for Nuclear Energy, and the National Committee for Science and Technology), existing in many countries under various names (*e.g.* Department of Energy and National Science Foundation in USA, CEA and CNRS in France) have been disbanded. Their counterparts in Poland are still in operation and are funding research projects (0.47% of GNP was allocated in 1998, compared to 0.13% in Romania). Moreover, our Ministry of Research was last year transformed into a lower ranking Agency. A much smaller country, Croatia, does have a Ministry of Research. We hope that our politicians, who believe that Romania will join the European Union, will realize the importance and strategic role of research activity, and will rectify this mistake sooner or later.

We decided to mark the 50th Anniversary of Institutional Research in Romania by organizing an International Symposium, intended to provide a forum for discussing the present status and the perspectives of different experimental and theoretical fields of Nuclear-, Particle-, and Atomic Physics, as well as some particular applications. The former directors of IFIN-HH, A. Berinde, A. Calboreanu, M. Caprini, M. Ivaşcu, M. Oncescu, G.Pascovici, M. Petraşcu, M. Petrovici, and F. Scarlat were our local guests of honour.

We are proud to see that despite the financial difficulties one has to face at home these years, our scientists are active members of an international community, which is able to maintain and increase via an enhanced scientific cooperation a high efficiency of research work, according to the well-established standards in operation. The new PACS number *23.70.+j Heavy-particle decay* was introduced following the development of a new field of nuclear physics, in which an international Romanian-German team had a leading role, starting with the predictions in the most cited paper, by Săndulescu, Poenaru and Greiner, published in 1980, four years before the first experimental confirmation. Several Departments of the Institute fulfil the conditions required for a Center of Excellence. In fact the analysis of scientometric parameters during the last ten years shows a significant increase in the quality and quantity of the Institute basic research activities, which is beyond doubt the result of an international effort.

This is one of the reasons why prestigious, very active scientists, enthusiastically responded to our invitation to come to Bucharest and present in their invited talks the status and perspectives of very exciting theoretical and experimental fields of research. The number of the oral- and poster contributions is so large that we will publish the corresponding papers in a dedicated issue of the *Romanian Journal of Physics.* We are very grateful to our guests.

We would also like to thank the International Advisory Committee consisting of Y. Abe (Kyoto), A. Antonov (Sofia), J. Aysto (Jyväskylä), P. von Brentano (Köln), D. G. Cacuci (Karlsruhe), R. F. Casten (Yale), A. Faessler (Tübingen), E. Gadioli (Milano), S. Galés (Orsay), M. Gavrilă (Harvard), A. Gelberg (Köln), W. Gelettly (Surrey), W. Greiner (Frankfurt), D. Guerreau (GANIL), J. H. Hamilton (Nashville), W. Henning (GSI), H. Horiuchi (Kyoto), E. Hourany (Orsay), M. Itkis (Dubna), C. Kalfas (Demokritos), H.V. Klapdor-Kleingrothaus (Heidelberg), K. L. Kratz (Mainz), R. J. Liotta (Stockholm), G. Münzenberg (GSI), W.von Oertzen (Berlin), H. Rebel (Karlsruhe), R. A. Ricci (Padova), G. Royer (Nantes), A. Sobiczewski (Warsaw), Sir A. Wolfendale (Durham), N. Zamfir (Yale), and A. Zeldes (Jerusalem), for their excellent recommendations on the choice of speakers.

Special thanks are due to UNESCO (UVO-ROSTE) Venice Office, JINR Dubna, the Romanian Commercial Bank, and ANSTI (the National Agency for Science, Technology and Innovation) — the main sponsors of the Symposium, which could not have been organized without this financial support. We also acknowledge the initiatives of our colleagues Prof. A. A. Răduţă and I. Vâţă of seeking financial support from the Venice regional Office for Science and Technology of UNESCO, and JINR Dubna, respectively.

On behalf of the Organizing Committee: Gh. Mateescu (Chairman), D. N. Poenaru (Co-chairman), V. Grecu, E. Drăgulescu, A. Enulescu, and S. Stoica, I would like to thank the technical staff (A. Olteanu, C. Ţucă, A. Aniţoaie, I. Ştefănescu, D. Sandu, P. Taină, M. Iordache, and P. Marian).

Dorin N. Poenaru

CONTENTS

OPENING CEREMONY

INAUGURAL TALK

SUPERHEAVY NUCLEI

NEW FISSION MODES

ASTROPHYSICS AND COSMIC RAYS

PARTICLE AND HIGH ENERGY PHYSICS

HADRONIC MATTER

NUCLEAR STRUCTURE AND REACTIONS

ATOMIC PHYSICS

APPLICATIONS AND HISTORY

MESSAGE FROM THE PRESIDENT OF ROMANIA, HIS EXCELLENCY EMIL CONSTANTINESCU, TO THE PARTICIPANTS AT SYMPOSIUM

Dear Professors, Researchers and Physicists,
Dear Guests, Distinguished Audience,

I wish to tell you first of all how deeply sorry I am for being unable to attend your meeting. As you know, we are right now preparing our participation in the summit of the European Union member and candidate countries that will be taking place in Helsinki. I have mentioned this event, because ever since its establishment, the Institute of Physics and Nuclear Engineering has been one of the most active links connecting Romania to that Europe of spiritual values to which our country now has a chance to integrate once and for all.

Under the dictatorship, your institute, aside from conducting an impressive scientific activity, has been one of the few privileged bulwarks of free thought where communist ideology had to keep on the defensive.

Now, the Institute of Physics and Nuclear Engineering has turned fifty. It is the age of full-blown ripeness in an individual. To a scientific community, it is both a tradition and a challenge.

Unfortunately, many people still think one does not need more than a pencil and a sheet of paper to work in modern physics; that a physicist nothing but sits in his or her laboratory and compares the filament of a bulb with the atomic nucleus. Should this ridiculous preconception be true, should such a scientist exist, he would turn out to be a prisoner in a jail of his own making — the kind of jail of which it is hardest of all to break out.

Being a scientist at the end of this millennium means above all communicating; keeping abreast with the latest results in one's field and being able to check them; publishing the results of one's own researches and receiving feedback. Individual limits thus sim-

ply vanish and give way to a marvelous sense of freedom, of mutual understanding, and of the universal nature of knowledge. The international symposium "Advances in Nuclear Physics" you are holding today to mark the anniversary of the Institute of Physics and Nuclear Engineering, therefore, is not only the well deserved crowning of a career, but also a new step forward on the road of truth.

Romania has been through a difficult time over the past few years, and sectors such as education, culture, scientific research, and health-care unfortunately bear the brunt. I am well aware of the hardships Romanian researchers have to face by being paid only a tiny fraction of what their western counterparts earn for the same amount of work. I am aware of the Romanian's frustration with the much awaited economic recovery being so slow to kick off. I am painfully aware of the despondency of many Romanian intellectuals saying to themselves, "There's nothing left to do".

A Romanian philosopher of our times was right to ask himself, "What can we do when there is nothing left to do?" I would relate this question to a remark Werner Heisenberg made thirty years ago with respect to the changes that thinking patterns suffered with the advance of science. As the prominent scientist put it, a theory by simply being true is not enough to transform thinking patterns. Real revolutions are when, in the novel context, there's something new to do, when a fruitful activity emerges.

We are all of us in a situation in which "nothing left to do" is tantamount to going back to a time where everything was known and set beforehand, and time was turning in a circle. The alternative lies in Heisenberg's words: "There is something to do." To physicists, I know, these words have a particular significance.

I am therefore convinced that the Horia Hulubei Institute of Physics and Nuclear Engineering will go on being the same privileged place on the map of Romanian spirituality into the next millennium.

Emil Constantinescu

ADDRESS BY THE PRESIDENT OF THE EUROPEAN PHYSICAL SOCIETY

SIR ARNOLD WOLFENDALE FRS
President of the European Physical Society

Ladies and Gentlemen,

It is a pleasure, and an honour, to bring the congratulations of the European Physical Society to the Horia Hulubei National Institute of Physics and Nuclear Engineering on this, the 50th Anniversary of its founding. The Institute is well known for the quality of its work in both the academic and applied areas of its work.

My personal research is in the field of Cosmic Ray Astrophysics and, as such, I have an interest in Nuclear Physics; indeed, a few years ago I chaired the United Kingdom committee which funded this area, as well as Particle Physics and Astronomy. It is evident to me that Nuclear Physics is still an area of considerable interest and one that is an ideal training ground for young physicists going on to other areas, too. Returning to the EPS, I must report on some of its activities which have relevance to the Institute that we are honouring. An important one is the production of Position Papers which should be of value to those politicians and other policy formers who are concerned with the programmes and funding of Research Institutes. There are special problems for Eastern Europe which are related to the ending of the Cold War. There seems to be a view that Research Institutes - and Physics researchers elsewhere, too - can have their funds reduced without much effect on the nation in question. This is a pernicious view, and one that must be fought. Except, perhaps, for Institutes which were previously engaged almost entirely in Defence work, there is need for more funds, not less, as can be demonstrated.

In the fundamental area, one where the IFIN-HH has such international prestige, the rest of the world is not standing still. Thus, nations must try harder to keep up so as to do their share of the world's research and thereby to reap the benefits of prestige and to provide superbly trained researchers who can interpret and apply the research done elsewhere. The training element cannot be overstressed. Without adequately trained physicists a nation will be unable to found the 'high-tech' industries which are so necessary in the modern world. The needs in the applied area are equally great. Environmental problems require urgent solutions and organizations such as IFIN-HH have an important role to play. New technology for State-run enterprises is also a must.

The EPS has also turned its attention to the question of a 'Scientist's Oath', and to the production of biographies of famous European physicists. These biographies which are directed at youngsters in school - differ from the normal ones in that they give interesting personal details as well as the science. Romania will have its share. An important function here is to enable youngsters in Europe to know who are famous in countries other than their own. Finally, I must return to the question of national support for Physics. I was disturbed to learn how low such support was here. I realise that there are severe economic problems but the way ahead is not to destroy the seed-corn. With the impending 10th anniversary of the downfall of the previous regime the possibility arises for the Government to mark its stewardship by a celebratory increase in funding to a more realistic level.

Good luck for the future.

Sir Arnold Wolfendale

OPENING REMARKS

LÁNYI SZABOLCS
President of National Agency for Science, Technology and Innovation (NASTI)

Ladies and Gentlemen,

I feel personally honoured and pleased to welcome all the participants to the International Symposium "Advances in Nuclear Physics" on the behalf of National Agency for Science, Technology and Innovation. I would like also to thank you for your effort to join us in Bucharest and to add value to our symposium.

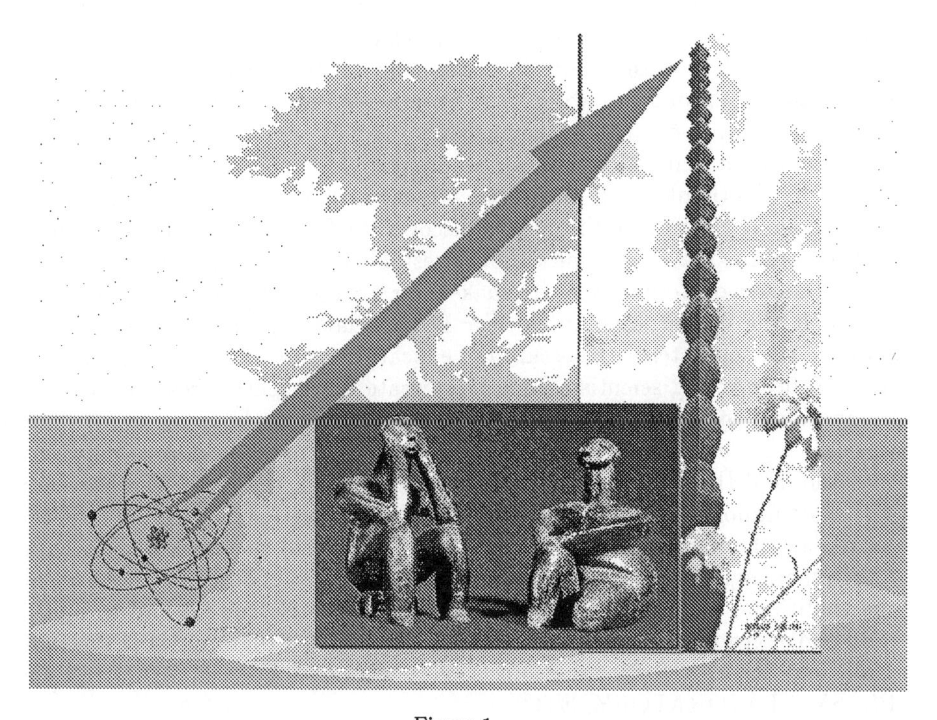

Figure 1.

National Agency for Science, Technology and Innovation has the great responsibility to lead the destiny of scientific research in a difficult period for Romania. Due to the lack of finance and decrease of the economy, the science which can drive the technology, is placed today a little bit in the shadow. But my intention is not to make now a progress effort on scientific research in Romania.

There are in this room a lot of people who were part of this report in the last fifty years in a special branch of science: nuclear physics. Recently, the main institute involved in this field, Institute of Physics and Nuclear Engineering (IFIN), has celebrated 50 years of activity.

On the large scale, this means almost nothing, but taking into account the short time from the birth of nuclear physics, this means history. And this history started earlier, in the years '30 with representative personalities as Alexandru Proca, Horia Hulubei, Şerban Ţiţeica and others precursors who has had as a sacred duty the desire to know the deepest secrets of the universe. During the time important effort was done in the universities of Bucharest, Cluj, Iasi and Timisoara to establish a Romanian Physics School, but the nucleus of nuclear research was built in Bucharest in the Institute of Atomic Physics (IFA). The most famous Romanian physicists, working in the labs and universities all around the world, were born here. Part of them are present in this symposium to confirm my sentences.

I don't want to mention names, because it could happen to forget someone, but I'd like to thank to all physicists, present or not present here, who were engaged along the time in this aim to discover the fundamental laws of nature and to confirm the Romanian tradition established thousands years ago by "The Thinkers of Cernavoda". To make their dream come true, in Cernavoda was built the first Romanian Nuclear Power Plant. Today our effort is to preserve the competence, to assure the education and training in the nuclear field, to bear young scientists and to integrate our scientific research in the international effort for the progress of science.

To be modest I'll establish the Romanian limits in research area between the atom and the "Infinity Column" passing through the roots (see Fig. 1). The most important value at present is the researcher itself, with his respect for life and human tresors, with his desire to perform the work and to find future challenges. Talking about future we try to keep good connections with the traditional partners (IAEA, CERN, NATO, IUCN Dubna) and to build new bridges to reach the European Union Research Programmes (FP5 and JRC) and the new Regional Projects as AUSTRON, DENSE MAGNETIC PLASMA LABORATORY, ETC.

As regarding FP5 and JRC our interest is to join both Fission and Fusion

Figure 2.

Projects and to find new ways to collaborate with different institute from JRC in specific interest area, to be part of international research community. Finally, I'd like to add my personally desire to sustain the scientists to dream the world leaded by the values inspiring science: love for creation, respect for life and human dignity.

I also wish you a successful, fruitful and pleasant meeting and a nice staying in Romania. Allow me to use this opportunity to wish all the best for you and for your family with the occasion of the Christmas Day, and the 2000 New Year (see Fig 2).

Lányi Szabolcs

THE 50 YEARS JUBILEE OF ROMANIAN NUCLEAR PHYSICS

DR. GH. MATEESCU, DIRECTOR GENERAL OF IFIN-HH

Distinguished Guests, Dear Colleagues, Ladies and Gentlemen,

I am honored to welcome in our midst Mr. Justin Tanase, the Counsellor of His Excellency the President of Romania, Mr. Emil Constantinescu, an academic and researcher, whom we regard as one of us. I am pleased to say that we also have here some members of Parliament: Senator Mihai Balanescu, Senator Oliviu Gherman, Deputy Aurel Sandulescu. I would also like to salute and to thank for being here with us the president of the National Agency for Science, Technology and Inovation, Professor Dr. Lànyi Szabolcs, and Dr. Dan Cutoiu, president of the National Commission for Nuclear Activities Control. A warm welcome as well to our foreign distingueshed guests and to the venerable veterans of our Romanian physics community, who have joined us for this anniversary. Also, let me, please to take the advantage of this opportunity to express my whole gratitude to Sir Arnold Wolfandale, the president of the European Physical Society, for his presence here. At last but not least, many thanks to the mass-media representatives which are present today at nuclear physics celebration.

The jubilee that we are marking here today is a milestone not only for the Horia Hulubei National Institute of Physics and Nuclear Engineering (IFIN-HH), but also for Romanian physics as a whole, for our research community, and, I dare say, for the entire community of intellectuals of this country. We all share the same interests and destiny, which the Institute of physics has assumed since its birth 50 years ago through the initiative of its founding father, Professor Horia Hulubei, a personality of European standing, a genuine school leader and a patriot. It has been expected of us, since the beginning, that we should make a contribution to the progress of human knowledge and bolster the economy, culture and education of this country. The pioneers who, a halfcentury ago, set off on this often arduous journey led by Hulubei, made this commitment quite explicitly.

What were the circumstances back in 1949 when Hulubei first established the Institute of Physics, then the Institute of Atomic Physics (IFA), which begot us all? It had been just four years since the end of the Second World War, the most devastating war in which Romania had ever been engaged. A major armament effort and operations on two distinct fronts had shattered the economy and the infrastructure was in tatters. Besides, a terrible famine had ravaged the country during the three years of peace. Between them, war and mass starvation, had taken a huge toll. Peace itself had come as a mixed blessing: Romania had slipped under the domination of a foreign power and military occupation. To cap it all, the country had to pay heavy damages to her erstwhile aggressor that, in 1940, had snatched large plots of her territory and which was then forcing on her its own political regime.

Consequently a considerable part of the political, military and academic elite was either in prison, or standing political trials, or, at best, sidelined. It was only the beginning of a hard trying period during which several generations of Romanians were sacrificed on behalf of foreign interests. In these difficult conditions, a number of leading intellectuals from different walks of science pleaded with the authorities and managed to squeeze the go-ahead for the establishment of modern research institutes, patterned after those that were operating abroad, particularly in the Soviet Union, a mandatory reference at the time. Anyway, it's the historians' job to examine the specific circumstances in which Romania's research institutes, including the Magurele Institute of Atomic Physics, were set up. The fact is that the newborn institutes - whether under control of the Academy, or of government bodies, as was IFA's case - gave Romania a chance of coming close to international scientific standards.

It is important to note that those initial research investments represented quite some money, especially in view of postwar hardships. The pioneers of Romanian research were quick to realize the opportunity and jumped at it, eager to serve not the political regime but scientific and technological progress and, of course, the Country's interests. However, owing to the regime rules, the Romanian scientists saw themselves isolated from the international scientific community, stifled by the all-powerful ideology, trammeled by redtape, baffled by spurious goals.

Still, Hulubei's dedication and diplomatic skill prevailed and the Institute stood up successfully. His team was soon able to produce remarkable achievements in a broad range of areas. They developed Romania's first laser and electronic computing machine; research groups were set up in atomic and nuclear physics, in particle physics, etc. The country's first cyclotrone and first research reactor, both of them expensive acquisitions from the Soviet Union, were mounted, adapted and developed at IFA; later on, a North+American tandem accelerator has also been installed and heavy ions accelerated; other facilities were added: the center for radioisotope research and production, the nuclear medicine center, etc.

Credit is also due to Hulubei for his promotion 30 years ago of the coun-

try's first nuclear power program (including the indigeneous production of nuclear fuel and of heavy water along with dedicated equipments for CANDU type NPP which entered into operation in 1996). Far from losing its interest and despite the country's curent difficulties, nuclear power remains a high-potential solution as the only way to be a country energetically independent, and to assert Romania's key role as a potential energy exporter in the region, particularly when its neighbors decide to close their Chernobyl-like reactors.

IFA, then IFIN-HH involved themselves in the country's economic development, providing specialized consultancy, non-conventional instrumentation and nuclear technical solutions in various fields of activity, such as medicine, farming, steel making, and the petroleum industry. The management teams that followed tried to continue the founder's policy of encouraging advanced research. Their persistent efforts even succeeded to some extent in breaking the isolation of Romanian physics by forging institutionalized ties with outstanding centers in western countries like Germany, France and Italy, but this came at the cost of concessions which carried many of them away from research into small-series production and services activity. However, this obedience to social command, as they perceived it at the time, ensured the survival of real scientific research while waiting for better times.

Dozens of our researchers from different fields of physics, as well as chemists, mathematicians, and engineers served on the faculty of the University of Bucharest and of the Bucharest Polytechnical Institute. Along the decades, hundreds of junior researchers from inside and outside the Institute earned their PhDs in the IFA system and with the help of IFA supervisors. A great number of physics students prepared their graduation theses at the Magurele center. In this respect, IFA and later on IFIN-HH successfully functioned as an institute of advanced studies, stimulating young people to further improve their knowledge by means of graduate studies. This major contribution IFA has made to the development of higher education was unfortunately neglected when the educational law was recently revised.

The political turnabout that took place in December 1989 drew cheers from the Magurele scientists and everyone expected the new authorities would reverse the previous neglect of scientific research and give it the better treatment that it certainly deserved. Except for a couple of years in the early 1990s these hopes never came true. The economic downturn, the lack of a long-term policy in science during our endless transition period had serious consequences: researchers number shrank, and so did the number of youths attracted to a scientific career; vast sectors of research and development went into disarray, and so on. The Institute has not been spared such mishaps. As a unit of advanced basic and applied research, IFIN-HH's projects of developing its experimental and applicative capacities were hit by a severe funding shortage. The Institute responded by revising its organizational structure to better meet research requirements and our offer as a supplier of nuclear applications and services has been reconsidered; we amplified also international R&D cooperation which our Government recognize to be Romanian first proved key factor on the way of European integration.

A jubilee is a good opportunity for planning for the future. Science, which substantially contributes to shaping the world of tomorrow and even of after tomorrow, cannot do without long-term plans. The country's future cannot be conceived without a developed Romanian science, with its own competitive facilities and an intense international cooperation giving us access to the highest level of world science and technology. Nuclear Physics is a forefront sector of research, not only as a trailblazer of contemporary science, but also thanks to its potential of providing valuable instruments and solutions to other research fields and to economy. Once its importance has been recognized, the next step should consist of earmarking adequate funding for its needs. Hoping that our decision-makers may change their minds about science, we are planning to further diversify our offer and prove that science is a profitable enterprise, in fact, the most profitable of all our long-term investments. In spite of our current plight, we are confident that we will succeed. We owe it not only to the founding fathers of fifty years ago, but also to the coming generations that will be right to blame us if we don't rise up to our forerunner's mark. It is in this spirit, distinguished Colleagues, that I now declare open our Symposion on "Advances in Nuclear Physics", wishing you all to make the best of it.

Have a good time and a good Conference! Thank you all for listening.

Gheorghe Mateescu

DEVELOPMENTS IN FISSION, FUSION, CLUSTER RADIOACTIVITY AND THE EXTENSION OF THE PERIODIC SYSTEM OF ELEMENTS

WALTER GREINER
Institut für Theoretische Physik, J.W. Goethe-Universität,
D-60054 Frankfurt, Germany

The extension of the periodic system into various new areas is investigated. Experiments for the synthesis of superheavy elements and the predictions of magic numbers are reviewed. Different ways of nuclear decay are discussed like cluster radioactivity, cold fission and cold multifragmentation, including the recent discovery of the tripple fission of ^{252}Cf. Furtheron, investigations on hypernuclei and the possible production of antimatter–clusters in heavy–ion collisions are reported. Various versions of the meson field theory serve as effective field theories at the basis of modern nuclear structure and suggest structure in the vacuum which might be important for the production of hyper– and antimatter. A perspective for future research is given.

There are fundamental questions in science, like e. g. "how did life emerge" or "how does our brain work" and others. However, the most fundamental of those questions is "how did the world originate?". The material world has to exist before life and thinking can develop. Of particular importance are the substances themselves, i. e. the particles the elements are made of (baryons, mesons, quarks, gluons), i. e. elementary matter. The vacuum and its structure is closely related to that. On this I want to report today. I begin with the discussion of modern issues in nuclear physics.

The elements existing in nature are ordered according to their atomic (chemical) properties in the **periodic system** which was developped by Mendeleev and Lothar Meyer. The heaviest element of natural origin is Uranium. Its nucleus is composed of $Z = 92$ protons and a certain number of neutrons ($N = 128 - 150$). They are called the different Uranium isotopes. The transuranium elements reach from Neptunium ($Z = 93$) via Californium ($Z = 98$) and Fermium ($Z = 100$) up to Lawrencium ($Z = 103$). The heavier the elements are, the larger are their radii and their number of protons. Thus, the Coulomb repulsion in their interior increases, and they undergo fission. In other words: the transuranium elements become more instable as they get bigger.

In the late sixties the dream of the superheavy elements arose. Theoretical nuclear physicists around S.G. Nilsson (Lund)[1] and from the Frankfurt school[2,3,4] predicted that so-called closed proton and neutron shells should counteract the repelling Coulomb forces. Atomic nuclei with these special **"magic" proton and neutron numbers** and their neighbours could again be rather stable. These magic proton (Z) and neutron (N) numbers were thought to be $Z = 114$ and $N = 184$ or 196. Typical predictions of their life times varied between seconds and many thousand years. Fig.1 summarizes the expectations at the time. One can see the islands of superheavy elements around $Z = 114$, $N = 184$ and 196, respectively, and the one around $Z = 164$, $N = 318$.

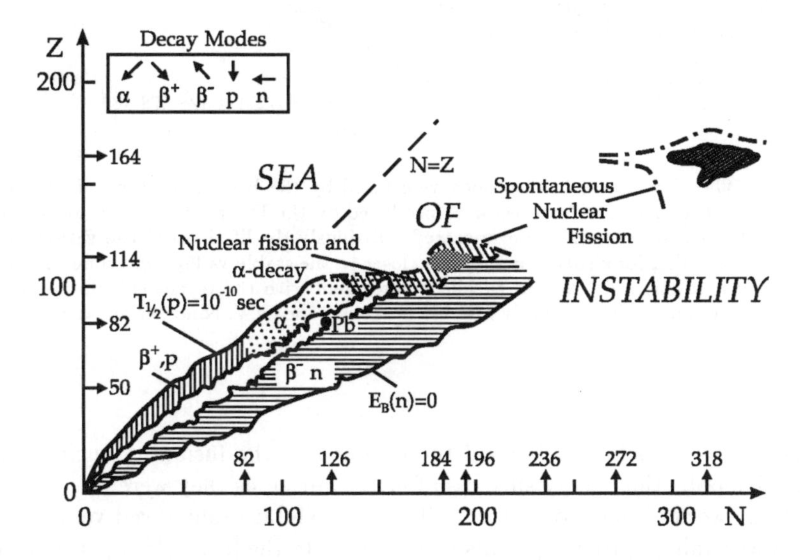

Figure 1. The periodic system of elements as conceived by the Frankfurt school in the late sixties. The islands of superheavy elements ($Z = 114$, $N = 184$, 196 and $Z = 164$, $N = 318$) are shown as dark hatched areas.

The important question was how to produce these superheavy nuclei. There were many attempts, but only little progress was made. It was not until the middle of the seventies that the Frankfurt school of theoretical physics together with foreign guests (R.K. Gupta (India), A. Sandulescu (Romania))[5] theoretically understood and substantiated the concept of bombarding of double magic lead nuclei with suitable projectiles, which had been proposed intuitively by the russian nuclear physicist Y. Oganessian[6]. The two-center shell model, which is essential for the description of fission, fusion and nuclear molecules, was developped in 1969-1972 together with my then students U. Mosel and J. Maruhn[7]. It showed that the shell structure of the two final

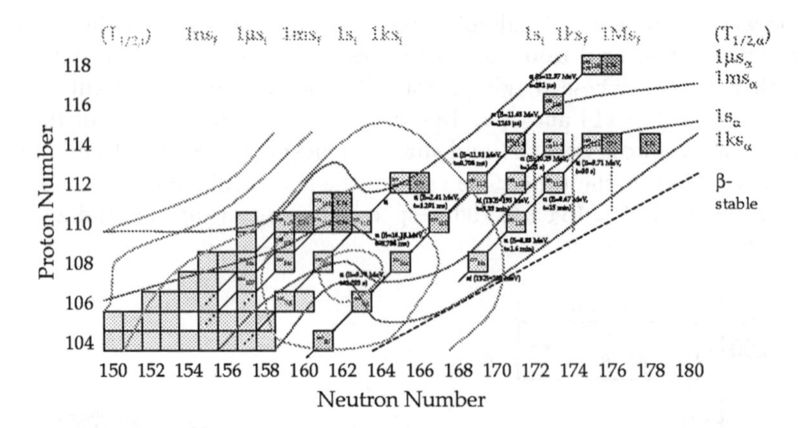

Figure 2. The $Z = 106 - 112$ isotopes were fused by the Hofmann–Münzenberg (GSI)-group. The two $Z = 114$ isotopes were produced by the Dubna–Livermore group. It is claimed that three neutrons are evaporated. Obviously the lifetimes of the various decay products are rather long (because they are closer to the stable valley), in crude agreement with early predictions [3,4] and in excellent agreement with the recent calculations of the Sobicevsky–group [11]. The recently fused $Z = 118$ isotope by V. Ninov et al. at Berkeley is the heaviest one so far.

fragments was visible far beyond the barrier into the fusioning nucleus. The collective potential energy surfaces of heavy nuclei, as they were calculated in the framework of the two-center shell model, exhibit pronounced valleys, such that these valleys provide promising doorways to the fusion of superheavy nuclei for certain projectile-target combinations (Fig. 3). If projectile and target approach each other through those **"cold" valleys**, they get only minimally excited and the barrier which has to be overcome (fusion barrier) is lowest (as compared to neighbouring projectile-target combinations). In this way the correct projectile- and target-combinations for fusion were predicted. Indeed, Gottfried Münzenberg and Sigurd Hofmann and their group at GSI [8] have followed this approach. With the help of the SHIP mass-separator and the position sensitive detectors, which were especially developped by them, they produced the pre-superheavy elements $Z = 106, 107, \ldots 112$, each of them with the theoretically predicted projectile-target combinations, and only with these. Everything else failed. This is an impressing success, which crowned the laborious construction work of many years. The before last example of this success, the discovery of element 112 and its long α-decay chain, is shown in Fig. 4. Very recently the Dubna–Livermore–group produced two isotopes of $Z = 114$ element by bombarding ^{244}Pu with ^{48}Ca (Fig. 2). Also this is a

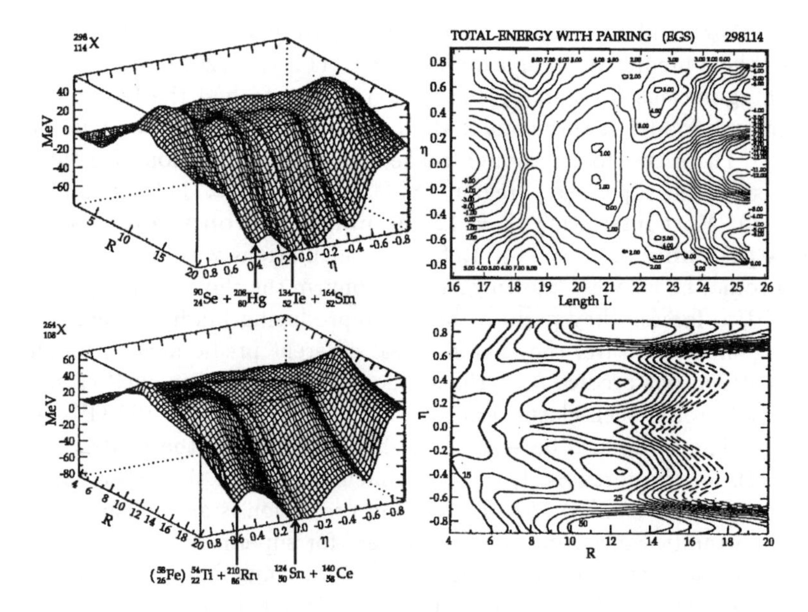

Figure 3. The collective potential energy surface of $^{264}108$ and $^{184}114$, calculated within the two center shell model by J. Maruhn et al., shows clearly the cold valleys which reach up to the barrier and beyond. Here R is the distance between the fragments and $\eta = \frac{A_1 - A_2}{A_1 + A_2}$ denotes the mass asymmetry: $\eta = 0$ corresponds to a symmetric, $\eta = \pm 1$ to an extremely asymmetric division of the nucleus into projectile and target. If projectile and target approach through a cold valley, they do not "constantly slide off" as it would be the case if they approach along the slopes at the sides of the valley. Constant sliding causes heating, so that the compound nucleus heats up and gets unstable. In the cold valley, on the other hand, the created heat is minimized. The colleagues from Freiburg should be familiar with that: they approach Titisee (in the Black Forest) most elegantly through the Höllental and not by climbing its slopes along the sides.

cold–valley reaction (in this case due to the combination of a spherical and a deformed nucleus), as predicted by Gupta, Sandulescu and Greiner [9] in 1977. There exist also cold valleys for which both fragments are deformed [10], but these have yet not been verified experimentally. The very recently reported $Z = 118$ isotope fused with the cold valley reaction [12] ^{58}Kr + ^{208}Pb by Ninov et al. [13] yields the latest support of the cold valley idea.

Studies of the shell structure of superheavy elements in the framework of the meson field theory and the Skyrme-Hartree-Fock approach have recently shown that the magic shells in the superheavy region are very isotope dependent [14] (see Fig. 5). Additionally, there is a strong dependency on the parameterset and the model. Some forces hardly show any shell structure, while others predict the magic numbers $Z = 114, 120, 126$. Using the heaviest known gg-nucleus Hassium $^{264}_{154}108$ as a criterium to find the best parameter sets in each model, it turns out that PL-40 and SkI4 produce best its binding energy. These two forces though make conflicting predictions for the magic number in the superheavy region: SkI4 predicts $Z = 114, 120$ and PL-40 $Z = 120$. Most interesting, $Z = 120$ **as magic proton number seems to be as probable as** $Z = 114$. Deformed calculations within the two models [15] for some the heaviest known gg-nuclei confirm the choice of the above mentioned two forces: they perform best in reproducing binding energies. Concerning shell structure, they again reveal different predictions: Though both parametrizations predict $N = 162$ as the deformed neutron shell closure, the deformed proton shell closures are $Z = 108$ (SkI4) and $Z = 104$ (PL-40) (see Fig. 6). There can also be seen some shell structure at the neutron number $N = 150$ for the force SkI4, the experimental data on the other hand show the shell closure at $N = 152$. This wrong prediction is an additional hint for the uncertainty concerning magic numbers for superheavy elements. Calculations of the potential energy surfaces [16] show single humped barriers, their heights and widths strongly depending on the predicited magic number. Furtheron, recent investigations in a chirally symmetric mean–field theory (see also below) result also in the prediction of these two magic numbers[40,42]. The corresponding magic neutron numbers are predicted to be $N = 172$ and - as it seems to a lesser extend - $N = 184$. Thus, this region provides an open field of research. R.A. Gherghescu et al. have calculated the potential energy surface of the $Z = 120$ nucleus. It utilizes interesting isomeric and valley structures (Fig. 7). The charge distribution of the $Z = 120, N = 184$ nucleus indicates a hollow inside. This leads us to suggest that it might be essentially a fullerene consisting of 60 α-particles and one binding neutron per alpha.

The determination of the chemistry of superheavy elements, i. e. the calculation of the atomic structure — which is in the case of element 112 the shell structure of 112 electrons due to the Coulomb interaction of the electrons and in particular the calculation of the orbitals of the outer (valence) electrons — has been carried out as early as 1970 by B. Fricke and W. Greiner[17]. Hartree-Fock-Dirac calculations yield rather precise results.

The potential energy surfaces, which are shown prototypically for $Z = 114$ in Fig 3, contain even more remarkable information that I want to mention cursorily: if a given nucleus, e. g. Uranium, undergoes fission, it moves in its potential mountains from the interior to the outside. Of course, this happens quantum mechanically. The wave function of such a nucleus, which decays by tunneling through the barrier, has maxima where the potential is minimal and minima where it has maxima. This is depicted in Fig. 8.

The probability for finding a certain mass asymmetry $\eta = \dfrac{A_1 - A_2}{A_1 + A_2}$ of the fission is proportional to $\psi^*(\eta)\psi(\eta)\mathrm{d}\eta$. Generally, this is complemented by a coordinate dependent scale factor for the volume element in this (curved) space, which I omit for the sake of clarity. Now it becomes clear how the so-called **asymmetric** and **superasymmetric** fission processes come into being. They result from the enhancement of the collective wave function in the cold valleys. And that is indeed, what one observes. Fig. 9 gives an impression of it.

For a large mass asymmetry ($\eta \approx 0.8,\ 0.9$) there exist very narrow valleys. They are not as clearly visible in Fig. 3, but they have interesting consequences. Through these narrow valleys nuclei can emit spontaneously not only α-particles (Helium nuclei) but also ^{14}C, ^{20}O, ^{24}Ne, ^{28}Mg, and other nuclei. Thus, we are lead to the **cluster radioactivity** (Poenaru, Sandulescu, Greiner [18]).

By now this process has been verified experimentally by research groups in Oxford, Moscow, Berkeley, Milan and other places. Accordingly, one has to revise what is learned in school: there are not only 3 types of radioactivity (α-, β-, γ-radioactivity), but many more. Atomic nuclei can also decay through spontaneous cluster emission (that is the "spitting out" of smaller nuclei like carbon, oxygen,...). Fig. 10 depicts some nice examples of these processes.

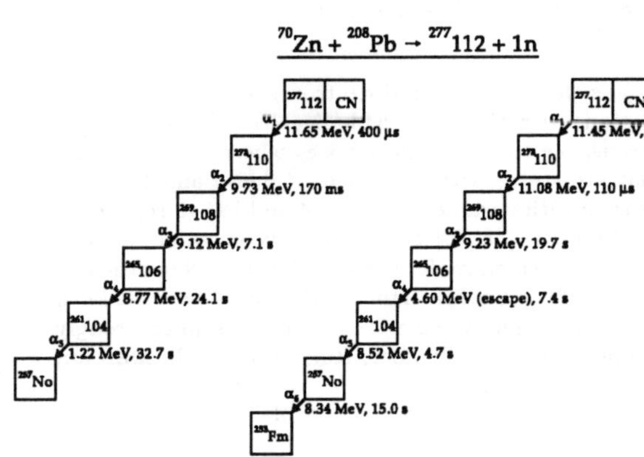

Figure 4. The fusion of element 112 with ^{70}Zn as projectile and ^{208}Pb as target nucleus has been accomplished for the first time in 1995/96 by S. Hofmann, G. Münzenberg and their collaborators. The colliding nuclei determine an entrance to a "cold valley" as predicted as early as 1976 by Gupta, Sandulescu and Greiner. The fused nucleus 112 decays successively via α emission until finally the quasi-stable nucleus ^{253}Fm is reached. The α particles as well as the final nucleus have been observed. Combined, this renders the definite proof of the existence of a $Z = 112$ nucleus.

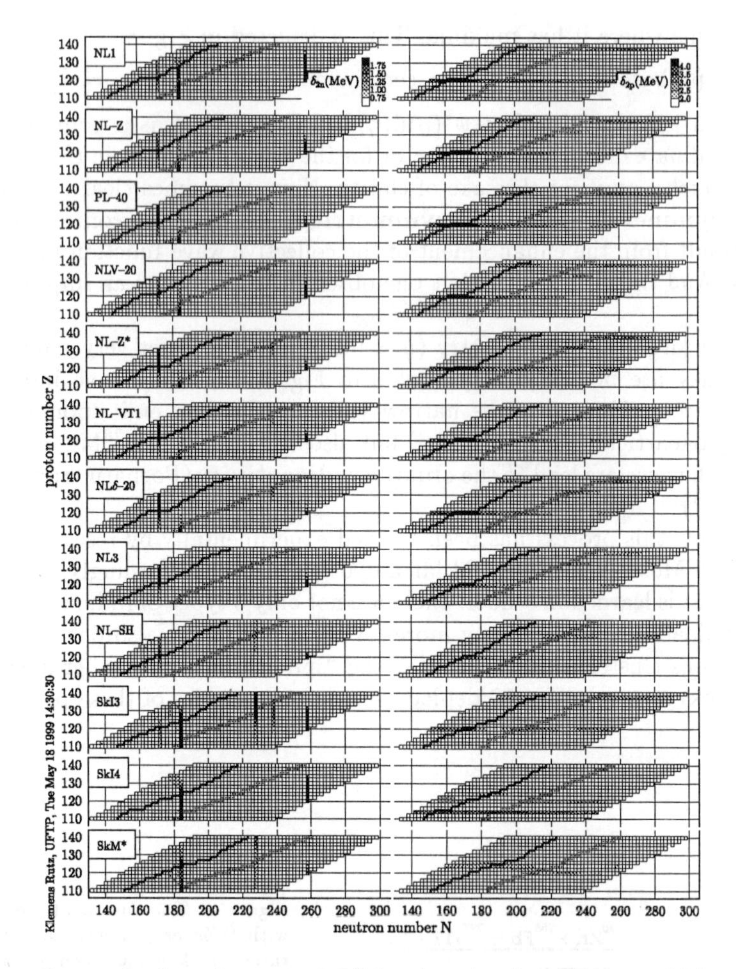

Figure 5. Grey scale plots of proton gaps (left column) and neutron gaps (right column) in the N-Z plane for spherical calculations with the forces as indicated. The assignment of scales differs for protons and neutrons, see the uppermost boxes where the scales are indicated in units of MeV. Nuclei that are stable with respect to β decay and the two-proton dripline are emphasized. The forces with parameter sets SkI4 and PL-40 reproduce the binding energy of ${}^{264}_{156}108$ (Hassium) best, i.e. $|\delta E/E| < 0.0024$. Thus one might assume that these parameter sets could give the best predictions for the superheavies. Nevertheless, it is noticed that PL-40 predicts only $Z = 120$ as a magic number while SkI4 predicts both $Z = 114$ and $Z = 120$ as magic numbers. The magicity depends — sometimes quite strongly — on the neutron number. These studies are due to Bender, Rutz, Bürvenich, Maruhn, P.G. Reinhard et al. [14].

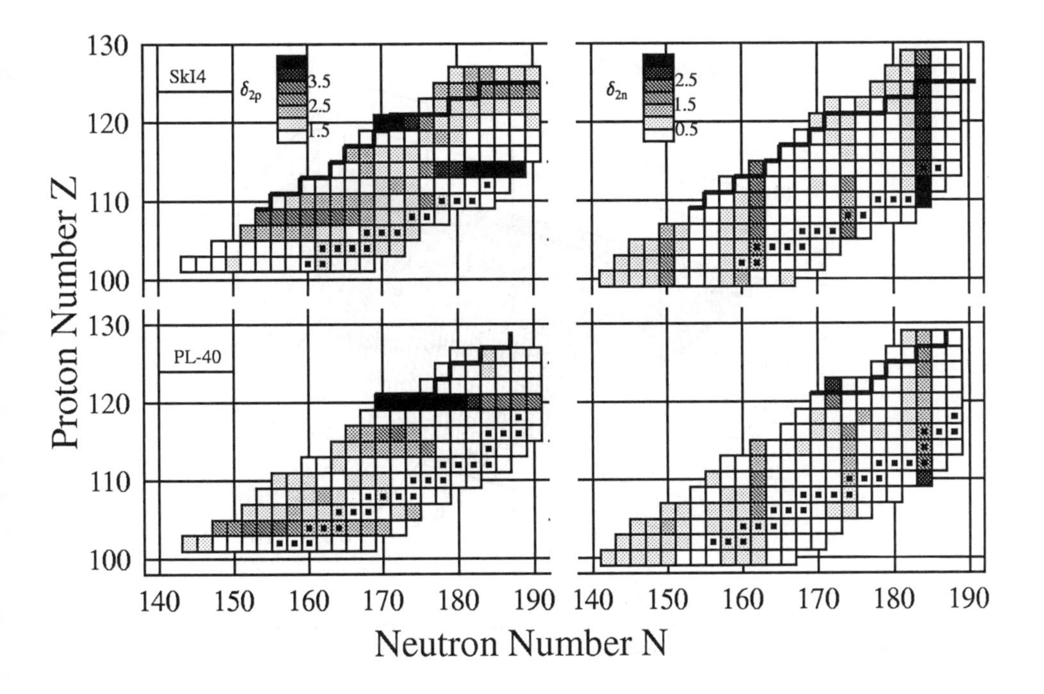

Figure 6. Grey scale plots of proton gaps (left column) and neutron gaps (right column) in the N-Z plane for deformed calculations with the forces SkI4 and PL-40. The assignment of scales is the same as in Fig 5. Besides the spherical shell closures one can see the deformed shell closures for protons at $Z = 104$ (PL-40) and $Z = 108$ (SkI4) and the ones for neutron at $N = 162$ for both forces.

The knowledge of the collective potential energy surface and the collective masses $B_{ij}(R, \eta)$, all calculated within the Two-Center-Shell-Modell (TCSM), allowed H. Klein, D. Schnabel and J. A. Maruhn to calculate lifetimes against fission in an "ab initio" way [19].
Utilizing a WKB-minimization for the penetrability integral

$$\mathcal{P} = e^{-\mathrm{I}}, \quad \mathrm{I} = \min_{\forall\,\text{paths}} \frac{2}{\hbar} \int_S \sqrt{2m(V(R,\eta) - E)}\, ds$$

$$= \min_{\forall\,\text{paths}} \frac{2}{\hbar} \int_0^1 \sqrt{\underbrace{2mg_{ij}}_{B_{ij}}(V(x_i(t) - E) \frac{dx_i}{dt}\frac{dx_j}{dt}}\, dt \qquad (1)$$

where $ds^2 = g_{ij}dx_idx_j$ and g_{ij} – the metric tensor – is in the well-known fashion related to the collective masses $B_{ij} = 2mg_{ij}$, one explores the minimal

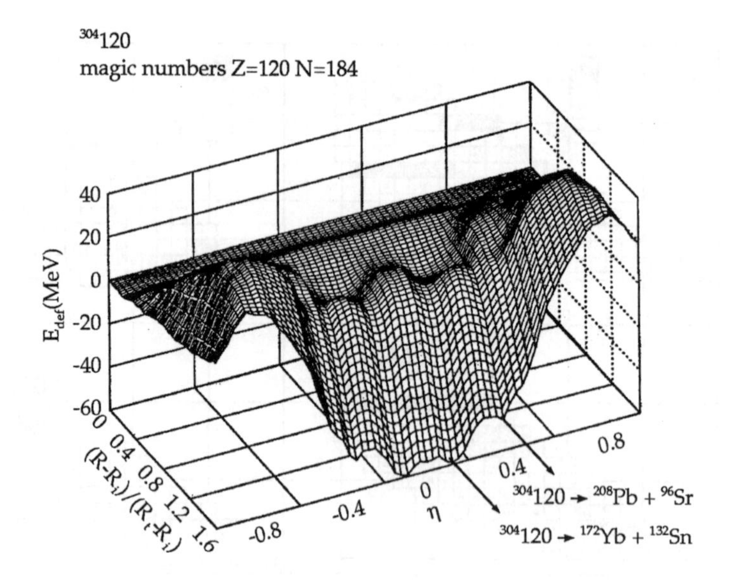

Figure 7. Potential energy surface as a function of reduced elongation $(R - R_i)/(R_t - R_i)$ and mass asymmetry η for the double magic nucleus $^{304}120$. $^{304}120_{184}$.

paths from the nuclear ground state configuration through the multidimensional fission barrier (see Fig. 11).

The thus obtained fission half lives are depicted in the lower part of figure 11. Their distribution as a function of the fragment mass A_2 resembles quite well the asymmetric mass distribution. Cluster radioactive decays correspond to the broad peaks around $A_2 = 20,\ 30\ (200,\ 210)$. The confrontation of the calculated fission half lives with experiments is depicted in Fig. 12. One notices "nearly quantitative" agreement over 20 orders of magnitude, which is – for an ab-initio calculation – remarkable!

Finally, in Fig. 13, we compare the lifetime calculation discussed above with one based on the Preformation Cluster Model by D. Poenaru et al. [20] and recognize an amazing degree of similarity and agreement.

The systematics for the average total kinetic energy release for spontaneously fissioning isotopes of Cm and No is following the Viola trend, but ^{258}Fm and ^{259}Fm are clearly outside. The situation is similar also for ^{260}Md, where two components of fission products (one with lower and one with higher kinetic energy) were observed by Hulet et al. [21]. The explanation of

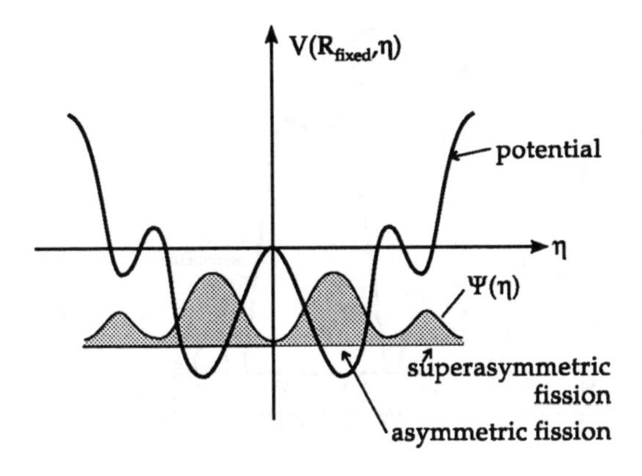

Figure 8. The collective potential as a function of the mass asymmetry $\eta = \frac{A_1 - A_2}{A_1 + A_2}$. A_i denotes the nucleon number in fragment i. This qualitative potential $V(R_{\text{fixed}}, \eta)$ corresponds to a cut through the potential landscape at $R = R_{\text{fixed}}$ close to the scission configuration. The wave function is drawn schematically. It has maxima where the potential is minimal and vice versa.

these interesting observations lies in two different paths through the collective potential. One reaches the scission point in a stretched neck position (i.e. at a lower point of the Coulomb barrier - thus lower kinetic energy for the fragments) while the other one reaches the scission point practically in a touching-spheres-position (i.e. higher up on the Coulomb barrier and therefore highly energetic fragments) [22]. The latter process is cold fission; i.e., the fission fragments are in or close to their ground state (cold fragments)and all the available energy is released as kinetic energy. Cold fission is, in fact, typically a cluster decay. The side-by-side occurence of cold and normal (hot) fission has been named bi-modal fission [22]. There has now been put forward a phantastic idea [23] in order to study cold fission (Cluster decays) and other exotic fission processes (ternary-, multiple fission in general) very elegantly: By measuring with e. g. the Gamma–sphere characteristic γ-transitions of individual fragments in coincidence, one can identify all these processes in a direct and simple way (Fig. 14). First confirmation of this method by J. Hamilton, V. Ramaya et al. worked out excellently [24]. This method has high potential for revolutionizing fission physics! With some physical intuition one can imagine that **triple** - and **quadriple** fission processes and even the process of **cold multifragmentation** will be discovered - absolutely fascinating! We have thus seen that fission physics (cold fission,

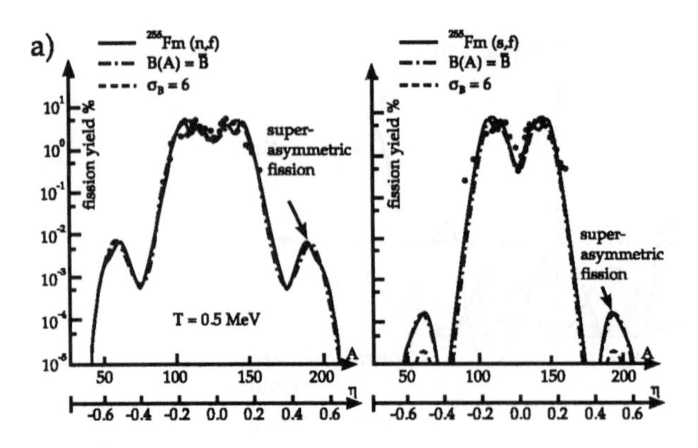

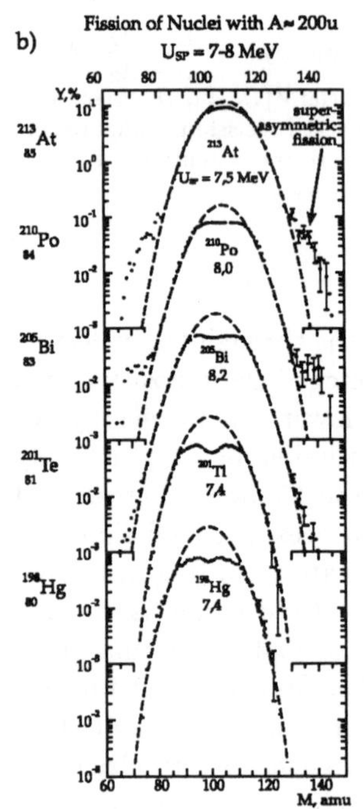

Figure 9. Asymmetric (a) and symmetric (b) fission. For the latter, also superasymmetric fission is recognizable, as it has been observed only a few years ago by the russian physicist Itkis — just as expected theoretically.

cluster radioactivity, ...) and fusion physics (especially the production of superheavy elements) are intimately connected.
Indeed, very recently, tripple fission of

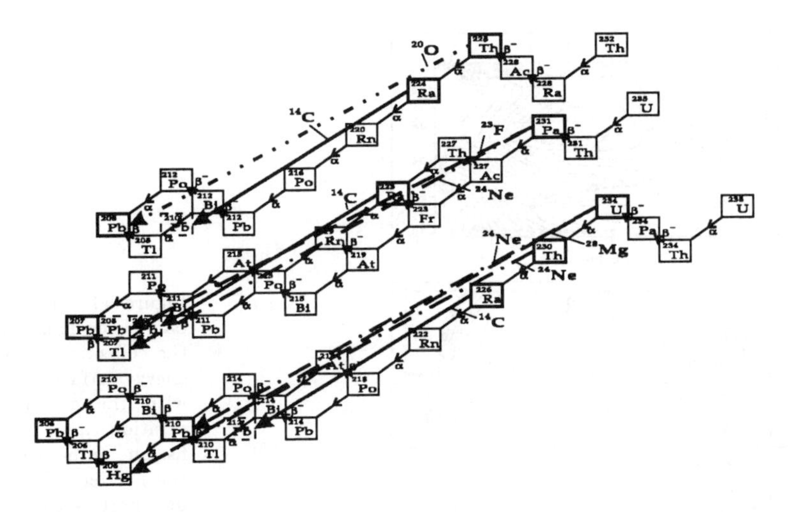

Figure 10. Cluster radioactivity of actinide nuclei. By emission of ^{14}C, ^{20}O,... "big leaps" in the periodic system can occur, just contrary to the known α, β, γ radioactivities, which are also partly shown in the figure.

$$\begin{aligned} {}^{252}\mathrm{Cf} &\rightarrow {}^{146}\mathrm{Ba} + {}^{96}\mathrm{Sr} \;\; + {}^{10}\mathrm{Be} \\ &\rightarrow {}^{112}\mathrm{Ry} + {}^{130}\mathrm{Sn} + {}^{10}\mathrm{Be} \\ &\rightarrow \ldots \end{aligned}$$

has been identified by measuring the various γ–transitions of these nuclei in coincidence (see Fig. 15). Even though the statistical evidence for the ^{10}Be line is small (≈ 50 events) the various coincidences seem to proof that spontaneous tripple fission out of the ground state of ^{252}Cf with the heavy cluster ^{10}Be as a third fragment exists. Also other tripple fragmentations can be expected. One of those is also denoted above. In fact, there are first indications, that this break–up is also observed. The most amazing observation is, however, the following: The cross coincidences seem to suggest that one deals with a simultaneous three–body breakup and not with a cascade process. For that one expects a configuration as shown in Fig. 16. Consequently the ^{10}Be will obtain kinetic energy while running down the combined Coulomb barrier of ^{146}Ba and ^{96}Sr and, therefore, the 3368 keV line of ^{10}Be should be Doppler–broadened. Amazingly, however, it is not and, moreover, it seems to be about 6 keV smaller than the free ^{10}Be γ–transition. If this turns out to be true, the only explanation will be that the Gamma is emitted while the nuclear molecule of the type shown in Fig. 16 holds. The molecule has to live longer than about 10^{-12} sec. The nuclear forces from the ^{146}Ba and ^{96}Sn cluster to

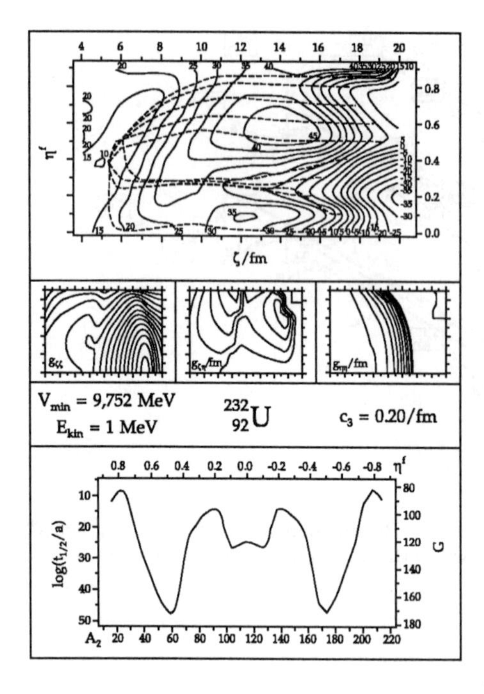

Figure 11. The upper part of the figure shows the collective potential energy surface for $^{232}_{92}U$ with the groundstate position and various fission paths through the barrier. The middle part shows various collective masses, all calculated in the TCSM. In the lower part the calculated fission half lives are depicted.

the left and right from ^{10}Be lead to a softening of its potential and therefore to a somewhat smaller transition energy. Thus, if experimental results hold, one has discovered long living ($\approx 10^{-12}$ sec) complex nuclear molecules. This is phantastic! Of course, I do immediately wonder whether such configurations do also exist in e.g. U + Cm soft encounters directly at the Coulomb barrier. This would have tremendous importance for the observation of the spontaneous vacuum decay [28], for which "sticking giant molecules" with a lifetime of the order of 10^{-19} sec are needed. The nuclear physics of such heavy ion collisions at the Coulomb barrier (giant nuclear molecules) should indeed be investigated!

As mentioned before there are other tri–molecular structures possible; some with ^{10}Be in the middle and both spherical or deformed clusters on both sides of ^{10}Be. The energy shift of the ^{10}Be–line should be smaller, if the outside clusters are deformed (smaller attraction $\Leftrightarrow$ smaller softening of the potential) and bigger, if they are spherical . Also other than ^{10}Be–clusters are expected to be in the middle. One is lead to the molecular doorway picture. Finally, these tri–body nuclear molecules are expected to perform themselves rotational and vibrational (butterfly, whiggler, β–, γ–type) modes. The energies

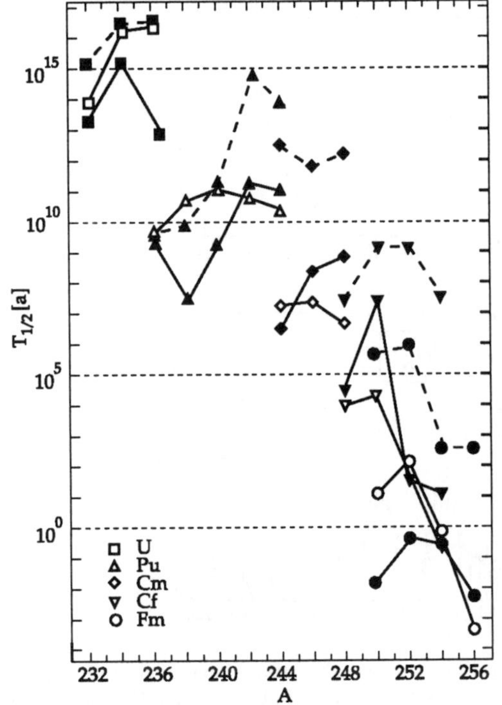

Figure 12. Fission half lives for various isotopes of $Z = 92$ (□), $Z = 94$ (△), $Z = 96$ (◇), $Z = 98$ (▽) and $Z = 100$ (○). The black curves represent the experimental values. The dashed and dotted calculations correspond to a different choice of the barrier parameter in the Two Center Shell Model ($c_3 \approx 0.2$ and 0.1 respectively).

were estimated by P. Hess et al [26]; for example rotational energies typically of the order of a few keV (4 keV, 9 keV, ...). A new molecular spectroscopy seems possible!

The "cold valleys" in the collective potential energy surface are basic for understanding this exciting area of nuclear physics! It is a master example for understanding the **structure of elementary matter**, which is so important for other fields, especially astrophysics, but even more so for enriching our "Weltbild", i.e. the status of our understanding of the world around us.

Nuclei that are found in nature consist of nucleons (protons and neutrons) which themselves are made of u (up) and d (down) quarks. However, there also exist s (strange) quarks and even heavier flavors, called charm, bottom, top. The latter has just recently been discovered. Let us stick to the s quarks. They are found in the 'strange' relatives of the nucleons, the so-called hyperons (Λ, Σ, Ξ, Ω). The Λ-particle, e. g., consists of one u, d and s quark, the Ξ-particle even of an u and two s quarks, while the Ω (sss) contains strange quarks only. Fig. 17 gives an overview of the baryons, which are of interest

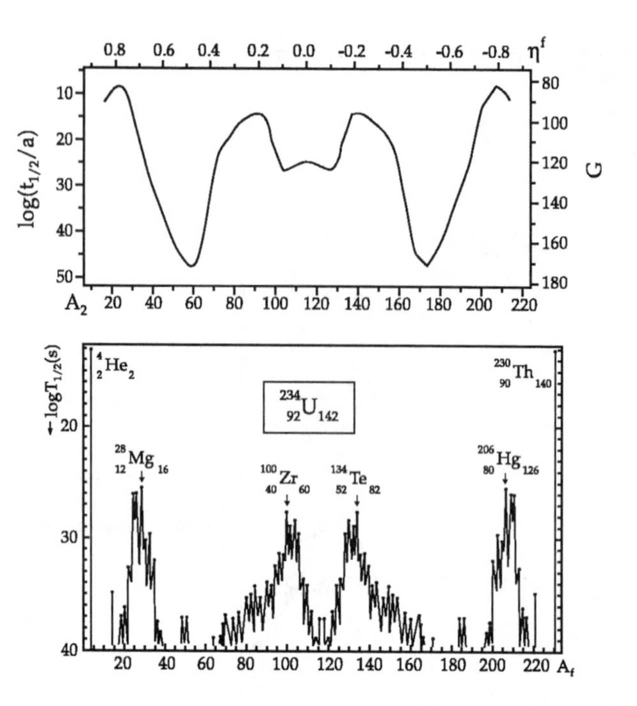

Figure 13. Comparison of the fission half lives calculated in the fission model (upper figure – see also Fig. 11) and in the Preformation Cluster Model [20]. In both models the deformation of the fission fragments is not included completely.

here, and their quark content.

If such a hyperon is taken up by a nucleus, a **hyper-nucleus** is created. Hyper-nuclei with one hyperon have been known for 20 years now, and were extensively studied by B. Povh (Heidelberg)[29]. Several years ago, Carsten Greiner, Jürgen Schaffner and Horst Stöcker[30] theoretically investigated nuclei with many hyperons, **hypermatter**, and found that the binding energy per baryon of strange matter is in many cases even higher than that of ordinary matter (composed only of u and d quarks). This leads to the idea of extending the periodic system of elements in the direction of strangeness.

One can also ask for the possibility of building atomic nuclei out of **anti-matter**, that means searching e. g. for anti-helium, anti-carbon, anti-oxygen. Fig. 18 depicts this idea. Due to the charge conjugation symmetry antinuclei should have the same magic numbers and the same spectra as ordinary nuclei. However, as soon as they get in touch with ordinary matter, they annihilate

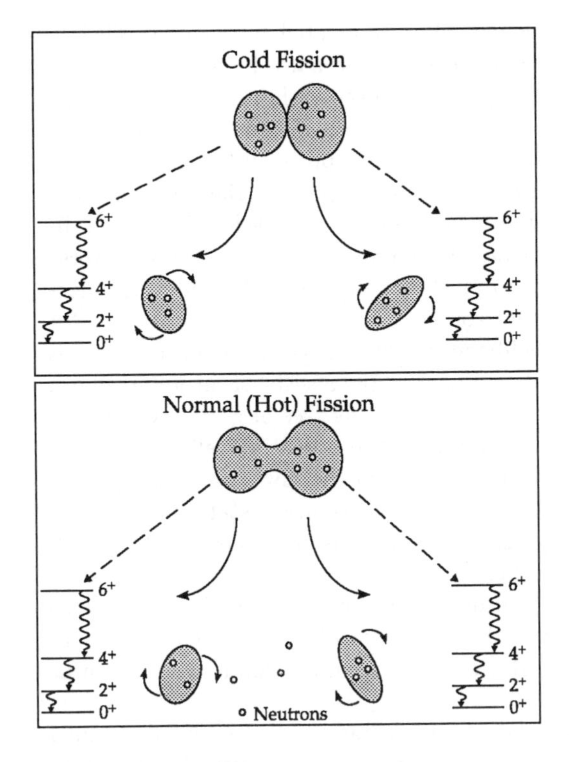

Figure 14. Illustration of cold and hot (normal) fission identification through multiple γ-coincidencs of photons from the fragments. The photons serve to identify the fragments.

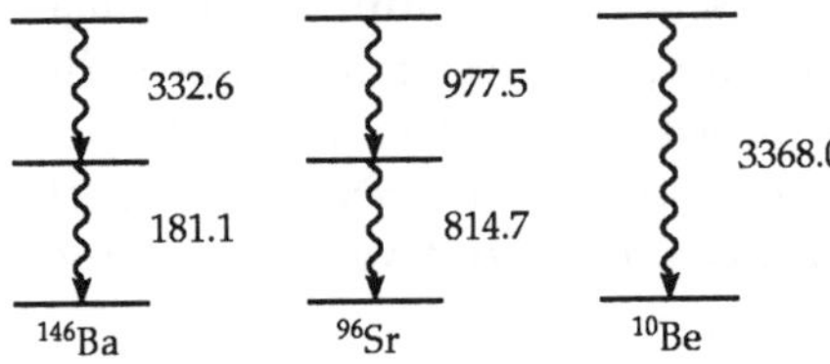

Figure 15. The γ–transitions of the three fission products of ^{252}Cf measured in coincidence. Various combinations of the coincidences were studied. The free 3368 keV line in ^{10}Be has recently been remeasured by Burggraf et al.[25], confirming the value of the transition energy within 100 eV.

with it and the system explodes.

Now the important question arises how these strange matter and antimatter clusters can be produced. First, one thinks of collisions of heavy nuclei, e. g. lead on lead, at high energies (energy per nucleon $\geq$ 200 GeV). Calculations with the URQMD-model of the Frankfurt school show that through **nuclear shock waves** [31,32,33] nuclear matter gets compressed to 5–10 times of its usual value, $\rho_0 \approx 0.17$ fm^3, and heated up to temperatures of $kT \approx 200$ MeV. As a consequence about 10000 pions, 100 Λ's, 40 Σ's and Ξ's and about as many antiprotons and many other particles are created in a single collision. It seems conceivable that it is possible in such a scenario for some Λ's to get captured by a nuclear cluster. This happens indeed rather frequently for one or two Λ-particles; however, more of them get built into nuclei with rapidly decreasing probability only. This is due to the low probability for finding the right conditions for such a capture in the phase space of the particles: the numerous particles travel with every possible momenta (velocities) in all directions. The chances for hyperons and antibaryons to meet gets rapidly worse with increasing number. In order to produce multi-Λ-nuclei and antimatter nuclei, one has to look for a different source.

In the framework of meson field theory the energy spectrum of baryons has a peculiar structure, depicted in Fig. 19. It consists of an upper and a lower continuum, as it is known from the electrons (see e. g. [28]). Of special interest in the case of the baryon spectrum is the potential well, built of the scalar and the vector potential, which rises from the lower continuum. It is known since P.A.M. Dirac (1930) that the negative energy states of the lower continuum have to be occupied by particles (electrons or, in our case, baryons). Otherwise our world would be unstable, because the "ordinary" particles are found in the upper states which can decay through the emission of photons into lower lying states. However, if the "underworld" is occupied, the Pauli-principle will prevent this decay. Holes in the occupied "underworld" (Dirac sea) are antiparticles.

The occupied states of this underworld including up to 40000 occupied bound states of the lower potential well represent the **vacuum**. The peculiarity of this strongly correlated vacuum structure in the region of atomic nuclei is that — depending on the size of the nucleus — more than 20000 up to 40000 (occupied) bound nucleon states contribute to this polarization effect. Obviously, we are dealing here with a **highly correlated vacuum**. A pro-

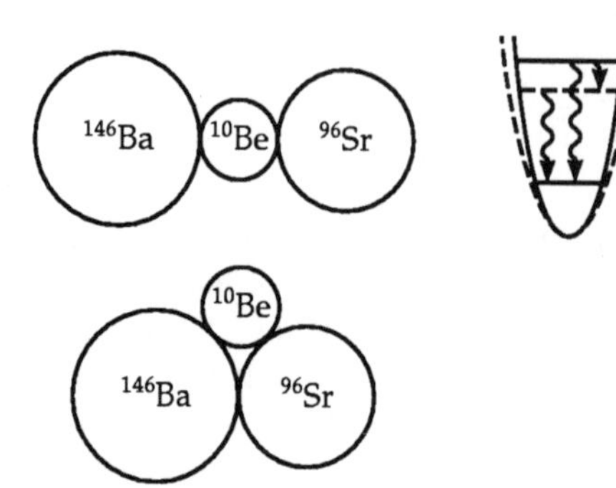

Figure 16. Typical linear cluster configuration leading to tripple fission of ^{252}Cf. The influence of both clusters leads to a softening of the ^{10}Be potential and thus to a somewhat smaller transition energy. Some theoretical investigations indicate that the axial symmetry of this configuration might be broken (lower lefthand figure).

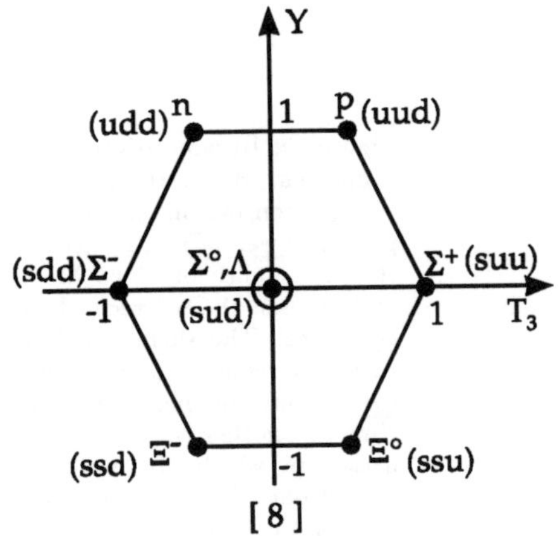

Figure 17. Important baryons are ordered in this octet. The quark content is depicted. Protons (p) and neutrons (n), most important for our known world, contain only u and d quarks. Hyperons contain also an s quark. The number of s quarks is a measure for the strangeness.

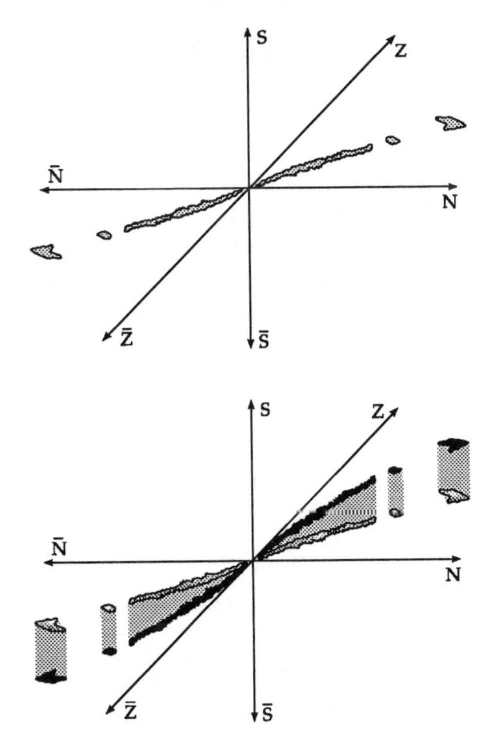

Figure 18. The extension of the periodic system into the sectors of strangeness (S, $\bar{S}$) and antimatter ($\bar{Z}$, $\bar{N}$). The stable valley winds out of the known proton (Z) and neutron (N) plane into the S and $\bar{S}$ sector, respectively. The same can be observed for the antimatter sector. In the upper part of the figure only the stable valley in the usual proton (Z) and neutron (N) plane is plotted, however, extended into the sector of antiprotons and antineutrons. In the second part of the figure it has been indicated, how the stable valley winds out of the Z-N-plane into the strangeness sector.

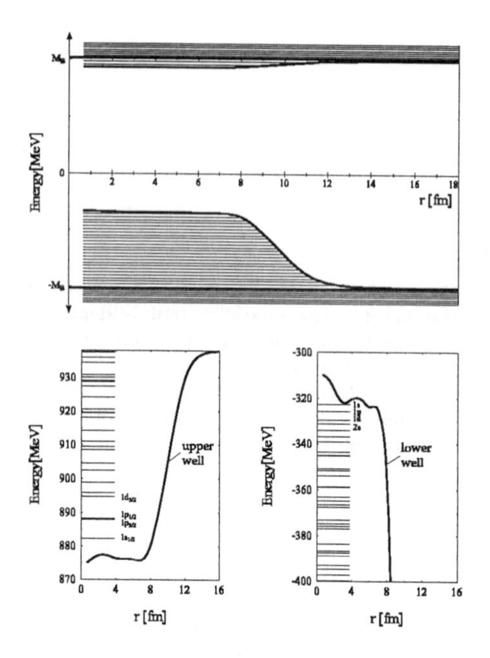

Figure 19. Baryon spectrum in a nucleus. Below the positive energy continuum exists the potential well of real nucleons. It has a depth of 50-60 MeV and shows the correct shell structure. The shell model of nuclei is realized here. However, from the negative continuum another potential well arises, in which about 40000 bound particles are found, belonging to the vacuum. A part of the shell structure of the upper well and the lower (vacuum) well is depicted in the lower figures.

nounced shell structure can be recognized [34,35,36]. Holes in these states have to be interpreted as bound antinucleons (antiprotons, antineutrons). If the primary nuclear density rises due to compression, the lower well increases while the upper decreases and soon is converted into a repulsive barrier (Fig. 20).

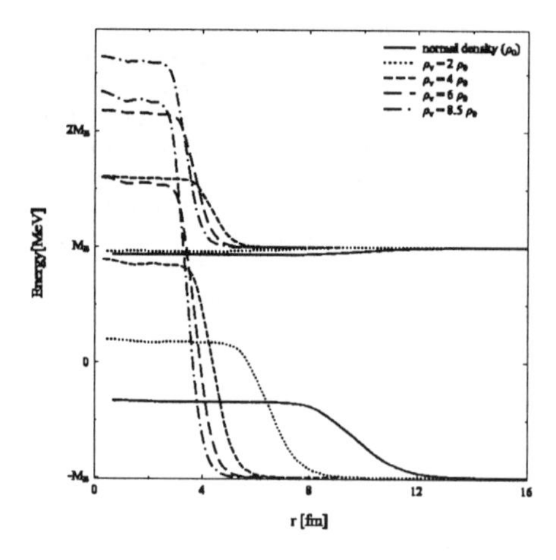

Figure 20. The lower well rises strongly with increasing primary nucleon density, and even gets supercritical (spontaneous nucleon emission and creation of bound antinucleons). Supercriticality denotes the situation, when the lower well enters the upper continuum.

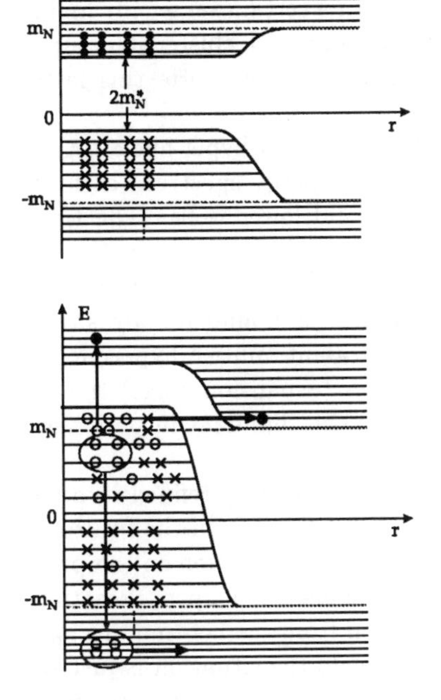

Figure 21.
Due to the high temperature and the violent dynamics, many bound holes (antinucleon clusters) are created in the highly correlated vacuum, which can be set free during the expansion stage into the lower continuum. In this way, antimatter clusters can be produced directly from the vacuum. The horizontal arrow in the lower part of the figure denotes the spontaneous creation of baryon-antibaryon pairs, while the antibaryons occupy bound states in the lower potential well. Such a situation, where the lower potential well reaches into the upper continuum, is called supercritical. Four of the bound holes states (bound antinucleons) are encircled to illustrate a "quasi-antihelium" formed. It may be set free (driven into the lower continuum) by the violent nuclear dynamics.

This compression of nuclear matter can only be carried out in relativistic nucleus-nucleus collision with the help of shock waves, which have been proposed by the Frankfurt school[31,32] and which have since then been confirmed extensively (for references see e. g. [37]). These **nuclear shock waves** are accompanied by heating of the nuclear matter. Indeed, density and temperature are intimately coupled in terms of the hydrodynamic Rankine-Hugoniot-equations. Heating as well as the violent dynamics cause the creation of many holes in the very deep (measured from $-M_B c^2$) vacuum well. These numerous bound holes resemble antimatter clusters which are bound in the medium; their wave functions have large overlap with antimatter clusters. When the primary matter density decreases during the expansion stage of the heavy ion collision, the potential wells, in particular the lower one, disappear.

The bound antinucleons are then pulled down into the (lower) continuum. In this way antimatter clusters may be set free. Of course, a large part of the antimatter will annihilate on ordinary matter present in the course of the expansion. However, it is important that this mechanism for the production of antimatter clusters out of the highly correlated vacuum does not proceed via the phase space. The required coalescence of many particles in phase space suppresses the production of clusters, while it is favoured by the direct production out of the highly correlated vacuum. In a certain sense, the highly correlated vacuum is a kind of cluster vacuum (vacuum with cluster structure). The shell structure of the vacuum levels (see Fig. 19) supports this latter suggestion. Fig. 21 illustrates this idea.

The mechanism is similar for the production of multi-hyper nuclei (Λ, Σ, Ξ, Ω). Meson field theory predicts also for the Λ energy spectrum at finite primary nucleon density the existence of upper and lower wells. The lower well belongs to the vacuum and is fully occupied by Λ's.

Dynamics and temperature then induce transitions ($\Lambda\bar{\Lambda}$ creation) and deposit many Λ's in the upper well. These numerous bound Λ's are sitting close to the primary baryons: in a certain sense a giant multi-Λ hypernucleus has been created. When the system disintegrates (expansion stage) the Λ's distribute over the nucleon clusters (which are most abundant in peripheral collisions). In this way multi-Λ hypernuclei can be formed.

Of course this vision has to be worked out and probably refined in many respects. This means much more and thorough investigation in the future. It is particularly important to gain more experimental information on the properties of the lower well by (e, e' p) or (e, e' p p') and also ($\bar{p}_c p_b$, $p_c\bar{p}_b$) reactions at high energy ($\bar{p}_c$ denotes an incident antiproton from the continuum, p_b is a proton in a bound state; for the reaction products the situation is just the opposite)[38]. Also the reaction (p, p' d), (p, p' ^{3}He), (p, p' ^{4}He) and others of similar type need to be investigated in this context. The systematic scattering of antiprotons on nuclei can contribute to clarify these questions. Problems of the meson field theory (e. g. Landau poles) can then be reconsidered. An effective meson field theory has to be constructed. Various effective theories, e. g. of Walecka-type on the one side and theories with chiral invariance on the other side, seem to give different strengths of the potential wells and also different dependence on the baryon density [39]. The Lagrangians of the Dürr-Teller-Walecka-type and of the chirally symmetric mean field theories look quite differently. We exhibit them — without further discussion — in the following equations:

$$\mathcal{L} = \mathcal{L}_{\text{kin}} + \mathcal{L}_{\text{BM}} + \mathcal{L}_{\text{vec}} + \mathcal{L}_0 + \mathcal{L}_{\text{SB}}$$

Non-chiral Lagrangian:

$$\mathcal{L}_{\text{kin}} = \frac{1}{2}\partial_\mu s \partial^\mu s + \frac{1}{2}\partial_\mu z \partial^\mu z - \frac{1}{4}B_{\mu\nu}B^{\mu\nu} - \frac{1}{4}G_{\mu\nu}G^{\mu\nu} - \frac{1}{4}F_{\mu\nu}F^{\mu\nu}$$

$$\mathcal{L}_{\text{BM}} = \textstyle\sum_B \overline{\psi}_B \big[i\gamma_\mu \partial^\mu - g_{\omega B}\gamma_\mu \omega^\mu - g_{\phi B}\gamma_\mu \phi^\mu - g_{\rho B}\gamma_\mu \tau_B \rho^\mu$$

$$-e\gamma_\mu \frac{1}{2}(1+\tau_B)A^\mu - m_B^*]\psi_B$$

$$\mathcal{L}_{\rm vec} = \frac{1}{2}m_\omega^2 \omega_\mu\omega^\mu + \frac{1}{2}m_\rho^2 \rho_\mu\rho^\mu + \frac{1}{2}m_\phi^2\phi_\mu\phi^\mu$$

$$\mathcal{L}_0 = -\frac{1}{2}m_s^2 s^2 - \frac{1}{2}m_z^2 z^2 - \frac{1}{3}bs^3 - \frac{1}{4}cs^4$$

Chiral Lagrangian:

$$\mathcal{L}_{\rm kin} = \frac{1}{2}\partial_\mu\sigma\partial^\mu\sigma + \frac{1}{2}\partial_\mu\zeta\partial^\mu\zeta + \frac{1}{2}\partial_\mu\chi\partial^\mu\chi - \frac{1}{4}B_{\mu\nu}B^{\mu\nu} - \frac{1}{4}G_{\mu\nu}G^{\mu\nu} - \frac{1}{4}F_{\mu\nu}F^{\mu\nu}$$

$$\mathcal{L}_{\rm BM} = \sum_B \overline{\psi}_B \left[i\gamma_\mu\partial^\mu - g_{\omega B}\gamma_\mu\omega^\mu - g_{\phi B}\gamma_\mu\phi^\mu - g_{\rho B}\gamma_\mu\tau_B\rho^\mu\right.$$

$$\left. -e\gamma_\mu\frac{1}{2}(1+\tau_B)A^\mu - m_B^*\right]\psi_B$$

$$\mathcal{L}_{\rm vec} = \frac{1}{2}m_\omega^2\frac{\chi^2}{\chi_0^2}\omega_\mu\omega^\mu + \frac{1}{2}m_\rho^2\frac{\chi^2}{\chi_0^2}\rho_\mu\rho^\mu + \frac{1}{2}m_\phi^2\frac{\chi^2}{\chi_0^2}\phi_\mu\phi^\mu + g_4^4(\omega^4 + 6\omega^2\rho^2 + \rho^4)$$

$$\mathcal{L}_0 = -\frac{1}{2}k_0\chi^2(\sigma^2+\zeta^2) + k_1(\sigma^2+\zeta^2)^2 + k_2(\frac{\sigma^4}{2}+\zeta^4) + k_3\chi\sigma^2\zeta - k_4\chi^4$$

$$+\frac{1}{4}\chi^4\ln\frac{\chi^4}{\chi_0^4} + \frac{\delta}{3}\ln\frac{\sigma^2\zeta}{\sigma_0^2\zeta_0^2}$$

$$\mathcal{L}_{\rm SB} = -\left(\frac{\chi}{\chi_0}\right)^2\left[m_\pi^2 f_\pi\sigma + \left(\sqrt{2}m_K^2 f_K - \frac{1}{\sqrt{2}}m_\pi^2 f_\pi\right)\zeta\right]$$

The non-chiral model contains the scalar-isoscalar field s and its strange counterpart z, the vector-isoscalar fields ω_μ and ϕ_μ, and the the ρ-meson ρ_μ as well as the photon A_μ. For more details see [39]. In contrast to the non-chiral model, the $SU(3)_L \times SU(3)_R$ Lagrangian contains the dilaton field χ introduced to mimic the trace anomaly of QCD in an effective Lagrangian at tree level (For an explanation of the chiral model see [39,40]).

The connection of the chiral Lagrangian with the Walecka-type can be established by the substitution $\sigma = \sigma_0 - s$ (and similarly for the strange condensate ζ). Then, e.g. the difference in the definition of the effective nucleon mass in both models (non-chiral:$m_N^* = m_N - g_s s$, chiral:$m_N^* = g_s\sigma$) can be removed, yielding:

$$m_N^* = g_s\sigma_0 - g_s s \equiv m_N - g_s s \qquad (2)$$

for the nucleon mass in the chiral model.

Nevertheless, if the parameters in both cases are adjusted such that ordinary nuclei (binding energies, radii, shell structure,...) and properties of infinite nuclear matter (equilibrium density, compression constant K, binding energy) are well reproduced, the prediction of both effective Lagrangians

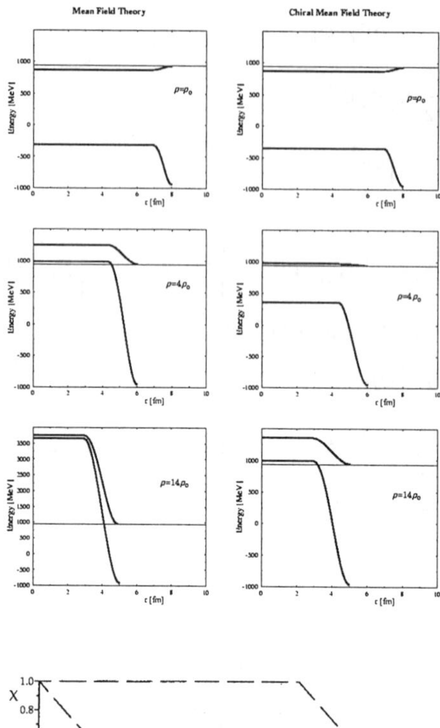

Figure 22. The potential structure of the shell model and the vacuum for various primary densities $\rho = \rho_0$, $4\rho_0$, $14\rho_0$. At left the predictions of ordinary Dürr-Teller-Walecka-type theories are shown; at right those for a chirally symmetric meson field theory as develloped by P. Papazoglu, S. Schramm et al. [39,40]. Note however, that this particular chiral mean–field theory does contain ω^4 terms. If introduced in both effective models, they seem to predict quantitatively similar results.

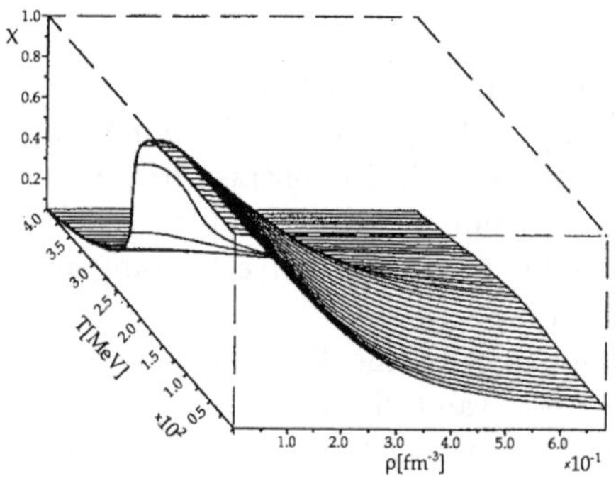

Figure 23. The strong phase transition inherent in Dürr-Teller-Walecka-type meson field theories, as predicted by J. Theis et al. [41]. Note that there is a first order transition along the ρ-axis (i.e. with density), but a simple transition along the temperature T-axis. Note also that this is very similar to the phase transition obtained recently from the Nambu-Jona-Lasinio-approximation of QCD [43].

for the dependence of the properties of the correlated vacuum on density and temperature is remarkably different. This is illustrated to some extend in Fig. 22. Accordingly, the chirally symmetric meson field theory predicts much higher primary densities (and temperatures) until the effects of the correlated vacuum are strong enough so that the mechanisms described here become effective. In other words, according to chirally symmetric meson field theories the antimatter-cluster-production and multi-hypermatter-cluster production out of the highly correlated vacuum takes place at considerably higher heavy ion energies as compared to the predictions of the Dürr-Teller-Walecka-type meson field theoories. This in itself is a most interesting, quasi-fundamental question to be clarified. Moreover, the question of the nucleonic substructure (form factors, quarks, gluons) and its influence on the highly correlated vacuum structure has to be studied. The nucleons are possibly strongly polarized in the correlated vacuum: the Δ resonance correlations in the vacuum are probably important. Is this highly correlated vacuum state, especially during the compression, a preliminary stage to the quark-gluon cluster plasma? To which extent is it similar or perhaps even identical with it? It is well known for more than 10 years that meson field theories predict a phase transition qualitatively and quantitatively similar to that of the quark-gluon plasma [41] — see Fig. 23.

The extension of the periodic system into the sectors hypermatter (strangeness) and antimatter is of general and astrophysical importance. Indeed, microseconds after the big bang the new dimensions of the periodic system, we have touched upon, certainly have been populated in the course of the baryo- and nucleo-genesis. Of course, for the creation of the universe, even higher dimensional extensions (charm, bottom, top) come into play, which we did not pursue here. It is an open question, how the depopulation (the decay) of these sectors influences the distribution of elements of our world today. Our conception of the world will certainly gain a lot through the clarification of these questions.

For the Gesellschaft für Schwerionenforschung (GSI), which I helped initiating in the sixties, the questions raised here could point to the way ahead. Working groups have been instructed by the board of directors of GSI, to think about the future of the laboratory. On that occasion, very concrete (almost too concrete) suggestions are discussed — as far as it has been presented to the public. What is missing, as it seems, is a **vision on a long term basis**. The ideas proposed here, the verification of which will need the **commitment for 2–4 decades of research**, could be such a vision with considerable attraction for the best young physicists. The new dimensions of the periodic system made of hyper- and antimatter cannot be examined in the "stand-by" mode at CERN (Geneva); a dedicated facility is necessary for this field of research, which can in future serve as a home for the universities. The GSI — which has unfortunately become much too self-sufficient — could be such a home for new generations of physicists, who are interested in the **structure of elementary matter**. GSI would then not develop just into a detector laboratory for CERN, and as such become obsolete. I can already see the enthusiasm in the eyes of young scientists, when I unfold these ideas to them — similarly as it was 30 years ago, when the nuclear physicists in the state of Hessen initiated the construction of GSI.

References

1. S.G: Nilsson et al. Phys. Lett. 28 B (1969) 458
   Nucl. Phys. A 131 (1969) 1
   Nucl. Phys. A 115 (1968) 545
2. U. Mosel, B. Fink and W. Greiner, Contribution to "Memorandum Hessischer Kernphysiker" Darmstadt, Frankfurt, Marburg (1966).
3. U. Mosel and W. Greiner, Z. f. Physik 217 (1968) 256, 222 (1968) 261
4. a) J. Grumann, U. Mosel, B. Fink and W. Greiner, Z. f. Physik 228 (1969) 371
   b) J. Grumann, Th. Morovic, W. Greiner, Z. f. Naturforschung *26a* (1971) 643
5. A. Sandulescu, R.K. Gupta, W. Scheid, W. Greiner, Phys. Lett. *60*B (1976) 225
   R.K. Gupta, A. Sandulescu, W. Greiner, Z. f. Naturforschung *32*a (1977) 704
   R.K. Gupta, A.Sandulescu and W. Greiner, Phys. Lett. *64*B (1977) 257
   R.K. Gupta, C. Parrulescu, A. Sandulescu, W. Greiner Z. f. Physik A283 (1977) 217
6. G. M. Ter-Akopian et al., Nucl. Phys. A*255* (1975) 509
   Yu.Ts. Oganessian et al., Nucl. Phys. A*239* (1975) 353 and 157
7. D. Scharnweber, U. Mosel and W. Greiner, Phys. Rev. Lett *24* (1970) 601
   U. Mosel, J. Maruhn and W. Greiner, Phys. Lett. *34*B (1971) 587
8. G. Münzenberg et al. Z. Physik A309 (1992) 89
   S.Hofmann et al. Z. Phys A*350* (1995) 277 and 288
9. R. K. Gupta, A. Sandulescu and Walter Greiner, Z. für Naturforschung *32*a (1977) 704
10. A. Sandulescu and Walter Greiner, Rep. Prog. Phys 55. 1423 (1992); A. Sandulescu, R. K. Gupta, W. Greiner, F. Carstoin and H. Horoi, Int. J. Mod. Phys. E1, 379 (1992)
11. A. Sobiczewski, Phys. of Part. and Nucl. 25, 295 (1994)
12. R. K. Gupta, G. Münzenberg and W. Greiner, J. Phys. G: Nucl. Part. Phys. 23 (1997) L13
13. V. Ninov, K. E. Gregorich, W. Loveland, A. Ghiorso, D. C. Hoffman, D.

M. Lee, H. Nitsche, W. J. Swiatecki, U. W. Kirbach, C. A. Laue, J. L. Adams, J. B. Patin, D. A. Shaughnessy, D. A. Strellis and P. A. Wilk, preprint

14. K. Rutz, M. Bender, T. Bürvenich, T. Schilling, P.-G. Reinhard, J.A. Maruhn, W. Greiner, Phys. Rev. C *56* (1997) 238.
15. T. Bürvenich, K. Rutz, M. Bender, P.-G. Reinhard, J. A. Maruhn, W. Greiner, EPJ A 3 (1998) 139-147.
16. M. Bender, K. Rutz, P.-G. Reinhard, J. A. Maruhn, W. Greiner, Phys. Rev. C 58 (1998) 2126-2132.
17. B. Fricke and W. Greiner, Physics Lett *30*B (1969) 317
    B. Fricke, W. Greiner, J.T. Waber, Theor. Chim. Acta (Berlin) *21* (1971) 235
18. A. Sandulescu, D.N. Poenaru, W. Greiner, Sov. J. Part. Nucl. 11(6) (1980) 528
19. Harold Klein, thesis, Inst. für Theoret. Physik, J.W. Goethe-Univ. Frankfurt a. M. (1992)
    Dietmar Schnabel, thesis, Inst. für Theoret. Physik, J.W. Goethe-Univ. Frankfurt a.M. (1992)
20. D. Poenaru, J.A. Maruhn, W. Greiner, M. Ivascu, D. Mazilu and R. Gherghescu, Z. Physik A*328* (1987) 309, Z. Physik A*332* (1989) 291
21. E. K. Hulet, J. F. Wild, R. J. Dougan, R. W.Longheed, J. H. Landrum, A. D. Dougan, M. Schädel, R. L. Hahn, P. A. Baisden, C. M. Henderson, R. J. Dupzyk, K. Sümmerer, G. R. Bethune, Phys. Rev. Lett. *56* (1986) 313
22. K. Depta, W. Greiner, J. Maruhn, H.J. Wang, A. Sandulescu and R. Hermann, Intern. Journal of Modern Phys. A*5*, No. 20, (1990) 3901
    K. Depta, R. Hermann, J.A. Maruhn and W. Greiner, in "Dynamics of Collective Phenomena", ed. P. David, World Scientific, Singapore (1987) 29
    S. Cwiok, P. Rozmej, A. Sobiczewski, Z. Patyk, Nucl. Phys. A491 (1989) 281
23. A. Sandulescu and W. Greiner in discussions at Frankfurt with J. Hamilton (1992/1993)
24. J.H. Hamilton, A.V. Ramaya et al. Journ. Phys. G **20** (1994) L85 - L89
25. B. Burggraf, K. Farzin, J. Grabis, Th. Last, E. Manthey, H. P. Trautvetter, C. Rolfs, *Energy Shift of first excited state in* ^{10}Be *?*, accepted for publication in Journ. of. Phys. G
26. P. Hess et al., *Butterfly and Belly Dancer Modes in* $^{96}Sr + ^{10}Be + ^{146}Ba$, in preparation
27. E.K. Hulet et al. Phys Rev C *40* (1989) 770.

28. W. Greiner, B. Müller, J. Rafelski, QED of Strong Fields, Springer Verlag, Heidelberg (1985). For a more recent review see W. Greiner, J. Reinhardt, *Supercritical Fields in Heavy-Ion Physics*, Proceedings of the 15th Advanced ICFA Beam Dynamics Workshop on Quantum Aspects of Beam Physics, World Scientific (1998)
29. B. Povh, Rep. Progr. Phys. *39* (1976) 823; Ann. Rev. Nucl. Part. Sci. *28* (1978) 1; Nucl. Phys. A*335* (1980) 233; Progr. Part. Nucl. Phys. *5* (1981) 245; Phys. Blätter *40* (1984) 315
30. J. Schaffner, Carsten Greiner and H. Stöcker Phys. Rev. C*45* (1992) 322; Nucl. Phys. B*24B* (1991) 246; J. Schaffner, C.B. Dover, A. Gal, D.J. Millener, C. Greiner, H. Stöcker: Annals of Physics*235* (1994) 35
31. W. Scheid and W. Greiner, Ann. Phys. *48* (1968) 493; Z. Phys. *226* (1969) 364
32. W. Scheid, H. Müller and W. Greiner Phys. Rev. Lett. 13 (1974) 741
33. H. Stöcker, W. Greiner and W. Scheid Z. Phys. A 286 (1978) 121
34. I. Mishustin, L.M. Satarov, J. Schaffner, H. Stöcker and W.Greiner Journal of Physics G (Nuclear and Particle Physics) *19* (1993) 1303
35. P.K. Panda, S.K. Patra, J. Reinhardt, J. Maruhn, H. Stöcker, W. Greiner, Int. J. Mod. Phys. E 6 (1997) 307
36. N. Auerbach, A. S. Goldhaber, M. B. Johnson, L. D. Miller and A. Picklesimer, Phys. Lett. B182 (1986) 221
37. H. Stöcker and W. Greiner, Phys. Rep. 137 (1986) 279.
38. J. Reinhardt and W. Greiner, to be published.
39. P. Papazoglou, D. Zschiesche, S. Schramm, H. Stöcker, W. Greiner, J. Phys. G 23 (1997) 2081; P. Papazoglou, S. Schramm, J. Schaffner-Bielich, H. Stöcker, W. Greiner, Phys. Rev. C 57 (1998) 2576.
40. P. Papazoglou, D. Zschiesche, S. Schramm, J. Schaffner–Bielich, H. Stöcker, W. Greiner, nucl–th/9806087, accepted for publication in Phys. Rev. C.
41. J. Theis, G. Graebner, G. Buchwald, J. Maruhn, W. Greiner, H. Stöcker and J. Polonyi, Phys. Rev. D 28 (1983) 2286
42. P. Papazoglou, PhD thesis, University of Frankfurt, 1998; C. Beckmann et al., in preparation
43. S. Klimt, M. Lutz, W. Weise, Phys. Lett. B*249* (1990) 386.

DISCOVERING SUPERHEAVY ELEMENTS

GOTTTFRIED MÜNZENBERG

Gesellschaft für Schwerionenforschung, GSI mbH, Planckstr. 1, 64291 Darmstadt and Johannes Gutenberg-Universität Mainz

Significant progress has been made approaching Superheavy Elements predicted for Z = 114. The heaviest element identified unambiguously at present, element 112, is already close to that region. The exciting physics result of heavy-element research is the discovery of a shell stabilised region located at Z = 108 and N = 162 which interconnects the trans-uranium region and the spherical superheavy elements. Recently the creation of elements 114, and 118 has been reported. In this paper a brief overview on experimental results on heavy element research will be given and discussed in the light of the experimental data.

1 Introduction

The motivation of heavy-element research is, to which extent shell stabilisation can extend the number of elements beyond the macroscopic limit determined by fission. Element 112 [1], the heaviest element safely identified at present is already far beyond that limit. It exists only by microscopic stabilisation. The shell stabilisation is due to a region of deformed nuclei centred at Z = 108 and N = 162. This has been explained in the frame of macrosopic- microscopic microscopic models[2,3]

The heaviest elements have been produced by complete fusion of heavy ions with production cross sections down to picobarns and rates as low as one nucleus per month. The identification is based on the in - situ observation of the heavy nuclei implanted in the silicon detectors. Characteristic for the heavy nuclei are long α-chains or α-fission sequences. For a safe identification the sequences must end in known transitions to be independent from theoretical predictions - an essential pre-requisite for the exploration of far unknown regions. To apply this technique it is neccssary to built-up such decay chains from the bottom, e.g. to create a reliable set of data for daughter nuclei, a strategy followed by GSI for safe identification of new elements[4].

2 The Production of Superheavy Elements

All of the heaviest elements have been synthesised by complete fusion of heavy ions. Two types of reactions have been used so far successfully: the cold heavy-ion fusion with lead or bismuth targets and appropriate projectile beams of the most neutron rich stable isotopes: ^{64}Ni, ^{70}Zn, or ^{86}Kr for the production

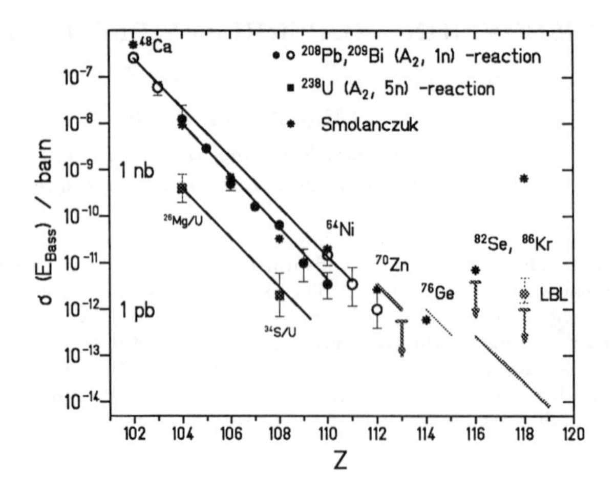

Figure 1. Production cross sections for the elements nobelium and beyond. The lines connect reactions with the same projectile isospin. The figure includes the prediction of Smolanczuk [9] and the Berkeley cross section given for the production of element 118, together with the upper limit of the GSI experiment.

of elements 110,111,112 or 118, respectively. The second way starts from actinide targets such as ^{235}U or ^{244}Pu and uses beams of ^{48}Ca to produce elements 112, and 114.

There are two limiting factors for heavy element production. The fusion process of heavy nuclear systems suffers from the large Coulomb forces involved. In a macroscopic picture asymmetric combinations will be favoured. This motivates the actinide way. The survival of the compound nucleus which in general is created as a highly excited system, is favoured for cold systems. This is the motivation for the cold fusion. It has been experimentally stated that shell effects play a role in the production yield of heavy elements. The theoretical motivation for the importance of shell effects in superheavy element production was discussed first in frame of the fragmentation model[5].

The strong Coulomb forces in the fusion of massive nuclear systems hinder the fusion process. In the exit channel they reduce the survival probability of the highly fissile heavy compound nuclei. For the fusion with lead or bismuth targets beyond Z = 108 only one-neutron evaporation has been observed. The excitation energy of the compound system of about 10 - 12 MeV is already close to the neutron evaporation energy, the natural lower limit for this reac-

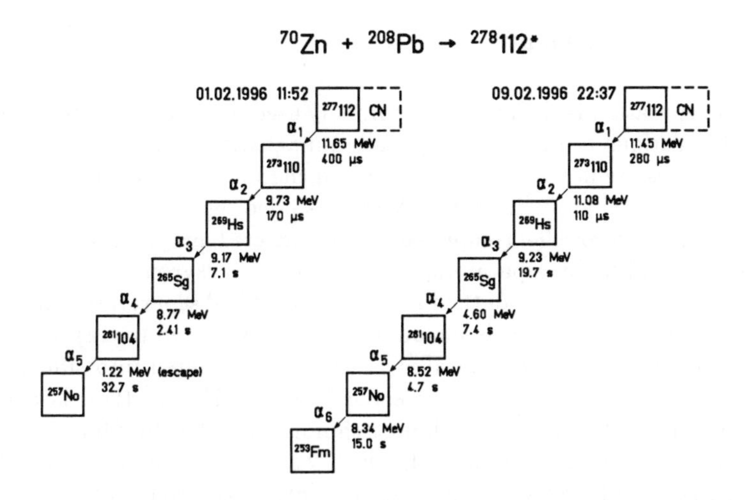

Figure 2. The two α - decay sequences observed for element 112.

tion channel. The production cross-sections for the heaviest elements are of the order of $10^{-36}cm^{-2}$. corresponding to production rates of 1 atom per 11 days for a cross section of 1 pb with exisiting, and 1 atom per 2 hours with future accelerators. Fig. 1 displays the production cross-sections for the elements nobelium and beyond. Projectiles with the same isospin projection are connected by solid lines[6,7]. It could be shown for element 110 that the production cross section profits from the isospin enhancement when using ^{64}Ni instead of ^{62}Ni. This effect could not be confirmed for element 112 , possibly because of the lack of a complete excitation function.

Theoretical models to predict the production cross sections for superheavy elements are under development, but up to now have only limited predictive power[8]. It is generally accepted that clusters, e. g. shell effects in target and projectile, as proven for the fusion with lead targets, enhance the fusion probability, whereas the stabilisation of shell effects against fission of the compound system could not been proven up to now. A recent prediction[9] predicts an increase of the production cross section up to several hundreds of picobarns for the production of elements beyond Z = 114 with the cold fusion (see Fig. 1). It should be noted here, that this model has been parametrised only with data from the heavy element production in cold fusion, a test with other reactions such as light symmetric systems is still lacking.

3 Recent Experiments

In-flight separation is the method commonly used at present in heavy-element research. Various types such as gas filled magnetic separators, energy-, and velocity filters are in use [10]. The separation characteristics of the various separator types are slightly different. While gas filled separators have the largest transmission and excellent suppression of scattered projectiles as compared to the other in-flight separators, they produce background from energetic projectile-scattered nuclei of the filling gas. Gas-filled separators transport all particle-evaporation channels including (x,n), (p,xn), and (α,xn).

For identification the separated nuclei are implanted into silicon detectors where their decay can be observed in situ. This method is a further development of the parent-daughter technique invented by Ghiorso et al.[11]. In contrast to the Ghiorso method, which was used in combination with the helium gasjet and well suited for the correlation for groups of particles and short α sequences, the implantation technique allows to identify single atoms and long chains and is therefore best suited for rare events. It allows the identification of new elements even on the basis of a single atom as has been demonstrated with the discovery of element 109[12]. Characteristic for the decay of a heavy nucleus are long α decay chains as well as sequences of α decays terminated by spontaneous fission. Recently the use of this method has been extended to lighter nuclei[13], e. g. in the identification of ^{100}Sn. Fig. 2 displays one out of the two α-decay chains observed for element 112 [1]. It should be noted that presently all new element investigation relies on in-flight separation and the single-atom correlation technique developed at GSI[14].

The unambiguous identification of new elements is based on the connection of the observed decay sequences to known nuclides. In the example of the identification of element 112 the isotopes of element 106 and below were safely identified before.

Recently two exciting discoveries were published. The Dubna laboratory reported the observation of α-fission sequences[15] in irradiations of ^{244,242}Pu with ^{48}Ca assigned to decays of the isotopes with masses 287, and 289 of the superheavy element with Z = 114. These data were completed by some new results including the isotope with mass 288. Berkeley announced the discovery of superheavy nuclei[16] in the reaction ^{86}Kr + ^{208}Pb = ^{293}118 + n by the observation of three long α-chains. The conclusive identification of the observed decays remains a problem, as the decay sequences end far in the unknown neutron rich trans-actinide region. Their assignment is primarily based on general considerations, consistency checks, and comparisons of the decay chains with theoretical models.

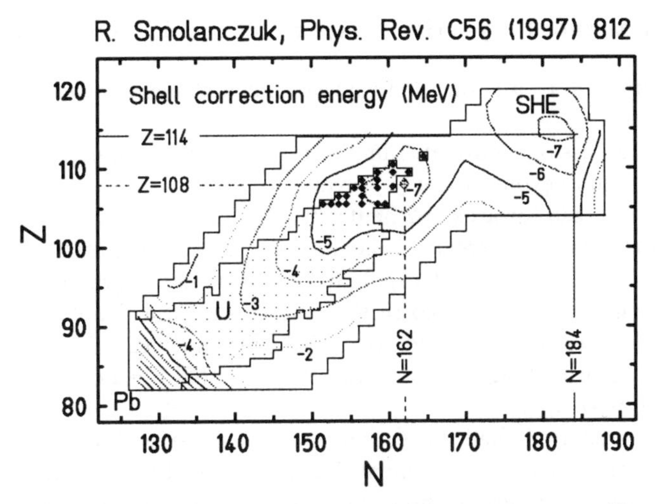

Figure 3. Shell correction energies form a macroscopic-microscopic model [18] for the region from lead to the spherical superheavy nuclei. The solid symbols mark the nuclides of the heaviest elements synthesised at GSI close to the deformed shell at Z=108 and N=162.

The irradiation of ^{208}Pb with ^{86}Kr to create element 118 was repeated at GSI with the projectile energy of the Berkeley experiment, adjusted by test reactions performed at both laboratories. The Berkeley result could not be confirmed[7,14] down to a level of 1 pb. Very recently the experiment was also repeated without positive result at GANIL, using the LISE velocity filter. An irradiation of ^{208}Pb with ^{84}Kr was made at RIKEN using the GARIS separator, also without positive result[17]. The more neutron deficient krypton isotope was chosen for two reasons: the decay chain of this isotope has some chance to end in a known region and the production cross section according to predictions of Abe should not be much less than that for ^{86}Kr.

It should be noted that the positive identification of the recently reported new elements will need some new experimental developments. These include the production of daughter nuclides where possible, chemical identification, and mass identification in isotope separators or with new kind of detector systems.

4 Ground-state Properties - Experiment and Predictions

The experimental results on the heaviest known elements reveal two remarkable features:
- spontaneous fission dominates in the region around element 104. For the heavier elements beyond α decay prevails.
- half-lives increase towards 162 neutrons.

These results were explained theoretically by a strong microscopic stabilisation, attributed to a hexadecapole deformation in the ground-state. Fig. 3 shows the calculated[18] shell- correction energies for the elements between the two shell closures at ^{208}Pb and the superheavy nucleus ^{298}114. In between there is the deformed region at Z = 108 and N = 162. Shell correction energies are as high as -7 MeV.

These calculated shell correction energies agree well with experimental data. The decay chains of element 112 pass close by the maximum of the shell correction and clearly show the step in decay energies and correlation times in the 110 to 108 region as a direct confirmation of the shell passing (Fig 2). The figure also reveals the character of the spherical superheavy shell which is much less pronounced as the much stronger lead shell.

The principal problems for the theoretical treatment of the heaviest nuclei are the high level density of the large nuclear shells with large angular momenta and the strong Coulomb field of the many protons which influences the diffuseness of the nuclear proton surface. Precise knowledge of spin-orbit coupling and the proton density at the nucleon surface are essential to calculate the location of the proton shell closure. Magic proton numbers of 114, 120, and 126 are currently predicted [19,20], depending on the models and forces. The neutron shell at N = 184 is fairly stable. Macroscopic-microscopic models tend to predict the magic proton number Z = 114, relativistic mean field models prefer Z = 120, whereas Skyrme-Hartree Fock models also predict Z = 126. A detailed comparative investigation of a great number of forces has been carried out by Bender et al.[19,21].

5 New Developments

With the elements seaborgium and beyond the predicted new species of shell nuclei has been discovered. They close the gap between the trans-uranium elements and the spherical superheavy shell closures. This region will offer a rich field of nuclear structure research as unexpectedly long half-lives have been observed experimentally. New types of experiments become possible such as the extension of heavy-element chemistry beyond the present

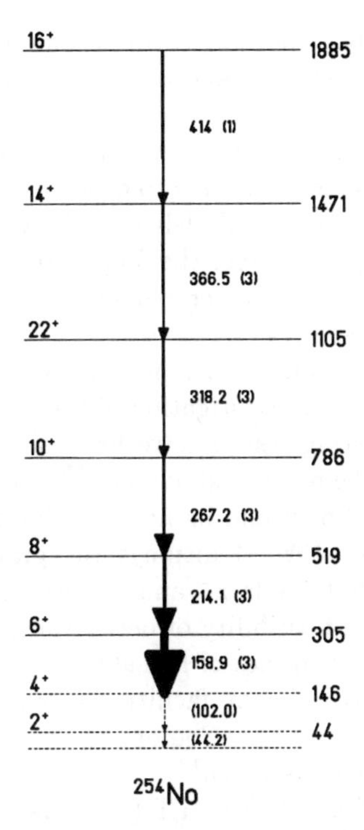

Figure 4. Level scheme [22] as measured with recoil- decay- tagging for ^{254}No

limit of nielsbohrium (Z = 107) and the application of ion traps for precision mass determination or the investigation of atomic properties of the heaviest elements.

The transition region between the deformed N = 162 shell and the spherical N = 184 shell is of specific interest for nuclear structure research. The high level density and large angular momenta will contribute to fast changes in structure as high spin and low spin states are close together. Already in the known isotopes a number of isomeric states have been found.

A key problem is the understanding of the production of superheavy nuclei, especially the importance of nuclear structure in the entrance and exit channels. New types of reactions need exploration. The fragmentation model predicts three paths to superheavy elements[5]: the cold fusion with lead and bismuth targets, the hot fusion with actinides, and as the third path the symmetric fusion or reversed fission. Presently only the cold fusion has been explored systematically. The hot, actinide based fusion data also are available, whereas the symmetric fusion just started to be explored at sufficiently high sensitivity[23].

High-current accelerators and new ion sources will provide beam intensities of more than 10^{14} ions/s allowing to proceed towards Z = 114 and beyond. Low-charge state accelerators will deliver still larger intensities. These developments will allow to proceed into the femtobarn cross sections and to extend our knowledge towards still heavier elements into the up to now inaccessible region of neutron rich species.

New detector developments such as the new generation of efficient 4π-Ge-arrays will allow a detailed investigation of the structure of the trans-actinide elements. First interesting results have been obtained with recoil tagged in-beam-spectroscopy. The rotational band of ^{254}No was observed which allowed to extract the deformation[22] and in addition the limiting angular momentum for the fusion process. Fig. 4 displays the ground-state band. It should be noted that the lowest levels 2^+ and 4^+ are converted. Intense projectile beams will open up the possibility of decay spectroscopy to explore in detail the character of the new species of hell stabilised trans-actinides.

The new radioactive beam facilities providing intense exotic nuclear beams with energies above the Coulomb barrier will produce neutron rich beams for heavy-element research[13]. Existing and planned laboratories as SPIRAL of GANIL (France), EXCYT (Catania, Italy), RIKEN (Japan) and its upgrade, MSU and its upgrade (Michigan State, USA) and the American RIA poroject, deliver already or will provide beams with intensities up to 10^7 /s for the running, and up to 10^{12} /s for the projected facilities. This will allow to explore the transition region between the deformed neutron shell at N = 162 and the spherical N = 184 region. The still open problems of isospin enhanced production cross section and the reversed fission are ideally explored with radioactive beams.

The combination of recoil separators and ion traps will open up new possibilities such as high precision mass measurements, laser spectroscopy to determine for instance deformations or ionisation potentials or chemical investigations. Such a trap could also serve as ion source to inject the in - flight separated nuclei into the GSI accelerator system. SHIPTRAP, a trap attached to SHIP is being planned, with a broad scientific program is under discussion.

6 Conclusions

In-flight separation is now the established method for heavy-element research. It permits the unambiguous identification of new elements and allows to obtain gross nuclear properties of a new nuclide from the decay of only single atoms. Recent heavy element research led to the discovery of deformed shell stabilised nuclei creating an extended region of enhanced stability which connects the trans-uranium region to the magic proton number 114, already closely approached with the discovery of element 112. First evidence for the superheavy elements 114 and 118 has been reported. Their positive identification will need still some efforts. New experimental developments will not only allow to extend the region of known elements but also give new access to their nuclear, atomic and chemical nature.

New high-current ion sources for stable beams will be essential for the production and investigation of new elements near and beyond element 114. The use of neutron rich beams of radioactive nuclei will open new perspectives for the investigation of heavy and superheavy elements in still inaccessible regions. Reaction studies to investigate the isospin dependence of the fusion cross sections will be the naturally first generation experiments with radioactive beams and give new insights in the production of superheavy nuclei. Trans-uranium elements up to the region around element 104 are expected to be accessible with secondary beams available from the presently planned facilities. The new region of neutron rich species near the deformed neutron shell and beyond, where the transition towards the spherical N = 184 shell is expected, is a rich and interesting field to explore. New insights in the shell stabilisation and the related existence of heavy elements in nature are expected.

The long half-lives of these nuclei will allow the application of new experimental techniques, giving new information about nuclear, atomic, and chemical properties of the elements at the top of the nuclear table.

Acknowledgements

The experiments at GSI were carried out together with S. Hofmann, F.P. Heßberger, V. Ninov, P. Armbruster, A.Yu. Lawrentew, A.G. Popeko, A.V. Eremin, S. Saro, and M.E. Leino. Fruitful discussions with A. Antonenko, W. Greiner, W. Nazarewicz, R. Smolanczuk, and A. Sobiczewski are highly appreciated.

References

1. S. Hofmann et al., Z. Phys. A354(1996)229
2. S. Cwiok et al., Nucl. Phys. A410(1983)254
3. P. Möller et al., Z. Phys. A323(1986)41
4. G. Münzenberg, Rep. Prog. Phys. 51 (1988) 57
5. A. Sandulescu and W. Greiner, J. Phys. G3(1977)189
6. S. Hofmann, Rep. Prog. Phys. 61(1998)639
7. S. Hofmann, Priv. comm.
8. G.G. Adamian et al., Nucl. Phys. A633(1998)409
9. R. Smolanczuk, Phys. Rev. C59(1999)2634
10. G. Münzenberg, in Nuclear Physics, eds. D. N. Poenaru and W. Greiner, Walter de Gruyter, Berlin, 1997
11. A. Ghiorso et al., Phys. Rev. Lett. 33(1974)1490
12. G. Münzenberg et al., Z. Phys. A315(1984)145
13. H. Geissel, G. Münzenberg, and K. Riisager, Ann. Rev. Nucl. Part. Sci. 45(1995)163
14. S. Hofmann and G. Münzenberg submitted to Rev. Mod. Phys.
15. Yu.Ts. Oganessian et al. Nature, 400(1999)242
16. V. Ninov et al. Phys. Rev. Lett. 83(1999)1104
17. I. Tanihata, priv. comm. 2000
18. R. Smolanczuk et al. Proc. EPS XV Nucl.Phys. Conf. St. Petersburg, Scientific Publ. Comp. Singapore (1995)313
19. K. Rutz et al., Phys. Rev. C56 (1997) 238
20. S. Cwiok et al. Nucl. Phys.A611(1996)211
21. M. Bender et al. to be published
22. P. Reiter, T. Khoo, C.J. Lister et al. Phys. Rev. C82(1999)509
23. C. Stodel et al., GSI Annual Report 1996; GSI-97-1

THE PROCESS OF COMPLETE FUSION OF NUCLEI WITHIN THE FRAMEWORK OF THE DINUCLEAR SYSTEM CONCEPT

V.V.VOLKOV

Joint Institute for Nuclear Research, Dubna, Russia

A non-traditional approach to the description of the process of complete fusion of nuclei — the dinuclear system concept (DNSC) is discussed. The DNSC reveals two important peculiarities of complete fusion of massive nuclei: the existence of a specific inner fusion barrier B^*_{fus} and the competition between the complete fusion and quasi-fission channels in the dinuclear system, which is formed on the capture stage. The DNSC was applied to the analysis of reactions used for the synthesis of transfermium and superheavy elements.

1 Introduction

In my talk I would like to discuss the process of fusion of two nuclei into a compound nucleus. Complete fusion is one of the main nuclear processes in low energy nucleus-nucleus collisions. Complete fusion reactions are used successfully for the synthesis of transfermium and superheavy elements, for the production of nuclei far from β-stability, for the production of super and hyper-deformed nuclei. Nevertheless, we cannot say that we have a true, realistic picture of the mechanism of compound nucleus formation. This situation is the result of the closed character of the complete fusion process. Fusing nuclei do not send any signals which allow one to reveal the mechanism of compound nucleus formation. Experimentalists detect products of compound nucleus decay. But it is well known that a compound nucleus forgets the history of its formation. Theoretical analysis of the transformation of two multinucleon systems into a new one is a very complicated problem and for the description of complete fusion process different models have been created.

2 Theoretical approaches to the description of the complete fusion process

2.1 A traditional approach to the description of the complete fusion process

Many theorists use an approach which may be called traditional. The basic idea of this approach is that complete fusion and fission may be regarded as a direct and a reverse process. Therefore the liquid drop model, which successfully describes the fission process can be applied to the description of the complete fusion process. The traditional approach is used in the macroscopic dynamical model by Swiatecki et al. [1], in the surface friction model by Gross [2] and Fröbrich [3], in the fluctuation — dissipation dynamics by Abe et al. [4]. Using these models many valuable results were obtained. However, serious difficulties arose in attempts to apply these models to the description of the synthesis of superheavy elements.

2.2 A non-traditional approach to the description of the complete fusion process — the Dinuclear System Concept

At Dubna a non-traditional approach to the description of the complete fusion process was proposed [5,6]. The basic idea of this approach is the assumption that complete fusion and deep inelastic transfer reactions (DITR) are similar nuclear processes. Indeed, in both processes a full dissipation of the collision kinetic energy occurs. In both processes the same forces act between colliding nuclei: the Coulomb, nuclear and centrifugal conservative forces and the dissipative forces of the nuclear friction. There is an interval on the angular momentum collision axis, where both reactions may be realized. What does this assumption give us? In contrast to fission, DITR are open reactions. They provide unique information on the interaction of two nuclei which appear to be in close contact after full dissipation of the collision kinetic energy. It is this unique information that has been used for revealing the mechanism of compound nucleus formation. This approach has received the name of the "Dinuclear System Concept" (DNSC) because the main content of the complete fusion process is formation and evolution of a dinuclear system. The DNSC was elaborated by G.G.Adamian, N.V.Antonenko, E.A.Cherepanov, A.K.Nasirov, W.Scheid, V.V.Volkov.

3 The main features of the dinuclear system concept

According to the DNSC, the main features of the complete fusion process are the following. At the capture stage, after full dissipation of the collision kinetic energy a dinuclear system (DNS) is formed. Complete fusion is an evolutionary process in which the nucleons of one nucleus gradually, shell by shell, are transferred to another nucleus. The nuclei of the DNS retain their individuality until the end of the fusion process. This important peculiarity of the DNS evolution is the consequence of the shell structure of nuclei.

Fig.1 illustrates the principal difference between the pictures of the com-

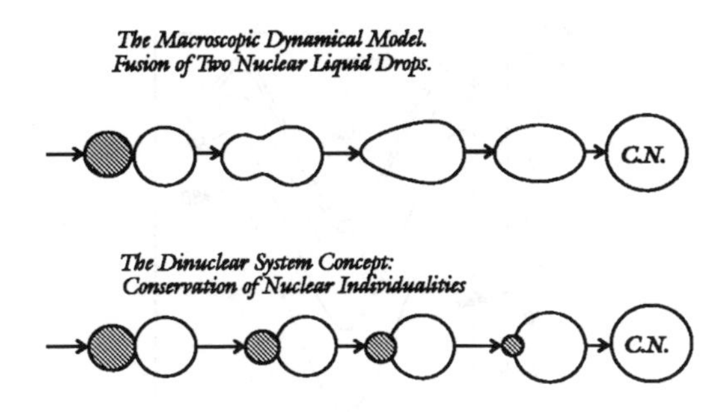

Figure 1. Illustration of compound nucleus formation process within the framework of the MDM and DNSC

pound nucleus formation process offered by the traditional approach and that offered by the Dubna approach. In the macroscopic dynamical model [1] fusing nuclear drops rapidly lose their individuality as a result of the neck formation. Complete fusion is a dynamical process, which is developing in the deformation space. In the DNSC the fusing nuclei retain their individuality until the end of the fusion process. Complete fusion is mainly a statistical process in which two initial nuclei are rebuilt into a compound nucleus.

4 Peculiarities of complete fusion of massive nuclei within the framework of the DNSC

The DNSC reveals two important peculiarities of complete fusion of massive nuclei:

— appearance of a specific inner fusion barrier $B^*_{\rm fus}$ and

— competition between the complete fusion and quasi-fission channels in the DNS formed at the capture stage.

As it is known, the DNS evolution is determined by the potential energy of the system, which is a function of the charge (mass) asymmetry and the collision angular momentum. Fig.2 shows potential energy V (Z, L) of the DNS formed in the reaction ^{110}Pd$+^{110}$Pd. At the horizontal axis the atomic number of one of the DNS nuclei is plotted. The figures on the curves indicate the collision angular momentum L. The potential energy is normalized to the potential energy of the compound nucleus, which is taken as zero. The initial DNS is situated in the minimum of the potential energy. It looks like a gigantic nuclear molecule. To realize complete fusion and to form a compound nucleus, the initial DNS has to overcome a potential barrier. It was called

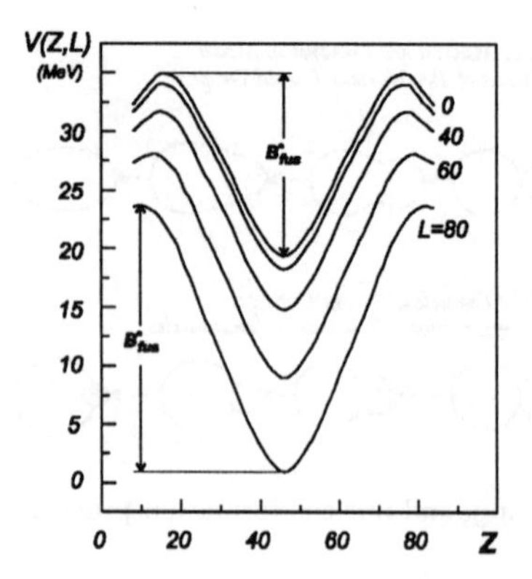

Figure 2. Potential energy of the DNS which is formed in the reaction $^{110}\mathrm{Pd} + ^{110}\mathrm{Pd}$.

"the inner fusion barrier — B^*_{fus}". The asterisk symbolizes that the energy for overcoming the barrier is taken from the DNS excitation energy — E^*. The inner fusion barrier — B^*_{fus} reflects the endothermic character of the fusion process for massive nuclei during the first stage of the DNS evolution. The value of B^*_{fus} depends on the angular momentum of collision.

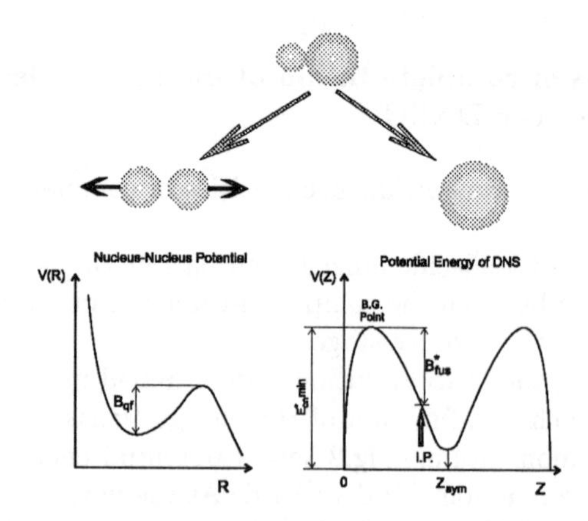

Figure 3. Two ways of evolution of a massaive asymmetric DNS. The nucleus-nelues potential (left) and potential energy of a DNS (right) are indicated.

The asymmetric initial DNS has two ways of evolution (Fig.3). It may increase its charge asymmetry and after overcoming the barrier — B^*_{fus} transform into a compound nucleus. Or it may evolve into a symmetric form. In the symmetric form the Coulomb repulsion between the DNS nuclei riches its maximum value and the DNS decays into two nearly equal fragments. It means that quasi-fission occurs. In the quasi-fission the DNS has to overcome the quasi-fission barrier B_{qf}. The competition between complete fusion and quasi-fission in the DNSC arises naturally as a consequence of the statistical nature of the DNS evolution.

5 The analysis of nuclear reactions used for the synthesis of transfermium and superheavy elements within the framework of the DNSC

5.1 The minimum of the excitation energy of compound nuclei in the cold fusion reactions

In the cold fusion reactions ^{208}Pb and ^{209}Bi targets are used and the products of reaction channel (HI, 1n) are detected. Models of complete fusion based

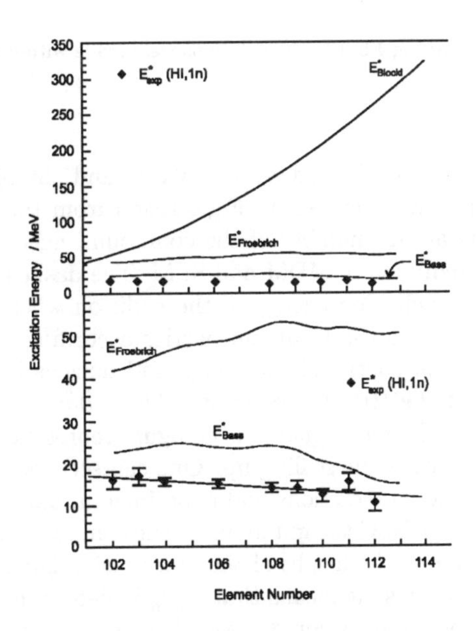

Figure 4. The minimal excitation energy of compound nuclei of the 102-112 elements which have been synthesized in cold fusion reaction; experimental data and theoretical predicitions.

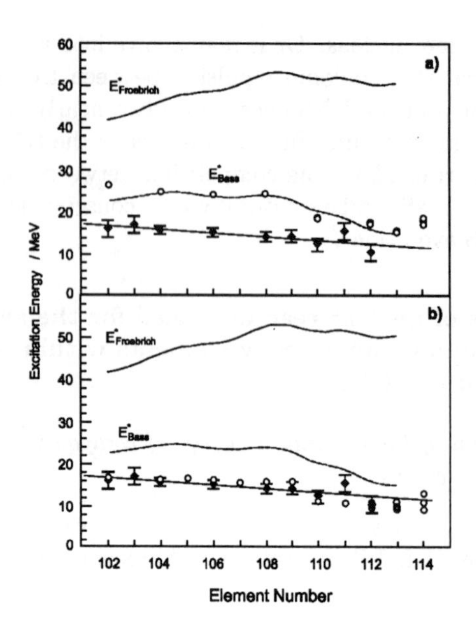

Figure 5. The same as in Fig.4 but the calculations were made within the framework of the DNSC.

on the traditional approach, namely the MDM and the SFM, predict very high excitation energy. It is seen in Fig.4 taken from the work [7]. At the horizontal axis the atomic number of the compound nucleus is plotted. The curves show the prediction of MDM and SFM. The dashed line indicates the compound nucleus excitation energy in the collision when kinetic energy of the projectile is equal to the Coulomb barrier according to Bass. The dots indicate the excitation energy of the compound nucleus in the maximum of the cross section of the (HI, 1n) reaction. These data were experimentally obtained at the GSI. The lower part of the figure represents the same data as the upper one but with some scaling up. One can see dramatic disagreement between the theoretical predictions and experimental data.

According to the DNSC the minimum of the compound nucleus excitation energy is determined by the height of the inner fusion barrier $B^*_{\rm fus}$ in relation to the compound nucleus. It means that $E^*_{\rm min}$ is determined by the shape of the potential energy curve. A comparison of the calculated and experimental data for $E^*_{\rm min}$ is shown in Fig.5. At the top of the figure the calculation results are represented in which the deformation of heavy nucleus of the DNS was taken into account. The calculated data turned out to be about 5 MeV higher than the experimental data. This difference disappears if one takes into account possible deformation of the light nucleus of the DNS. The deformation of the light nucleus was taken in the excited state 2^+.

So the DNSC make it possible to estimate the minimal excitation energy of the compound nucleus and the optimal value of the bombarding energy in the cold synthesis of SHE.

5.2 The role of quasi-fission in the synthesis of superheavy elements

Fig.6. shows the calculated production cross section of elements 104, 108 and 110 synthesized in cold fusion reactions. The calculations were made by B.Pustylnik [8] within the framework of the traditional approach. The optical model was used for the calculation of the capture cross section. The statistical model — for the calculation of the survival probability of the excited compound nucleus. The experimental data were obtained at the GSI [9]. One can see satisfactory agreement between the calculated and experimental data for element 104. But there is dramatic disagreement in the case of elements 108 and 110. The cause of this disagreement is quasi-fission. However in the traditional approach quasi-fission is not taken into account. The calculations were made according to the ratio (1):

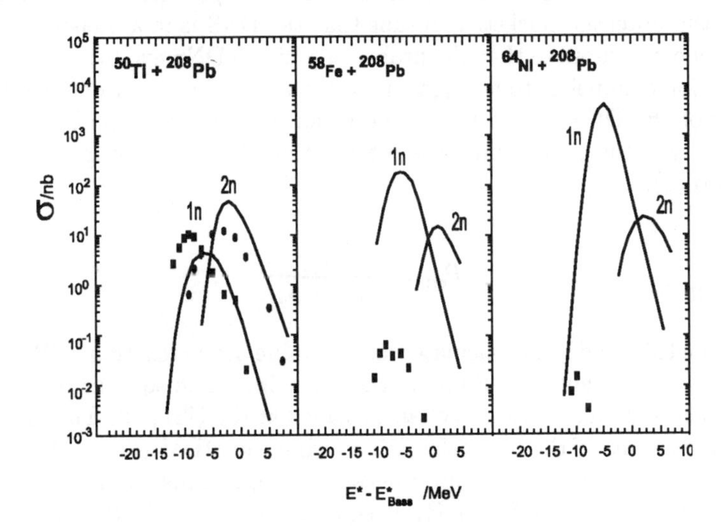

Figure 6. The production cross-section of elements 104, 108 and 110 in cold fusion reactions; • — the experimental data, the lines are the results of calculations.

$$\sigma_{\rm ER} = \sigma_c \cdot W_{\rm sur}, \tag{1}$$

where σ_c — is the capture cross section and $W_{\rm sur}$ — the survival probability of the compound nucleus during its deexcitation. There is no term which reflects the influence of quasi-fission. According to the DNSC, the production cross section of a heavy element is defined by the ratio (2):

$$\sigma_{\rm ER} = \sigma_c \cdot P_{cn} \cdot W_{\rm sur}. \tag{2}$$

Here P_{cn} — is the probability of the compound nucleus formation in the competition with quasi-fission. It should be emphasized that only the DNSC provides a basis for the creation of a realistic model of competition between the complete fusion and quasi-fission channels.

5.3 The model of competition between complete fusion and quasi-fission for the symmetric nuclear reactions

Our first model of competition between complete fusion and quasi-fission was created for the symmetric nuclear reactions between massive nuclei. In these reactions the initial DNS is situated in the minimum of the potential energy V(Z, J) and in the minimum of the nucleus-nucleus potential V(R), at the bottom of the potential pocket. It means that the DNS is in a quasi-equilibrium state. One can suppose that the probability of the DNS evolution to the complete fusion channel or to the quasi-fission channel is proportional to the level density of the DNS on the top of the fusion and quasi-fission barriers. The probability of the compound nucleus formation after the capture is defined by the ratio (3):

$$P_{cn} = \frac{\rho_{B^*_{\rm fus}}}{\rho_{B^*_{\rm fus}} + \rho_{qf}}. \tag{3}$$

Using this model, we calculate the $\sigma_{\rm ER}$ value in the reaction ^{110}Pd+^{110}Pd. The result is presented in Fig.7 by the solid line. The squares represent the experimental data by Maravek et al. obtained at the GSI [10]. The dashed line shows the results of the $\sigma_{\rm ER}$ calculation, using the MDM. The MDM does not take into account the competition between complete fusion and quasi-fission. But in this reaction the factor P_{cn} is equal to $10^{-3} - 10^{-4}$. It means that the quasi-fission channel strongly dominates over the complete fusion channel.

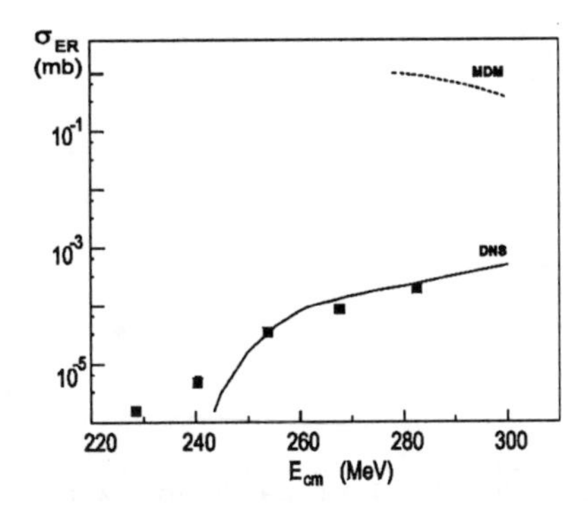

Figure 7. The evaporation residue cross-section $\sigma_{\rm ER}$ in the reaction ^{110}Pd $+^{110}$ Pd; • — the experimental data, the solid line is the results of calculations within the DNSC, the dashed line — within the MDM.

5.4 The models of competition between complete fusion and quasi-fission for the asymmetric nuclear reactions

In the asymmetric nuclear reactions the initial DNS is in a non-equilibrium state and the model for the symmetric reactions cannot be used. We proposed two models for the asymmetric nuclear reactions. In the first model the Monte-Carlo method was used for calculating the DNS evolution [11]. In the second model the evolution of the DNS was considered as a diffusion process proceeding along two collective coordinates. Diffusion along the mass asymmetry coordinate $\eta = (A_1 - A_2)/(A_1 + A_2)$ leads to complete fusion. Diffusion along the R-coordinate leads to quasi-fission (R is the distance between the two centers of the DNS nuclei) [12]. A quasi-stationary solution to the two-dimensional Fokker-Planck equation is used for describing the DNS evolution. The parameter of the model is the DNS viscosity.

The second model was applied to the calculation of P_{cn} in the reactions of synthesis of transfermium and superheavy elements [13]. Fig.8 presents the dependence of P_{cn} on the atomic number of the compound nucleus for cold fusion reactions. For element 104, P_{cn} is equal to $5 \cdot 10^{-2}$. However, for elements 112 and 114 P_{cn} drops to $10^{-}6$ and $10^{-}7$, respectively. The calculations of the P_{cn} values indicate that quasi-fission is the main factor responsible for decreasing the production cross section in the cold fusion reactions.

Using the data for P_{cn}, it is possible to reproduce the experimental data for the production cross section of elements 102-112 in cold fusion reactions. In Fig.9 the filled squares represent the experimental data for $\sigma_{\rm ER}$ [14], the

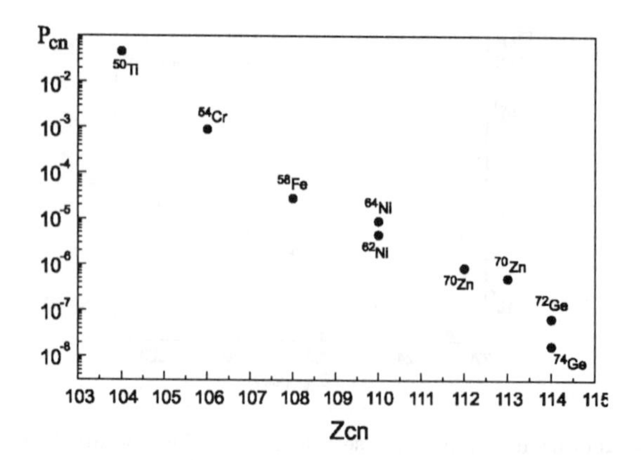

Figure 8. The probability of the complete fusion P_{cn} in the cold fusion reactions $HI, 1n)$

open circles — the calculated data [15]. One can see satisfactory agreement between the two sets of data.

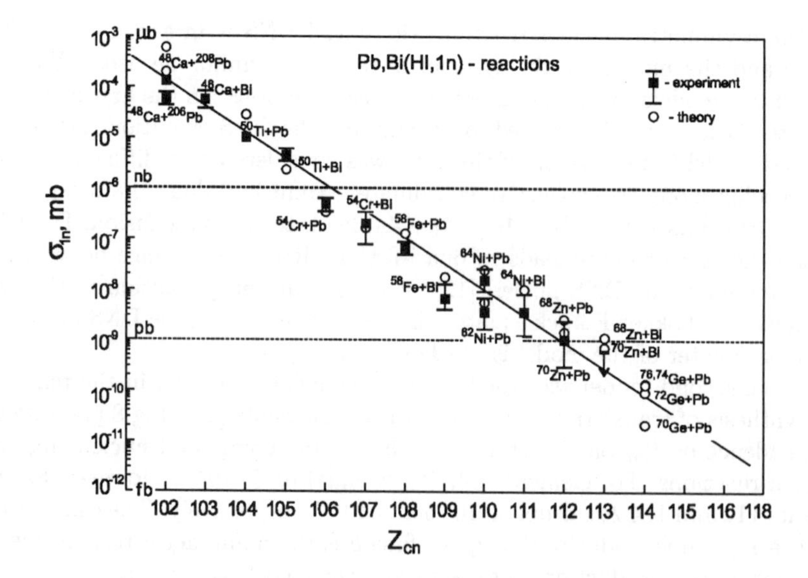

Figure 9. The production cross-sections of elements 102-114 synthesized in the cold fusion reactions (HI, 1n); filled squares are the experimental data, open circles — results of calculations with using the P_{cn} values.

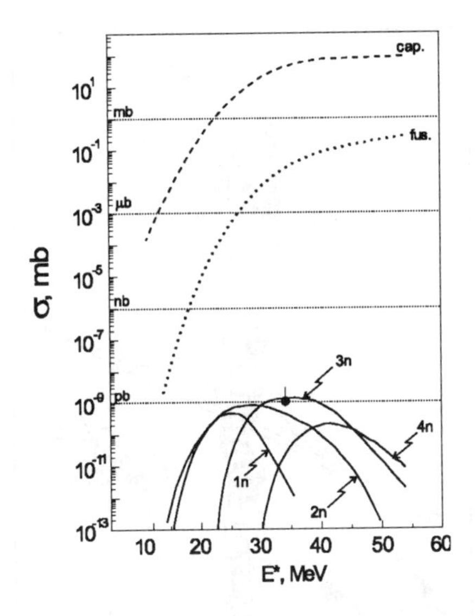

Figure 10. The calculation results of $\sigma_{\rm ER}$ of element 114 in the warm fusion ^{244}Pu+^{48}Ca [15]; the dot is experimental value of obatinedt in the reaction ^{244}Pu+^{48}Ca$\rightarrow^{289}$ $114 + 3n$ [16].

The DNSC makes it possible to reproduce the production cross section of transfermium and superheavy elements in warm fusion reactions, in which ^{48}Ca ions are used [15]. Fig.10 shows the results of calculations of the production cross section of element 114, which was synthesized at Dubna in the reaction $^{244}\mathrm{Pu} +^{48} \mathrm{Ca} \rightarrow^{289} 114 + 3n$ [16]. The first (the upper) curve is the capture cross section σ_c, which was calculated within the framework of the optical model. The second curve is the compound nucleus formation cross section. The difference between these curves reflects the influence of quasi-fission in this reaction. Quasi-fission decreases the probability of the compound nucleus formation by a factor of one thousand. The curves with indexes 1n, 2n, 3n and 4n reflect the competition between the fission and reaction channels with emission of different number of neutrons. The dot is the experimental production cross section. One can see rather good agreement between the calculations and the experimental data. It should be emphasized that the calculations were made before obtaining the experimental data.

For many years fission of excited compound nuclei has been considered the main obstacle on the way to SHE. The DNSC reveals a new serious threat — quasi-fission. Fig.11 shows the two real threats which experimenters encounter on the way to SHE. The synthesis of element 114 in warm fusion reaction and

Figure 11. The two threats which experimenter encounter on the way to superheavy elements

the failure to synthesize element 113 in a cold fusion reaction indicate that the tigers of quasi-fission are more dangerous for the synthesis of SHE than the crocodiles of fission!

6 Conclusion

Analyzing experimental data on the synthesis of transfermium and superheavy elements within the framework of the DNSC one may come to a conclusion that this concept gives today the most realistic description of the complete fusion process.

References

1. Swiatecki W.J., *Phys. Scripta* **24** (1981) p.113; Bjornholm S. and Swiatecki W.J., *Nucl. Phys.* **A319** (1982) p.471; Blocki J.P., Feldmeir H.,

Swiatecki W.J., *Nucl. Phys.* **A459** (1986) p.145.
2. Gross D.H.E., Nayak R.C., Satpathy L., *Z. Phys.* **A299** (1981) p.63.
3. Fröbrich P., *Phys. Reports* **116** (1984) p.338.
4. Abe Y., Ayik S., Reinhard P.G. and Suraud E., *Phys. Reports* **C275** (1996) p.49; Arimoto Y., Wada T., Ohta M. and Abe Y., *Phys. Rev.* **C59** (1999) p.796.
5. Volkov V.V. in: Proc. Int. School-Seminar on Heavy Ion Physics, Dubna, 1986 (JINR, D7-87-68, Dubna 1987) p.528; *Izv. AN SSSR ser. fiz.* **50** (1986) p.1879.
6. Antonenko N.V., Cherepanov E.A., Nasirov A.K., Permjakov V.P. and Volkov V.V., *Phys. Lett.* **B319** (1993) p.425; *Phys. Rev.* **C51** (1995) p.2635.
7. Popeko A.G., *Il Nuovo Cim.* **110A** (1997) p.1137.
8. Pustylnik B., in Dymanical aspects of nuclear fission, ed. by Jan Kliman, Boris I. Pustylnik (JINR, Dubna 1996) p.121.
9. Hofmann S. in Proc. on XV Nucl. Phys. Conf. LEND-95, St.-Petersburg, April 18-22, 1995, World Sci. Singapure, 1995, p.305.
10. Morawek W., Ackermann T., Brohn T., Clerk H.G., Collerthon V., Hanlet E., Horz M., Schwab W., Voss B., Scmidt K.H. and Hessberger F.P., *Z. Phys.* **A341** (1991) p.75.
11. Cherepanov E.A., Volkov V.V., Antonenko N.V., Nasirov A.K. in Heavy Ion Physics and its Application, Lanzhou, China, 1995 ed. Y.X.Luo, G.M.Jin and J.Y.Liu (World scientific, Singapure) 1996, p.272.
12. Adamian G.G., Antonenko N.V., Scheid W., *Nucl. Phys.* **A618** (1997) p.176.
13. Adamian G.G., Antonenko N.V., Scheid W., Volkov V.V., *Nucl. Phys.* **A633** (1998) p.154.
14. Hofmann S., *Rep. Prog. Phys.* **61** (1998) p.639.
15. Cherepanov E.A., Volkov V.V. in 1st Intern. Conf. on the Chemistry and Physics of the Transactinidc Elements, Sep. 26-30, 1999, Seeheim, Germany, Extended Abstracts.
16. Oganessian Yu.Ts., Utenkov V.K., Lobanov Yu.V. et al., *Phys. Rev. Lett.* **83** (1999) p.3154.

FUSION REACTIONS FOR PRODUCING SUPERHEAVY NUCLEI

R. SMOLAŃCZUK

Nuclear Theory Department, Soltan Institute for Nuclear Studies, Hoża 69, PL-00-681 Warszawa, Poland
and
Nuclear Science Division, Lawrence Berkeley National Laboratory, 1 Cyclotron Road, Berkeley, CA 94720, USA
E-mail: smolan@fuw.edu.pl

Cold fusion reactions for the production of new elements are indicated. Decay properties of the resulting isotopes are discussed.

In Ref. [1], we proposed a model for the description of cold fusion reactions [2,3,4,5] with the emission of only one neutron. The model is based on the concept of the quantal tunneling through the Coulomb barrier and statistical description of the competition between neutron emission and fission in the exit channel.

We investigated theoretically the bombardments of ^{208}Pb, ^{207}Pb and ^{209}Bi target nuclei with various heavy ions. Within our model, we reproduced [1,6,7] the measured formation cross section of deformed transactinide nuclei produced in cold fusion reactions and predicted [1,6,7,8] the optimal bombarding energies and the formation cross sections for many deformed and spherical nuclei with $Z \leq 120$. The main result is that the optimal reactions with stable projectiles for producing superheavy nuclei are ^{208}Pb(^{86}Kr,1n)^{293}118, ^{208}Pb(^{87}Rb,1n)^{294}119 and ^{208}Pb(^{88}Sr,1n)^{295}120. Pretty large formation cross sections of ^{293}118, ^{294}119 and ^{295}120 of the order of hundreds of picobarns were obtained. [1,7,8] The fusion hindrance in the entrance channel for these reactions is reduced due to the magicity of the reaction partners. The next reason for the large cross sections is the choice of the projectiles which are the most neutron-rich stable isotopes of Kr, Rb and Sr. The neutron-rich projectiles might lead to superheavy nuclei with higher fission barriers and, consequently, with larger survival probability in the exit channel.

Decay properties of ^{293}118, ^{294}119 and ^{295}120 and their descendants were calculated in Refs. [9,8] by using the macroscopic-microscopic model. [10,11] As the macroscopic energy, the Yukawa-plus-exponential potential [12] was taken. The Strutinsky method [13] was used to calculate the microscopic energy and the single-particle spectra were obtained by means of the Woods-Saxon potential. [14]

The initial part of the predicted [9] α-decay chain of the nucleus $^{293}118$ is the sequence of high energy α particles with energy higher than 10 MeV and with half-lives equal to $\approx$ 1 ms. For $^{294}119$ and $^{295}120$, the initial parts of their decay chains are predicted [9,8] to be the sequences of $\approx$ 100 μs and $\approx$ 20-50 μs α particles, respectively. The calculated decay properties may help in the identification of the synthesized nuclei in cases were all decay products were not observed so far.

An experiment on the synthesis of element 118 was performed [15] recently by the Berkeley group. A very good agreement between the predicted [9] and the measured [15] decay properties of the nuclei constituting the observed α-decay chain provide strong evidence that element 118 was synthesized.

It is an open question if the discrepancy between the measured and the predicted cross section arises from possible model uncertainties discussed in Ref. [7] or forbidden α decays as explained by the authors of Ref. [16] Such decays might be responsible for the measurement, through α decay, only a part of the formation cross section of $^{293}118$.

Acknowledgments

This work was supported by the Director, Office of Science, Office of High Energy and Nuclear Physics, Division of Nuclear Physics under U.S. Department of Energy Contract No. DE-AC03-76SF00098 and Grant of the Polish Committee for Scientific Research (KBN) No. 2 P03B 099 15.

References

References

1. R. Smolańczuk, *Phys. Rev.* C **59**, 2634 (1999).
2. Yu.Ts. Oganessian, in *Classical and Quantum Mechanical Aspects of Heavy Ion Collisions*, Vol. 33 of *Lecture Notes in Physics* (Springer, Heidelberg, 1975), p. 221.
3. G. Münzenberg, *Rep. Prog. Phys.* **51**, 57 (1988).
4. S. Hofmann, *Rep. Prog. Phys.* **61**, 639 (1998).
5. P. Armbruster, *Rep. Prog. Phys.* **62**, 465 (1999).
6. R. Smolańczuk, *Phys. Rev. Lett.* **83**, 4705 (1999).
7. R. Smolańczuk, *Phys. Rev.* C **61**, 011601(R) (2000).
8. R. Smolańczuk, *Phys. Rev. Lett.* (submitted).
9. R. Smolańczuk, *Phys. Rev.* C **60**, 021301(R) (1999).
10. R. Smolańczuk, *Phys. Rev.* C **56**, 812 (1997).

11. R. Smolańczuk, J. Skalski, and A. Sobiczewski, *Phys. Rev.* C **52**, 1871 (1995).
12. P. Möller and J.R. Nix, *At. Data Nucl. Data Tables* **26**, 165 (1981).
13. V.M. Strutinsky, *Nucl. Phys.* **A95**, 420 (1967); **A122**, 1 (1968).
14. S. Ćwiok, J. Dudek, W. Nazarewicz, J. Skalski, and T. Werner, Comput. Phys. Commun. **46**, 379 (1987).
15. V. Ninov, K.E. Gregorich, W. Loveland, A. Ghiorso, D.C. Hoffman, D.M. Lee, H. Nitsche, W.J. Swiatecki, U.W. Kirbach, C.A. Laue, J.L. Adams, J.B. Patin, D.A. Shaughnessy, D.A. Strellis, and P.A. Wilk, *Phys. Rev. Lett.* **83**, 1104 (1999).
16. S. Ćwiok, W. Nazarewicz, and P.H. Heenen, *Phys. Rev. Lett.* **83**, 1108 (1999).

WAYS FOR FUSION OF HEAVY NUCLEI

G.G. ADAMIAN[1,2], N.V. ANTONENKO[2], A. DIAZ-TORRES[1], S.P. IVANOVA[2], W. SCHEID[1], V.V. VOLKOV[2]

[1] *Institut für Theoretische Physik der Justus-Liebig-Universität, Giessen, Germany*
[2] *Joint Institute for Nuclear Research, Dubna, Russia*

It is shown that the compound nucleus is formed by a transfer of nucleons (dinuclear system concept) and not by a melting of the nuclei.

1 Introduction

The experimental synthesis of superheavy elements and heavy nuclei far from the line of stability stimulates the study of the mechanism of fusion in heavy ion collisions at low energies. Models for calculation of fusion cross sections can be discriminated by the assumptions on the dynamics of the fusion process. The important collective degrees of freedom are the relative internuclear distance R between the nuclear centers and the mass asymmetry coordinate η defined as $\eta = (A_1 - A_2)/(A_1 + A_2)$ where A_1 and A_2 are the mass numbers of the clusters. Here, η=0 means a symmetric system and η=$\pm$ 1 the fused system. Depending on the main degree of freedom used for the description of fusion two different types of models for fusion can be distinguished. The first type of models assumes a melting of nuclei along the internuclear distance[1]. The second type is the dinuclear model [2,3]. Here, the fusion proceeds in the mass asymmetry degree of freedom by transferring nucleons or clusters between the nuclei.

2 Evaporation residue cross section for fusion

The evaporation residue cross section can be written as a sum over the contributions of partial waves

$$\sigma_{ER}(E_{c.m.}) = \sum_J \sigma_c(E_{c.m.}, J) P_{CN}(E_{c.m.}, J) W_{sur}(E_{c.m.}, J). \tag{1}$$

The first factor is the partial capture cross section for the transition of colliding nuclei over the entrance (Coulomb) barrier with a probability $T(E_{c.m.}, J)$. The second factor in (1) is the probability of complete fusion that the system approaches an excited compound nucleus, starting from the touching configuration. This probability includes the competition with quasifission after capture. The last factor in (1) is the surviving probability of the fused system and regards the fission and neutron evaporation of the excited compound nucleus. The contributing angular momenta in σ_{ER} are limited by W_{sur} with $J_{max} \approx$ 10-20 in the case of highly fissile superheavy nuclei. For small angular momenta we can approximate this cross section as follows

$$\sigma_{ER}(E_{c.m.}) = \sigma_c(E_{c.m.})P_{CN}(E_{c.m.}, J=0)W_{sur}(E_{c.m.}, J=0) \qquad (2)$$

with $\sigma_c(E_{c.m.}) = \pi\lambda\!\!\!^{-2}(2J_{max}+1)T(E_{c.m.}, J=0)$. For reactions leading to superheavy nuclei, values of J_{max}=10 and $T(E_{c.m.}, J=0)$=0.5 are chosen for bombarding energies $E_{c.m.}$ near the Coulomb barrier.

3 Adiabatic treatment of internuclear motion

Models describing the fusion process as an internuclear melting of nuclei often use adiabatic potential energy surfaces (PES). These potentials are calculated with the macroscopic-microscopic method of the Strutinsky formalism. Here we apply the microscopic two-center shell model with the following coordinates: elongation $\lambda = \ell/(2R_0)$, where ℓ is the length of the system and $2R_0$ the diameter of the compound system, neck parameter ε with ε=0 showing no neck and $\varepsilon \approx 1$ showing necked-in shapes, mass asymmetry η and deformation parameters $\beta_i = a_i/b_i$ which are ratios of semiaxes of the clusters i=1 and 2. The adiabatic potential is obtained as

$$U(\lambda, \varepsilon, \eta, \eta_z, \beta_i) = U_{LDM} + \delta U_{shell} + \delta U_{pairing}, \qquad (3)$$

where U_{LDM} is the liquid drop potential, δU_{shell} the shell correction part originating from the TCSM and $\delta U_{pairing}$ the pairing energy part.

Fig. 1 shows the adiabatic potential energy surface for the system $^{110}Pd+^{110}Pd$ [4]. The fission-type valley along λ can be recognized at small ε. The potential energy surface is used for classical dynamical calculations of trajectories. In addition to the PES a kinetic energy $T = \frac{1}{2}\sum_{i,j} M_{ij}(q)\dot{q}_i\dot{q}_j$ and a Rayleigh dissipation function $\Phi = \frac{1}{2}\sum_{i,j}\gamma_{ij}(q)\dot{q}_i\dot{q}_j$ are applied, where the masses are calculated in Werner-Wheeler or hydrodynamical approximation and the friction coefficients with the expression $\gamma_{ij} = \Gamma M_{ij}/\hbar$ following from the linear response theory. The quantity Γ is an average width of single particle states. Trajectories of the descent to the fission-type valley are given in Fig. 1. They start from a DNS shape in a touching configuration with λ=1.59 and ε=0.75 and end in the fission valley.

In order to calculate the fusion probability we started from a value of $\lambda = \lambda_v$

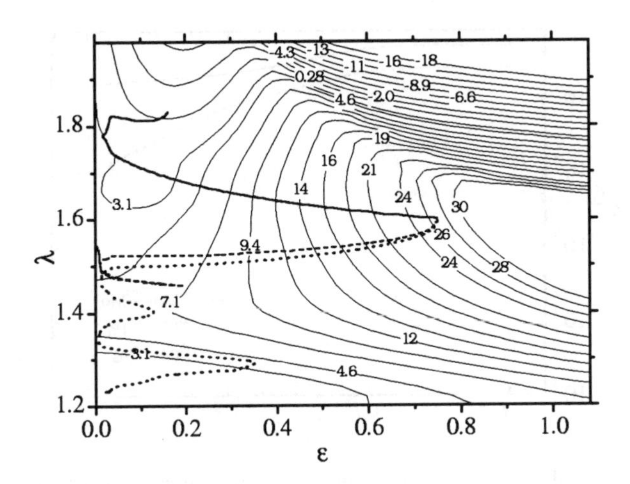

Figure 1. Adiabatic potential energy contours drawn in the (λ, ε)-plane for the system $^{110}Pd+^{110}Pd$ with shell corrections and β_i=1. The dynamical trajectories starting from the touching configuration with initial kinetic energies of 0, 40 and 60 MeV are shown by solid, dashed and dotted curves, respectively.

in the fission valley obtained from the dynamical calculation of the descent into this valley. The fusion probability is defined by the rate $\Lambda_{fus}(t)$ according to an one-dimensional Kramers expression which is a quasistationary solution of the Fokker-Planck equation.

$$P_{CN} = \int_0^{t_0} \Lambda_{fus}(t)dt, \tag{4}$$

where t_0 is the lifetime of the DNS and $\Lambda_{fus} \sim \exp(-B_{fus}/T)$. Here, B_{fus} is the barrier for fusion and $T = \sqrt{E^*/a}$ the local thermodynamic temperature with the excitation energy E^* of the system. Fig. 2 shows that the fusion probabilities obtained with this model are much larger than the values found from experimental data. We conclude that an adiabatic calculation of the potential energy surface is not adequate for the description of fusion of heavy nuclei.

4 Microscopic inertia tensor

The hydrodynamical mass for the neck coordinate is small and lets increase the neck of the DNS rapidly to a shape of a strongly deformed compound nucleus.

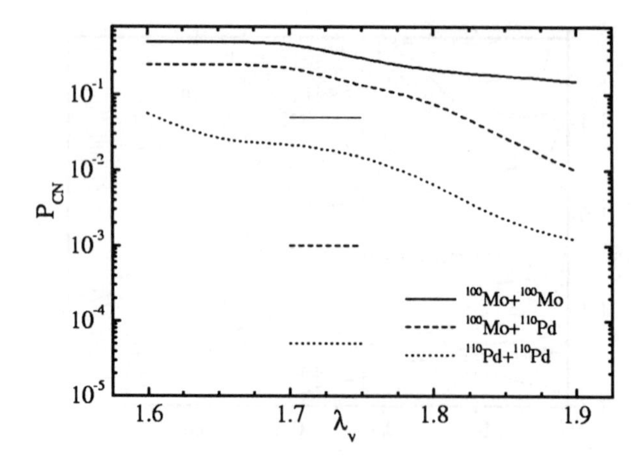

Figure 2. Fusion probabilities depending on the starting value λ_v in the adiabatic PES for various reactions. The values extracted from experimental data are given by horizontal lines.

Howerver, crossings of single particle levels lead to a large inertia of the system which hinders the growth of the neck. Using the single-particle spectrum and wave functions of the TCSM one can obtain the mass parameters with an extented cranking formula

$$M_{ij}^{cr} = \hbar^2 \sum_{\alpha,\beta} \frac{<\alpha|\frac{\partial \hat{H}}{\partial q_i}|\beta><\beta|\frac{\partial \hat{H}}{\partial q_j}|\alpha>}{(\epsilon_\beta - \epsilon_\alpha)^2 + \frac{1}{4}(\Gamma_\beta + \Gamma_\alpha)^2} \frac{n_\alpha - n_\beta}{\epsilon_\beta - \epsilon_\alpha}, \tag{5}$$

where Γ_α and Γ_β are the widths of the single particle states. The contributions to the mass parameters (5) can be classified as those coming from nondiagonal and diagonal elements, respectively. The main contribution to M_{ij}^{cr} arises from the diagonal parts at $\epsilon_\beta \to \epsilon_\alpha$. Then expression (5) can be approximated as

$$M_{ij}^{cr} \approx \hbar^2 \sum_{\alpha} \frac{\partial \epsilon_\alpha}{\partial q_i} \frac{\partial \epsilon_\alpha}{\partial q_j} \frac{1}{\Gamma_\alpha^2} (\frac{-dn_\alpha}{d\epsilon_\alpha}), \tag{6}$$

where the derivative of the occupation number, $-dn_\alpha/d\epsilon_\alpha$, has a bell-like shape with a width T and is peaked at the Fermi energy.
Comparing our results with M_{ij}^{WW} obtained in the Werner-Wheeler approximation for a touching configuration with excitation energy 30 MeV (T=1.3 MeV), we find $M_{\lambda\lambda}^{cr} = M_{\lambda\lambda}^{WW}$, $M_{\varepsilon\varepsilon}^{cr} \approx$ (20-30)$M_{\varepsilon\varepsilon}^{WW}$ and $M_{\lambda\varepsilon}^{cr} \approx 0.4\ M_{\lambda\varepsilon}^{WW}$, practically independent of the mass number of the system [4,5]. The neck parameter calculated with the Werner-Wheeler and cranking mass formulas is

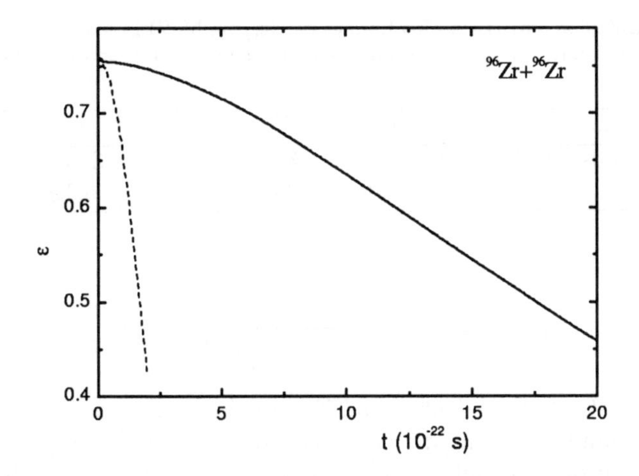

Figure 3. The dynamics of the neck parameter ε in the system ^{96}Zr+^{96}Zr calculated with mass parameters of the extended cranking model (solid curve) and in Werner-Wheeler approximation (dashed curve).

compared in Fig. 3. As result we conclude that the initial neck coordinate is nearly kept due to the large microscopical mass parameter for a time comparable with the reaction time. Therefore, the initial neck coordinate stands approximately fixed, which is assumed in the dinuclear model. However, an adiabatic treatment[5] of fusion in λ also yields fusion probabilities which are considerably overestimated in comparison to the experimental data.

5 Dynamical diabatic description

Because the adiabatic PES is not adequate for the description of fusion, we investigated the dynamics of the transition between an initially diabatic (sudden) interaction potential and an adiabatic potential during the fusion process. The main question is the survival of the dinuclear configuration evolving in the mass asymmetry.

In the calculations we applied the maximum symmetry method for the Hamiltonian of the diabatic TCSM [6]. Diabatic TCSM levels do not show avoided level crossings. Avoided level crossings can be removed from the adiabatic TCSM by eliminating the symmetry-violating parts from the adiabatic Hamiltonian of the TCSM. The diabatic potential can be written

$$U_{diab}(\lambda, \varepsilon) = U_{adiab}(\lambda, \varepsilon) + \delta U_{diab}(\lambda, \varepsilon), \quad (7)$$

Table 1. Quasifission barrier (in MeV), inner barriers (in MeV) in η and λ calculated with the dynamical diabatic potential, fusion probabilities in λ and η, and experimental fusion probability.

Reaction	B^{λ}_{qf}	B^{η}_{fus}	B^{λ}_{fus}	P^{λ}_{CN}	P^{η}_{CN}	P^{exp}_{fus}
$^{110}Pd+^{110}Pd\rightarrow^{220}U$	1.3	12	36	$3\cdot 10^{-15}$	$3\cdot 10^{-4}$	$\sim 10^{-4}$
$^{86}Kr+^{160}Gd\rightarrow^{246}Fm$	0.2	12	65	$4\cdot 10^{-26}$	$7\cdot 10^{-5}$	$5\cdot 10^{-5}$

where

$$\delta U_{diab} = \sum_{\alpha} \epsilon_{\alpha}^{diab} n_{\alpha}^{diab} - \sum_{\alpha} \epsilon_{\alpha}^{adiab} n_{\alpha}^{adiab}. \tag{8}$$

Here, ϵ_{α}^{diab}, $\epsilon_{\alpha}^{adiab}$ and n_{α}^{diab}, n_{α}^{adiab} are the single particle energies and occupation numbers of the diabatic and adiabatic levels, respectively. The initial occupation probabilities n_{α}^{diab} are determined by the configuration of the separated nuclei. They depend on time since the excited diabatic levels get deexcited:

$$\frac{dn_{\alpha}^{diab}(\lambda,\varepsilon,t)}{dt} = -\frac{1}{\tau(\lambda,\varepsilon,t)}(n_{\alpha}^{diab}(\lambda,\varepsilon,t) - n_{\alpha}^{adiab}(\lambda,\varepsilon)), \tag{9}$$

where τ is a relaxation time determined by a mean single particle width depending on the diabatic occupation numbers

$$(\tau(\lambda,\varepsilon,t))^{-1} = \sum_{\alpha} \bar{n}_{\alpha}^{diab}\Gamma_{\alpha}/(2\hbar\sum_{\alpha}\bar{n}_{\alpha}^{diab}). \tag{10}$$

Diabatic potentials show a strong increase with decreasing elongation λ. Their repulsive character screens smaller values of the elongation and hinders the DNS to melt into the compound nucleus. Fig. 4 shows the diabatic potential for the system $^{110}Pd+^{110}Pd$ as a function of the elongation λ for the initial time and the lifetime t_0 of the DNS. The nuclei are considered as spherical with a neck parameter ε=0.75. In Table 1 examples for barriers and probabilities for forming compound nuclei are listed [7]. The probabilities were calculated with Kramers expressions similar to Eq.(4).

As in Ref. [8], our analysis with the diabatic dynamics [7] yields the result that the fusion of heavy nuclei along the internuclear distance in the coordinates R or λ is very unprobable, especially in the case of fusion of equal nuclei in contrast to experimental data. These facts, demonstrated with the examples of Table 1, strongly support our standpoint that the correct model for fusion of heavy nuclei is the dinuclear system model where fusion is described by the transfer of nucleons in the touching configuration.

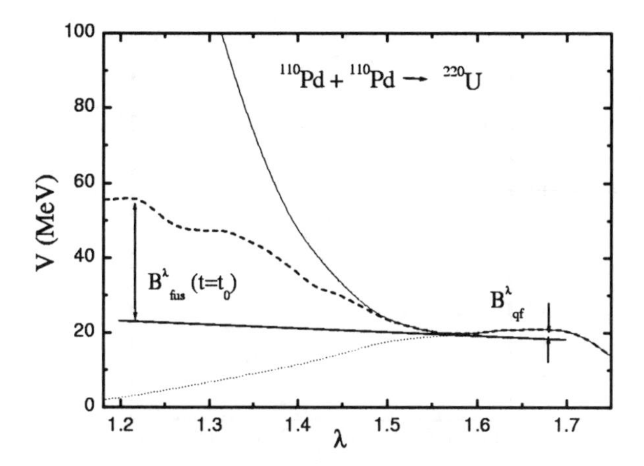

Figure 4. The diabatic time-dependent potentials (solid and dashed curves) and the adiabatic potential (dotted curve) for ^{110}Pd+^{110}Pd as a function of λ.

6 Cross sections for superheavy nuclei and summary

The main result of this study is that fusion of heavy nuclei to superheavy nuclei is correctly described by the dinuclear model. This model assumes that the compound nucleus is formed by nucleon or cluster transfer between two touching nuclei in the dinuclear configuration. The available experimental data can be well reproduced with the dinuclear system concept with exception of the reaction ^{86}Kr+^{208}Pb→^{293}118+1n. We calculated evaporation residue cross sections according to Eq.(1) for lead- and actinide-based reactions. The strong decrease of the fusion cross sections with the charge number of the fused system is caused by the decrease of the probability P_{CN} for forming the compound nucleus (see Fig. 5). For example, for ^{70}Zn+^{208}Pb→^{277}112+1n we found P_{CN}=1$\cdot$ 10^{-6}, σ_{ER}=1.8 pb (exp. 1.0 pb) and for ^{86}Kr+^{208}Pb→^{293}118+1n we obtained P_{CN} = 1.5$\cdot$ 10^{-10}, σ_{ER}=5.1 fb [9]. The surviving probability varies moderately with the shell structure of the compound nucleus between 10^{-4} and 10^{-2}.

A further proof for the dinuclear system concept would be the reproduction of experimental quasifission distributions. These cross sections could give further important information on the dynamics of the fusion process. Work on this topic is in progress.

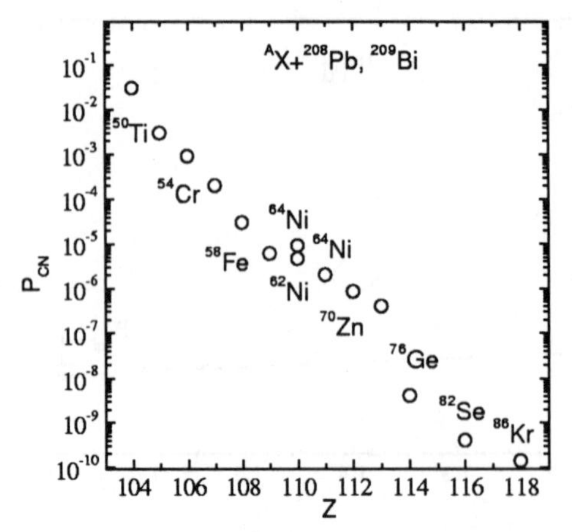

Figure 5. Calculated fusion probabilities P_{CN} for cold fusion in (HI,1n) lead-based reactions for the projectiles indicated.

References

1. A. Sandulescu, R.K. Gupta, W. Scheid, W. Greiner, *Phys. Lett.* B **60**, 225 (1976).
2. V.V. Volkov, *Izv. AN SSSR ser. fiz.* **50**, 1879 (1986).
3. N.V. Antonenko, E.A. Cherepanov, E.K. Nasirov, V.B. Permjakov, V.V. Volkov, *Phys. Lett.* B **319**, 425 (1993); *Phys. Rev.* C **51**, 2635 (1995).
4. G.G. Adamian, N.V. Antonenko, S.P. Ivanova, W. Scheid, *Nucl. Phys.* A **646**, 29 (1999).
5. G.G. Adamian, N.V. Antonenko, A. Diaz-Torres, W. Scheid, *Nucl. Phys.* A , (2000).
6. A. Diaz-Torres, N.V. Antonenko, W. Scheid, *Nucl. Phys.* A **652**, 61 (1999).
7. A. Diaz-Torres, G.G. Adamian, N.V. Antonenko, W. Scheid, Melting or nucleon transfer in fusion of heavy nuclei?, preprint, Giessen 1999.
8. G.G. Adamian, N.V. Antonenko, Yu.M. Tchuvil'sky, *Phys. Lett.* B **451**, 289 (1999).
9. G.G. Adamian, N.V. Antonenko, W. Scheid, Isotopic dependence of fusion cross sections in reactions with heavy nuclei, preprint, Giessen 2000.

NEW DATA ON THE FRAGMENT ANGULAR MOMENTUM IN SPONTANEOUS FISSION OF ^{252}CF

G.M.TER-AKOPIAN, G.S.POPEKO, A.V.DANIEL, YU.TS.OGANESSIAN
Joint Institute for Nuclear Research, Dubna,141980, Russia

J.H.HAMILTON, J.KORMICKI*, A.V.RAMAYYA, K.HWANG
Physics Department, Vanderbilt University, Nashville, TN 37235 USA

W.GREINER
Institut für Theoretishe Physik der J.W. Goethe Universitat, D-60054, Frankfurt am Main, Germany

A.SANDULESCU, A.FLORESCU
Institute for Atomic Physics, Bucharest, P.O. Box MG-6, Rumania

J.O.RASMUSSEN
Lawrence Berkeley National Laboratory, Berkeley CA 94720, U.S.A.

M.A.STOYER
Lawrence Livermore National Laboratory, Livermore CA 94550, U.S.A.

J.D.COLE
Idaho National Engineering Laboratory, Idaho Falls, ID 83415 U.S.A.

Average angular momentum for primary fission fragments as a function of neutron multiplicity and neutron-to-proton ratio of the primary fission fragments was extracted for the first time for the Mo-Ba and Zr-Ce charge splits of ^{252}Cf. The data are discussed in terms of the energy balance occurring at the scission point

1 Introduction

Though a considerable progress has been made in understanding the process of low-energy fission [1] still only a little is known about the energy balance occurring at the scission point. Regarding the value of the energy that released at the descent of the nucleus from saddle to scission only theory estimates exist. For the case of ^{252}Cf spontaneous fission a value of 30 MeV could be assumed for this energy release [2]. However, the partition of this free energy

---

*ALSO AT UNISOR, ORISE, OAK RIDGE, TN 37831; ON LEAVE FROM INSTITUTE OF NUCLEAR PHYSICS, CRACOW, POLAND.

remains so far the least-understood aspect of nuclear fission [1,3]. Evidently, the fragment pre-scission collective motion and internal excitation degrees of freedom share a part of this energy release resulting in the pre-scission kinetic energy and pre-scission excitation of the fragments. However there is so far no direct way to acquire data showing the amount of energy that is stored at the scission point in these two forms. The pre-scission kinetic energy makes a small addition to the total kinetic energy (TKE) of the fragments obtained at infinity. There is no way to estimate this small addition on the background of a dominating term in TKE coming from the fragment electrostatic interaction at the scission point. A value of about 8 MeV was deduced for the pre-scission kinetic energy from the analysis of angular and kinetic energy distributions of α particles emitted in ternary fission [4]. Lacking in something better one should take this indirect estimation as a reference point. Data on the number of prompt fission neutrons are useless for estimating the fragment intrinsic excitation at scission because the main contribution in the excitation energy of the separated primary fragments comes from the energy stored in the deformation acquired by the fragments at the scission point. Analysis of the odd-even Z staggering in fragment yields indicates that there is only a small chance for pair breaking at the descent to the scission point. Following this way an upper limit of 11 MeV was estimated for the intrinsic excitation of ^{252}Cf at its scission point (see Ref. [1]). Remaining are the excitations of collective degrees of freedom at scission. One can not exclude that the major part of the energy released at descent goes to the excitation of collective vibrations at scission. It is generally recognized that the angular momentum that fission fragments acquire in low-energy fission is a probe for the states of excitation for some of the orthogonal angular-momentum bearing modes at the scission point. These are the so-called "bending" and "wriggling" oscillation modes [5]. In previous investigations (see Ref. [6]) integrated information was extracted on the fragment angular momentum. Such integrated data were deduced for the first time from the rates of discrete γ transitions occurring between levels of ground state rotational bands in the fission fragments of ^{252}Cf [7]. The authors [7] also pointed at the advantages of this approach compared to other ones used for the fragment angular momentum study.

In this paper we present new data on the spontaneous fission of ^{252}Cf where for the first time the fragment angular momentum values were measured in function of the prompt neutron multiplicity and neutron-to-proton ratio of primary fission fragments.

2 Experimental method and results

Fission fragment angular momenta were obtained from the analysis of two-dimensional coincidence spectra of prompt γ rays emitted by secondary fragments. A hermetically sealed source of ^{252}Cf giving 6×10^4 fission events per second was placed at the center of Gammasphere when it had 36 Ge detectors. Approximately, 2.5×10^9 $\gamma-$ ray coincidence events with a multiplicity of $\geq$2 were recorded. Intensities of $\gamma - \gamma$ coincidence peaks were extracted from two-dimensional spectra as described in Ref. [8]. The presently available new data on fission fragment level schemes [9,10] played an essential role in the realization of the new method.

Independent yields of fission fragment pairs of three charge splits of the ^{252}Cf nucleus were obtained from the data of this experiment (see Ref. [11]). For the fragment pairs that involved two even-even nuclei, the yields were estimated from the peak areas resulting from the coincidences of $2^+ \rightarrow 0^+$ transitions in the ground state bands of both fragments. In the case of odd mass fragments, the pair yields were obtained from the sums of all $\gamma - \gamma$ peaks originating from transitions to the fragment ground states. After correcting relative peak areas for the detection efficiency, $\gamma-$ray conversion probability and effects caused by isomeric transitions, independent yields (in percents to the total number of fission events) for secondary fragment pairs were estimated. For the charge splits Z_L/Z_H =44/45 (Ru–Xe), 42/56 (Mo–Ba) and 40/58 (Zr–Ce), the yields were obtained for 33, 52 and 43 fragment pairs, respectively. The most extensive data sets involve secondary fragment pairs formed after neutron evaporation from primary fragment pairs with prompt neutron numbers (ν_{tot}) ranging from zero to ten (eight) neutrons for the Mo–Ba (Zr-Ce) splits. The mass, excitation energy and TKE distributions of primary fission fragments resulted after unfolding the yield patterns of secondary fragment pairs (the unfolding procedure is described in Ref. [8]). Least square fits to the data of Ru-Xe and Zr-Ce splits showed the contribution of one fission mode with $\langle TKE \rangle$ $\cong$193 and 183 MeV, respectively, for the Ru–Xe and Zr–Ce splits. These values are close to $\langle TKE \rangle$ =189 MeV known for the ^{252}Cf spontaneous fission [12]. For the Mo–Ba split, a reasonable fit could be obtained only at an assumption that, in addition to the "normal" mode ($\langle TKE \rangle$ $\cong$189 MeV), a second, peculiar mode ($\langle TKE \rangle$ $\cong$153 MeV) contributes in this split (see details in Ref. [8,11,13].

For a given secondary fragment, we measured level populations when it was obtained in pairs with different complementary secondary fragments. To calculate the average angular momentum we determined, from γ transition intensities, populations P_i for the levels of spin J_i, and calculated average

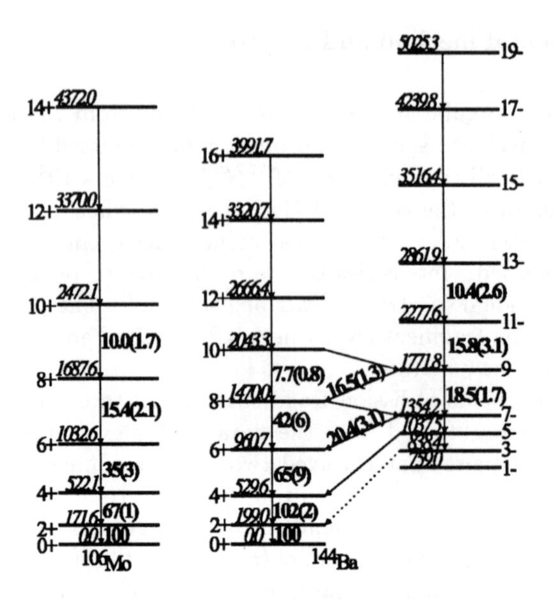

Figure 1. Level diadrams of ^{106}Mo and ^{146}Ba The numbers printed in boldface show the γ transition intensities obtained when the two fragments are emitted in one secondary pair.

spin values $\langle J \rangle = \Sigma_i(P_i \times J_i)/\Sigma_i P_i$ for each chosen secondary fragment. For the even-even fragments, where level schemes and excited level spins are more accurately known, we measured the intensities of γ transitions in the ground state band and side bands, and transitions feeding the ground state band from the side bands. These measurements were made for any individual secondary even-even fragment (A1, Z1) when this fragment was formed in pairs with different partners (A2, Z2). For the even-even fragment A2, Z2, the γ ray intensities in fragment A1, Z1 were derived from the areas of the coincidence peaks of the $2^+ \rightarrow 0^+$ transitions. For odd mass partner fragments A2, Z2, the sum of the intensities of all transitions leading to the ground state of the partner fragment were determined. Transition intensities were calculated from the peak areas corrected for internal conversion. As an example, the measured transition intensities are presented in Fig. 1 for one secondary fragment pair, ^{106}Mo-^{146}Ba. For fragment pairs showing a strong population of side bands, the intensities of the side bands were taken into account. For instance, for the fragment pair ^{144}Ba-^{106}Mo taking into account the side band population, which in ^{144}Ba is 30 %, changes the average spin value from $\langle J \rangle = 5.5$ to $\langle J \rangle = 7.0$. For low-yield pairs the contribution of side bands is generally less

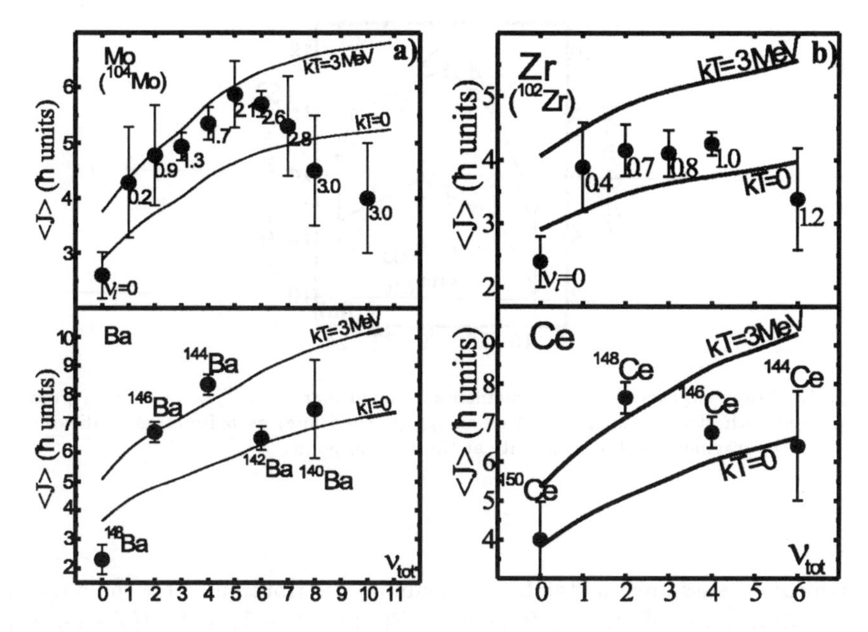

Figure 2. a) Average angular momentum of Mo and Ba primary fragments as a function of the prompt neutron number ν_{tot}. Results are deduced from values $\langle J \rangle$ obtained for the pairs of secondary fragments involving ^{104}Mo and different complementary Ba fragments. Numbers ν_l for prompt neutrons emitted by light fragments are presented at corresponding data points in upper panel. b) The same for the Zr-Ce pairs. Results are deduced from values $\langle J \rangle$ obtained for the pairs of secondary fragments involving ^{102}Zr and different complementary Ce fragments.

important, and neglecting the side band intensities causes only small ($\leq 8\%$) underestimate for the $\langle J \rangle$.

The average angular momenta of primary fragments are obtained from the average spin values of secondary fragments corrected for the angular momentum removed by neutron evaporation. For each secondary fragment pair with a fixed ν_{tot}, the mean numbers of neutrons evaporated by the light and heavy fragments, ν_l and ν_h, were extracted from the unfolding results [8]. These ν_l and ν_h values and the mass numbers of the "average" primary fragments, for which the average angular momenta are estimated, are fractional because several primary fragments contribute to the yield of each secondary fragment (see Fig. 2 and 4).

The average spin j removed by one evaporated neutron was calculated with evaporation code PACE2. To fit the measured $\langle J \rangle$ values for secondary

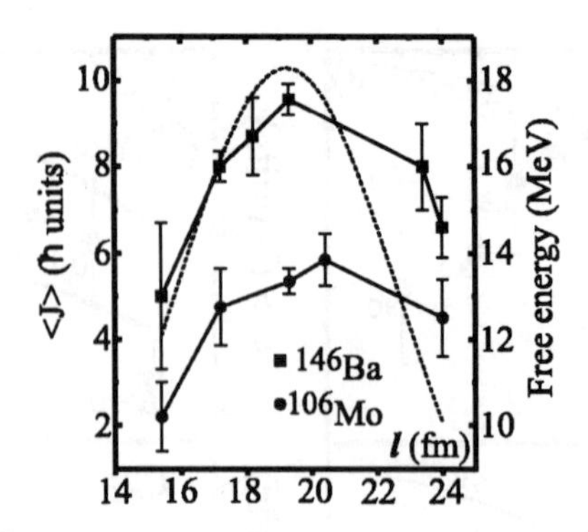

Figure 3. Fragment average angular momenta in the primary pair $^{106}Mo - ^{146}Ba$ (symbols connected with straight lines) and free energy (dashed line) as a function of distance l between the fragment centers of gravity at the scission point.

fragments we assumed a standard probability distribution [14] for the angular momentum states of primary fragments: $P(J) \propto (2J+1)exp\{-J(J+1)/B2\}$ and fitted parameter B in this distribution (B is approximately equal to the rms value of $(J+1/2)$).

More than 70 angular momentum values were determined for the primary Ba-Mo and Ce-Zr fragment pairs. Some of these data are presented in Fig. 2 as a function of ν_{tot}. Average angular momenta deduced for light primary fragment from the average spins of ^{104}Mo and ^{102}Zr are shown in the upper panels. The lower panels show the angular momenta of the primary Ba and Ce fragments, which resulted from the average spins observed for even-even secondary Ba and Ce fragments emitted in pairs with ^{104}Mo and ^{102}Zr.

In Fig. 3 the average angular momenta of two primary fragments, ^{106}Mo and ^{146}Ba, one of the most abundant primary pair in the ^{252}Cf spontaneous fission, are shown in function of the distance l between the centers of gravity of the fragments at the scission point. The distance l was derived from the fragment excitation energy, which, in its turn, was derived from relevant values ν_l, ν_h assuming that the fragment excitation obtained at infinity was stored in their scission-point deformations. These results were derived from the angular momenta of secondary pairs that were formed mostly after neutron evaporation from the $^{106}Mo - ^{146}Ba$ primary pair.

In Fig. 4 the sums of the angular momenta of the primary light and heavy Mo-Ba and Zr-Ce pairs are shown as a function of the mass number of the heavy fragment.

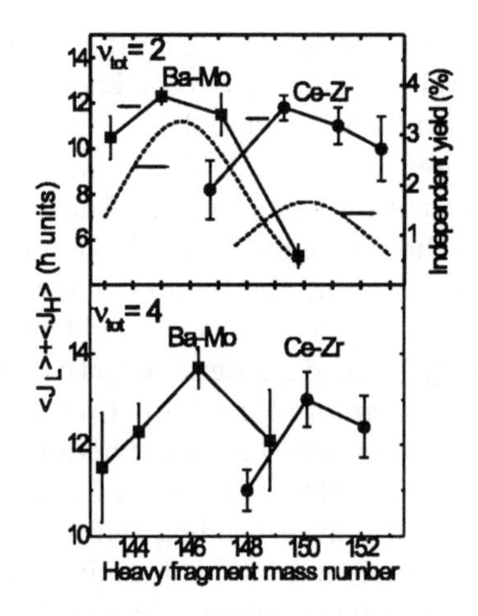

Figure 4. Sums of fragment average angular momenta obtained for the Mo-Ba (squares) and Zr-Ce (circles) splits. The data are presented in function of the heavy fragment mass number for primary fragment pairs associated with the evaporation of ν_{tot} =2 and 4 neutrons. Dashed lines in upper panel show the yields of primary fragment pairs [8].

3 Discussion

It will be appropriate at first to consider the results presented in Fig. 2 and 3 in the light of the so called model of the scission point bending oscillation. The authors of Ref. [15] formulated this model assuming the case of one spherical fragment at the scission point. Later the consideration was extended in Ref. [16] to the case of two deformed fragments at the scission point. Both papers assumed a strong damping of the ordered motion along the fission path, leading from the saddle to scission, by coupling to other collective degrees of freedom. Therefore the presence of statistical equilibrium only between collective degrees of freedom was assumed. This kind of equilibrium in fission was first discussed in Ref. [17]. From Refs. [15,16] one could anticipate that the primary fragment angular momentum rises with the increase of the bending temperature and/or the fragment deformation at the scission point. It is little known about how the bending temperature could vary with changing the scission-point deformation. Therefore we took fixed bending temperatures and, using formulae of Ref. [16], calculated the values of angular momentum, which depend on the mass, charge and deformation of both fragments. The scission point configuration was approximated as a system of two coupled spheroids performing bending oscillation. Quadrupole deformation parameters were calculated using the liquid-drop formula [2] with fragment excitation energies deduced from the known numbers of evaporated neutrons. Curves obtained for kT=0 and 3 MeV are shown in Fig. 2. Rather high bending temperatures (kT=2-3 MeV) follow from these figures for the most probable fission events associated with 2-5 neutrons evaporation. This is in agreement with conclusions made in Ref. [16] about integrated data on the fragment angular momentum from the thermal neutron fission of ^{235}U. Such high temperatures seem to be possible in the light of the basic assumptions of thermal equilibrium established between collective degrees of freedom and weak coupling to intrinsic degrees of freedom. However, one should not take too seriously these absolute temperature values because inclusion of some coupling between collective and intrinsic degrees of freedom (as it was suggested in Ref. [18], for instance) will reduce bending temperatures inferred with the help of this model from our angular momentum data. Leaving aside the issue of the absolute scission-point temperature, one could infer instructive conclusions from its relative variations considered from the point of view of bending oscillation model. Experimental points in Fig. 2 and 3 show systematically that the mean angular momentum values are lower for light fragments as compared to the heavy ones. However, this difference, which sometimes is close to 100 %, can be well explained in terms of bending oscillations.

One could explain the decrease of the angular momentum seen for the case of the Mo–Ba splitting at $\nu_{tot} > 5$ in Fig. 2 invoking a low bending temperature (essentially kT=0) for the second, peculiar fission mode, which was found in the Ba-Mo split of ^{252}Cf [8]. This feature of the peculiar fission mode, together with its large scission-point deformation and low TKE, is reminiscent of cold deformed fission [1], cold because the energy at scission is in deformation. However, while the peculiar fission mode is not seen in the case of the Zr–Ce split of ^{252}Cf [8], relevant data in Fig. 2 suggest a similar dependence of the bending temperature on the scission point deformation. Higher bending temperature is obtained at "normal" scission point deformations resulting in $\nu_{tot} = 2 - 5$, while lower temperature is associated with large ($\nu_{tot} > 5$) and small ($\nu_{tot} = 0$) deformations. Correlation between this temperature course and behavior of the free energy release obtained at the scission point becomes visible in Fig. 3 where the free energy (E_{free}) is shown in function of the scission-point deformation together with the average angular momenta of ^{106}Mo and ^{146}Ba. Here $E_{free} = Q - V_{int} + V_{def}$; Q, V_{int}, and V_{def} are, respectively, the reaction Q value, the sum of the Coulomb and nuclear interaction energies, and the fragment deformation energy at the scission point. It appears that the bending oscillations become cooler when the free energy is low, and they attain their maximum value when the free energy is maximum. Such temperature behavior appears to be reasonable in the light of a similar regularity seen for the intrinsic excitation (see Ref. [1]).

An interesting correlation between the sums of the angular momenta and the yields of Mo–Ba and Zr–Ce primary fragment pairs is seen in Fig. 4 where these values are shown in function of the heavy primary fragment mass number. The fragment pairs lying on the left and right wings of the yield distributions are associated with the dipole oscillation occurring at the descent from saddle to scission. The dipole oscillation takes away some 2.5-3.0 MeV from the descent energy release [1]. This correlation shows that the subtraction of 2.5-3.0 MeV from the total energy balance results in a remarkable reduction of the bending temperature. The observed correlation gives a clear indication that there is a coupling between the two collective degrees of freedom, i.e. when energy is pumped into the dipole oscillation, the bending mode receives a lower excitation. Recently, the so called "orientation pumping" has been discussed [19] as a mechanism bearing the fragment angular momentum. It appears to us, however, that one can not explain the regularities, that are revealed for the fragment angular momentum values shown in Figs. 2-4 at the expense of the "orientation pumping" alone. Perhaps, the observed fragment angular momentum is a result of a combined action of bending oscillation and "orientation pumping". One can not exclude that the "orientation pumping" could be responsible for the creation of the relatively large obtained angular momentum values whereas, the regularities seen in Figs. 2-4 are caused by the bending oscillation.

Acknowledgments

G.M.T.-A., A.V.D., A.S., A.F. and W.G. thank Vanderbilt and JIHIR for their hospitality and financial support. Work is supported at VU, INEL, LBNL, LLNL by the DOE grant and contracts DE-FG05-88ER40407, DE-AC07-76ID01570, DE-AC03-76SF00098 and W-7405-ENG48, at JINR (Dubna) and University of Frankfurt by the grants of RBSF (98-02-04134) and DFG (436RUS 113/484/O(R)). JIHIR is supported by member institutions the UTN, VU and ORNL and DOE grant DE-FG05-87ER40361.

References

1. F. Gönnenwein, In: *The Nuclear Fission Process*, C.Wagemans, Ed. (CRC Press, Boca Raton, 1991), p. 287.
2. U. Brosa, S. Grossmann and A. Müller, *Phys. Pep.*, **197** (1990) 167.
3. R. Vandenbosch, *Nucl. Phys.*, **A502** (1989) 1c.
4. C. R. Guet, In: *Physics and Chemistry of Fission*, Proc. Int. Symp., 1979, (IAEA, Vienna, 1980) vol II, p. 247.
5. J. R. Nix and W. J. Swiatecki, *Nucl. Phys.*, **71** (1965) 1.
6. D. De Frene, In: *The Nuclear Fission Process,* C. Wagemans, Ed. (CRC Press, Boca Raton, 1991), p. 475.
7. J. B. Wilhelmy et al., *Phys. Rev. C,* bf5 (1972)2041.
8. G. M. Ter-Akopian et al., *Phys. Rev. C,* **55** (1997) 1146.
9. J. H. Hamilton et al., *Prog. Part. Nucl. Phys.,* **35** (1995) 635.
10. J. H. Hamilton et al., *Prog. Part. Nucl. Phys.,* **38** (1997) 273.
11. G. M. Ter-Akopian et al., In: *Fission and Properties of Neutron-Rich Nuclei,* Proc. Int. Conf., Sanibel Island, Florida, USA, 1997, J. H. Hamilton and A. V. Ramayya, Eds. (World Scientific, Singapore, 1998) p. 165.
12. C. Budtz-Jorgensen and H.-H. Knitter, *Nucl. Phys.,* **A490** (1988) 307.
13. G. M. Ter-Akpoian et al., *Phys. Rev. Lett.,* **77** (1996) 32.
14. H. Warhanek and R. Vandenbosh, *J. Inorg. Nucl. Chem.,* **26** (1964) 669.
15. J. O. Rasmussen, W. Nörenberg and H. J. Mang, *Nucl. Phys.,* **A136** (1969) 465.
16. M. Zielinska-Pfabe and Dietrich K., *Phys. Lett.,* **B49** (1974) 123.
17. W. Nörenberg, In: *Physics and Chemistry of Fission,* Proc. 2nd Int. Symp., 1969 (IAEA, Vienna 1969) p. 51.
18. B. D. Wilkins, E. F. Steinberg and R. R. Chasman, *Phys. Rev. C,* **14** (1976) 1832.
19. I. N. Mikhailov and P. Quentin, Preprint CSNSM 99-01, Orsay, 1999.

MOLECULAR FINGERPRINTS IN THE TERNARY COLD FISSION

A.SANDULESCU AND Ş. MIŞICU

National Institute for Physics and Nuclear Engineering, Bucharest-Magurele, P.O.B. MG-6, ROMANIA

E-mail: misicu@theor1.theory.nipne.ro

W.GREINER

Institut für Theoretische Physik, Frankfurt am Main, Germany

Very recent experimental data suggests the existence of a trinuclear molecular structure before the cold ternary fragmentation of ^{252}Cf takes place. We present a theoretical prediction which encompasses only the calculation of the half-lives of the α-like trinuclar molecules, the collective modes for the ^{10}Be-like trinuclear molecules and the shift of the first 2^+ state of ^{10}Be in the quasi-molecular configuration.

1 Introduction

Nuclear cold fission is a rare phenomenon consisting in the disintegration of a large nucleus, such as ^{252}Cf, in two or more fragments with a very small dissipation of energy on degrees of freedom, other than the translational motion[1]. Before scission takes place, and after preformation from the mother nucleus was accomplished, there is a transient stage when the clusters are in close vicinity. A very recent experiment on ^{10}Be-accompanied cold fission of ^{252}Cf is supporting the idea that the time spent by the three clusters in this transient stage is larger than 10^{-13}s which brings into discussions the concept of a long living giant molecule. The present work is dedicated to the theoretical treatment of this topic.

2 Life-Times of Trinuclear Molecules

As has been showed in ref.[1], the introduction of a repulsive core in the nucleon-nucleon interaction will determine a typical molecular minima in the nucleus-nucleus potential, provided that at least one of them have a non-negligible deformation. *Mutatis mutandis*, the three-body potential, assumed to be the sum of all two-body components,

$$V = V_{12} + V_{13} + V_{23} \quad , \tag{1}$$

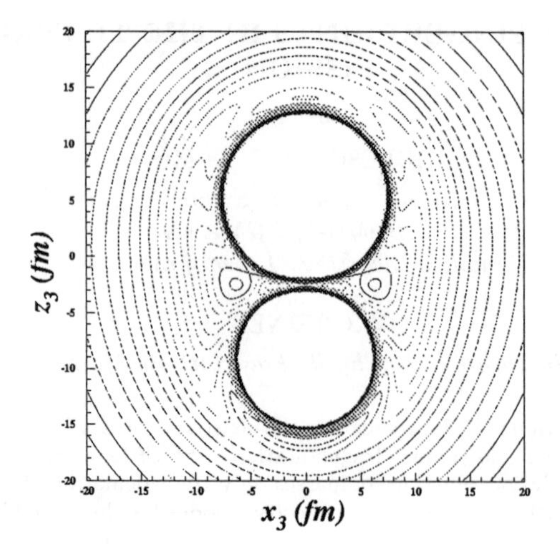

Figure 1. Contor lines of the total three-body potential.

displays a similar quasimolecular pattern with two minima in the equatorial region and two at the poles of the system (see Fig.1). Due to the axial symmetry, the minima in the equatorial region are equivalent, and in fact one could speak about a ring which represents the geometrical locus of the points where the three-body potential attains an absolute minimum(see Fig.1). In the case of α-like quasi-molecules such minima are formed in all two-body channles, e.g. α-^{92}Kr, α-^{156}Nd and ^{156}Nd-^{96}Kr. Based on the multidimensional tunneling approach for the ternary cold fission of ^{252}Cf[1], we assume that the motion of the three clusters is slow enough such that at each step of the tunneling process the system will be found in the configuration corresponding to the absolute minimum of the three-body potential. Thus, the trinuclear system, after the preformation, will perform an oscillatory motion in the generalized pocket, between the first two turning points (where $Q = V$), which provides an oscillation time $T_{\rm osc}$ or an assault frequency $\nu = \frac{1}{T_{\rm osc}}$. After this time, estimated to be in the range 10^{-21}-10^{-22}s, the system is passing from the second turning point and enters the barrier or the classicaly forbidden region. As showed elsewhere[1], in this region one have to solve a set of coupled differential equations derived from the minimization of the reduced action

$$S = \int \sqrt{2(V-Q)(m_1 + \sum_{i=2}^{n} m_i \frac{dx_i}{dx_1})} \tag{2}$$

where m_i are the masses of the fission degrees of freedom, x_1 is one of these coordinates, varying monotically, such as the elongation R of the binary subsystem. In this dynamical approach the third cluster has the tendency to get away from the molecular(fission) axis with large values of the x_3 coordinate at the third turning point. Once obtained the solutions of these equations, $x_i(t)$, we compute the penetrability factor $P = e^{-2S/\hbar}$, the relative production yields $Y_{\rm rel}$ and the half-life of the decaying molecule $T_{1/2} = \ln 2/\nu P$, assuming that the spectroscopic factors are the same for all cold splittings. The result was surprising : the half-live $T_{1/2}$ is huge, ranging between 1 and some tenths of seconds ! This conclusion is strongly supporting the molecular scenario, since as we mentioned earlier, one should expect times larger than 10^{-13}s for the ephemeral existence of the ^{10}Be-like trinuclear molecule. Another point supporting our scenario is given by the final mean kinetic energy of the α-cluster. Beyond the third turning point, when the system becomes unstable and eventually decays, we performed classical trajectory calculations using the method presented in [2] and the output of the dynamical calculation under the ternary barrier sketched above. The mean values for the splittings ^{132}Sn+α+^{116}Pd and ^{1156}Nd+α+^{92}Kr are 17.5 Mev and 19 MeV respectively, provided the α-particle is ejected in the band $4 \leq x_3 \leq 5$ fm. On the other hand, the experimental information obtained by the Darmstadt group is indicating an average kinetic energy of the $\alpha \approx 18.7$ MeV, when the cold limit, TXE$\rightarrow$0, is approached. Therefore our calculations are also confirming an increase of the average kinetic energy of the α particle, compared to the well known value of 15.9 MeV in hot fission.

3 Collective Modes of Trinuclear Molecules

In the previous section we discussed about the life-time of the trinuclear molecule which can be related to experimental observables such as the final kinetic kinetic energies and emission angles of the separated fragments. This is, so to speak, an indirect proof of the existence of the quasimolecular structure. Clearly, a better test would be to look for observables related to the still undecayed trinuclear system. Like in the case of dinuclear molecules one expects that the trinuclear molecules are likely to develope a rotational and vibrational spectrum. Knowing the excited states of the individual nuclei and their probability to decay by γ emission one could then look in the coincidence γ-ray spectrum for such new lines. We therefore proposed a way in which such vibrations and rotations could occur. For the time being we consider only a linear chain scenario, in which the light cluster, ^{10}Be is sandwiched between the two heavy clusters. An example would be the ternary cold splitting ^{96}Sr+^{10}Be+^{146}Ba[3]. In such a configuration the interaction between the heavy fragments is almost entirely given by the Coulomb term. However the interaction between the lighter fragment and the heavy fragments consists also of a noticeable nuclear component, which in fact is responsible for the nuclear bond. Like in the case of binary molecules, butterfly modes can occur, in which the fragments rotates in phase while the lighter fragment is approximately preserving its pole-pole configuration with the heavy fragments. The classical expression of the kinetic energy of the three-body system, after removing the center of mass contribution, is expressed as a sum of translational and rotational degrees of freedom :

$$T = \frac{1}{2}\mu_{12}\dot{\boldsymbol{R}}^2 + \frac{1}{2}\mu_{(12)3}\dot{\boldsymbol{\xi}}^2 + \frac{1}{2}\sum_{i=1}^{3} {}^t\boldsymbol{\omega}_i \boldsymbol{\mathcal{J}}_i \boldsymbol{\omega}_i \tag{3}$$

The first term describes the relative motion of the dinuclear sub-system (12), formed from the heavy clusters, whereas the second one corresponds to the relative motion of the third cluster with respect to the heavy fragments. The angular velocities of the rotational motion of the three clusters, referred to the laboratory frame, are denoted by $\boldsymbol{\omega}_{1,2,3}$, ${}^t\boldsymbol{\omega}$ being the transpose of $\boldsymbol{\omega}$. The inertia tensors $\boldsymbol{\mathcal{J}}_i$ are defined in the intrinsic frame such that in the absence of β and γ vibrations, the only non-vanishing components are the first two diagonal terms, $(\boldsymbol{\mathcal{J}}_i)_{11} = (\boldsymbol{\mathcal{J}}_i)_{22} \equiv J_i$, the quantum rotation around the symmetry axis of any of the two heavy fragments being discarded. When γ vibrations are included there will be a contribution to $(\boldsymbol{\mathcal{J}}_i)_{33}$ [4]. We also assume that the heavy fragments are constrained to rotate only around an axis perpendicularly to the axis joining their centers, this possibility being justified experimentally by the small forward anisotropy of the angular distribution of prompt γ radiation.

The molecular frame is defined in such a way that the z-axis coincides with the fission axis and the three-body plane is choosen to coincide with the $x-z$ molecular plane. The Cluster's Euler angles are $(\chi_i, \varphi_i, \phi_i)$, where φ_i describes the angle between the fragment i symmetry axis and the molecular z-axis. The geometry of our problem, with the heavy fragments symmetry axes lying in the same plane, makes the Euler angles χ equal , i.e. $\chi_1 = \chi_2$. They are combined in the variable $\theta_3 = (\chi_1 + \chi_2)/2$ which measures the rotation of the trinuclear aggregate with respect to the fission axis. Due to the symmetry of the problem the third Euler angle ϕ is a redundant variable. Eventually the total kinetic energy (3) can be expressed as a sum of three parts, the rotational energy T_{rot}, the internal kinetic energy T_{int} and the Coriolis coupling T_{cor}. In terms of the time derivatives of the Molecule's Euler angles, $(\dot{\theta}_1, \dot{\theta}_2, \dot{\theta}_3)$, specifying the rotation of the molecular frame, the classical expressions of the kinetic energy reads

$$T = \frac{1}{2} \sum_{ij} g_{ij} \dot{\theta}_i \dot{\theta}_j \tag{4}$$

where the various components of the metric tensor g_{ij} are listed elsewhere [3]. In the particular case of the spontaneous fission of ^{252}Cf, the spin $\boldsymbol{J}$ of the mother nucleus is 0, the total helicity K, i.e. the projection of the total angular momentum on the fission axis is also zero for both binary and ternary fragmentations, and therefore $\dot{\theta}_3 = 0$. To simplify even further one choose the molecular frame such that $\theta_3 = 0$. After quantizing the kinetic energy in three coordinates $(\varphi_1, \theta_1, \theta_2)$ and neglecting terms multiplied by the non-diagonal matrix-element $J_1 + J_2 \frac{R_1+R_3}{R_2+R_3} - \mu_{(12)3}(R_1 + R_3)\xi_0$, which prove to be small in the resulting metric tensor, we arrive to a form of the kinetic energy in which the rotations are decoupled from the butterfly vibrations

$$\hat{T} = -\frac{\hbar^2}{2(J_0 + J_1 + J_2)} \left(\frac{1}{\sin\theta_2} \frac{\partial^2}{\partial\theta_1^2} + \cot\theta_2 \frac{\partial}{\partial\theta_1} + \frac{\partial^2}{\partial\theta_2^2} \right) - \frac{\hbar^2}{2\mathcal{J}_\varphi} \frac{\partial^2}{\partial\varphi_1^2} \tag{5}$$

where

$$\mathcal{J}_\varphi = J_1 + J_2 \left(\frac{R_1 + R_3}{R_2 + R_3} \right)^2 + \mu_{(12)3}(R_1 + R_3)^2$$

$$J_0 = \mu_{12}(R_1 + R_2 + 2R_3)^2 + \mu_{(12)3}\xi_0^2$$

For small non-axial fluctuations(bendings), the potential in the neighborhood of the scission, or "molecular equilibrium" point $R_0 \equiv R_1 + R_2 + 2R_3$, $\xi_0 = \frac{A_1(R_1+R_3)-A_2(R_2+R_3)}{A_1+A_2}$ gets a simplified form, provided we keep terms up to the second power in angle :

$$V = V(R_0, \xi_0) + \frac{1}{2} C_\varphi \varphi^2 \tag{6}$$

Thus we separated the quantum motion into a two-dimensional rotation and a one-dimensional vibration, of a butterfly type, whose spectrum is simply given by

$$E_\varphi = \left(n_\varphi + \frac{1}{2} \right) \hbar \sqrt{\frac{C_\varphi}{J_\varphi}} \tag{7}$$

The values of $\hbar\omega_\varphi$ are ranging between 2 and 2.5 MeV.

4 Polarization of the light cluster in Trinuclear Molecules

Another observable belonging to the quasimolecular system is the shift of the first 2_1^+ state in ^{10}Be. The existence of such a shift was claimed very recently [5]. Using the Gammasphere with 72 detectors, the γ-ray from the first 2^+ state in ^{10}Be, was measuread in coincidence with the γ rays of the fission partners ^{146}Ba and ^{96}Sr. It was observed that the spectrum corresponding to these γ quanta is not Doppler broadened as if the quanta are emitted from a resting source, and secondly the energy value of the 2_1^+ state in such a quasi-bound configuration is lowered by 6 keV compared to the value compiled for the free configuration (E_γ = 3368.03 keV). Most probable, if the measurement was correct, the shift of the 2_1^+ level of ^{10}Be is a consequence of its polarization induced by the other two clusters in the quasi-molecular configuration. In order to evaluate this polarization we make recourse to the following assumptions :

a) For a a given configuration of the three clusters, which should correspond to a precise point on the fission path, the static energy of the quasi-molecule must have a minimum with respect to variations in the deformations [6], i.e.

$$\frac{\partial E_{\text{ternary}}}{\partial \beta_i} = 0, \qquad (i = 1, 2, 3) \tag{8}$$

b) The deformation energy for the light cluster 3 resembles very much a vibrator with a small cubic anharmonicity for not to large departures from the spherical equilibrium position:

$$E_{\text{def}}(\beta_3) = \frac{1}{2} C_2 \beta_3^2 + C_3 \beta_3^3 \tag{9}$$

where the stiffness parameters have the numerical values: C_2 =7.688 MeV and C_3 =-2.855 MeV, obtained upon interpolating the Hartree-Fock deformation curve.

c) The interaction between the light cluster 3 and the heavy clusters, 1 and 2, can be expanded with respect to the deformation of the third cluster β_3, keeping in the same time the deformations of the heavy clusters, β_1 and β_2, freezed.

Accordingly, discarding the constant terms, the vibrational Hamiltonian of ^{10}Be reads

$$H_3^{(\beta)} = -\frac{\hbar^2}{2B} \frac{\partial^2}{\partial \beta_3^2} + \frac{1}{2} C_2' \beta_3^2 + C_3' \beta_3^3 \tag{10}$$

where $C_{2,3}'$ are the modified stiffness parameters of the quadratic and cubic potential terms and B, the effective mass, is computed from the experimental

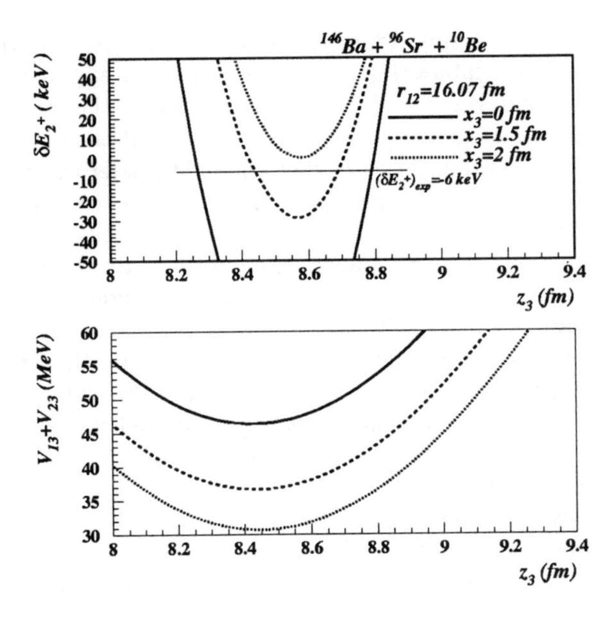

Figure 2. Three-body potential.

values of $B(E2, 0_1^+ \rightarrow 2_1^+)$ and the energy of the 2_1^+ state. The corresponding spectrum can be deduced by applying the stationary perturbation theory in the second order approximation [7].

$$E'_n = \hbar\omega \left\{ n + \frac{1}{2} - (\hbar\omega)^2 \frac{15C_3'^2}{4C_2'^3} \left(n^2 + n + \frac{11}{30} \right) \right\} \tag{11}$$

where $\hbar\omega = \sqrt{\frac{C_2'}{B}}$. Then, the shift can be readily obtained by taking the difference between the energies of the 1-phonone state in the quasi-bound configuration (with primed $C_{2,3}$) and the free case (with unprimed $C_{2,3}$)

$$\delta E_{2+} = E'_1 - E_1 \tag{12}$$

In Fig.2 we represented the energy shift of the 2_1^+ state, in the case of the splitting ^{146}Ba + ^{96}Sr + ^{10}Be, as a function of the location of the light cluster on the molecular z-axis (upper panel). In the lower panel the sum $V_{13} + V_{23}$ of the interactions between clusters 1 and 3 and between clusters

2 and 3, is plotted. The heavy fragments were supposed to have their symmetry axes aligned. An interesting fact, which becomes apparent from the inspection of these figures, is the close proximity between the location of the minima of the two sets of curves, which may suggest that the lowest value of the energy shift is obtained in the case when the light cluster is located on, or very close from, the electro-nuclear saddle curve[a]. The curves represented in Fig.2 were obtained for a tip distance between the heavy clusters, $d = 2$ fm, which translated in inter-cluster distances is $r_{12} = 16.07$ fm. If we increase or we decrease this distance, the negative energy shifts will gradually disappear. If we make the assumption that the light cluster is located on the bottom of the potential pocket, or in quantum terms, is filling-up the first state of the quantum well produced by the interaction of the three fragments, then the experimental shift, δE_{2^+} =-6 keV can be reproduced if the tip distance between the heavy fragments is confined in the interval 1.5 - 2.5 fm, and the cluster ^{10}Be is located off the fission axis at a height x_3 not larger than 2 fm. The present calculations are therefore predicting a polarization of ^{10}Be, consistent with the observed 6 keV shift, to arise in a triangular configuration or a linear configuration of the giant molecule. However, in view of the conclusions reached in the section dedicated to the molecular life-time, a linear configuration seems to be unlikely. Moreover, in order to get a negative shift in a linear configuration one have to suppose a somewhat to large overlap between the clusters.

References

1. A.Săndulescu, Ş. Mişicu, F.Carstoiu and W.Greiner, Phys.Part.Nucl. **30**, 386 (1999).
2. Ş.Mişicu, A.Sandulescu, F.Carstoiu, M.Rizea and W.Greiner, Il Nuovo Cimento **A112**, 313 (1999).
3. Ş.Mişicu, P.O.Hess, A.Sandulescu and W.Greiner J.Phys.G **25**, L147 (1999).
4. J.M.Eisenberg and W.Greiner, *Nuclear Theory I: Nuclear Models*, (North-Holland, Amsterdam, 1987).
5. A.V.Ramayya, J. K. Hwang, J. H. Hamilton, A. Săndulescu, W. Greiner and GANDS Collaboration Phys.Rev.Lett. **81**, 1998 (947)
6. F.Dickman and K.Dietrich, Nucl.Phys.A **129**, 241 (1969)
7. S.Flügge, *Practical Quantum Mechanics I* (Springer Verlag, Berlin, 1971).

[a]Moving off the fission axis we defined the electro-nuclear saddle curve as the geometrical locus of points where the z-component of the force exerted by cluster 1 on cluster 3 is compensated by the forces exerted by cluster 2 on cluster 3 [1].

THE QUASI-MOLECULAR STAGE OF TERNARY FISSION

D. N. POENARU AND B. DOBRESCU

Horia Hulubei National Institute of Physics and Nuclear Engineering,
P.O. Box MG-6, RO-76900 Bucharest, Romania
E-mail: poenaru@ifin.nipne.ro

W. GREINER

Institut für Theoretische Physik der Universität, Postfach 111932,
D-60054 Frankfurt am Main, Germany

J. H. HAMILTON AND A. V. RAMAYYA

Department of Physics, Vanderbilt University, Nashville, Tennessee, USA

We developed a three-center phenomenological model, able to explain qualitatively the recently obtained experimental results concerning the quasimolecular stage of a light-particle accompanied fission process. It was derived from the liquid drop model under the assumption that the aligned configuration, with the emitted particle between the light and heavy fragment, is reached by increasing continuously the separation distance, while the radii of the heavy fragment and of the light particle are kept constant. In such a way, a new minimum of a short-lived molecular state appears in the deformation energy at a separation distance very close to the touching point. This minimum allows the existence of a short-lived quasi-molecular state, decaying into the three final fragments. The influence of the shell effects is discussed. The half-lives of some quasimolecular states which could be formed in the ^{10}Be and ^{12}C accompanied fission of ^{252}Cf are roughly estimated to be the order of 1 ns, and 1 ms, respectively.

1 Introduction

The light particle accompanied fission was discovered [1] in 1946, when the track of a long-range particle (identified by Farwell *et al.* to be ^{4}He) almost perpendicular to the short tracks of heavy and light fragments was observed in a photographic plate. The fission was induced by bombarding ^{235}U with slow neutrons from a Be target at a cyclotron. The largest yield in such a rare process (less than one event per 500 binary splittings) is measured for α-particles, but many other light nuclei (from protons to oxygen or even calcium isotopes) have been identified [2] in both induced- and spontaneous fission phenomena. If A_1 and A_2 are the mass numbers of the heavy fragments (assume $A_1 \geq A_2$), then usually the mass of the light particle $A_3 << A_2$. The "true" ternary fission, in which $A_1 \simeq A_2 \simeq A_3$, has not yet been experimentally detected.

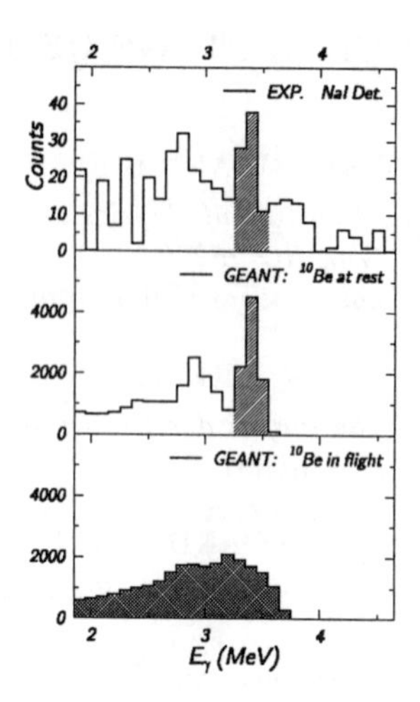

Figure 1. Gamma-ray energy spectrum of ^{10}Be accompanying fission of ^{252}Cf (top) and the simulated (with the code GEANT) pulse-height distributions for a (not-Doppler-broadened) 3.37 MeV γ emission from ^{10}Be at rest (middle) as well as for ^{10}Be in flight (bottom) from the source to the detector (duration of about 1 ns). The flight-path of the order of 10 cm. One uses NaI scintillator detectors. Data from Fig. 6 of Ref. 13.

Many properties of the binary fission process have been explained [3] within the liquid drop model (LDM); others like the asymmetric mass distribution of fragments and the ground state deformations of many nuclei, could be understood only after adding the contribution of shell effects. [4,5] As it was repeatedly stressed (see [6] and the references therein), shell effects proved also to be of vital importance for cluster radioactivities predicted [7] in 1980.

The total kinetic energy (TKE) of the fragments, in the most frequently detected binary or ternary fission mechanism, is smaller than the released energy (Q) by about 25–35 MeV, which is used to produce deformed and excited fragments. These then emitt neutrons (each with a binding energy of about 6 MeV) and γ-rays. From time to time a "cold" fission mechanism is detected, in which the TKE exhausts the Q-value, hence no neutrons are emitted, and the fragments are produced in or near their ground-state. The first experimental evidence for cold binary fission in which its TKE exhaust Q was reported [8] in 1981. Larger yields were measured [9] in trans-Fm ($Z \geq 100$) isotopes, where the phenomenon was called bimodal fission.

The correlated fragment pairs in cold ternary (α- and ^{10}Be accompanied spontaneous fission of ^{252}Cf) processes were only recently discovered, [10,11]

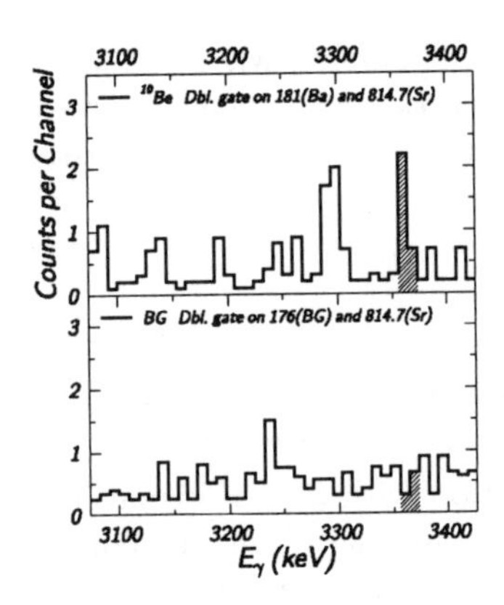

Figure 2. The cumulated γ-ray energy spectrum of ^{10}Be accompanying fission of ^{252}Cf (top) and the corresponding background (bottom). The spectrum of the 3.362 γ is not-Doppler-broadened, suggesting an emission from ^{10}Be at rest. The stopping time of ^{10}Be in the absorber mounted around the source is of the order of 1 ps. Data from Fig. 2 of Ref. 11. Coincidence spectra obtained with Ge detectors in GAMMASPHERE.

by measuring triple γ coincidences in a modern large array of γ-ray detectors (GAMMASPHERE). The fragments are identified by their γ-ray spectra. Among other new aspects of the fission process seen for the first time with this new technique, [10,12] one should mention the double fine structure, and the triple fine structure in binary and ternary fission.

A particularly interesting feature, observed [11,13] both in ^{10}Be- and ^{12}C accompanied cold fission of ^{252}Cf is related to the width of the light particle γ-ray spectrum (see Figs. 1 and 2). For example, the 3.368 MeV γ line of ^{10}Be, with a lifetime of 125 fs is not Doppler-broadened, as it should be if it would be emitted when ^{10}Be is in flight (taking about 1 ns to reach the detector). A plausible suggestion was made, that the absence of Doppler broadening is related to a trapping of ^{10}Be in a potential well of nuclear molecular character. [11]

Quasi-molecular configurations of two nuclei have been suggested as a natural explanation for the resonances measured [14] in ^{12}C+^{12}C scattering and reactions. There are also other kinds of such binary molecules (see [15] and references therein), like spontaneously fissioning shape-isomers. The above mentioned experiments can be considered as the first evidence for a more complex quasi-molecular configuration of three nuclei. The purpose of the present lecture is to show, within a phenomenological three-center model, that a minimum which could explain the existence of these quasi-molecules is produced in the potential barrier, when the formation of the light particle occurs in the neck between the two heavier fragments. In this way we extend to ternary fission our unified approach of cold fission, cluster radioactivities, and α-decay. [6]

2 Shape Parametrization

The shape parametrization with one deformation parameter as follows has been suggested from the analysis [16] of different aligned and compact configurations of fragments in touch. A lower potential barrier for the aligned cylindrically-symmetric shapes with the light particle between the two heavy fragments, is a clear indication that during the deformation from an initial parent nucleus to three final nuclei, one should arrive at such a scission point. In order to reach this stage we shall increase continuously the separation distance, R, between the heavy fragments, while the radii of the heavy fragment and of the light particle are kept constant, $R_1 =$ constant, $R_3 =$ constant. Unlike in the previous work, we now adopt the following convention: $A_1 \geq A_2 \geq A_3$. The hadron numbers are conserved: $A_1 + A_2 + A_3 = A$.

At the beginning (the neck radius $\rho_{neck} \geq R_3$) one has a two-center evolution (see Fig. 3) until the neck between the fragments becomes equal to the radius of the emitted particle, $\rho_{neck} = \rho(z_{s1}) |_{R=R_{ov3}} = R_3$. This Eq. defines R_{ov3} as the separation distance at which the neck radius is equal to R_3. By placing the origin in the center of the large sphere, the surface equation

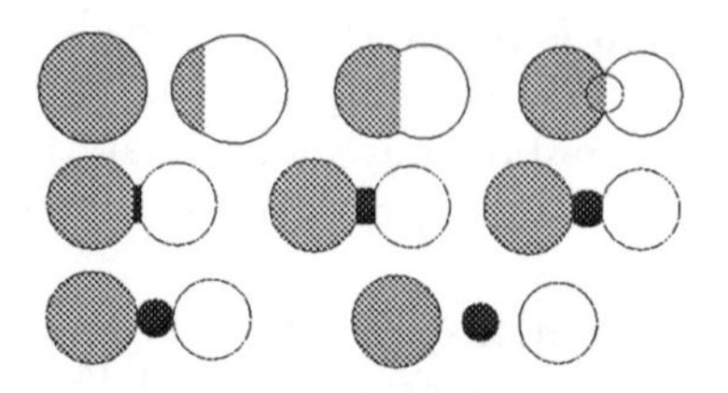

Figure 3. Evolution of nuclear shapes during the deformation process from one parent nucleus ^{252}Cf to three separated fragments ^{146}Ba, ^{10}Be, and ^{96}Sr. In the upper part the binary stage is illustrated; the separation distance increases from R_i to R_{ov3}, passing through R_{min1b} and R_{min2b} values. In the middle, the ternary stage of the process develops by forming the third particle in the neck. The quasi-molecular shape, at which $R = R_{min-t}$ is the intermediate one in this row. At the bottom the fragments are separated.

in cylindrical coordinates is given by:

$$\rho_s^2 = \begin{cases} R_1^2 - z^2 & , \; -R_1 \le z \le z_{s1} \\ R_2^2 - (z-R)^2 & , \; z_{s1} \le z \le R + R_2 \end{cases} \tag{1}$$

Then for $R > R_{ov3}$ the three center starts developing by decreasing progressively with the same amount the two tip distances $h_1 + h_{31} = h_{32} + h_2$. Besides this constraint, one has as in the binary stage, volume conservation and matching conditions. The R_2 and the other geometrical quantities are determined by solving numerically the corresponding system of algebraic equations. By assuming spherical nuclei, the radii are given by $R_j = 1.2249 A_j^{1/3}$ fm ($j = 0, 1, 3$), $R_{2f} = 1.2249 A_2^{1/3}$ with a radius constant $r_0 = 1.2249$ fm, from Myers-Swiatecki's variant of LDM. Now the surface equation can be written as

$$\rho_s^2 = \begin{cases} R_1^2 - z^2 & , \; -R_1 \le z \le z_{s1} \\ R_3^2 - (z-z_3)^2 & , \; z_{s1} \le z \le z_{s2} \\ R_2^2 - (z-R)^2 & , \; z_{s2} \le z \le R + R_2 \end{cases} \tag{2}$$

and the corresponding shape has two necks and two separating planes. Some of the important values of the deformation parameter R are the initial distance $R_i = R_0 - R_1$, and the touching-point one, $R_t = R_1 + 2R_3 + R_{2f}$. There is also R_{ov3}, defined above, which allows one to distinguish between the binary and ternary stage.

3 Deformation Energy

According to the LDM, by requesting zero energy for a spherical shape, the deformation energy, $E^u(R) - E^0$, is expressed as a sum of the surface and Coulomb terms

$$E^u_{def}(R) = E_s^0[B_s(R) - 1] + E_C^0[B_C(R) - 1] \tag{3}$$

where the exponent ^{u} stands for uniform (fragments with the same charge density as the parent nucleus), and ^{0} refers to the initial spherical parent. In order to simplify the calculations, we initially assume the same charge density $\rho_{1e} = \rho_{2e} = \rho_{3e} = \rho_{0e}$, and at the end we add the corresponding corrections. In this way we perform one numerical quadrature instead of six. For a spherical shape $E_s^0 = a_s(1 - \kappa I^2)A^{2/3}$; $I = (N-Z)/A$; $E_C^0 = a_c Z^2 A^{-1/3}$, where the numerical constants of the LDM are: $a_s = 17.9439$ MeV, $\kappa = 1.7826$, $a_c = 3e^2/(5r_0)$, $e^2 = 1.44$ MeV·fm.

The shape-dependent, dimensionless surface term is proportional to the surface area:

$$B_s = \frac{E_s}{E_s^0} = \frac{d^2}{2} \int_{-1}^{+1} \left[y^2 + \frac{1}{4} \left(\frac{dy^2}{dx} \right)^2 \right]^{1/2} dx \tag{4}$$

where $y = y(x)$ is the surface equation in cylindrical coordinates with -1, +1 intercepts on the symmetry axis, and $d = (z'' - z')/2R_0$ is the seminuclear length in units of R_0. Similarly, for the Coulomb energy [17] one has

$$B_c = \frac{5d^5}{8\pi} \int_{-1}^{+1} dx \int_{-1}^{+1} dx' F(x, x') \tag{5}$$

$$\begin{aligned} F(x,x') = \{ & yy_1[(K-2D)/3] \cdot \\ & \left[2(y^2+y_1^2) - (x-x')^2 + \frac{3}{2}(x-x') \left(\frac{dy_1^2}{dx'} - \frac{dy^2}{dx} \right) \right] + \\ & K \left\{ y^2 y_1^2/3 + \left[y^2 - \frac{x-x'}{2}\frac{dy^2}{dx} \right] \left[y_1^2 - \frac{x-x'}{2}\frac{dy_1^2}{dx'} \right] \right\} \} a_\rho^{-1} \end{aligned} \tag{6}$$

K, K' are the complete elliptic integrals of the 1st and 2nd kind

$$K(k) = \int_0^{\pi/2} (1 - k^2 \sin^2 t)^{-1/2} dt; \quad K'(k) = \int_0^{\pi/2} (1 - k^2 \sin^2 t)^{1/2} dt \tag{7}$$

and $a_\rho^2 = (y + y_1)^2 + (x - x')^2$, $k^2 = 4yy_1/a_\rho^2$, $D = (K - K')/k^2$.

The new minimum, which can be seen in Fig. 4 at a separation distance $R = R_{min-t} > R_{ov3}$, is the result of a competition between the Coulomb- and surface energies. At the beginning ($R < R_{min-t}$) the Coulomb term is stronger, leading to a decrease in energy, but later on ($R > R_{min-t}$) the light particle formed in the neck posses a surface area increasing rapidly, so there is also an increase in energy up to $R = R_t$.

Now let us analyse the influence of various corrections, which could in principle alter this image. After performing numerically the integrations, we add the following corrections: for the difference in charge densities reproducing the touching point values; for experimental masses reproducing the Q_{exp}-value at $R = R_i$, when the origin of energy corresponds to infinite separation distances between fragments, and the phenomenological shell corrections δE

$$E_{LD}(R) = E_{def}^u(R) + (Q_{th} - Q_{exp}) f_c(R) \tag{8}$$

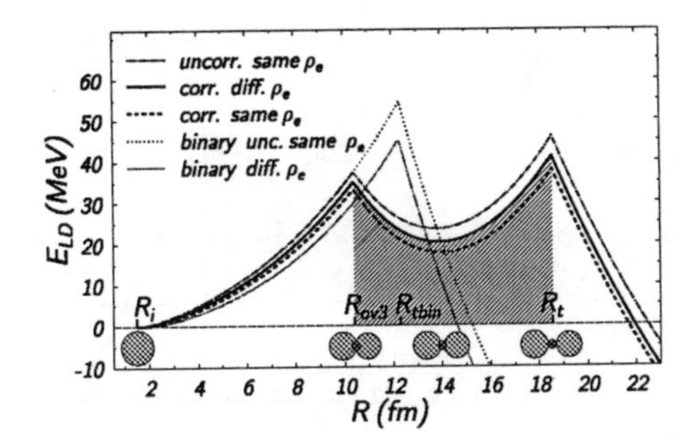

Figure 4. The liquid drop model deformation energy versus separation distance for the ^{20}O accompanied cold fission of ^{252}Cf with ^{132}Sn and ^{100}Zr heavy fragments. In order to simplify the numerical calculations we start by assuming the same charge density of the fragments. One can see the effect of successive corrections taking into account the experimental Q-value and the difference in charge density. Similar curves for the binary fission posses a narrower fission barrier. The new minimum appears in the shaded area from R_{ov3} to R_t.

where $f_c(R) = (R - R_i)/(R_t - R_i)$, and

$$Q_{th} = E_s^0 + E_C^0 - \sum_1^3 (E_{si}^0 + E_{Ci}^0) + \delta E^0 - \sum_1^3 \delta E^i \tag{9}$$

The correction increases gradually (see Fig. 4 and Fig. 5) with R up to R_t and then remains constant for $R > R_t$. The barrier height increases if $Q_{exp} < Q_{th}$ and decreases if $Q_{exp} > Q_{th}$. In this way, when one, two, or all final nuclei have magic numbers of nucleons, Q_{exp} is large and the fission barrier has a lower height, leading to an increased yield. In a binary decay mode like cluster radioactivity and cold fission, this condition is fulfilled when the daughter nucleus is ^{208}Pb and ^{132}Sn, respectively.

4 Shell Corrections and Half-lives

Finally we also add the shell terms

$$E(R) = E_{LD}(R) + \delta E(R) - \delta E^0 \tag{10}$$

Presently there is not available any microscopic three-center shell model reliably working for a long range of mass asymmetries. This is why we use a phenomenological model, instead of the Strutinsky's method, to calculate the shell corrections. The model is adapted after Myers and Swiatecki. [5]

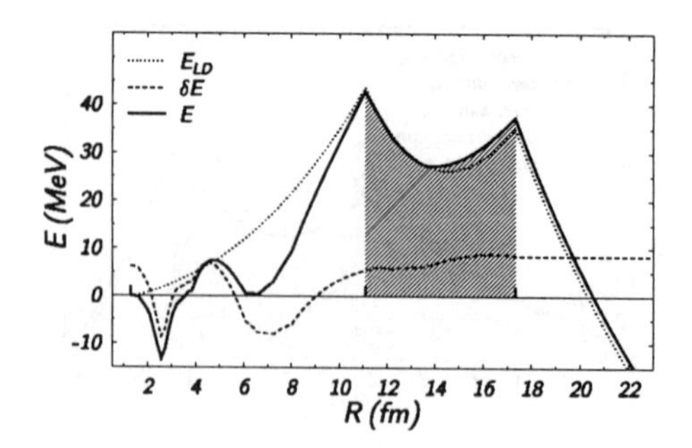

Figure 5. The liquid drop model, E_{LD}, the shell correction, δE, and the total deformation energies, E, for the ^{10}Be accompanied cold fission of ^{252}Cf with ^{146}Ba and ^{96}Sr heavy fragments. The new minimum appears in the shaded area from R_{ov3} to R_t.

At a given R, we calculate the volumes of fragments and the corresponding numbers of nucleons $Z_i(R)$, $N_i(R)$ $(i = 1, 2, 3)$, proportional to the volume of each fragment. Fig. 3 illustrates the evolution of shapes and of the fragment volumes. Then we add for each fragment the contribution of protons and neutrons

$$\delta E(R) = \sum_i \delta E_i(R) = \sum_i [\delta E_{pi}(R) + \delta E_{ni}(R)] \tag{11}$$

which are given by

$$\delta E_{pi} = Cs(Z_i); \quad \delta E_{ni} = Cs(N_i) \tag{12}$$

where

$$s(Z) = F(Z)/[(Z)^{-2/3}] - cZ^{1/3} \tag{13}$$

$$F(n) = \frac{3}{5}\left[\frac{N_i^{5/3} - N_{i-1}^{5/3}}{N_i - N_{i-1}}(n - N_{i-1}) - n^{5/3} + N_{i-1}^{5/3}\right] \tag{14}$$

in which $n \in (N_{i-1}, N_i)$ is either a current Z or N number and N_{i-1}, N_i are the closest magic numbers. The constants $c = 0.2$, $C = 6.2$ MeV were determined by fit to the experimental masses and deformations. The variation with R is calculated [18] as

$$\delta E(R) = \frac{C}{2}\left\{\sum_i [s(N_i) + s(Z_i)]\frac{L_i(R)}{R_i}\right\} \tag{15}$$

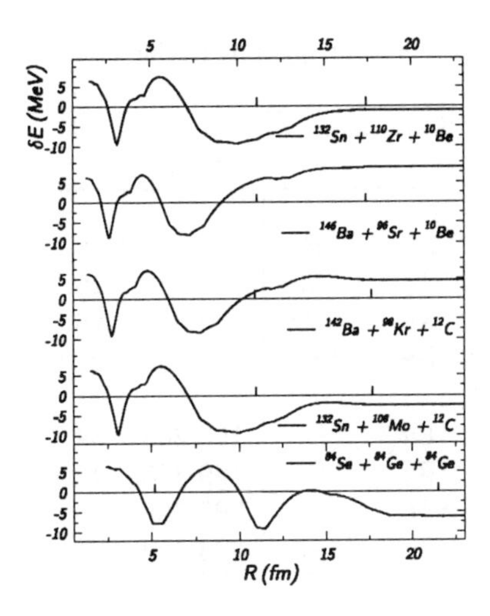

Figure 6. The shell correction energy variation with the separation distance for two examples of ^{10}Be-, two of ^{12}C accompanied cold fission, compared to the "true" ternary fission (in nearly three identical fragments) of ^{252}Cf. The three partners are given. The two vertical bars on each plot show the positions of R_{ov3} and of R_t.

where $L_i(R)$ are the lengths of the fragments along the axis of symmetry, at a given separation distance R. During the deformation, the variation of separation distance between centers, R, induces the variation of the geometrical quantities and of the corresponding nucleon numbers. Each time a proton or neutron number reaches a magic value, the correction energy passes through a minimum, and it has a maximum at midshell (see Fig. 5 and Fig. 6). The first narrow minimum appearing in the shell correction energy δE in Fig. 5, at $R = R_{min1b} \simeq 2.6$ fm, is the result of almost simultaneously reaching the magic numbers $Z_1 = 20$, $N_1 = 28$, and $Z_2 = 82$, $N_2 = 126$. The second, more shallower one around $R = R_{min2b} \simeq 7.2$ fm corresponds to a larger range of R-values for which $Z_1 = 50$, $N_1 = 82$, $Z_2 = 50$, $N_2 = 82$ are not obtained in the same time. In the region of the new minimum, $R = R_{min-t}$, for light-particle accompanied fission, the variation of the shell correction energy is very small, hence it has no major consequence. One can say that the quasimolecular minimum is related to the collective properties (liquid-drop like behavior). On the other side, for "true" ternary process (see the bottom part of Fig. 6) both minima appear in this range of values, but no such LDM effect was found there. In order to compute the half-life of the quasi-molecular state, we have first to search for the minimum E_{min} in the quasimolecular well, from R_{ov3} to R_t, and then to add a zero point vibration energy, E_v: $E_{qs} = E_{min} + E_v$.

Table 1. Calculated half-lives of some quasi-molecular states formed during the ternary fission of ^{252}Cf.

Particle	Fragments		Q_{exp} (MeV)	K	$\log T(s)$
^{10}Be	^{132}Sn	^{110}Ru	220.183	19.96	-11.17
	^{138}Te	^{104}Mo	209.682	25.23	-8.89
	^{138}Xe	^{104}Zr	209.882	26.04	-8.54
	^{146}Ba	^{96}Sr	201.486	22.98	-9.86
^{12}C	^{147}La	^{93}Br	196.268	39.80	-2.56
	^{142}Ba	^{98}Kr	199.896	42.71	-1.30
	^{140}Te	^{100}Zr	209.728	38.21	-3.25
	^{132}Sn	^{108}Mo	223.839	31.46	-6.18

The half-life, T, is expressed in terms of the barrier penetrability, P, which is calculated from an action integral, K, given by the quasi-classical WKB approximation

$$T = \frac{h \ln 2}{2E_v P}; \quad P = exp(-K) \tag{16}$$

where h is the Planck constant, and

$$K = \frac{2}{\hbar} \int_{R_a}^{R_b} \sqrt{2\mu[E(R) - E_{qs}]}\, dR \tag{17}$$

in which R_a, R_b are the turning points, defined by $E(R_a) = E(R_b) = E_{qs}$ and the nuclear inertia is roughly approximated by the reduced mass $\mu = m[(A_1A_2 + A_3A)/(A_1 + A_2)]$, where m is the nucleon mass, $\log[(h \ln 2)/2] = -20.8436$, $\log e = 0.43429$ and $\sqrt{8m/\hbar^2} = 0.4392$ MeV$^{-1/2}\times$fm^{-1}.

The results of our estimations for the half-lives of some quasimolecular states formed in the ^{10}Be- and ^{12}C accompanied fission of ^{252}Cf are given in Table 1. They are of the order of 1 ns and 1 ms, respectively, if we ignore the results for a division with heavy fragment ^{132}Sn, which was not measured due to very high first excited state. Consequently the new minimum we found can qualitatively explain the quasimolecular nature of the narrow line of the ^{10}Be γ-rays.

It is interesting to note that the trend toward a split into two, three, or four nuclei (the lighter ones formed in a long neck between the heavier fragments) has been theoretically demonstrated by Hill, [19] who investigated the classical dynamics of an incompressible, irrotational, uniformly charged liquid drop. No mass asymmetry was evidenced since any shell effect was ignored.

In conclusion, we should stress that a quasimolecular stage of a light-particle accompanied fission process, for a limited range of sizes of the three partners, can be qualitatively explained within the liquid drop model.

Acknowledgments

We are grateful to M. Mutterer for enlightening discussions.

References

1. L. W. Alvarez, as reported by G. Farwell, E. Segrè, and C. Wiegand, *Phys. Rev.* **71**, 327 (1947). T. San-Tsiang *et al.*, *C. R. Acad. Sci. Paris* **223**, 986 (1946).
2. M. Mutterer and J.P. Theobald, in *Nuclear Decay Modes*, ed. D.N. Poenaru (IOP Pub., Bristol, 1996), p. 487.
3. N. Bohr and J. Wheeler, *Phys. Rev.* **55**, 426 (1939).
4. V.M. Strutinsky, *Nucl. Phys.* A **95**, 420 (1966).
5. W.D. Myers and W.J. Swiatecki, *Nucl. Phys.* A **81**, 1 (1966).
6. D.N. Poenaru and W. Greiner, in *Nuclear Decay Modes*, (IOP Publishing, Bristol, 1996), p. 275.
7. A. Săndulescu, D.N. Poenaru and W. Greiner, *Sov. J. Part. Nucl.* **11**, 528 (1980).
8. C. Signarbieux *et al.*, *J. Phys. Lett. (Paris)* **42**, L437 (1981).
9. E.K. Hulet *et al.*, *Phys. Rev. Lett.* **56**, 313 (1986).
10. A.V. Ramayya *et al.*, *Phys. Rev.* C **57**, 2370 (1998).
11. A.V. Ramayya *et al.*, *Phys. Rev. Lett.* **81**, 947 (1998).
12. J.H. Hamilton *et al.*, *Prog. Part. Nucl. Phys.* **35**, 635 (1995). G.M. Ter-Akopian *et al.*, *Phys. Rev. Lett.* **77**, 32 (1996).
13. P. Singer *et al.*, in *Dynamical Aspects of Nuclear Fission*, Proc. 3rd Int. Conf., Častá-Papiernička, Slovakia, 1996 (JINR, Dubna, 1996), p. 262.
14. D.A. Bromley *et al.*, *Phys. Rev. Lett.* **4**, 365 (1960).
15. W. Greiner, J.Y. Park, W. Scheid, *Nuclear Molecules* (World Sci., Singapore, 1995).
16. D.N. Poenaru *et al.*, *Phys. Rev.* C **59**, 3457 (1999).

17. D.N. Poenaru and M. Ivaşcu, *Comput. Phys. Commun.* **16**, 85 (1978).
18. H. Schultheis and R. Schultheis, *Phys. Lett.* B **37**, 467 (1971).
19. D.L. Hill in *Proc. of the 2nd U N Int. Conf. on the Peaceful Uses of Atomic Energy* (United Nations, Geneva, 1958), p. 244.

REALISTIC FISSION SADDLE-POINT SHAPES AND HEIGHTS

P. MÖLLER

P. Moller Scientific Computing and Graphics, Inc., P. O. Box 1440, Los Alamos, NM 87544, USA

D. G. MADLAND

Theoretical Division, Los Alamos National Laboratory, Los Alamos, NM 87545, USA

A. IWAMOTO

Department of Materials Sciences, Japan Atomic Energy Research Institute, Tokai-mura, Naka-gun, Ibaraki, Japan

We calculate complete fission potential-energy surfaces for five shape coordinates: elongation, neck radius, light-fragment deformation, heavy-fragment deformation, and mass asymmetry for even nuclei in the range $82 \leq Z \leq 100$. The potential energy is calculated in terms of the macroscopic-microscopic model with a folded-Yukawa single-particle potential and a Yukawa-plus-exponential macroscopic model in the three-quadratic-surface parameterization.

1 Introduction

Because nascent fragment shell effects strongly influence the structure of the fission potential-energy surface long before scission, usually already in the outer saddle region, it is crucial to include in calculations the nascent fragment deformations as two independent shape degrees of freedom. In addition, elongation, neck radius, and mass-asymmetry shape-degrees of freedom are required to properly describe the complete fission potential-energy surface. For nascent-fragment deformations we choose spheroidal deformations characterized by Nilsson's quadrupole ϵ parameter. This single fragment-deformation parameter is sufficient because higher-multipole shape-degrees of freedom are usually of minor importance in the fission-fragment mass region below the rare-earths. Our potential-energy model is the macroscopic-microscopic finite-range liquid-drop model as defined in Ref. [1] with shape-dependent Wigner and A^0 terms as defined in Ref. [2]. Additional details about our current approach are given in Ref. [3].

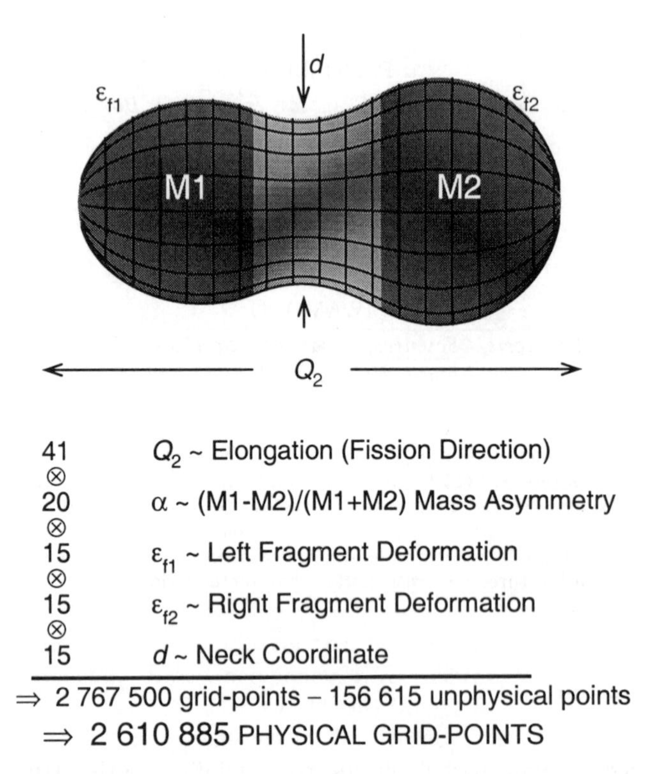
Five Necessary Saddle Shape Coordinates
d
ε f1
ε f2
M1
M2
Q2
41 Q2 ~ Elongation (Fission Direction)
⊗
20 α ~ (M1-M2)/(M1+M2) Mass Asymmetry
⊗
15 ε f1 ~ Left Fragment Deformation
⊗
15 ε f2 ~ Right Fragment Deformation
⊗
15 d ~ Neck Coordinate
⇒ 2 767 500 grid-points – 156 615 unphysical points
⇒ 2 610 885 PHYSICAL GRID-POINTS

Figure 1. Five-dimensional shape parameterization used in potential-energy calculation. Different shades of gray indicate the three different quadratic surfaces. The first derivative is continuous at where the surfaces meet.

2 Five-Dimensional Potential-Energy Surfaces

We consider here the physically relevant part of the full 5-dimensional space of the three-quadratic-surface parameterization in terms of 41 values of the charge quadrupole moment Q_2[a], 15 steps in the neck size, 15 values of the fragment deformation ϵ, where $-0.2 \leq \epsilon \leq 0.50$ for each of the two nascent fragments, and 20 values of the mass asymmetry $\alpha_g = (M_1 - M_2)/(M_1 + M_2)$,

[a] Q_2 is always given in term of ^{240}Pu with the same shape as the nucleus considered, so that the nuclear size effect is eliminated

where M_1 and M_2 are the volumes of the left and right nascent fragments were they completed to closed shapes, and where $\alpha_g = -0.02(0.02)0.36$. The various shape coordinates are enumerated in Fig. 1 where also an example of a shape is shown. We have earlier [3] emphasized that it is important to consider a *dense* grid in ϵ and mass asymmetry because fragment shell corrections vary rapidly in a narrow range of these deformation coordinates. For example, near ^{132}Sn the microscopic corrections vary by one MeV for a change of the nucleon number A by one unit.

3 Analysis of Calculated Surfaces by Imaginary Waterflow in Five Dimensions

We investigate the structure of the multidimensional surface by considering imaginary water flows [4] in the 5-dimensional potential-energy surface. For example, we imagine we stepwise flood, in intervals of 1 MeV, the second minimum with water. During the flooding process we check at what water level a preselected "exit" grid-point that is clearly in the fission valley near scission gets "wet". When this happens, the water level has passed the threshold energy for fission. We can determine the saddle-point energy to desired accuracy by repeating the filling procedure with successively smaller stepwise increases of the water level. The saddle-point shape can also be obtained from this procedure.

Once the threshold energies for fission have been identified, it is also of interest to establish if structure in the potential-energy surface leads to multimode fission, such as that of the well-known three-peaked mass distribution in ^{228}Ra fission [5]. To look for such structure we investigate as a first step if there are valleys of distinctly different character running in the fission direction of increasing Q_2. That is, for 10 or more fixed Q_2 values beyond the outer saddle region, we determine all minima in the remaining 4-dimensional space of the two fragment deformations, neck size and mass asymmetry.

We find that there are often two distinct valleys in the region beyond the second saddle region, one corresponding to a mass asymmetry α_g of about $(140-(A-140))/A$ and one corresponding to mass symmetry. To understand the significance of these valleys it is necessary to determine more details about their structure.

A slight modification of the flooding method shows that separate saddle points provide entry to the two valleys. In the standard flooding strategy an exit point in either of the two valleys gets "wet" as soon as the same specific saddle threshold is exceeded. This is because the water will flow over the lowest saddle point and down the corresponding valley and then,

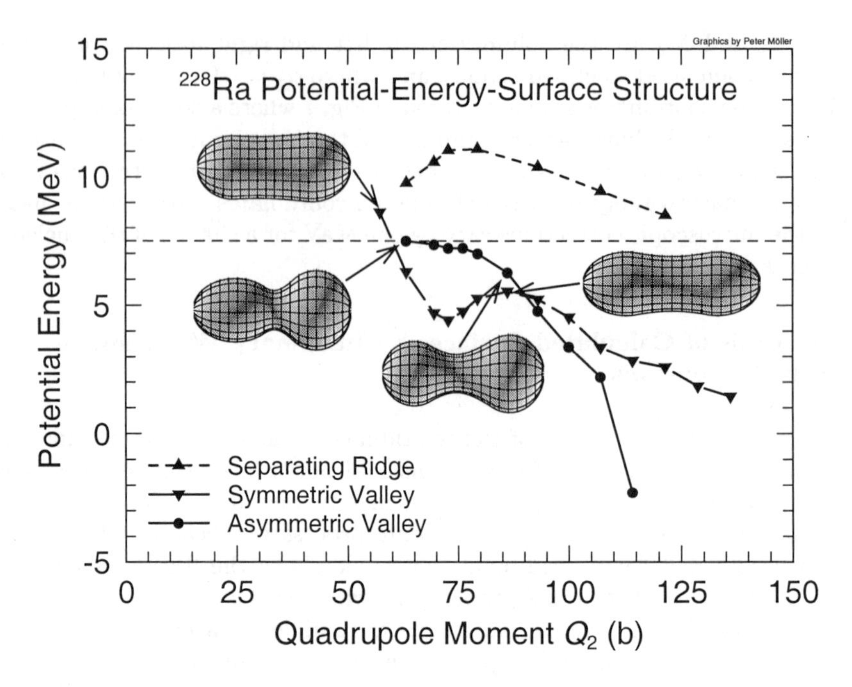

Figure 2. Potential-energy valleys and ridges and corresponding nuclear shapes for ^{228}Ra.

when sufficiently far along in this valley it will flow backwards up the other valley. Therefore we block the flow at a certain Q_2 value, say 86 b so that the water does not flow any further. This can be simply accomplished by ignoring points with higher Q_2. In such an approach one valley gets flooded through one saddle point corresponding to a distinct energy and shape and the other valley through another saddle with a distinctly different shape and energy. If the blockage is moved to successively higher Q_2 then the two saddle points remain unchanged until a critical Q_2 is reached, at which the higher of the two saddles disappears and a new saddle appears just at the blocking wall. This happens when we are sufficiently far down the valley that is entered through the lower saddle point so that the ridge separating the two valley becomes lower than the higher of the two saddle points seen when the valleys were blocked higher up. To determine the height of the ridge between the two valleys along their entire length we study for for each fixed Q_2 the the remaining 4-dimensional space in which the two valleys correspond to two minima and the ridge to the saddle separating them. We use the flooding algorithm in four dimensions to localize this saddle.

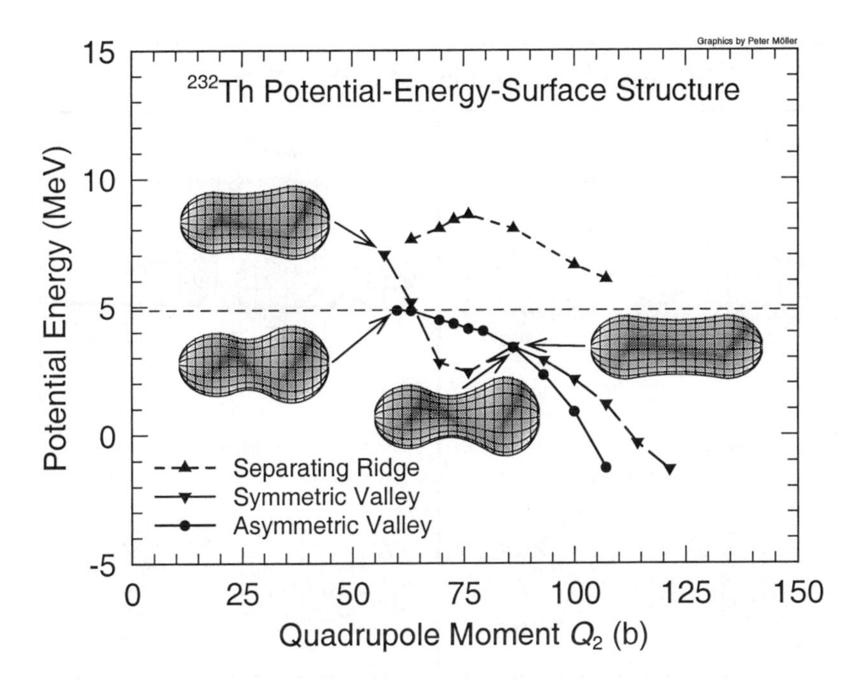

Figure 3. Potential-energy valleys and ridges and corresponding nuclear shapes for ^{232}Th.

As examples of the structures we have found in the calculated 5-dimensional potential-energy surfaces we show in Figs. 2 and 3 some fission-valley and separating-ridge features obtained for ^{228}Ra and ^{232}Th. The first point on the fission-valley curves is the saddle point for entry into the particular valley. The nuclear shapes corresponding to the saddle points are shown to the left in the figure. Shapes corresponding to the symmetric and mass-asymmetric valleys at $Q_2 = 86$ b are shown to the right. Note that the shape corresponding to the *entry* to the mass-*symmetric* valley is slightly mass-*asymmetric*. The thin dashed line is the calculated threshold potential energy for fission, which to be consistent with the other curves is given relative to the spherical macroscopic energy. To obtain the threshold energy for fission the calculated ground-state microscopic correction has to be subtracted.

The calculated structure of the potential-energy surface therefore leads to the observed bimodal fission features in this region of nuclei [5,6]. The high ridge separating the two valleys for ^{228}Ra is peaked at 2.47 MeV above the entrance saddle to the symmetric valley. It therefore keeps the mass-symmetric and mass-asymmetric modes well separated until scission, whereas

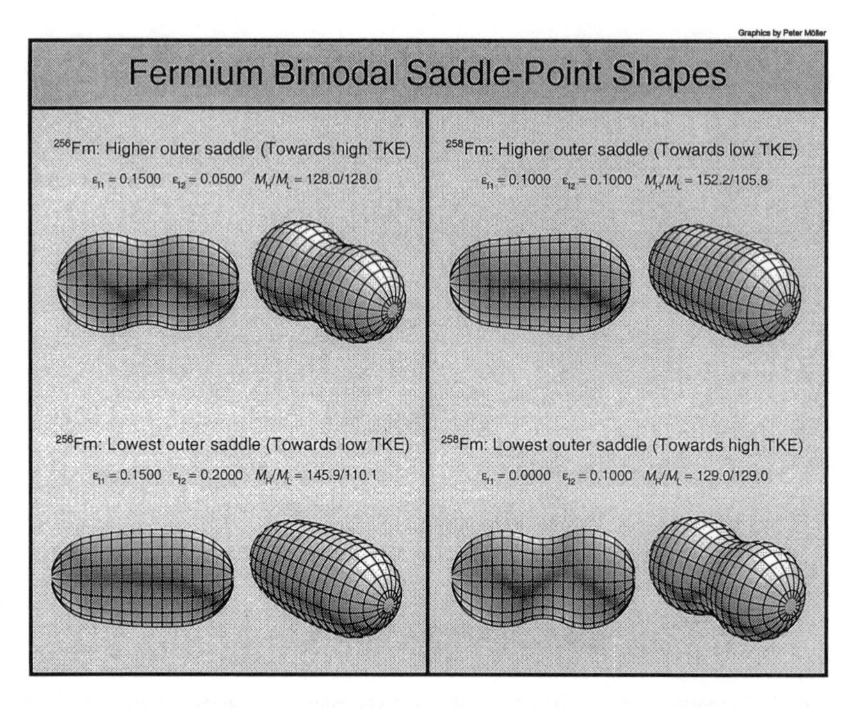

Figure 4. Multiple, bimodal saddle-point shapes for ^{256}Fm and ^{258}Fm based on the grid in Ref. [3].

the lower separating ridge, peaked at 1.56 MeV above the entrance saddle to the symmetric valley, allows the symmetric component to partially revert back to the asymmetric valley before scission for ^{232}Th.

Also, nuclei in the region near ^{258}Fm exhibit bimodal features in fission as discussed in [7]. We have earlier tentatively identified bimodal structures in calculated *two-dimensional* potential-energy surfaces [8,2], but it remained until now to verify that these interpretations are still valid when the calculation is taken from two to five dimensions. In the Fm region we have used a modification of the water-flow analysis to find alternative saddle points that are higher in energy than the lowest threshold saddle point. We build a dam across the lowest saddle point in the following manner: Once the lowest saddle-point has been identified, we raise the energy of this saddle to a high value, for example 10000 MeV. Then we search again for a saddle point. We then normally find that the lowest saddle is now next to this first dam element. We then raise also the energy of this point to 10000 MeV. The procedure is repeated. When we find a saddle that is not next to a dam element we tag this as an alternative saddle point, possibly relevant to bimodal fission. For ^{256}Fm and ^{258}Fm we find the two distinct classes of saddle points shown in Fig. 4. For ^{256}Fm the shape of the lowest saddle indicates it corresponds to normal, low-TKE fission similar to what is observed in fission of slightly lighter actinides. However, another saddle point exists, which we calculate to be 0.30 MeV higher than the lower saddle point. This may correspond to fission into compact scission configurations with high kinetic energies. For ^{258}Fm the latter type of saddle-point becomes the lowest saddle point. Thus, we reproduce the experimentally observed transition point between asymmetric low-TKE fission and symmetric high-TKE fission as observed experimentally [7].

This research is supported by the US DOE under contract W-7405-ENG-36.

References

1. P. Möller, J. R. Nix, W. D. Myers, and W. J. Swiatecki, Atomic Data Nucl. Data Tables **59** (1995) 185.
2. P. Möller, J. R. Nix, and W. J. Swiatecki, Nucl. Phys. **A492** (1989) 349.
3. P. Möller and A. Iwamoto, Phys. Rev. Lett. to be published (1999).
4. A. Mamdouh, J. M. Pearson, M. Rayet, and F. Tondeur, Nucl. Phys. **A644** (1998) 389.
5. E. Konecny, H. J. Specht, and J. Weber, Proc. Third IAEA Symp. on the physics and chemistry of fission, Rochester, 1973, vol. II (IAEA, Vienna, 1974) p. 3.
6. Y. Nagame, I. Nishinaka, K. Tsukada, S. Ichikawa, H. Ikezoe, Y. L. Zhao, Y Oura, K. Sueki, H. Nakahara, M. Tanikawa, T. Ohtsuki, K. Takamiya, K. Nakanishi, H. Kudo, Y. Hamajima, Y. H. Chung, Radiochimica Acta, **78** (1997) 3.
7. E. K. Hulet, J. F. Wild, R. J. Dougan, R. W. Lougheed, J. H. Landrum, A. D. Dougan, M. Schädel, R. L. Hahn, P. A. Baisden, C. M. Henderson, R. J. Dupzyk, K. Sümmerer, and G. R. Bethune, Phys. Rev. Lett. **56** (1986) 313.
8. P. Möller, J. R. Nix, and W. J. Swiatecki, Nucl. Phys. **A469** (1987) 1.

THE ORIGIN OF COSMIC RAYS

A. D. ERLYKIN
P. N. Lebedev Physical Institute, Leninsky Prospekt 53, 117924, Moscow, Russia

A. W. WOLFENDALE
Physics Department, University of Durham, DH1 3LE, UK

Discovered in 1912, cosmic rays are still proving an enigma. Where exactly do they come from? Although at 'low' energies, say below 10^{11} eV, it is pretty certain that they are generated in sources in our own Galaxy, at higher energies the problems start to multiply. Up to 10^{16} eV supernova remnants may well be responsible but above that it is almost anyone's guess. At the very highest energies detected so far ($\sim 3 \times 10^{20}$ eV), where a single particle carries as much energy as a mass of 1 kg falling 5 m, the problem of origin is so acute that alarmingly exotic processes have to be invoked. The present status of the search for the origin of cosmic rays will be described here.

1 Introduction

Cosmic rays were discovered by Viktor Hess in 1912 and it is true to say that today, 87 years later, the origin of all but the lowest energy part of the radiation, or, to be more specific, the manner and location of the acceleration of the particles, is still largely a matter of speculation. The purpose of this paper is to endeavour to set the stage by briefly describing the astrophysical aspects of cosmic rays — particle type, energy, etc, and, after reviewing possible origins, to give the authors own predictions.

It should be stated immediately that the cosmic radiation does not represent some tiny phenomenon on the cosmic scale; the energy density in the primary radiation above the atmosphere is about the same as that in starlight (and in other astronomical entities, too, as we shall see). At the rather parochial level of their interaction with human beings, one can remark that there are about five secondary cosmic rays (mainly muons) passing through our heads every second. We really ought to know where they are coming from.

2 Properties of the primary cosmic rays

2.1 Energy spectrum and mass composition

Although their origin is obscure, the properties of cosmic rays are reasonably well known and we start with a description.

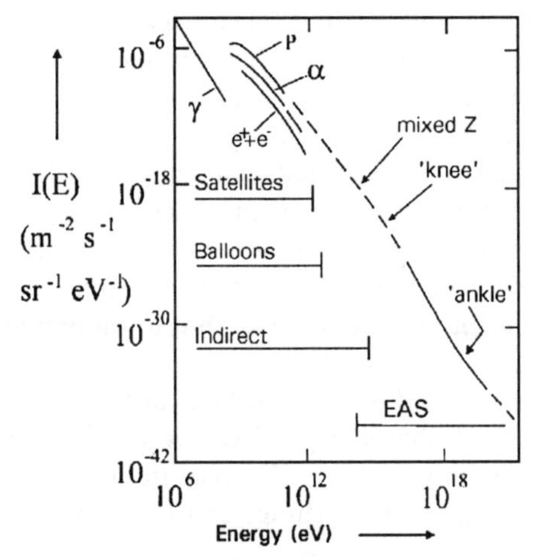

Figure 1. Energy spectra of the major components of the cosmic radiation. The mass composition is uncertain in the dashed part. From 10^{17}eV to 3×10^{18}eV iron nuclei of Galactic origin appear to be common; at even higher energies extragalactic particles start to predominate. The proton component is probably strong amongst the EG particles but there may also be heavier nuclei. [1] The methods of study are indicated.

The energy spectra of the major components of the cosmic radiation are shown in figure 1. Starting with nuclei, not surprisingly, protons and helium predominate and many other nuclei have also been identified, with lower intensities (see later). The shape of the energy spectrum carries with it information about particle production and propagation. Starting at the lowest energies, there is 'curvature' below about 10^{10} eV/nucleon due to the modulating effect of the solar wind and the intensity in this region is sensitive to solar activity. Continuing to higher energies there is evidence for a power law with constant exponent $(N(E)dE\alpha E^{-\gamma}dE$, with $(\gamma \sim 2.65)$ up to about 10^{15} eV/nucleon, above which γ increases to about 3.15. Finally, there is good evidence for a remarkable flattening of the spectrum above about 10^{19} eV.

The mass composition is known quite well below several times 10^{10} eV/nucleon and with modest accuracy to about 1 TeV and figure 2 shows a comparison of the elemental abundances of the cosmic rays with the 'Solar System' or 'universal abundances'. It will be noticed that there are strong similarities, especially when it is realized that the large excesses in CR in the region of Li, Be, B and Sc, Ti and V (and some others) can be easily explained in terms of the fragmentation of nearby, heavier nuclei in their passage through the interstellar medium (see section 2.3). However, there is a strange underabundance of H and He which is not easy to understand.

Turning to electrons, both particles and anti-particles have been seen; indeed anti-protons have also been observed, these particles having been generated in CR interactions with the gas in the interstellar medium (ISM).

Gamma rays have been detected, both a general continuum and fluxes from specific discrete sources. It is interesting to recall that initially it was considered by many that the 'radiation' comprised some form of ultra-γ-radiation (hence the use of the term cosmic *radiation*) but in fact, in the region up to a few times 10^9 eV (where most of the energy of CR resides) the ratio of γ-intensity to particle intensity is only $\simeq 10^{-6}$. Despite its low intensity, it turns out that the γ-radiation has important things to say about the 'origin problem'.

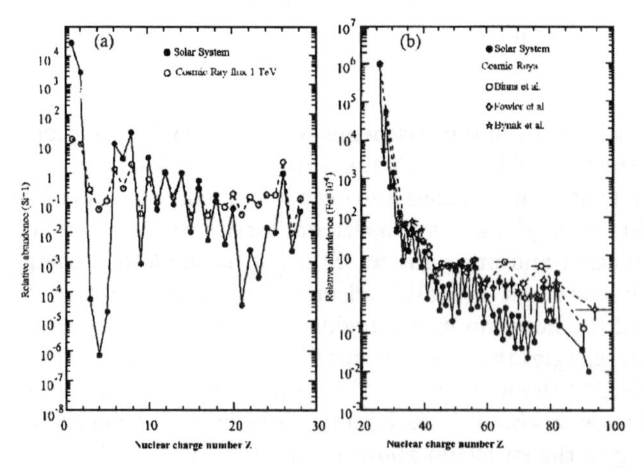

Figure 2. Elemental abundances of cosmic rays (open circles), compared with the Solar System abundaces (joined by full lines) from the summary of Ref. 2. (a) Cosmic rays at 1TeV; the results at lower energies are similar, but more precise, and there is a wealth of isotopic information below 1 GeV. (b) Cosmic rays above 1.5 GeV per nucleon for $Z \geq 26$. The remarkable excesses in CR in some cases can be understood in terms of these nuclei being the products of heavier nuclei which fragment during collisions with gas nuclei in the interstellar medium.

2.2 Spatial extent of cosmic rays: interactions in the ISM

The question of the interactions of CR with the ISM, and their effect on the characteristics of the detected particles, can be considered by moving directly to a basic problem: do cosmic rays represent a local phenomenon in the Galaxy?

For the electron component the answer is immediately no because the distribution of a radio continuum, due almost certainly to synchrotron radiation from CR electrons spiralling in the magnetic field in the Galaxy, is widespread. In fact, radio data indicate the generation of electrons in specific sources in various parts of the Galaxy and of course emission has been detected from other galaxies, too.

The situation with the predominant nuclear component (p, α, ...) is more difficult and the answer to the question is clearly bound up with the general origin problem. Here we can make some preliminary comments. The scale of distance in the Galaxy from the CR point of view is set by the Larmor radius ($\rho[cm] = pc[eV]/(300H[G]Z)$) and insofar as a typical field in the ISM is $\simeq 3\mu G$ then $\rho \simeq p[10^{15}eV/c]/3Z$ parsec (1 parsec $= 3 \times 10^{18}cm \simeq 3$ light years). The observation of particles with momenta as high as 10^{20} eV/c thus implies that these particles at least are likely to be widespread ($\rho \simeq$ kpc even if Z $\gg$ 1), and to cover the Galaxy. This argument can clearly be extended downwards to perhaps 10^{16} eV or so, but in the important region $10^{9} - 10^{10}$ eV, where $\rho \simeq 10^{-6}$ pc (for protons), it has no validity of course, and it could be that these particles are generated locally. However, a number of arguments can be made which give the contrary view and these can be itemized:

(i) The long-term near-constancy of the radiation (a variation of less than $\simeq 2$ over 10^{9} y, for example), during which time the Solar System has made about one-quarter of an orbit round the Galaxy, with respect to the general star population.

(ii) The likelihood of sources of nuclei being distributed in a similar fashion to those of electrons,even if the sources are not identical.

(iii) The possibility of explaining a number of features of the radiation in terms of the interaction of CR nuclei with the nuclei of the ISM during long passage ($\simeq 10^{7}$ y) through the tenuous gas.

3 Origin of low-energy particles

3.1 Galactic or extragalactic?

It is the phenomenon mentioned at the end of the last section, namely the interaction of cosmic rays with gas in the ISM, that has led to the elucidation of the origin of the lowest-energy CR (E $< 10^{10}$ eV or so) — or at least a distinction between particles of Galactic and extragalactic origin. (Hereafter we write 'extragalactic' as EG).

Concerning the EG possibility, although the energy requirements would be large there would be no objection to having CR protons filling the Universe quite uniformly. The same cannot be said for electrons because EG ones are absorbed by inverse Compton (IC) interactions on the cosmic microwave background (CMB).

The Durham group [3] claimed the identification of a radial gradient of CR protons in the Galaxy (a fall-off in intensity with increasing galactocentric distance) using the 'gamma-ray technique'. The principle of the method is to divide the measured gamma-ray intensity along a particular line of sight by the column density of gas along that line. Insofar as many of the gamma rays come from interactions of the CR protons with the gas, the result of the division is the average CR proton intensity along the line. After initial scepticism most contemporary workers [4] now agree with the conclusion.

Another method was suggested by Ginzburg, [5] this being to determine the average proton intensity in the Large Magellanic Cloud (LMC) by the gamma-ray technique and to compare it with the local CR intensity. Equality would indicate an EG origin but if the LMC intensity were much lower a Galactic origin for the local CR protons would follow.

Until the last few years, satellite data of sufficient sensitivity were not available but with the advent of the Compton Gamma Ray Observatory (CGRO), or more particularly the EGRET instrument of the CGRO, [6] the test can now be made. The first study by [7] gave equality for the Galaxy and the LMC, but Chi and Wolfendale [8] found a clear difference, the LMC proton intensity being much less than locally. The reason for the discrepancy is not fully understood. It is a relief, however, that both groups agree that the Small Magellanic Cloud (SMC) cosmic-ray intensity is much lower than locally. A Galactic origin for most rays is therefore assured. Our most recent analysis gives the results shown in figure 3.

3.2 Nature of the sources

Although we have put forward evidence for supernova remnants (SNR) being responsible for the low-energy CR [10,11] the situation is still not quite clear. What is needed is the next generation of gamma-ray detectors (satellite-borne), which will give unambiguous evidence for gamma-ray excesses in the shells of known SNR.

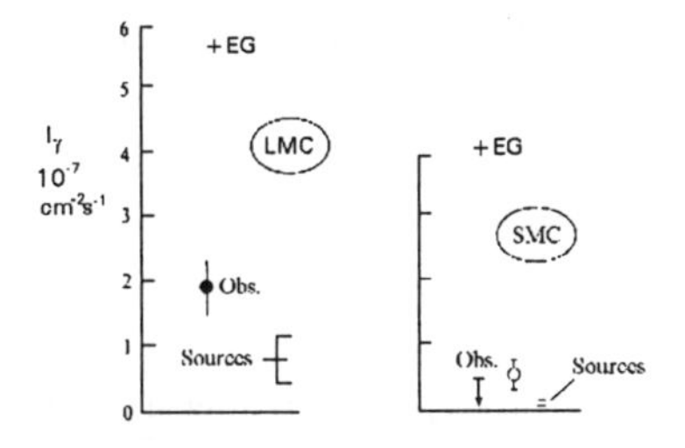

Figure 3. Flux of gamma rays from the Large Magellanic Cloud (LMC) and Small Magellanic Cloud (SMC), in comparison with what would have been expected if the cosmic-ray particle intensity were the same there as here, i.e. if CR were extragalactic (denoted EG). The results show that the CR intensity in the clouds is much less than that at Earth; thus, most CR that we detect come from Galactic sources and are not Universal. The gamma rays are those above 100 MeV. The data are from the Compton Gamma Ray Observatory. [9]The contribution from discrete sources, marked 'sources', is very uncertain.

3.3 The TeV region

Although the TeV region (1 TeV = 10^{12} eV) is not really a low-energy part of the spectrum, it can be considered briefly here. The use of detectors which respond to the Cherenkov radiation emitted by those secondary electrons in the upper atmosphere that travel faster than the velocity of light in air, has led to a breakthrough in this area. The technique allows sources of gamma rays of TeV energy to be identified. So far some four SNRs have been identified (SN1006, Vela, Crab and PSR 1706-44-with no pulsar modulation).

Before jumping to the conclusion that the gamma rays are secondary to protons (and other nuclei) accelerated in the sources it must be pointed out that energetic electrons can also produce gamma rays by way of synchrotron radiation and it is not at all clear which is responsible here.

The whole area of gamma-ray astronomy is, in fact, bedevilled by the uncertain contribution of electrons, the electrons generating gamma rays by a variety of processes. In a sense there is a similar problem with cosmic rays in other galaxies - we know that there are radio signals from these galaxies but these are certainly due to electrons. We have no direct evidence of cosmic-ray nuclei in other galaxies (the situation for the Magellanic Clouds, referred to earlier, is that we have upper limits to the proton flux there).

Returning to gamma rays, it is only in the $10^8 - 10^9$ eV region that we have the possibility of seeing a characteristic signal due to protons as distinct from electrons (the so-called pion peak). At higher energies the ambiguity remains.

4 Origin of CR in the $10^{15} - 10^{16}$ eV region

4.1 The problem

Inspection of figure 1 shows a 'knee' in the primary energy spectrum which, surely, is trying to tell us something about the origin — or propagation — of the particles of these energies.

The 'knee' feature has been known for over 40 years, ever since such a feature was found [12] in the size spectrum of extensive air showers (the proxy indicator of primary energy). Most workers have argued that it is a propagation effect and explained it in terms of a simple power law production spectrum with an increasing loss of particles from the Galaxy as the Galactic magnetic field starts to lose its trapping ability. However, we [13] have argued strenuously that this is not the case, but rather that the knee is indicative of another CR component which pokes through a smoothly falling 'background spectrum'. Insofar as this paper is devoted to unsolved problems we must include this topic — the reason being the remarkable lack of acceptance of our hypothesis (so far)!

4.2 The Erlykin and Wolfendale single-source model

Figure 4 indicates the manner in which we 'explain the knee'. The idea is that a single nearby local SNR is accelerating the particles and it is the characteristic sharply 'peaked' (in the manner in which it is plotted, namely E^3 I (E)) spectra of the important elements oxygen and iron which explain the spectral shape (the oxygen being largely responsible for the knee itself).

This beautiful theory which, in our view, explains a mass a cosmic-ray data, has, alas, a problem. This is: where is the actual supernova? There are a few candidates but so far it has not been possible to identify the actual one.

The 'smoking gun' remains to be identified. Nevertheless, we persist by analysing each and every piece of cosmic-ray data in the energy region in question as it reaches the public domain. Figure 5 is from a very recent paper of ours; [14] in our view, it shows the peak above the conventional Galactic modulation expectation rather well.

If our model is correct, there is the possibility of an increase in cosmic-ray intensity in the future as the SNR shell comes closer, if indeed we are *outside* the shell.

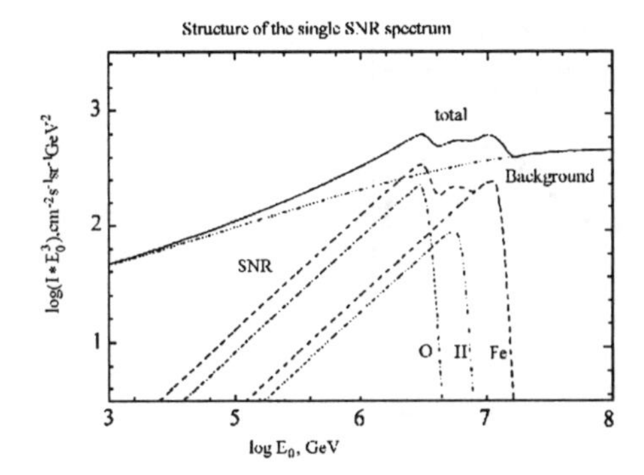

Figure 4. Energy spectra of CR from the 'single source' postulated in Ref. 13. Note: the ordinate is the logarithm of the intensity, I(E), multiplied by E^3; in real life the intensity plummets with increasing energy. 'SNR' denotes the contribution from a single supernova remnant. 'Background' is a slowly varying spectrum from unknown multiple sources.

5 Cosmic rays of the highest energies

5.1 The anisotropy problem

If all cosmic rays were to come from sources in our Galaxy, the anisotropy (i.e. the nonuniformity of arrival directions) would be much bigger than is observed — above 10^{19} eV, at least — whatever the nature of the particles. Thus, it seems very likely indeed that these ultra-high-energy particles, which extend to 3×10^{20} eV and possibly higher, [15] are coming from extragalactic sources. Just what these sources are is subject to considerable doubt.

Two situations can be distinguished for the acceleration mechanism — bottom-up or top-down. The former relates to particles which have been accelerated up from low energies in some way—just as, at low energies, SNR seem to behave. For top-down it is postulated that exotic, massive particles of some form 'decay down' yielding the ultra-energetic particles. The situations will be considered in turn, after examining the effect of the ubiquitous microwave background — which is a relic of the Big Bang.

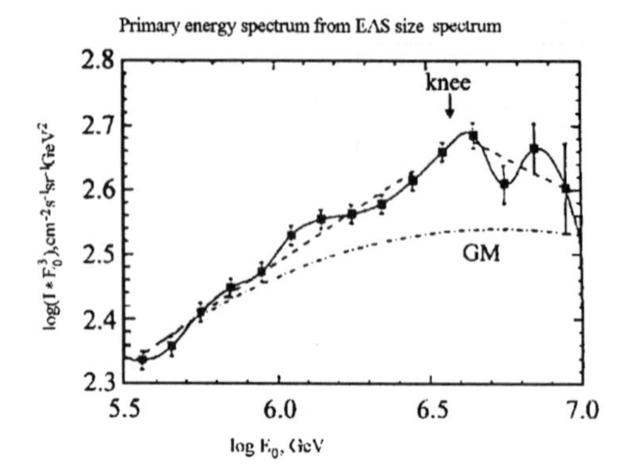

Figure 5. One exemple of very recent data which support the Earlykin and Wolfendale model. [14] The data shown there come from our analysis of results given by Ref. 16 based on measurements of Cherenkov radiation at La Palma (the 'HEGRA' array). 'GM' denotes the spectrum expected from the conventional model (galactic modulation) in which the galactic magnetic field ceases to trap particles efficiently after a certain energy is reached; it is clear that it does not fit the observations.

5.2 The cosmic microwave cut-off

The presence of the 2.7 cosmic microwave background (CMB) radiation causes a most interesting effect: if the ultra-high-energy cosmic rays (UHECR) are Universal in origin then the measured intensity should lurch downwards at about 6×10^{19} eV, which is where the interaction of protons with the CMB becomes serious. In fact, there is no evidence at all for this fall, indeed (figure 1) the intensity falls *less* slowly than expected, here. The interactions cause of loss of energy which is, eventually, converted into gamma rays. The result is that the particles of higher energy actually recorded must come from sources within a few tens of Mpc. A similar situation exists for heavy nuclei, which fragment over similar distances.

The fact that the sources should be 'local' (if they are discrete) means, in principle, that they have a bigger chance of being detected, the reason being the undoubted presence of an EG magnetic field which will randomize the directions of those further away.

Progress in the search for local sources will now be considered.

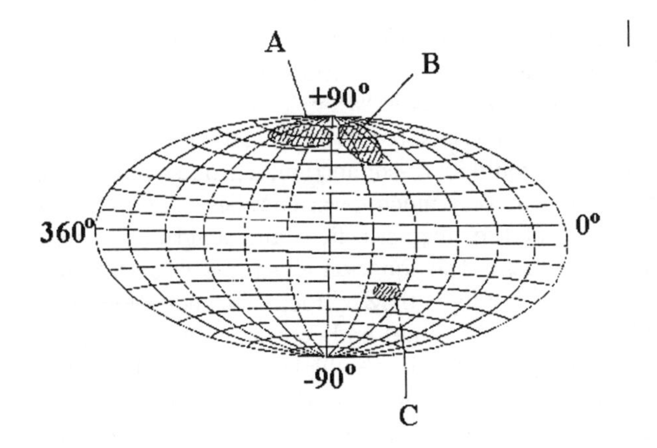

Figure 6. A map of the Galaxy. The line from longitude 0° to 360° represents the mid-plane of the Milky Way. Latitude +90° corresponds to the North Galactic Pole and −90° corresponds to the South Galactic Pole. Regions of the sky from which there *may* be a small excess of ultra-high-energy CR fluxes, above 10^{19} eV, are shwon shaded: A, B and C. The analysis , [17] relied on earlier works (Ref. 9, 18, 19). This is the nearest we have got to identifying source regions for ultra-high-energy cosmic rays. The evidence is admittedly, weak.

5.3 The evidence for specific EG sources: colliding galaxies?

The bottom-up mechanism would presumably give rise to apparent 'sources' of EG particles in rather specific regions of space (galaxies of particular types, etc).

The subject of searches for EG sources has had a chequered history; 'here today and gone tomorrow' sums it up. Many workers in the field would probably not object to figure 6 (from Ref. 17) which identifies regions of the sky from which there seem to be rather more arrival directions than would be expected for a purely uniform, i.e. isotropic, distribution. We remember here that at these energies (above 10^{19} eV) the deflections in the Galaxy and in EG space out to 20 Mpc, i.e. about 0.4% of the 'radius' of the Universe, should be small.

Having (perhaps) identified directions to putative sources — what are they? There are many galaxies in the error boxes so one has to be specific and look for reasonable, but rare, objects. The best we have been able to come up with, within 20 Mpc, is 'colliding galaxies'. Within 10 Mpc there are nine such pairs of galaxies and of these three fall in the regions A, B and

C (one in each). In the range 10–20 Mpc, 10 more have been identified, of which five fall in A or B. Statistically, then, the outlook is bright.

There is a problem, however, and that is 'what is the mechanism?' A possibility is that the magnetic fields associated with the galaxies fight-it-out in the collision. Specifically, there could be re-connection, a process that seems to work on the sun, but of course at very much lower energies. The difficulty concerns the likely magnitudes of the magnetic field and length of system over which the field is coherent. Using the equation given in section 2.2, $\rho = p[10^{15}eV/c]/3Z$ parsec, then $p[10^{20}eV/c] = 0.3Z\rho[\mathrm{Mpc}]$. This equation is for 3μG. If the field is 0.3μG (a more likely value) and converting momentum to energy, then $E[10^{20}eV] = 3Z\rho[\mathrm{Mpc}]$. Now for the highest reasonable charge, iron, Z = 26 and for ρ = 10kpc (i.e. $\rho[\mathrm{Mpc}] = 0.01$) the value of E ($10^{20}$ eV) is unity. This means that a particle of this energy would have a radius of curvature equal to 10 kpc, which is probably the maximum distance over which the magnetic field could be coherent. The converse is that an acceleration mechanism could, perhaps, just reach this energy, but it seems doubtful (we note that for SNR acceleration the maximum energy is quite a lot less than a simple theory like the above would predict).

Notwithstanding these remarks, the colliding galaxy idea is the best that we have been able to suggest. Others have suggested 'active galactic nuclei' (AGN) as being responsible at the very highest energies, [2] but a quantitative estimate of the particle energy expected is again difficult. Furthermore, there are so many AGN 'out there' that finding genuine coincidences is difficult. There are no obvious super-energetic objects in the region (some tens of Mpc) that is relevant to the search. The search goes on.

5.4 The top-down mechanism

5.4.1 Cosmic strings. Several exotic mechanisms have been suggested which could, in principle, be responsible for the UHECR. Firstly, we consider 'cosmic strings'. Topological defects in the universe may well exist and manifest themselves as cosmic strings . [20]

In Ref. 21 it was postulated that the strings contain extraordinarily condensed regions of 'X-particles' with masses $\sim 10^{15}$ GeV (the Grand Unified Theory energy scale). The X-particles can be 'shaken free' when the strings intersect or collapse.

The X-particles, which were previously stable in the potential well associated with their great mass density ($\sim 10^{22} gcm^{-1}$), are then able to decay into a variety of particles, including protons. It is plausible that the energy spectrum will be acceptable and the maximum energy of the protons will certainly be high enough to account for the observations.

A problem has been identified, however, and this concerns the decay scheme of the X-particles. It seems that gamma rays should be produced in abundance and, specifically, the gamma/proton ratio should [22] be about 30. The gamma rays should cascade through the universe and be detectable; they are not. A way out of the difficulty is to assume that the UHECR are not protons (or nuclei) but are, themselves, gamma rays! Such a situation, although not impossible, seems very unlikely.

5.4.2 Dark matter: 'cryptons'. In astronomy, when all else fails, one invokes black holes or dark matter — or both! So it is here. It has been suggested that the 'dark matter', which surrounds the Galaxy, contains 'cryptons'. [23,24] The cryptons are hypothesiyed to have mass $\sim 10^{12}$ GeV and lifetime $\geq 10^{16}$ y (i.e. longer than the 'age' of the Universe). They, too, can decay into protons and, in principle, explain the UHECR.

This beautiful idea is still accepted by some workers, but the present authors are not convinced. The problem concerns the expected anisotropy. If the halo were big enough — many times the 'radius' of the Galaxy — the predicted anisotropy would be small and acceptable, but it appears not to be. Thus, Ref. 25 conclude that even approaching 10^{20} eV only 10% of the UHECR can come from this mechanism.

6 Conclusions

It will be evident from what has been described that our knowledge about the origin of cosmic rays is surprisingly poor.

Discovered in 1912, and studied strenuously ever since, we have had modest success only to about 10^{16} eV, i.e. for seven orders of magnitude out of approaching 12 (the range of 'cosmic rays' at present is $10^9 - 10^{21}$ eV). Even below 10^{16} eV, the evidence is only really secure up to about 10^{11} eV. We (the authors) are quite sure up to 10^{16} eV, but most other workers are not! All would agree, however, that the SNR model itself is a good one, for this energy region; it is just the hard evidence that is lacking.

It is above 10^{16} eV where the problems are really severe. It is true to say that if no particles had been detected at these energies none would have been predicted!

Most would agree that Galactic sources of some sort — many of which are producing iron nuclei — predominate to a few 10^{18} eV.

Above 10^{19} eV, extragalactic particles are almost certainly in the majority. What the sources are and how they accelerate particles we just do not know.

References

1. T. Wibig and A. W. Wolfendale, *J. Phys. G: Nucl. Part. Phys.* **25**, 1099 (1999).
2. B. Wiebel-Sooth and P. L. Biermann , *Landolt-Börnstein* Group VI, vol.**3** sub-vol.C. *Astronomy and Astrophysics—ISM, Galaxy, Universe* (Springer, Berlin, 1999) p.37.
3. D. Dodds, A.W. Strong and A.W. Wolfendale, *Mon. Not. R. Astron. Soc.* **171**, 569 (1975).
4. A.W. Strong and J.R. Mattox, *Astron. Astrophys.* **308**, L21 (1996).
5. V.L. Ginzburg, *Nature (Phys. Sci. Suppl.)* **239**, 8 (1972).
6. C.E. Fichtel, *NASA Technical Memo* No 83957 (1982).
7. P. Sreekumar *et al, Astrophys. J.* **400**, L67 (1992).
8. X. Chi and A.W. Wolfendale, *J. Phys. G: Nucl. Part. Phys.* **19**, 795 (1993).
9. X. Chi *et al, J. Phys. G: Nucl. Part. Phys.* **18**, 539 (1992).
10. C.L. Bhat *et al, Nature* **314**, 515 (1985).
11. J.L. Osborne, A.W. Wolfendale and L. Zhang, *J. Phys. G: Nucl. Part. Phys.* **21**, 429 (1995).
12. G.V. Kulikov and G.B. Khristiansen, *Sov. Phys.-JETP* **35**, 635 (1958).
13. A.D. Erlykin and A.W. Wolfendale, *Astropart. Phys.* **8**, 265 (1998).
14. A.D. Erlykin and A.W. Wolfendale, *Astron. Astrophys.* in press.
15. D. J. Bird *et al, Preprint Astrophys. J.* **441**, 144 (1995).
16. F. Arqueros *et al, Astron. Astrophys.* in press.
17. S.S. Al-Dargazelli *et al, J. Phys. G: Nucl. Part. Phys.* **22**, 1825 (1996).
18. T. Stanev *et al, Phys. Rev. Lett.* **75**, 3056 (1995).
19. N. Hyashida *et al, Phys. Rev.* **77**, 1000 (1996).
20. A. Vilenkin, *Phys. Rep.* **121**, 263 (1985).
21. P. Bhattacharjee and N.C. Rana, *Phys. Lett. B* **246**, 365 (1990).
22. X. Chi *et al, Astropart. Phys.* **1**, 239 (1993).
23. V. Berezinsky, B. Pasquale and A. Vilenkin, *Preprint astro-ph/980327*, v**2** 8 May (1998).
24. M. Birkel and S. Sarkar, hep/ph 9804285, v**2** (1998).
25. A. Benson, A. Smialkowski and A. W. Wolfendale, *Astropart. Phys.* **10**, 313 (1999).

TEN YEARS OF HEIDELBERG–MOSCOW EXPERIMENT – A FRESH LOOK

HANS V. KLAPDOR-KLEINGROTHAUS† *

† *Max–Planck–Institut für Kernphysik*

P.O. Box 103980, D–69029 Heidelberg, Germany

The Heidelberg–Moscow double beta decay project operating now for almost ten years in the Gran Sasso Underground Laboratory in Italy, yields the world–wide most stringent limit on the Majorana neutrino mass. Simultaneously, new limits on other parameters of Beyond Standard Model physics have been obtained on the *TeV* scale, and also the best limits on cold dark matter in the universe using only raw data.

The two fundamental questions to all unified theories of modern particle physics, of neutrino mass and nature, and of the existence of a superworld predicted in supersymmetric theories, probably will be solved underground.

Double beta decay is the only process allowing to solve the question of the nature of the neutrino: Dirac or Majorana particle. Also, it is the common opinion of theorists in the field that the question of the neutrino mass matrix cannot be solved by neutrino oscillation experiments (solar, atmospheric, etc) alone but requires in addition a sufficiently sensitive double beta decay experiment.

Neutrinoless double beta decay is a hypothetical extremely rare radioactive decay mode, in which exchange of a neutrino between two nucleons triggers their decay under emission of two electrons. This process would violate conservation of lepton number L, and, equally important, of baryon number minus lepton number (B–L), which would imply immediately Beyond Standard Model physics.

This mode is only possible if the neutrino has a Majorana mass, which requires that the neutrino is its own antiparticle. – However, by far most Grand Unified Models of particle physics predict Majorana neutrinos. – The decay rate is proportional to an effective neutrino mass $< m >$ squared, which is a superposition of the different neutrino mass eigenstates. Consequently double beta decay yields important information on the parameters of the neutrino mass and mixing matrix. The information from double beta decay is particularly important to fix the absolute neutrino mass scale, since neutrino oscillation experiments measure only differences between mass eigenstates.

*SPOKESMAN OF THE HEIDELBERG–MOSCOW COLLABORATION

There are worldwide several double beta experiments running looking for this type of decay for various nuclei, which has, however, not been observed until now. The Heidelberg–Moscow experiment is the by far most sensitive one and is probing for the first time the sub–eV range for the Majorana neutrino mass.

The team from the Max Planck Institute for Nuclear Physics in Heidelberg and the Kurchatov Institute in Moscow constructed a setup of five high-purity Germanium detectors enriched in the isotope Germanium–76 to 86% (natural abundance 7.8%) of total mass of 11.5 *kg*. This results in the largest source strength ever used in a double beta decay experiment. The setup is operated in a heavy shielding by some ten tons of superclean lead and copper, under 1500 *meters* of rock (corresponding to 3500 m of water shielding) in the Gran Sasso Underground Laboratory in Italy. The background level reached by the experiment is the lowest worldwide for this kind of experiment - 0.06 *events* per year and kilogramm of detector mass in the energy range of the expected double beta decay signal, which would be a peak in the spectrum produced by the decay electrons, at 2038 keV.

In almost 10 years of measurement, the Heidelberg-Moscow team has obtained a lower limit for the half life of neutrinoless double beta decay of Germanium–76 of several 10^{25} years. This is the world record under all running investigations. With the deduced conservative limit for the Majorana neutrino mass of 0.38 eV, the Heidelberg–Moscow experiment is not only the first one exploring the sub-eV range of the neutrino mass, but also enters in a range of values, which has stringent consequences for neutrino mass models and for cosmological models assuming neutrinos as hot dark matter in the universe (see Fig. 1).

For example, in degenerate neutrino mass models, in which the neutrino mass eigenstates have very close values, the above value excludes models with cold and hot dark matter ($CHDM$ models), and also those including a non–vanishing cosmological constant Λ ($\Lambda CHDM$ models) for the case of the small mixing angle solution of the solar neutrino problem. This means, in this case neutrinos as hot dark matter would have to be of Dirac type. Assuming the bestfits of the large mixing angle solution or the vacuum oscillation solutions, still $CHDM$ models are excluded, while $\Lambda CHDM$ models are still marginally possible (requiring $< m >= 0.15 - 0.30\ eV$). In the case of inverse hierarchical neutrino mass scenarios, with only two neutrinos contributing to hot dark matter, only $LCDM$ models with a small Hubble constant h = 0.5 are marginally not yet excluded.

Already now, if assuming the small angle solution of the solar neutrino problem to be valid, the experiment would rule out the whole range of sensi-

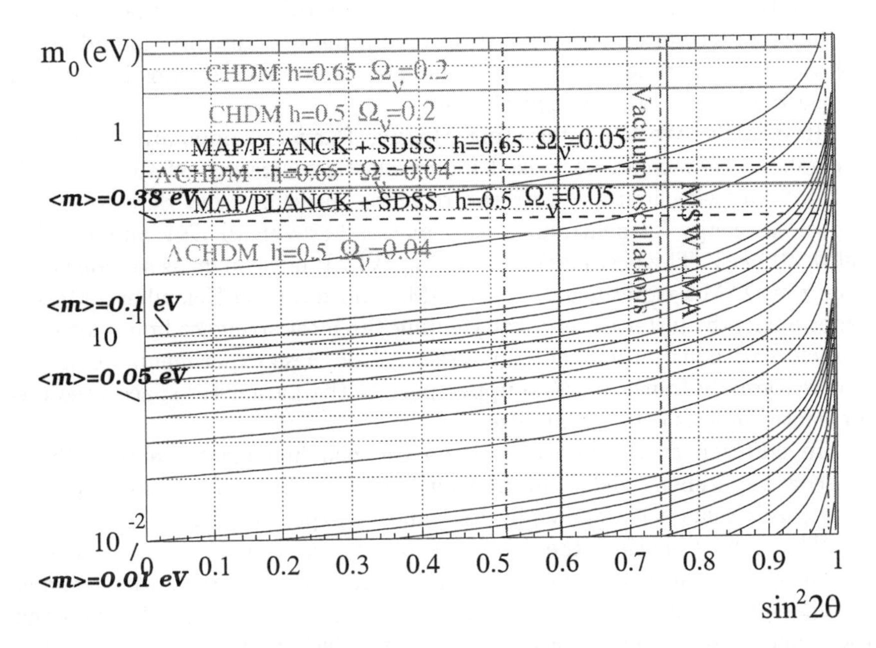

Figure 1. Isomass–lines for the effective neutrino mass $< m >$ measured in double beta decay in the $m_0 - \sin^2(2\theta)$ plane for the case of a degenerate three-neutrino scenario with mass m_o. Also shown are the bestfits for $CHDM$ and $\Lambda CHDM$ for different values of the Hubble constant, the sensitivity of MAP/PLANCK combined with SDSS, and the regions of the MSW LMA and of the vacuum oscillation solutions of the solar neutrino problem. The Heidelberg–Moscow experiment excludes already most of the cosmological models and also, assuming the MSW SMA solution, the whole range of sensitivity of the future satellite experiments MAP and PLANCK for cosmological models involving neutrinos as hot dark matter (from [Kla99b]).

tivity of the future satellite experiments MAP and PLANCK for cosmological models of the above types involving neutrinos as hot dark matter (see Fig. 1). Fig. 1 also demonstrates that, combined with the neutrino oscillation results obtained in the recent solar and atmospheric neutrino oscillation experiments and with precise determinations of cosmological parameters double beta decay is obviously the only way to obtain precise informations about the neutrino mixing and the absolute mass scale in partially degenerate and degenerate neutrino mass scenarios.

Looking into four–neutrino scenarios, including in addition to the three known neutrino flavours a fourth 'sterile' neutrino, there are as shown recently by Giunti et al., Bilenky et al., and others, only two schemes, that can accomodate the results of *a l l* neutrino oscillation experiments (including the Los Alamos experiment LSND). The first of these two scenarios is ruled out by the Heidelberg-Moscow experiment.

If WIMPs (Weakly Interacting Massive Particles) populate the halo of our galaxy, they could be looked for by elastic scattering off the Ge nuclei and the following ionization by the recoiling nucleus in our detectors. The deposited energy for neutralinos with masses between 10 GeV and 1 TeV is below 100 keV. The best current limits on WIMP nucleon cross sections come from the DAMA experiment, from CDMS and from the Heidelberg–Moscow experiment, the latter experiment yielding the most stringent limits for using raw data without pulse shape analysis. All of these experiments at present just marginally touch only the upper part of the parameter space predicted for neutralinos as cold dark matter.

In addition to the information on the neutrino mass the Heidelberg–Moscow experiment yields information on beyond standard model physics on the TeV scale, where new physics could be expected (Fig. 2).

This is possible since the $\Delta L = 2$ process of neutrinoless double beta decay could occur in *a n y* lepton-number violating theory, so, e.g. by exchange of supersymmetric particles like neutralinos, gluinos, sleptons, etc., and thus the process would yield information on the underlying theories and on properties of the involved particles. It is important to note, and has been proved theoretically already in the early 80's by Schechter and Valle, that independent of the mechanism underlying the double beta decay, its occurrence *a l w a y s* would imply a nonvanishing Majorana neutrino mass.

The half life limit measured in the Heidelberg-Moscow experiment yields *a l o w e r* limit on the the mass of a superheavy left–handed neutrino of $M > 8 \times 10^5$ GeV, a limit which could be reached only by a far–future 2 TeV Linear Electron–Electron Collider. It yields further an upper limit on the Yukawa coupling Λ'_{111} in the R–parity breaking part of the superpotential of the Minimal Supersymmetric Standard Model (MSSM), which is more stringent than present limits from the TEVATRON and HERA colliders, and which immediately excluded the possibility of squarks of first generation being produced in the recently discussed high–Q squared events at HERA.

The Heidelberg–Moscow result also restricts stringently products of higher generation Yukawa couplings, it yields the sharpest limits for a Majorana–type mass of the sneutrino, the supersymmetric partner of the neutrino, sets limits to compositeness (assuming a substructure of quarks and leptons), which are, as shown recently by Panella et al., more stringent than those from the LEPII accelerator,yields bounds on violation of special relativity (VLP)and equivalence principle (VEP) in the neutrino sector, in the range of small mixing, which cannot be constrained by other experiments, etc.

In summary the Heidelberg-Moscow experiment yields a lot of new information, including the worldwide most stringent information on the Majorana neutrino mass, important for fixing the neutrino mass matrix and cosmological models, including the most stringent limits for cold dark matter (on the basis of raw data), and the sharpest bounds on other beyond standard model physics parameters in the TeV range, where new physics can be expected, and which partly can be covered only by future colliders. For a recent summary of all of these results see [Kla98], [Kla99], [HM99], [HM98], [Kla99a], [Bha99], [Kla2000].

The Heidelberg–Moscow experiment will remain the most sensitive double beta decay experiment also for the next years.

The method of double beta decay could be pushed in another type of setup to a limit of 0.01 eV or ultimately 0.001 eV, which probably would be the ultimate value obtainable in double beta decay experiments. For this purpose the Heidelberg group has proposed the new project GENIUS [Kla98], [Kla99b], [Kla99c], which would have implications for a broad area of physics encompassing particle physics and astrophysics, and which would serve as an important bridge between the physics that will be gleaned from high energy accelerators such as LHC and NLC on the one hand and satellite experiments such as MAP and PLANCK on the other.

References

1. G. Bhattacharyya, H. V. Klapdor-Kleingrothaus and H. Päs *"Neutrino Mass and Magnetic Moment in Supersymmetry without R–Parity in the light of Recent Data"*, *Phys. Lett.* **B 463** (1999) 77 – 82 and *Preprint hep-ph/* **9907432** (1999)
2. HEIDELBERG–MOSCOW Collaboration, L. Baudis, J. Hellmig, G. Heusser, H. V. Klapdor–Kleingrothaus, S. Kolb, B. Majorovits, H. Päs, Y. Ramachers, H. Strecker, V. Alexeev, A. Balysh, A. Bakalyarov, S. T. Belyaev, V. I. Lebedev and S. Zhukov *"New limits on dark–matter interacting particles from the Heidelberg–Moscow experiment"*, *Phys. Rev.* **D 59** (1998) 022001-1 – 022001-5
3. HEIDELBERG–MOSCOW Collaboration, L. Baudis, A. Dietz, G. Heusser, H. V. Klapdor–Kleingrothaus, I. V. Krivosheina, S. Kolb, B. Majorovits, V. F. Melnikov, H. Päs, F. Schwamm, H. Strecker, V. Alexeev, A. Balysh, A. Bakalyarov, S. T. Belyaev, V. I. Lebedev and S. Zhukov

"Limits on the majorana neutrino mass in the 0.1–eV range", *Phys. Rev. Lett.* bf 83 (1999) 41 – 44

4. H. V. Klapdor–Kleingrothaus *"Status and Perspectives of Double Beta Decay – Window to New Physics Beyond the Standard Model of Particle Physics"*, *Intern. Journ of Modern Phys.* **A 13, No. 23** (1998) 3953 – 3992
5. H. V. Klapdor–Kleingrothaus *"Double Beta and Dark Matter Search – Window to New Physics Beyond the Standard Model of Particle Physics"*, in Proc. "Lepton and Baryon Number Violation in Particle Physics, Astrophysics and Cosmology", eds by *H.V. Klapdor–Kleingrothaus and I.V. Krivosheina, Intern. Workshop at ECT,* Trento, Italy, April 20 – April 25, 1998, *World Scientific* (1998), 251 – 301
6. H. V. Klapdor–Kleingrothaus, H. Päs and U. Sarkar *"Test of Special Relativity and Equivalence Principle from Neutrinoless Double Beta Decay"*, *Eur. Phys. J.* **A 5** (1996) 3 – 6
7. H. V. Klapdor–Kleingrothaus L. Baudis, G. Heusser, B. Majorovits, H. Päs "GENIUS - a Supersensitive Germanium Detector System for Rare Events", *Proposal* **August 1999 second draft**, *Preprint hep-ph/* **9910205** (1999)
8. H. V. Klapdor–Kleingrothaus *"Double Beta Decay with Ge–detectors – and future of Double Beta and Dark Matter Search (GENIUS)"*, "in Proc. of "NEUTRINO'98" Takayama, Japan, 4 – 9 June 1998, ed. by *Y. Suzuki and Y. Totsuka, Nucl. Phys.* **B 77** *Proc. Suppl.* (1999) 357 – 368
9. H. V. Klapdor–Kleingrothaus, H. Päs and U. Sarkar *"Effects of New Gravitational Interactions on Neutrinoless Double Beta Decay", to be published* (2000)

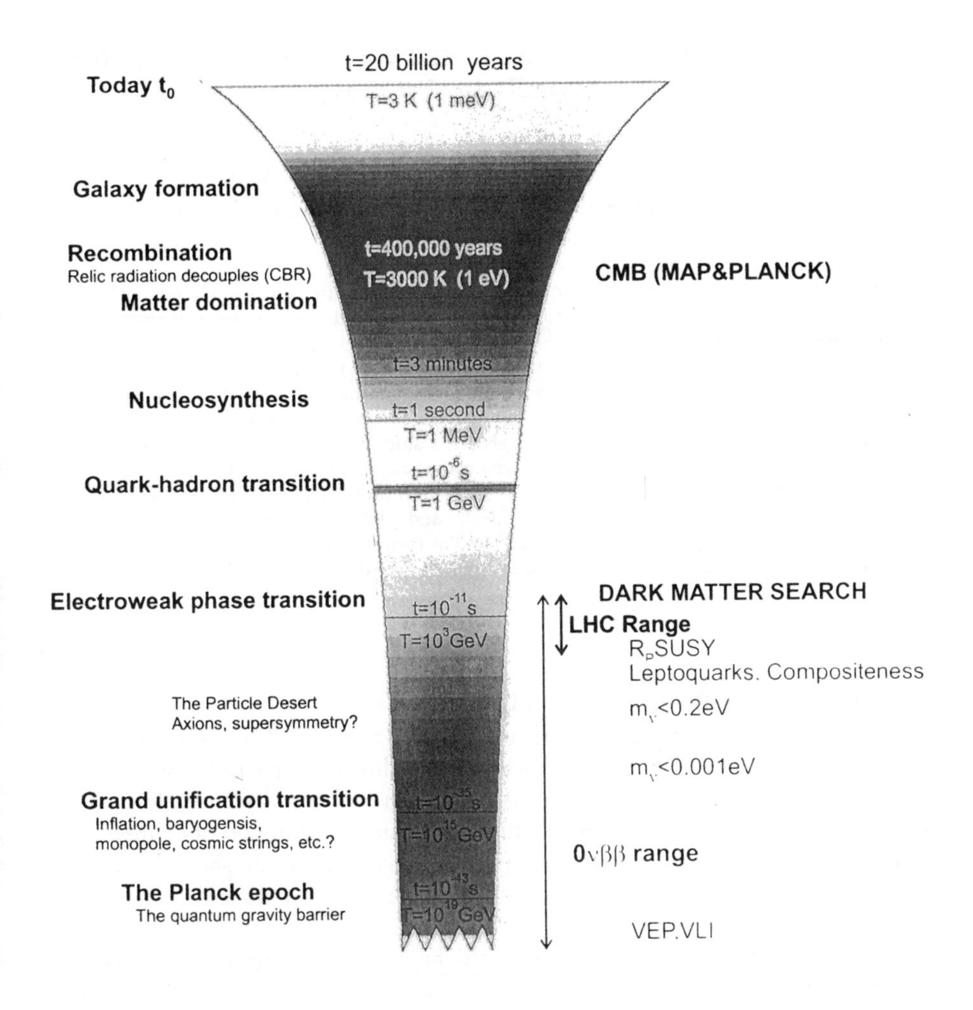

Figure 2. *The potential of double beta decay for research into particle physics and the related cosmology. Instead of the (light) neutrino masses the scale is indicated for the right-handed heavy counterparts according to the seesaw mechanism.*

COSMIC RAYS AS INTERFACE BETWEEN ASTROPHYSICS AND PARTICLE PHYSICS

H. REBEL FOR THE KASCADE COLLABORATION
Institut für Kernphysik, Forschungszentrum Karlsruhe, Germany
E-mail: rebel@ik3.fzk.de

The interpretation of extensive air shower (EAS) observations needs a sufficiently accurate knowledge of the interactions driving the cascade development in the atmosphere. While the electromagnetic and weak interaction parts do not provide principal problems, the hadronic interaction is a subject of uncertainties and debates, especially in the ultrahigh energy region extending the energy limits of man made accelerators and experimental knowledge from collider experiments. Since the EAS development is dominantly governed by soft processes, which are presently not accessible to a perturbative QCD treatment, one has to rely on QCD inspired phenomenological interaction models, in particular on string-models based on the Gribov-Regge theory like QGSJET, VENUS and SIBYLL. Recent results of the EAS experiments KASCADE are scrutinized in terms of such models used as generators in the Monte Carlo EAS simulation code CORSIKA.

1 Introduction

In cosmic ray investigations, in addition to the astrophysical items of origin, acceleration and propagation of primary cosmic rays, there is the historically well developed aspect of the interaction of high-energy particles with matter. Cosmic rays interacting with the atmosphere as target (on sea level it is equivalent to a lead bloc of 1m thickness) produce the full zoo of elementary particles and induce by cascading interactions intensive air showers (EAS) which we observe with large extended detector arrays distributed in the landscapes, recording the features of different particle EAS components. The EAS development carries information about the hadronic interaction (but it has to be disentangled from the unknown nature and quality of the beam). When realizing the present limits of man made accelerators, it is immediately obvious why there appears a renaissance of interest in cosmic ray studies from the point of view of particle physics. EAS observations of energies 10^{15} eV represent an almost unique chance to test theoretical achievements of very high energy nuclear physics.
Actually the astrophysicist is faced with the situation that reliable interpretations of the features of the secondary particle production, and of their relation to the characteristics of the primary particle are necessarily related to our understanding of the hadronic interactions. This aspect is particularly stimulating for high-energy physicists, since there is not yet an exact way to

calculate the properties of the bulk of hadronic interactions.
This lecture reviews some relevant aspects of hadronic interactions affecting the EAS development, illustrated with recent results of EAS investigations of the KASCADE experiment [1], especially of studies of the hadronic EAS component using the iron sampling calorimeter of the KASCADE central detector [2].

2 EAS Development and Hadronic Interactions

The basic ingredients for the understanding of EAS are the total cross sections of hadron air collisions and the differential cross sections for multiparticle production. Actually our interest in the total cross section is better specified by the inelastic part, since the elastic part does not drive the EAS development. Usually with ignoring coherence effects, the nucleon-nucleon cross section is considered to be more fundamental than the nucleus-nucleus cross section, which is believed to be obtained in terms of the first. Due to the short range of hadron interactions the proton will interact with only some, the so-called wounded nucleons of the target. The number could be estimated on basis of geometrical considerations, in which size and shape of the colliding nuclei enter. All this is mathematically formulated in the Glauber multiple scattering formalism, ending up with nucleon-nucleus cross sections.
Looking for the cross features of the particle production, the experiments show that the bulk of it consists of hadrons emitted with limited transverse momenta ($< P_t > \sim 0.3\,\mathrm{GeV/c}$) with respect to the direction of the incident nucleon. In these "soft" processes the momentum transfer is small. More rarely, but existing, are hard scattering processes with large P_t-production.

It is useful to remind that cosmic ray observations of particle phenomena are strongly weighted to sample the production in forward direction. The kinematic range of the rapidity distribution for the Fermilab proton collider for 1.8 TeV in the c.m. system is equivalent to a laboratory case of 1.7 PeV. Here the energy flow is peaking near the kinematical limit. That means, most of the energy is carried away longitudinally. This dominance of longitudinal energy transport has initiated the concept, suggested by Feynman: The inclusive cross sections are expressed by factorizing the longitudinal part with an universal transverse momentum distribution $G(P_t)$ and a function scaling with the dimensionless Feyman variable x_F, defined as the ratio of the longitudinal momentum to the maximum momentum. Though this concept, expressing the invariant cross sections by

$$E \cdot d^3\sigma/dp^3 \sim x_F \cdot d^3\sigma/dx_F dp_T \tag{1}$$

provides an orientation in extrapolating cross sections, it is not correct in reality, and the question of scaling violation is a particular aspect in context of modeling ultrahigh-energy interactions.

3 Hadronic Interaction Models as Generators of Monte-Carlo Simulations

Microscopic hadronic interaction models, i.e. models based on parton-parton interactions are approaches, inspired by the QCD and considering the lowest order Feyman graphs involving the elementary constituents of hadrons (quarks and gluons). However, there are not yet exact ways to calculate the bulk of soft processes since for small momentum transfer the coupling constant α_s of the strong interaction is so large that perturbative QCD fails. Thus we have to rely on phenomenological models which incorporate concepts from scattering theory.

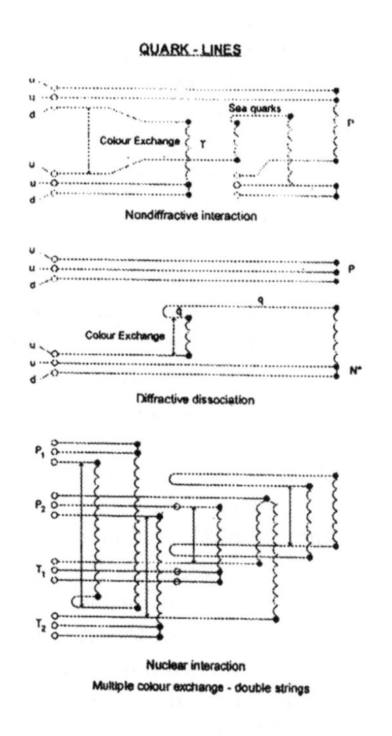

Figure 1. Parton interaction lines.

A class of successful models are based on the Gribov-Regge theory. In the language of this theory the interaction is mediated by exchange particles, so-called Reggeons. At high energies, when non-resonant exchange is dominating, a special Reggeon without colour, charge and angular momentum, the Pomeron, gets importance. In a parton model the Pomeron can be identified as a complex gluon network or generalised ladders i.e. a colourless, flavourless multiple (two and more) gluon exchange. For inelastic interactions such a Pomeron cylinder of gluon and quark loops is cut, thus enabling colour exchange ("cut cylinder") and a re-arrangement of the quarks by a string formation. Fig.1 recalls the principles by displaying some parton interaction diagram's.

- The interacting valence quarks of projectile and target rearrange by gluon exchange the color structure of the system (the arrow indicates the colour exchange by opening the cylinder). As a consequence, constituents of the projectile and target (a fast quark and slow di-quark e.g.) for a colour singlet string with partons of large relative momenta. Due to the confinement the stretched chains start to fragment (i.e. a spontaneous $q\bar{q}$-production) in order to consume the energy within the string. We recognize a target string (T) and a projectile string (P), which are the only chains in pp collisions. In multiple collision processes in a nucleus, sea quarks are additionally excited and may mediate nucleon-A interactions. While in the intermediate step the projectile diquark remains inert, chains with the sea quark of the projectile are formed.

- Most important are diffractive processes, signaled in the longitudinal momentum (x_F) distribution by the diffractive peak in forward directions. Here the interacting nucleon looks like a spectator, in some kind of polarisation being slowed down a little bit due to a soft excitation of another nucleon by a colour exchange with sea quarks (quark-antiquark pairs spontaneously created in the sea).

- There is a number of such quark lines, representing nondiffractive, diffractive and double diffractive processes, with single and multiple colour exchange.

The various string models differ by the types quark lines included. For a given diagram the strings are determined by Monte Carlo procedures. The momenta of the participating partons are generated along the structure functions. The models are also different in the technical procedures, how they incorporate hard processes, which can be calculated by perturbative QCD. With increasing energy hard and semihard parton collisions get important, in particular minijets induced by gluon-gluon scattering.

In summary, the string models VENUS [3], QGSJET [4] and DPMJET [5] which are specifically used as generators in Monte-Carlo EAS simulations are based on the Gribov-Regge theory.They describe soft particle interactions by exchange of one or multiple Pomerons. Inelastic reactions are simulated by cutting Pomerons, finally producing two color strings per Pomerons which subsequently fragment into color-neutral hadrons.All three models calculate detailed nucleus-nucleus collisions by tracking the participants nucleons both in target and projectile.The differences between the models are due to some technical details in the treatment and fragmentation of strings. An important difference is that QGSJET and DPMJET are both able to treat hard processes, whereas VENUS, in the present form, does not. VENUS on the other hand allows for secondary interactions of strings which are close to each other in space and time. That is not the case in QGSJET and DPMJET. SIBYLL [6] and HDPM [7] extrapolate experimental data to high energies guided by simple theoretical ideas. SIBYLL takes the production of minijets into account. These models are implemented in the Karlsruhe Monte Carlo simulation programm CORSIKA [7,8] to which we refer in the analyses of data. An extensive comparison of the various models and studies of their influence on the simulated shower development and EAS observables have been made in ref.[9]. There are distinct differences in the average multiplicities and the multiplicity distributions generated by different models. Nevertheless the variations in the average longitudinal development, though visible, appear to be relatively small. It should be noted that when inspecting the development of single showers with identical initial parameters, instead of average quantities, we get impressed by the remarkable fluctuations and sometimes unusual EAS developments. A further aspect which affects the accuracy of the simulations are the tracking algorithms propagating the particles through the atmosphere. In devising the CORSIKA code great care has been taken on this aspect, since the outcome for arrival time and lateral distributions could be significantly influenced by the tracking procedures.

4 The KASCADE Apparatus

From the very beginning, when planning the KASCADE experiment [1] the setup of an calorimeter for efficient studies of the hadronic component in the shower center has been foreseen with the intention of checking the predictions of hadronic interaction models.

The KASCADE detector array consists of an field array of 252 detector stations, arranged in a regular way in an area of $200 \cdot 200\,\mathrm{m}^2$, and of a complex central detector with a sampling calorimeter for hadron detection [2]. The field detectors identify the EAS event, they provide the principal trigger (a coincidence in at least eight stations), the basic characterisation (angle of incidence, shower axis and core location) and do sample the lateral distribution of the electron-photon and muon component from which the shower size and quantities characterising the intensity and muon content of the showers are determined. In the array stations the muon detectors are positioned directly below the scintillators of the electron-photon detectors, shielded by lead and iron corresponding to 20 radiation lengths, imposing a energy detection threshold of about 300 MeV.

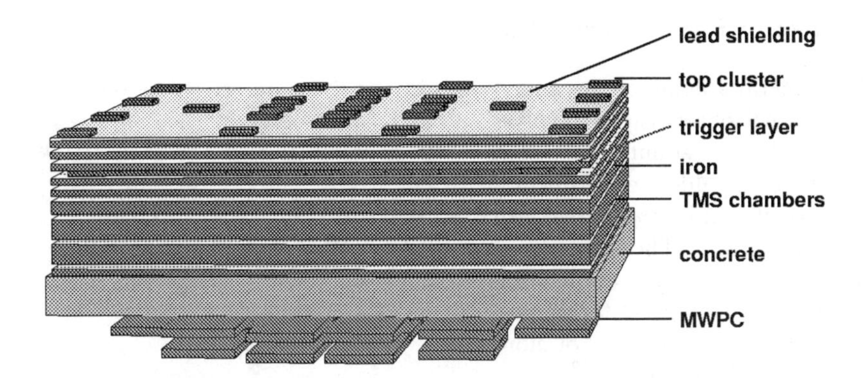

Figure 2. Scheme of the KASCADE central detector.

The central detector combines various types of detector installation with with an iron sampling calorimeter of eight layers of active detectors.
The iron absorbers are 12-36 cm thick, increasingly in the deeper parts of calorimeter. Therefore the energy resolution does not scale as $1/\sqrt{E}$, but is rather constant, slowly varying from $\sigma\sqrt{E} = 20\%$ at 100 GeV to 10% at 10 TeV. In total (including the concrete ceiling) the calorimeter thickness corresponds to 11 interaction lengths ($\lambda_I = 16.7\,\mathrm{cm}$ Fe) for vertical muons. On top, a 5 cm lead layer absorbs the electromagnetic component to a sufficiently low level.
The active detectors are 10.000 ionisation chambers using room temperature liquids tetramethylsilan (TMS) and tetramethylpentane (TMP) operated with a large dynamical range (5.10^4). This ensures that the calorimeter measures linearly the energy of single hadrons up to 15 TeV. The third layer of the calorimeter setup is an "eye" of 456 plastic scintillator, which deliver a fast trigger signal. Independently from hadron calorimetry, it is used as additional muon detector and as timing facility for muon arrival time measurements. In the basement of the iron calorimeter there are position sensitive multiwire proportional chamber (MWPC) installed for specific studies of the structure of the shower core and of the EAS muon component with an energy threshold of 2 GeV.
The energy calibration of the energy deposit of single ionisations chambers is made by means of the through-going muons, and the transition curves, i.e. the longitudinal profiles of the energy deposition are compared with simulations (using the detector simulation code GEANT [10] with the FLUKA description).

5 Test of EAS Observables

The general scheme of the analysis of EAS observations involves Monte Carlo simulations constructing pseudo experimental data which can be compared with the real data [11]. The king-way of the comparison is the application of advanced statistical techniques of multivariate analyses of non-parametric distributions [12].
The mass composition of cosmic rays in the energy region above 0.5 PeV is poorly known. Hence the comparison of simulation results based on different interaction models has to consider two extreme cases of the primary mass: protons and iron nuclei, and the criteria of our judgment of a model is directed to the question, if the data are compatible in the limits of the predicted extremes of protons and iron nuclei. We consider the hadronic observables [13], in dependence from shower parameters which characterize the registered EAS, in particular indicating the primary energy:

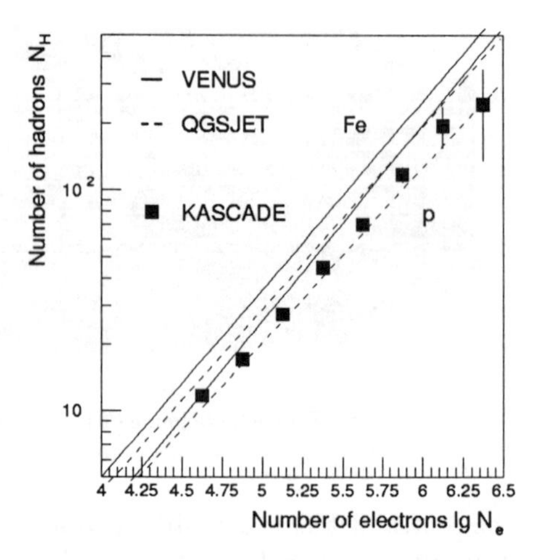

Figure 3. Hadron number N_H - shower size N_e correlation.

- The shower size N_e, i.e. the total electron number
- The muon content N_μ^{tr} which the number of muons obtained from an integration of the lateral distribution in the radial range from 40 to 200 m. It has been shown that this quantity is approximately an mass independent energy estimator for the KASCADE layout, conveniently used for a first energy classification of the showers [14].

First, the dependence of the average number of hadrons N_H with an energy $E_H > 100\,\text{GeV}$ from the shower size is shown and compared with the predictions of the VENUS and QGSJET model. The energy range covers the range from 0.2 PeV to 20 PeV. The result shows some preference for the QGSJET model, and such an indication is corroborated by other tests.
There is another feature obvious. When shower observables are classified along the electromagnetic shower sizes N_e, a proton rich composition is displayed. This effect is understood by the fact that at the same energy

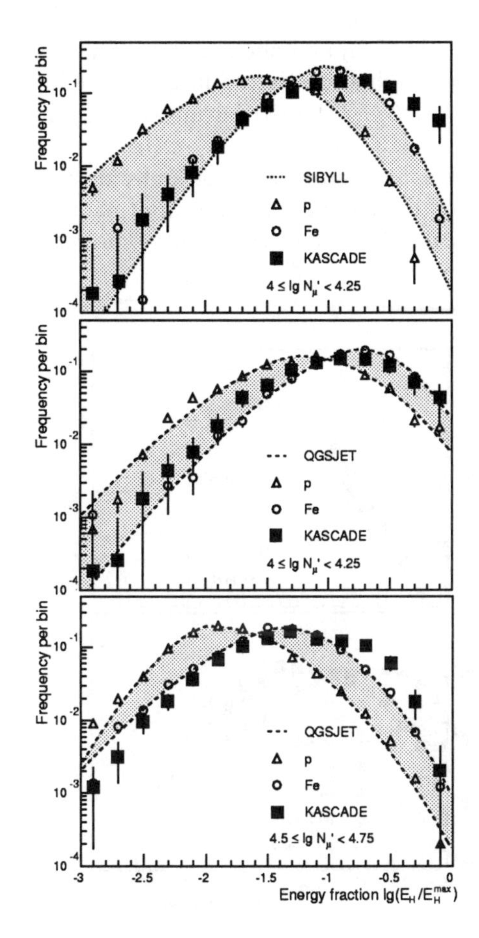

Figure 4. Distribution of the energy fraction of the EAS hadrons.

protons produce larger electromagnetic sizes than iron induced showers, i.e. with the same shower size iron primaries have higher energies, where the steeply falling primary induces the dominance of protons in the sample.

Another example considers the frequency distributions of the energy of each single hadron E_H with respect to the energy of the most energetic hadron E_H^{max}. The data are compared with predictions of SIBYLL and QGSJET for iron and proton induced showers.

- For a primary proton one expects that the leading particle is accompanied by a swarm of hadron of lower energies. For a primary iron nuclei the energy distribution appears narrowed.

- The two upper curves display the case for a primary energy below the knee (about 3 PeV). The deficiencies of SIBYLL are obvious and have been also evidenced by other tests, especially with the muon content [13]. SIBYLL seems to produce a wrong EAS muon intensity, and it fair to mention that just this observation has prompted the authors to start a revision of the SIBYLL model.

- At energies well above the knee (about 12 PeV) also the QGSJET exhibits discrepancies, at least in the energy distribution of the hadrons of the shower core. Other observables like lateral distribution and the total number of hadrons, however are appear more compatible with the model.

How to interpret this results? Tentatively we may understand that in the simulations E_H^{max}, the energy of the leading hadrons is too large. Lowering E_H^{max} would lead to a redistribution of the E/E_H^{max} distribution shifting the simulation curves in direction of the data.
A further test quantity is related to the spatial granularity of hadronic core of the EAS. The graph (Fig.5 left) shows the spatial distribution of hadrons for a shower induced by a 15 PeV proton. The size of the points represents the energy (on a logarithmic scale).

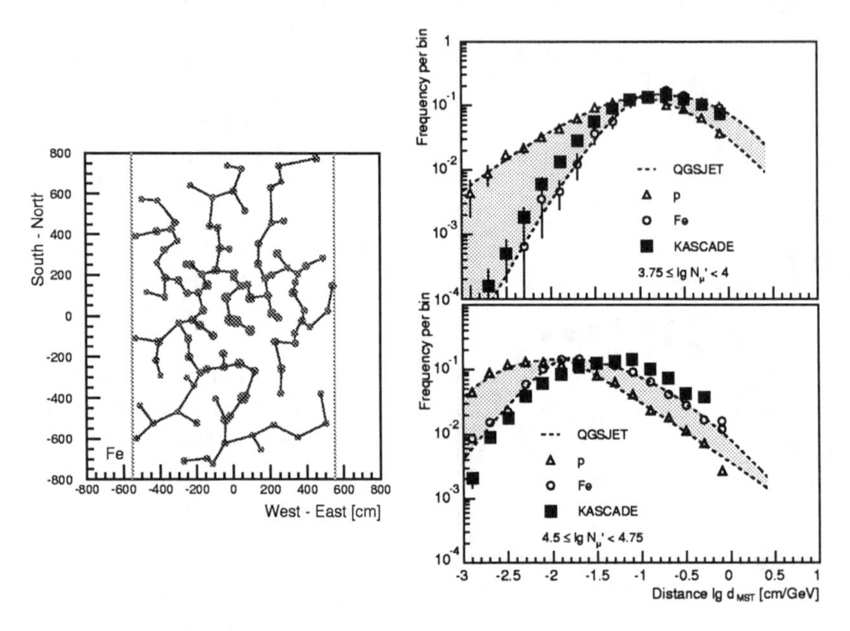

Figure 5. Left: Example of a hadronic core observed in the calorimeter (top view). Right: Frequency distributions of the distances of the minimum-spanning-tree.

For a characterisation of the pattern a minimum spanning tree is constructed. All hadron points are connected by lines and the distances are weighted by the inverse sum of energies. The minimum spanning tree minimizes the total sum of all weighted distances. The test quantity is the frequency distribution of the weighted distances d_{MST}. Results are shown for two different bins of the truncated muon size or of the primary energy (2 and 12 PeV), respectively (Fig.5 right). Again we are lead to the impression that either the distribution pattern is not reproduced or the high-energy hadrons are missing in the model.

Tentatively we may deduce from these indications, that the transfer of energy to the secondaries - what we phenomenologically characterize with the not very well defined concept of the inelasticity of the collision - appears to be underestimated.

In order to underline this feature we may inspect the variation of some other observables with the quantity $log_{10}(N_{\mu}^{tr}) \propto log_{10}(E_{prim})$: The so-called shower age s, which characterizes the stage of the EAS development, the number of observed hadrons N_h with $E_H > 100\,\mathrm{GeV}$, the energy sum $\sum E_h$ of this hadrons and the energy of the highest energy hadrons E_H^{max}. Fig.6 compares with predictions with the QGS model (with the limit log $N_{\mu}^{tr} < 4.6$). The

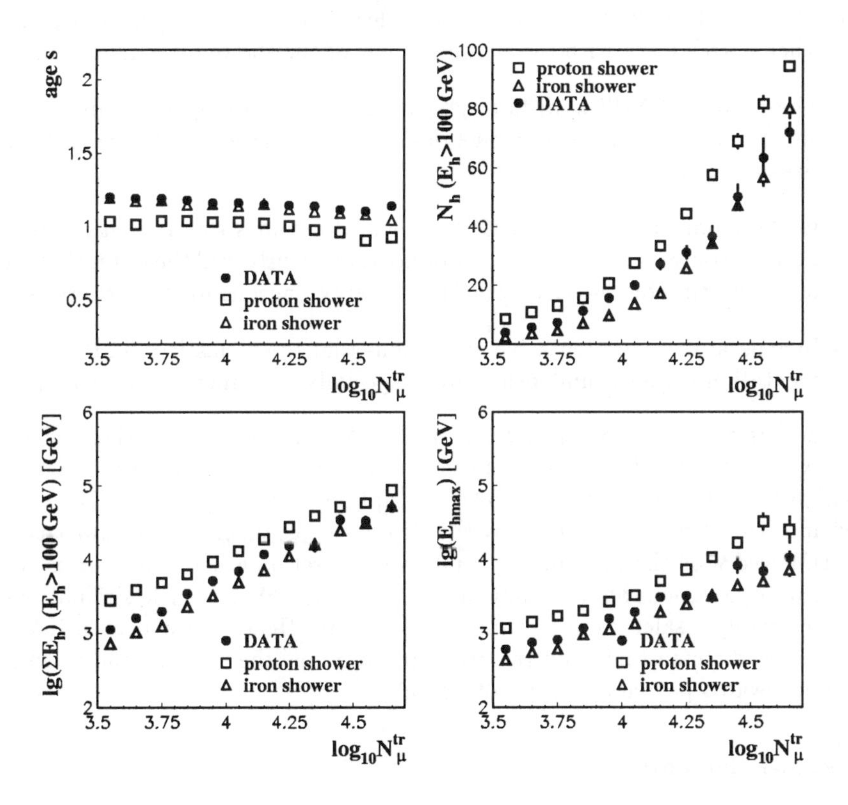

Figure 6. Comparison of various experimentally accessible EAS variables with predictions of the QGSJET model [15].

predictions of the VENUS display the same features. Globally we realize the tendency that the experimental data approach the predictions for iron induced showers, i.e. for faster developed EAS. But this may be hardly interpreted as consequence of a heavier mass composition, rather as arising from a larger inelasticity of the hadronic collisions e.g..

6 Concluding Remarks

From the investigation of a series EAS observables and comparisons with different hadronic interaction models, en vogue for ultrahigh energy collisions, we conclude with following messages:

- The model SIBYLL, in the present release, has problems, in particular when correlations with the muon content of the showers are involved.
- The model VENUS is in fair agreement with the data, but it indicates also some problems at high energies, when correlations with the shower sizes are considered.
- In the moment the model QGSJET, which includes the minijet production - in contrast to VENUS - reproduces sufficiently well the data, though it underestimates the number of high energy hadrons for high energies.
- In general there are tentative indications that the inelasticity in the fragmentation region is underestimated especially with increasing energy.

All models are in a process of refinements. Actually somehow triggered by the experimental indications, there is a common enterprise of VENUS and QGSJET towards a combined model descriptions: NEXUS [16]. That is a unified approach combining coherently the Gribov-Regge theory and perturbative QCD. Faced with the experimental endeavour to set up giant arrays for astrophysical observations at extremely high energies, the Monte Carlo simulations need certainly a safer ground of model generators. Hence our efforts in KASCADE are directed to extend the array and to refine the present studies with results towards primary energies of 10^{17} eV.

Acknowledgments

The experimental results are based on a KASCADE publication [13]. In particular, I would like to thank Dr. Andreas Haungs for contributions and clarifying discussions.

References

1. H.O. Klages et al. - KASCADE collaboration, *Nucl.Phys.B (Proc.Suppl.)* **52B**, 92 (1997).
2. J. Engler et al., *Nucl.Instr.Meth.A* **427** 528 (1999).
3. K. Werner, *Phys.Rep.* **232**, 87 (1993).
4. N.N. Kalmykov and S.S. Ostapchenko, *Phys. At. Nucl.* **56**, 346 (1993).
5. J. Ranft, *Phys. Rev. D* **51**, 64 (1995).
6. R.S. Fletcher el al., *Phys. Rev. D* **50**, 5710 (1994).
7. J.N. Capdevielle et al., KfK Report 4998, Kernforschungszentrum Karlsruhe (1992).
8. D. Heck et al., FZKA Report 6019, Forschungszentrum Karlsruhe (1998).
9. J. Knapp, D. Heck, G. Schatz, FZKA Report 5828, Forschungszentrum Karlsruhe (1996).
10. GEANT 3.15, Detector Description and Simulation Tool, CERN Program Library Long Writeup W5015, CERN (1993).
11. H. Rebel, Proc. of the Workshop ANI 98, eds. A.A. Chilingarian, H. Rebel, M.Roth, M.Z. Zazyan, FZKA 6215, Forschungszentrum, p.1 (1998).
12. A. Chilingarian, "ANI - Nonparametric Statistical Analysis of High Energy Physics and Astrophysics Experiments", Users Guide, 1998; M.Roth, FZKA Report 6262, Forschungszentrum Karlsruhe (1998).
13. T. Antoni et al. - KASCADE collaboration, *J. Phys. G: Part. Phys.* **25**, 2161 (1999).
14. R. Glasstetter for the KASCADE collaboration, *Proc. 25th ICRC (Durban)* 6, 157 (1997).
15. A. Haungs, private communication
16. H.J. Drescher et al., *preprint hep-ph/9903296* (March 1999).

NUCLEAR STRUCTURE NEAR THE NEUTRON DRIP-LINE AND R-PROCESS NUCLEOSYNTHESIS

K.–L. KRATZ[1], P. MÖLLER[2], B. PFEIFFER[1] AND W.B. WALTERS[3]
[1] *Institut für Kernchemie, Universität Mainz, D-55128 Mainz, Germany*
E-mail: klkratz@mail.kernchemie.uni-mainz.de
[2] *Theoretical Division, LANL, Los Alamos, NM, USA*
E-mail: moller@moller.lanl.gov
[3] *Department of Chemistry, University of Maryland, College Park, MD, USA*
E-mail: ww3@umail.umd.edu

A correct understanding and modeling of the production of the heaviest elements in nature via the r-process requires the knowledge of nuclear properties far from stability and a detailed prescription of the astrophysical environment. Experiments have played a pioneering role in exploring the characteristics of nuclear structure in terms of masses and β-decay properties. Initial examinations paid attention to far unstable nuclei with magic neutron numbers related to r-process peaks, while present activities are centered on the evolution of shell effects with the distance from the valley of stability. In this paper, the main focus is put on the region around the doubly-magic nucleus ^{132}Sn, where increasing experimental evidence for a weakening of the N=82 shell strength is observed. This new phenomenon leads to improved predictions of the solar-system isotopic r-abundances ($N_{r,\odot}$).

1 Introduction

Relative to other fields, explosive nucleosynthesis is propably unique in its requirements of a very large number of physical quantities in order to achieve a satisfactory description of the astrophysical phenomena under study. On the nuclear level, atomic masses, β-decay properties and cross sections are but a few examples that are of paramount importance in stellar models. Because nuclei of extreme N/Z composition, often quite different from what has so far been studied on earth, exist in explosive environments, an understanding of their nuclear-structure properties presents a continuing, stimulating challenge to the experimental and theoretical nuclear-physics community.

Only some 15 years ago, astrophysicists believed that nuclear-structure information on r-process nuclei would never become available from terrestrial experiments. However, already a few years later, in 1986 a new aera in nuclear astrophysics started with the identification of the first two classical, neutron-magic "waiting-point" isotopes [1], $^{80}Zn_{50}$ and $^{130}Cd_{82}$ [2,3,4]. As was shown in [4,5], with their β-decay properties first evidence for the existence of an $N_{r,\odot}(Z)\times\lambda_\beta(Z)\approx$const. correlation could be achieved, which immediately became very important to constrain the equilibrium conditions of an r-process.

And this success strongly motivated further experimental and theoretical nuclear structure investigations, as well as astrophysical r-process studies [6,7]. For example, the second N=50 waiting-point isotope ^{79}Cu could be identified [8]. Experimentally known strong P_n branches were shown to be the nuclear-structure origin of the odd-even staggering in the A$\simeq$80 $N_{r,\odot}$ peak [9]; and from the interpretation of the $^{80}Zn_{50}$ decay scheme, evidence for a vanishing of the spherical N=50 shell closure far from stability was obtained [10].

Since the late 1980's, considerable progress has been achieved in the study of nuclear-structure systematics (see, e.g. [11,12,13]) of neutron-rich medium- to heavy-mass nuclei at different laboratories using various production, separation and detection methods. However, due to the generally very low production yields, the restriction to chemically non-selective ionization modes, or the application of non-selective detection methods, no further information on isotopes lying **on** the r-process path could be obtained for quite some time. Only in recent years, further progress in the identification of r-process nuclides was achieved by considerably improving the **selectivity** in production and detection methods of rare isotopes, e.g. by applying Z-selective laser ion-source systems, or projectile-fragmentation TOF techniques.

Recent experiments

There are mainly three mass regions, where at present nuclear-structure information is of particular astrophysical interest. The first is the *r-process seed* region which seems to be not well understood (see, e.g. [14,15]), involving very neutron-rich Fe-group isotopes up to the doubly-magic nucleus $^{78}Ni_{50}$. Recent spectroscopic results can, for example, be found in [16,17] and references therein. The second region is that of far-unstable nuclei around A$\simeq$115. Here, some r-process calculations show a pronounced *r-abundance trough*, which is believed to be duc to nuclear-model deficiencies at and beyond N=72 mid-shell (see, e.g. [6,18]). Recent experimental information on that mass range can be found, for example, in [16,17,19,20] and references therein. The third region of interest is that around the *double-magic nucleus* $^{132}_{50}Sn_{82}$. Apart from astrophysical importance (formation of the A$\simeq$130 peak of the $N_{r,\odot}$ [6,21]), this area is of considerable shell-structure interest. The isotope ^{132}Sn itself, together with the properties of the nearest-neighbor single-particle ($^{133}_{51}Sb_{82}$ and $^{133}_{50}Sn_{83}$) and single-hole ($^{131}_{50}Sn_{81}$ and $^{131}_{49}In_{82}$) nuclides are essential for tests of the shell model, and as input for any reliable future microscopic nuclear-structure calculations towards the neutron-drip line.

The bulk of data so far known in this mass region have been obtained from β-decay spectroscopy at the mass-separator facilities OSIRIS (Sweden) and ISOLDE [12,13,16,17,19,20]. The structures of $^{131}Sn_{81}$ (ν-hole) and $^{133}_{51}Sb$ (π-particle) are fairly well known since more than a decade. More recently, several ν-particle states in $^{133}Sn_{83}$ have been identified at the General Purpose Separator (GPS) of the new PS-Booster ISOLDE facility (see, e.g. Hoff et al. in [17]). From these data, valuable information on the spin-orbit splitting of the 2f- and the 3p-orbitals was obtained. These results were compared to mean-field and HFB predictions, and it was found that none of the potentials currently used in *ab inito* shell-structure calculations was capable of properly reproducing the ordering and spacing of these states (see, e.g. [22], where also possible astrophysical consequences are given). Of particular interest in this context are the surprisingly low-lying $\nu p_{3/2^-}$ (854 keV) and $\nu p_{1/2^-}$ (1655 keV) states in ^{133}Sn. According to the standard Nilsson model [23], for example, they are expected at 2.89 MeV and 4.36 MeV, respectively. Such lowering of the energies of low-j orbitals has recently been predicted as a *neutron-skin* phenomenon to occur near the neutron drip-line (see, e.g. [24]), but not yet in (N_{mag}+1) ^{133}Sn which is still neutron-bound by $S_n \simeq 2.4$ MeV. Nevertheless, following the suggestion of Dobaczewski [24], we have modified the Nilsson potential by reducing the strength of the l^2-term, in order to study its effect on different orbitals [25]. And indeed, this procedure has led to the desired change in the position and even the ordering of the ν-particle states beyond N=82, and has thus allowed at least a qualitative reproduction of the experimental observation of low-lying low-j and high-lying high-j orbitals in ^{133}Sn. It is interesting to note in this context, that as a simultaneous consequence of reducing the l^2-strength, also the N=82 shell gap is reduced (*"quenched"*).

With these new data, the ^{132}Sn valence-nucleon region is nearly complete. The only missing information are the π-hole states in $^{131}_{49}In$, which can in principle be studied through β-decay of the exotic nucleus $^{131}Cd_{83}$. Recent laser-ionization studies at Mainz [26] have made possible first spectroscopic measurements on Cd isotopes in the A$\simeq$130 region at the ISOLDE-GPS. Although the data are not yet fully analyzed, we present here some new results obtained with the resonance-ionization laser ion source (RILIS), which again indicate that the structure around ^{132}Sn is not at all well understood.

For example, in this experiment it was possible to remeasure ^{130}Cd and to determine, for the first time, the $T_{1/2}$ and P_n values of ^{131}Cd and ^{132}Cd. For the N=83 isotope ^{131}Cd a half-life of $T_{1/2}$=(68±3) ms was obtained, which is surprisingly short when compared to current model predictions. For example, the QRPA calculations for pure Gamow-Teller (GT) decay [27] yield $T_{1/2}$(GT)=943 ms when using the β-decay energy Q_β=11.61 MeV from the FRDM mass model [28]. A similar picture arises for the N=84 nuclide ^{132}Cd,

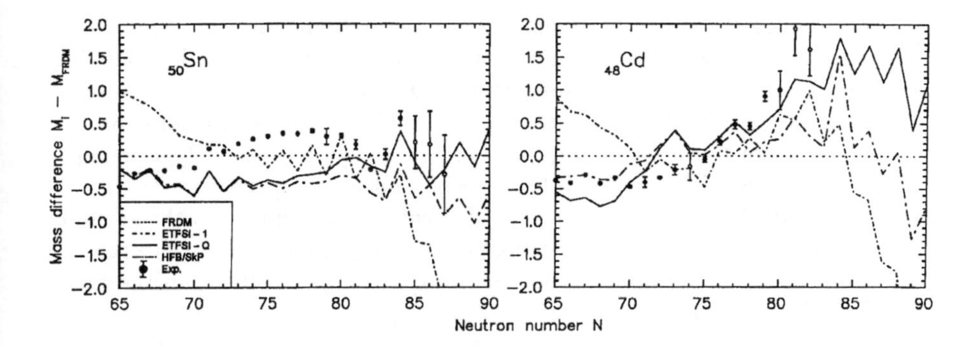

Figure 1. Experimental and theoretical mass differences for $_{50}$Sn and $_{48}$Cd isotopes normalized to FRDM values (M_i–M_{FRDM}). For discussion, see text.

where the experimental half-life of $T_{1/2}$=(95±12) ms is again considerably shorter than the predicted $T_{1/2}$(GT)=633 ms [27], when using Q_β=10.65 MeV from FRDM. In trying to understand the structure and decay of these Cd nuclides on the basis of the nuclear properties of neighboring nuclides, three quantities are of interest, the Q value for β-decay, an estimate of the log(ft) values for the allowed (GT) and forbidden (ff) decay channels, and an estimate of the positions of the states in the daughter nuclides that will be strongly populated. In the following, we will discuss the importance of these quantities at the example of $^{131}Cd_{83}$ decay.

The Q_β value of ^{131}Cd is predicted by recent global mass models to lie between 11.5 MeV and 12.6 MeV, with the general tendency in this mass region that those models which exhibit a strong N=82 shell closure (FRDM[28], GTNM[29], ETFSI-1[30], TFM[31]) give the "lower" values. Recent experimental trends in the ^{132}Sn area [32], however, seem to favor the Q_β predictions from mass models with shell quenching (HFB/SkP[33], ETFSI-Q[34], INM[35]). As can be seen from Fig. 1, between the sequence of $_{50}$Sn and $_{48}$Cd isotopes there is a significant change in the trend of the experimental and theoretical mass differences (here normalized to the FRDM predictions, $M_i - M_{FRDM}$). Clearly, for the Cd isotopic chain the best agreement with the measured masses is obtained with the quenched ETFSI-Q masses. For the case of $^{131}Cd_{83}$, the choice of Q_β=12.56 MeV from this latter model results in a reduction of the $T_{1/2}$(GT) prediction already by about a factor of 3, however still being too long by a factor of 5.

For the log(ft) values, the principal GT-decay branch in the ^{132}Sn region [36] is the $\nu g_{7/2} \rightarrow \pi g_{9/2}$ transition with a log(ft)$\simeq$4.4, whereas the also possible $\nu f_{7/2} \rightarrow \pi f_{5/2}$ is blocked according to our QRPA calculations [37] because of the complete occupancy of the $\pi f_{5/2}$ orbital. The main ff-transitions, on the other hand, are the $\nu f_{7/2}$ or $\nu h_{11/2} \rightarrow \pi g_{9/2}$ decays and the $\nu d_{3/2}$ or $\nu s_{1/2} \rightarrow \pi p_{1/2}$ decays with minimum log(ft)$\simeq$5.3 and 5.1, respectively [36]. In the case of ^{131}Cd, the above $\nu f_{7/2} \rightarrow \pi g_{9/2}$ ff-decay branch represents the ground-state (g.s.) transition; and its estimated partial half-life of $T_{1/2}$(ff)$\simeq$125 ms turns out to be more important for the total β-decay half-life than the main GT-branch to the $[\nu f^{+1}_{7/2}(\nu g^{-1}_{7/2}\pi g^{-1}_{9/2})]$ 3 quasi-particle (QP) state. Hence, ff-transitions cannot be neglected in the ^{132}Sn region. Therefore, any straight-forward comparisons of experimental β-decay quantities (like $T_{1/2}$ and P_n) with model predictions of only the GT-part of β-decay, or of models with different sophistication (like QRPA and Gross Theory) among each other – often used to "validate" the reliability and predictive power of such theories (see, e.g. [38,39]) – must be regarded as meaningless.

To describe the β-decay of $^{131}Cd_{83}$ correctly, we must now add the main decay branches described above. With this, our QRPA model predicts a total β-decay half-life of $T_{1/2}$(GT+ff)$\simeq$195 ms. However, the theoretical $T_{1/2}$ is still too long by a factor 3, and – even more important – the predicted P_n value is with roughly 60 % still much too high compared to P_n(exp)$\simeq$3.5 % [40]. As it is very unlikely that the log(ft) values in ^{131}Cd decay are considerably smaller than in nearby known decays [36] and predicted by our QRPA, and that another sizeable increase in Q_β beyond the predictions of [33,34] occurs, the only solution is a significantly lower energy for the states strongly populated in the ^{131}Sn daughter. In particular, the 3QP state fed by the major GT-branch, must have been shifted down in energy by at least 1 MeV (i.e. **below** $S_n\simeq$6.3 MeV) in order to explain the small experimental P_n value. Within our QRPA model, we can account for the required changes relative to the "normal" shell-structure description of ^{132}Sn by using the modified Nilsson potential for very neutron-rich nuclei, as suggested in [24,25]. With a reduction of the l^2-term in the Ragnarsson-Sheline parameterization [23] by 25 %, we indeed obtain the required downward-shifts of both the main GT-branch (to 5.65 MeV) and the $\nu h_{11/2} \rightarrow \pi g_{9/2}$ ff-transition (to about 4 MeV). In consequence, our QRPA model now calculates $T_{1/2}\simeq$85 ms and $P_n\simeq$5 %, in good agreement with the initially surprising experimental values.

Based on similar Q_β and level-systematics arguments, and following the above "stepwise" QRPA shell-model calculations for GT- and ff-transitions, also the basic $J^\pi=1^+$ and 1^- 2QP pattern of ^{130}In and ^{132}In populated in β-decay of $^{130}Cd_{82}$ and $^{132}Cd_{84}$, respectively, can be explained. Details on their nuclear-structure features will be published in a forthcoming paper.

It is interesting to note in this context, that with the above QRPA parameterization the N=82 shell gap in ^{131}Cd and ^{132}Cd (here defined as $E(\nu f_{7/2}) - E(\nu h_{11/2})$) is reduced from about 4.9 MeV (Folded-Yukawa, Nilsson) to about 3.9 MeV (Nilsson/$0.75\times l^2$), yielding another indication of an onset of shell quenching already in the vicinity of ^{132}Sn, in addition to the effect seen already in nuclear masses (see Fig. 1).

Another recent study of nuclear-structure developments towards N=82 concerns the βdn- and γ-spectroscopic measurements of neutron-rich Ag nuclides at the PS-Booster ISOLDE facility, using an improved version of the RILIS system together with new microgating procedures. This approach was of considerable assistance in minimizing the activities from surface-ionized In and Cs isobars. In this context, the additional "selectivity" of the spin- and moment-dependent hyperfine (HF) splitting was used to enhance the ionization of either the $\pi p_{1/2}$–isomer or the $\pi g_{9/2}$ g.s.-decay of the Ag isotopes. Details about first βdn- and γ-spectroscopic applications of this technique to short-lived isomers of ^{122}Ag up to the clasical N=82 r-process waiting-point nuclide ^{129}Ag can be found e.g. in [16,17].

Because of its importance for r-process calculations, in particular for the r-matter flow through the A$\simeq$130 $N_{r,\odot}$ peak (see, e.g. [21], and Kratz et al. in [16,17]), we briefly focus on the results obtained for this latter isotope. With an off-center laser frequency enhancing the ionization of the $\pi g_{9/2}$ level, an unambiguous identification of the βdn-decay from the g.s. of ^{129}Ag with a half-life of 46^{+5}_{-9} ms was possible. This value is in very good agreement with the recent QRPA prediction of 47 ms [27], but is lower than our old waiting-point requirement of about 130 ms [5,6]. However, that estimate was based on our earlier half-life measurement of $T_{1/2}\simeq$195 ms for ^{130}Cd, and on the older r-process residuals $N_{r,\odot}$ of [41]. Moreover, for non-equilibrium phases of the r-process, also the $T_{1/2}$ contribution from the $\pi p_{1/2}$ isomer would be of importance for the "stellar" half-life of ^{129}Ag. With the QRPA model of [37], we calculate the GT-decay for such an isomer to $T_{1/2}\simeq$320 ms. When including an estimate for the expected ff-strength (deduced from the decay of the $J^{\pi}=1/2^-$ isomer in isotonic ^{131}In), a minimum value of $T_{1/2}$(GT+ff)$\simeq$125 ms is suggested. And, indeed, a careful re-examination of the A=129 βdn-decay curve taken with the laser at central frequency gave a first indication of a weak, "longer-lived" ^{129}Ag component with $T_{1/2}\simeq$160 ms. Now, it will be important to use isomer-specific ionization in combination with isobar separation at ISOLDE-HRS to ascertain the existence of this $\pi p_{1/2}$ isomer.

The astrophysical half-life of ^{129}Ag for non-equilibrium r-process conditions is expected to be a mixture of the g.s. and isomer values. Based on our present knowledge on level systematics in the ^{132}Sn region, neutron-capture γ-decay from $S_n \simeq 5.6$ MeV in ^{129}Ag would populate the $\pi p_{1/2}$ isomer to roughly 35 % and the $\pi g_{9/2}$ g.s. to about 65 % [42]. For $T_9 \simeq 1$, this would result in an average stellar $T_{1/2} \simeq 80$ ms. Based on the new experimental half-lives for ^{130}Cd and ^{129}Ag, together with recently improved QRPA calculations for the other (Z<47) N=82 waiting-point isotopes, an excellent reproduction of the shape of the $N_{r,\odot}$ peak from A=124 to A=133 is obtained. From this result, we conclude that the effect of neutrino-processing of the initial r-abundances during freeze-out is considerably smaller than recently postulated by Qian et al. [43]. The main effect of "filling-up" the low-mass wing of the $A \simeq 130$ (as well as the $A \simeq 195$) r-peak is rather due to βdn-branching during the first 150 ms of the freeze-out [44], which has been ignored by the above authors.

In addition to the study of gross β-decay properties of neutron-rich Ag isotopes, also the level systematics of the Cd-daughters has been extended up to the r-process path (see, e.g. Kratz et al. in [16,17]). The new results are shown in Fig. 2, together with known $E(2^+)$ and $E(4^+)/E(2^+)$ level systematics of neighboring elements. It is quite evident from this figure, that the trend for the Z=48 isotopes clearly deviates from that of the Z=50 (Sn) and 52 (Te) isotones. For the Cd nuclides this indicates an extension of the well-known vibrational character of the lighter isotopes up to the magic shell with deduced values for $B(E2) \simeq 3950$ e^2fm^4 for $^{126,128}Cd_{78,80}$. For comparison, $^{132}Te_{80}$ has a $B(E2) \simeq 3000$ e^2fm^4. So far, no microscopic model can fully account for the observed trend of the level systematics near ^{132}Sn. Only a very recent deformed SMF approach with selfconsistent cranking by Reinhard [45] results in relatively low $E(2^+) \simeq 1$ MeV for N=78 and 80 Cd isotopes, with **no** increase in 2^+ energy in neutron-magic $^{130}Cd_{82}$. Experimentally, we do, however, have an indication of a moderate increase of $E(2^+)$ from about 650 keV in ^{126,128}Cd to approximately 1.4 MeV in ^{130}Cd. Moreover, the tentative ratio of $E(4^+)/E(2^+)=1.69$ would once more indicate a weakening of the N=82 spherical shell below ^{132}Sn.

2 Reproduction of r-process abundances

Summarizing the recent results from the ^{132}Sn region, we conclude that by now there is sufficient experimental evidence to confirm our earlier predictions concerning a weaker strength of the N=82 shell closure far from stability than described by the standard FRDM and ETFSI-1 mass models (for a more detailed discussion, see e.g. [46]; and Kratz et al. in [16,17]. Therefore, these new nuclear- structure signatures should no longer be ignored in astrophysical applications. Consequently, we prefer to use "quenched" mass models,

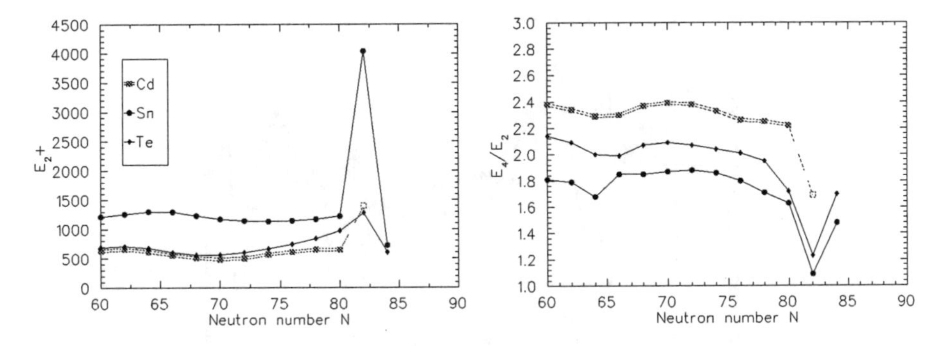

Figure 2. Systematics of the first 2^+ levels in neutron-rich $_{48}$Cd to $_{52}$Te isotopes (left part); and $E(4^+)/E(2^+)$ ratios (right part). For discussion, see text.

and short-range extrapolations of the new nuclear-structure signatures in our QRPA calculations of β-decay properties.

Although also on the astrophysical side there has been considerable improvement since the review by Thielemann and Kratz [47] in 1991 (for a recent review, see e.g. [48]), even today none of the presently favored r-process models, in particular (i) the supernova type II neutrino wind and (ii) neutron star mergers (see, e.g. [49,14,15], and Kratz et al. in [17]), were able to account for the production and ejection of r-process material with the correct distribution to explain the solar-system r-abundances in a realistic and selfconsistent manner. Therefore, our collaboration has continued to use the more deductive approach [47] to obtain further insight into the r-process mystery (see, e.g. [18,21]). In our new attempts, which still are independent of a specific stellar model or site, the above modern nuclear data base has been combined with most recent astrophysical observables [50,51] in order to derive the necessary conditions required to observe these features. As an example of the effect of the new neutron-dripline physics on the calculated isotopic r-abundance pattern, in Fig. 3 we show global N_r distributions from a superposition of 16 components with different, correlated neutron densities, process durations and weights at constant T_9=1.35 [18,21,15] for the two versions of the ETFSI nuclear mass model [30,34]. In both cases, identical conditions for the stellar parameters were used. With the ETFSI-1 mass model, apart from pronounced A$\simeq$115 and A$\simeq$175 abundance troughs, too little r-process material is observed in the whole region beyond the A$\simeq$130 peak. As mentioned above, the solution to the A$\simeq$115 r-abundance deficiency clearly lies in the prediction of nuclear

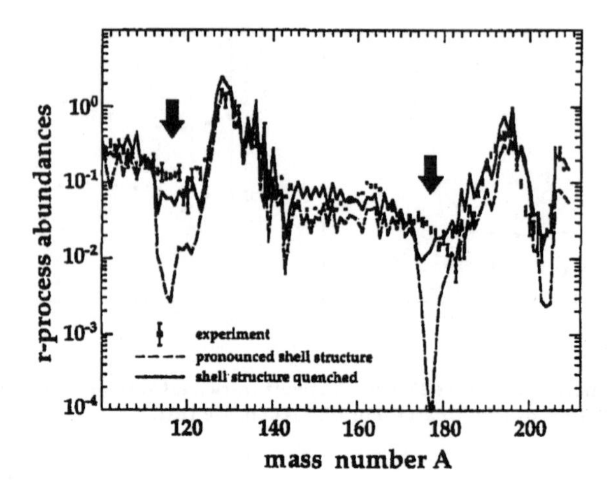

Figure 3. Global r-abundance fits for the ETFSI-1 mass model [30] with pronounced shell structure and the ETFSI-Q formula [34] with quenched shell gaps. For discussion, see text and [18,21].

properties along the r-process path in this mass range up to the N=82 shell closure below ^{132}Sn. Analoguous deficiencies in describing shape-transitions around N$\simeq$104 midshell and the N=126 shell strength seem to be the origin of the deep A$\simeq$175 r-abundance trough [18,21].

With the "quenched" mass model ETFSI-Q, however, a considerable improvement of the overall $N_{r,\odot}$ fit is observed. In particular, the prominent abundance troughs in the A$\simeq$115 and 175 regions are eliminated to a large extent, and the r-matter flow is speeded up at A$\geq$130. Hence, with our new experimental data the application of a "quenched" mass model [33,34,35,52] for r-abundance calculations beyond A$\simeq$150 seems to be highly recommended, in particular if predictions for the ^{203}Tl to ^{209}Bi region are required. Moreover, also extrapolations up to the progenitors of the long-lived actinites ^{232}Th and ^{235,238}U should now be more reliable [18], since they are for the first time based on an internally consistent, modern nuclear-physics input.

3 Summary

We have discussed in this paper which nuclear properties are of prime importance in rapid neutron-capture nucleosynthesis. Although our knowledge of nuclear structure near the neutron drip-line has considerably improved in recent years, there are still many open questions, such as the magnitudes of shell quenching at N=50, 82 and 126, or the nature and stellar site of the r-process. We therefore believe that continued progress in the understanding of the coupling between nuclear physics at the drip-lines and explosive nucleosynthesis scenarios will remain an exciting challenge also in the future.

References

1. Burbidge, E.M. et al., *Rev. Mod. Phys.* **29**, 547 (1957).
2. Lund, E. et al., *Phys. Scr.* **34**, 614 (1986).
3. Gill, R.L. et al., *Phys. Rev. Lett.* **56**, 1874 (1986).
4. Kratz, K.-L. et al., *Z. Phys.* **A325**, 489 (1986).
5. Kratz, K.-L., *Rev. Mod. Astr.* **1**, 184 (1988).
6. Kratz, K.-L. et al., *Ap. J.* **403**, 216 (1993).
7. Thielemann, F.-K. et al., *Nucl. Phys.* **A570**, 329c (1994).
8. Kratz, K.-L. et al., *Z. Phys.* **A340**, 419 (1991).
9. Kratz, K.-L. et al., *Z. Physik* **A336**, 357 (1990).
10. Kratz, K.-L. et al., *Phys. Rev.* **C38**, 278 (1988).
11. *Nuclear Structure of the Zr Region, Res. Rep. in Phys.*, Springer (1988).
12. *Nuclei Far From Stability / Atomic Masses and Fundamental Constants 1992*, *Inst. of Phys. Conf. Ser.* **132** (1993).
13. *Exotic Nuclei and Atomic Masses - ENAM95*, Ed. Frontières (1995).
14. Takahashi, K. et al., *Astron. Astrophys.* **286**, 857 (1994).
15. Freiburghaus, C. et al., *Ap. J.* **516**, 381 (1999).
16. *Exotic Nuclei and Atomic Masses - ENAM98*, *AIP Conf. Proc.* **455** (1998).
17. The ISOLDE Laboratory Portrait, *Hyperfine Interactions*, in print.
18. Pfeiffer, B. et al., *Z. Phys.* **A357**, 235 (1997).
19. *Fission and Properties of Neutron-Rich Nuclei*, World Scientific (1999).
20. *Fission and Properties of Fission-Product Nuclides*, *AIP Conf. Proc.* **447** (1998).
21. Kratz, K.-L. et al., *Nucl. Phys.* **A630**, 352c (1998).
22. Rauscher, T. et al., *Phys. Rev.* **C57**, 2031(1998).
23. Ragnarsson, I. and Sheline, R.K., *Phys. Scr.* **29**, 385 (1984).
24. Dobaczewski, J. et al., *Phys. Rev. Lett.* **72**, 981 (1994).
25. Pfeiffer, B. et al., *Acta Physica Polonica* **B27**, 475 (1996).
26. Erdmann, N. et al., *Appl. Phys.* **B66**, 431 (1998).
27. Möller, P. et al., *At. Data Nucl. Data Tables* **66**, 131 (1997).
28. Möller, P. et al., *At. Data Nucl. Data Tables* **59**, 185 (1995).
29. Hilf, E.R. et al., CERN-Rep. 76-13, 142 (1976).

30. Aboussir, Y. et al., *At. Data Nucl. Data Tables* **61**, 127 (1995).
31. Myers, W.D. and Swiatecki, W.J., Report LBL-36803 (1994); and *Nucl. Phys.* **A601**, 141 (1996).
32. Audi, G. et al., *Nucl. Phys.* **A624**, 1 (1997).
33. Dobaczewski, J., HFB/SkP Mass Table, priv. comm.
34. Pearson, J.M. et al., *Phys. Lett.* **B397**, 455 (1996); and ETFSI-Q Mass Table, priv. comm.
35. Satpathy, L. and Nayak, R.C., *At. Data Nucl. Data Tables* **39**, 241 (1988); and *J. Phys. G: Nucl. Part. Phys.* **24**, 1527 (1998).
36. Firestone, R.B. et al., Table of Isotopes, 8th Edition, Wiley, New York (1996).
37. Möller, P. and Randrup. J., *Nucl. Phys.* **A514**, 1 (1990).
38. Goriely, S. and Clerbaux, B., *Astron. Astrophys.* **346**, 798 (1999).
39. Arnould, M. and Takahashi, K., *Rep. Prog. Phys.* **62**, 393 (1999).
40. Hannawald, M. et al., to be subm. to *Phys. Rev. C.*
41. Walter, G. et al., *Astron. Astrophys.* **167**, 186 (1986).
42. Rauscher, T., priv. comm.
43. Qian, Y.-Z. et al., *Ap. J.* **494**, 285 (1998).
44. Kratz, K.-L. et al., to be subm. to *Phys. Rev. Lett.*
45. Reinhard, P.G., priv. comm. (1999).
46. Kratz, K.-L. et al., Proc. 10th Capture Gamma-Ray Spectroscopy and Related Topics - CGS10, Santa Fe, NM, 1999; AIP Conf. Proc., in press.
47. Thielemann, F.-K. and Kratz, K.-L., Proc. 22nd Masurian Lakes Summer School 1991; IOP, 187 (1992).
48. Wallerstein, G. et al., *Rev. Mod. Phys.* **69**, 995 (1997).
49. S.E. Woosley et al.; *Ap. J.* **433**, 229 (1994).
50. Beer, H. et al., *Ap. J.* **474**, 843 (1997).
51. C. Arlandini et al., *Ap. J.* **525**, 886 (1999).
52. Brown, B.A., *Phys. Rev.* **C58**, 220 (1998).

NUCLEAR ASTROPHYSICS WITH RADIOACTIVE BEAMS

L. TRACHE, A. AZHARI, H. L. CLARK, C. A. GAGLIARDI,
Y.-W. LUI, A. M. MUKHAMEDZHANOV, X. TANG, AND R. E. TRIBBLE
Cyclotron Institute, Texas A&M University, College Station, Texas, USA

V. BURJAN, J. CEJPEK, V. KROHA, S. PISKOR AND J. VINCOUR
Institute for Nuclear Physics, Czech Academy of Sciences, Prague-Rez, Czech Republic

F. CARSTOIU
"Horia Hulubei" Institute of Physics and Nuclear Engineering, Bucharest-Magurele, Romania

A major contribution in nuclear astrophysics is expected now and in the near future from the use of radioactive beams. This paper presents an indirect method to determine the astrophysical S-factor at the very low energies relevant in stellar processes (tens and hundreds of keV) from measurements at energies more common to the nuclear physics laboratories (5-30 MeV/nucleon, e.g.). The Asymptotic Normalization Coefficient method consists in the determination from peripheral transfer reactions of the single particle wave function of the outermost charged particle (proton or alpha particle) around a core in its asymptotic region only, as this is the part contributing to nuclear reactions at very low energies. It can be applied to the study of radiative proton or alpha capture reactions, a very important class of stellar reactions. The method is briefly presented along with our recent results in the determination of the astrophysical factor S_{17} for the proton capture reaction $^7Be(p,\gamma)^8B$. The reaction is crucial for the understanding of the solar neutrino production. Our study was done at the superconducting cyclotron K500 of Texas A&M University, using the ^7Be beam produced with MARS. Two proton transfer reactions with radioactive beams were measured $^{10}B(^7Be,^8B)^9Be$ and $^{14}N(^7Be,^8B)^{13}C$, as well as proton transfer reactions involving stable partners. We present the experiments, discuss the results and the uncertainties arising from the use of calculated optical potentials between loosely bound radioactive nuclei. A test case for our method is included.

1 Introduction

Let me (L.T.) begin by saying, in only two short sentences, that I am very glad to be here, in Bucharest, where I spent most of my formative years, for the 50-th anniversary of what we all knew and still know by the acronym of IFA (Institutul de Fizica Atomica) despite the changes of official names it went through during the years. It is a pleasure for me to see these places again, but especially to meet people I was working with for so many years, either Romanian scientists from "our institute", or guests "from abroad".

We know for some time now that nuclei are the fuel of the stars, but in many cases we do not have yet enough information, or enough precise, about the nuclear reactions taking place in stars to make reliable quantitative predictions about astrophysical processes. The recent major efforts and breakthroughs in astrophysical observations are matched by an increased interest in nuclear astrophysics. In short, one can say that the goal of nuclear astrophysics is to provide the knowledge and data needed to understand how nuclear processes influence astrophysical phenomena, past and present. A recent town meeting organized by the American Physical Society at the University of Notre Dame ended with a white paper that summarizes many of the remaining challenges for the nuclear physics community and sketches possible ways to go [1]. In the determination of the nuclear reaction rates relevant for astrophysics, we need cross sections at very low energies, in the range of tens and hundreds of keV. One approach is to determine the cross sections from direct measurements of the same reaction in the nuclear physics laboratory. These cross sections, especially when charged particles are involved, are usually very low, and are therefore very difficult to measure directly. To overcome the limitations of this approach, a range of few indirect methods is currently employed [2]. The use of radioactive beams will open enormous new possibilities for all of the indirect methods. However, this paper, despite its rather general title, will refer only to the last of these indirect methods, developed and used in our group at the Cyclotron Institute of the Texas A&M University. The method relies on the fact that at low energies a charged particle radiative capture reaction to a loosely bound state is a surface process, due to the Coulomb repulsion in the entrance channel. Its cross section is determined by the tail of the radial overlap integral between the bound state wave function of the final nucleus and those of the initial colliding nuclei. This overlap integral is asymptotically proportional to a well known Whittaker function, and therefore the knowledge of its asymptotic normalization alone determines the cross section. This asymptotic normalization, in turn, can be determined from the measurement of a transfer reaction involving the same vertex, provided that this second reaction is also peripheral. We call it the Asymptotic Normalization Coefficient (ANC) method. All the experiments were done at or around 10 MeV/nucleon, energies much easier to handle in the nuclear physics laboratories. Part of the results were published before ([4,5,6,7,8,9]).

2 ANCs from Proton Transfer Reactions

Traditionally spectroscopic factors have been obtained from proton transfer reactions by comparing experimental cross sections to DWBA predictions.

For peripheral transfer, we show below that the ANC is better determined and is the more natural quantity to extract. Consider the proton transfer reaction $a + A \rightarrow c + B$, where $a = c + p$, $B = A + p$. As was previously shown [3,4] we can write the transfer cross section in the form

$$\frac{d\sigma}{d\Omega} = \sum_{j_B j_a} \frac{(C^B_{Apl_Bj_B})^2}{b^2_{Apl_Bj_B}} \frac{(C^a_{cpl_aj_a})^2}{b^2_{cpl_aj_a}} \sigma^{DW}_{l_Bj_Bl_aj_a}, \tag{1}$$

where $\sigma^{DW}_{l_Bj_Bl_aj_a}$ is the calculated DWBA cross section and j_i, l_i are the total and orbital angular momenta of the transferred proton in nucleus i. The factors $b_{cpl_aj_a}$ $b_{Apl_Bj_B}$ are the ANCs of the bound state proton wave functions in nuclei a and B which are related to the corresponding ANC of the overlap function by

$$(C^a_{cpl_aj_a})^2 = S^a_{cpl_aj_a} b^2_{cpl_aj_a}, \tag{2}$$

where $S^a_{cpl_aj_a}$ is the spectroscopic factor. We have used this formulation to extract ANCs from three peripheral proton transfer reactions involving stable beam and target nuclei, ^{9}Be(^{10}B,^{9}Be)^{10}B (Ref. [4]), ^{13}C(^{14}N,^{13}C)^{14}N (Ref. [5]) and ^{16}O(^{3}He,d)^{17}F (Ref. [6]). For surface reactions the cross section is best parametrized in terms of the product of the square of the ANCs of the initial and final nuclei $(C^B)^2(C^a)^2$ rather than spectroscopic factors, because of the strong dependence of the latter on the parameters of the core-proton potential used in the calculations. This is shown in Fig. 1 where the extracted spectroscopic factors vary with up to a factor two, whereas the corresponding C^2 are very stable (2%) against the reduced radius and diffuseness (r_0, a) of the potential used, when values r_0=1.0 to 1.3 fm and a=0.5 to 0.7 fm were used.

2.1 Using ANCs to Predict Astrophysical S-Factors: A Test Case

The ANCs found from the proton transfer reactions can be used to determine direct capture rates at astrophysical energies. Astrophysical S-factors have been determined for ^{16}O(p,γ)^{17}F as a test of the technique. The relation of the ANCs to the direct capture rate at low energies is straightforward to obtain. The cross section for the direct capture reaction $A + p \rightarrow B + \gamma$ can be written as

$$\sigma = \lambda |< I^B_{Ap}(\mathbf{r}) \mid \hat{O}(\mathbf{r}) \mid \psi^{(+)}_i(\mathbf{r}) >|^2, \tag{3}$$

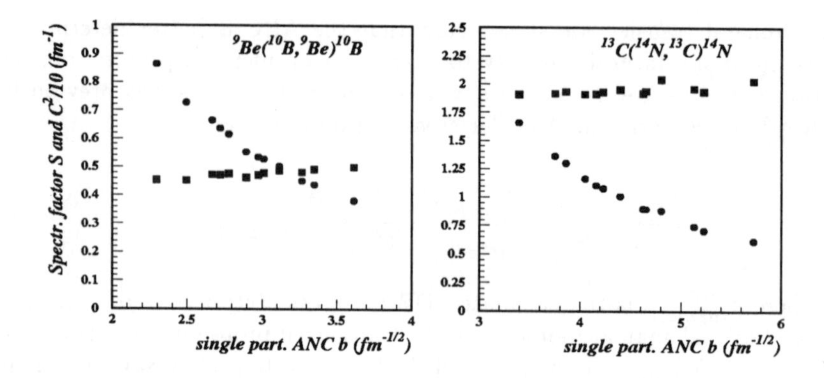

Figure 1. Comparison between the spectroscopic factor S (dots)and the ANC C^2 extracted from experimental data for ^{10}B (left) and ^{14}N (right), as a function of the single particle ANC b.

where λ contains kinematical factors, I^B_{Ap} is the overlap function for $B \rightarrow A+p$, $\hat{O}$ is the electromagnetic transition operator, and $\psi_i^{(+)}$ is the scattering wave in the incident channel. If the dominant contribution to the matrix element comes from outside the nuclear radius, the overlap function may be replaced by

$$I^B_{Ap}(r) \approx C\frac{W_{-\eta,l+1/2}(2\kappa r)}{r}, \tag{4}$$

where C is the ANC as above, W is the Whittaker function, η is the Coulomb parameter for the bound state $B = A+p$, and κ is the bound state wave number. Thus, the direct capture cross sections are directly proportional to the squares of these ANCs.

For ^{16}O(p,γ)^{17}F, the required C's are just the ANCs found above from the transfer reaction ^{16}O(^{3}He,d)^{17}F. Using the results outlined above, the S-factors describing the capture to both the ground and first excited states for ^{16}O(p,γ)^{17}F were calculated, with no additional normalization constants, with the standard definition of the astrophysical S-factor [?]. The results obtained compare very well [6] to the two previous measurements of ^{16}O(p,γ)^{17}F.

3 Optical Model Potentials for loosely bound p-shell nuclei

Another source of uncertainty in calculations can be the optical model potentials used. Parameters for Woods-Saxon type potentials were obtained from the fit of the elastic scattering angular distributions measured in 7 combinations projectile-target measured at TAMU. In addition, nucleus-nucleus potentials were calculated by a double folding procedure using six different effective nucleon-nucleon interactions. The nuclear densities calculated for each partner in the Hartree-Fock approximation were folded with six different nucleon-nucleon interactions. The resulting nucleus-nucleus potentials were later renormalized [12] to obtain a fit of the elastic scattering data. The normalization constants have similar values in all systems for each effective interaction used, which makes it appear likely that the procedure can be extended to the calculation of optical potentials for other similar nucleus-nucleus systems. The procedure and renormalization coefficients needed for the analysis of elastic data with these double folding potentials are extracted and discussed in [9].

From all effective interactions used, we conclude that of Jeukenne et al. [11] gives the best results. It provides us with an imaginary part that has a geometry which is independent from that of the real part of the potential. We find that while the depth of the real potential needs a substantial renormalization ($\langle N_V \rangle = 0.366 \pm 0.014$), the imaginary part needs no such renormalization ($\langle N_W \rangle = 1.000 \pm 0.087$). This also suggests that the imaginary part of the effective interaction is well accounted for. The renormalized double folded potentials obtained were also used in the DWBA analysis of the proton transfer reactions with stable nuclei, and the results were found to be in excellent agreement with those given by the phenomenological Woods-Saxon potentials.

The procedure found was applied to extract the optical model potentials for the ^{7}Be and ^{8}B radioactive projectiles needed in the description of the ^{7}Be+^{10}B and ^{7}Be+^{14}N experiments.

4 Transfer Reactions with Radioactive Beams

We have measured the (^{7}Be,^{8}B) reaction on a 1.7 mg/cm^2 ^{10}B target [7] and a 1.5 mg/cm^2 Melamine target [8] in order to extract the ANC for $^{8}B \rightarrow {}^{7}Be + p$. The radioactive ^{7}Be beam was produced at 12 MeV/nucleon by filtering reaction products from the ^{1}H(^{7}Li,^{7}Be)n reaction in the recoil spectrometer MARS, starting with a primary ^{7}Li beam at 18.6 MeV/nucleon from the TAMU K500 cyclotron. The beam was incident on an H_2 cryogenic gas target, cooled by LN_2, which was kept at 1 atmosphere (absolute) pressure. Reaction products were measured by 5 cm $\times$ 5 cm Si detector telescopes consisting of a 100 μm ΔE strip detector, with 16 position sensitive strips, followed by a 1000 μm E counter.

A single 1000 μm Si strip detector was used for initial beam tuning. This detector, which was inserted at the target location, allowed us to optimize

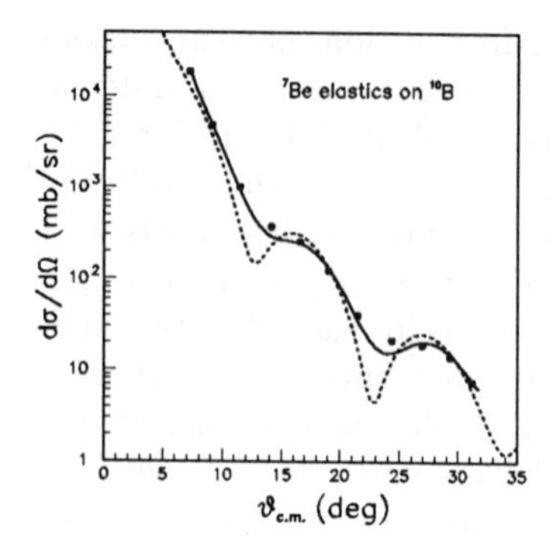

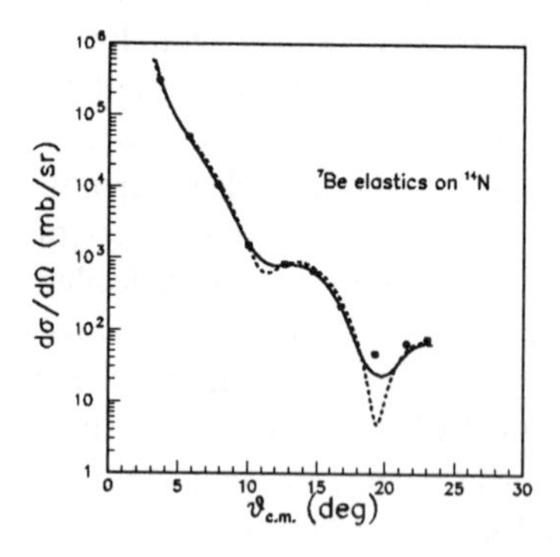

Figure 2. Angular distributions for elastic scattering from the ^{10}B and ^{14}N targets. The dashed curves are from optical model calculations of the target components and the solid curves are smoothed over the angular acceptance of each bin.

the beam shape and to normalize the ^{7}Be flux relative to a Faraday cup that measured the intensity of the primary ^{7}Li beam. Following optimization, the approximate ^{7}Be beam size was 6 mm × 3 mm (FWHM), the energy spread was ≈ 1.5 MeV, the full angular spread was $\Delta\theta \approx 28$ mrad and $\Delta\phi \approx 62$ mrad, and the purity was ≥99.5% ^{7}Be for the experiment with the ^{10}B target. The beam size and angular spread were improved for the experiment with the ^{14}N target to 4 mm × 3 mm (FWHM), $\Delta\theta \approx 28$ mrad and $\Delta\phi \approx 49$ mrad. Periodically during the data acquisition, the beam detector was inserted to check the stability of the secondary beam tune. The system was found to be quite stable over the course of the experiment with maximum changes in intensity observed to be less than 5%. The typical rate for ^{7}Be was $\approx$ 1.5 kHz/pnA of primary beam on the production target. Primary beam intensities of up to 80 pnA were obtained on the gas cell target during the experiments.

Results for the elastic scattering angular distributions from the two targets are shown in Fig. 2. A Monte Carlo simulation was used to generate the solid angle factor for each angular bin and the smoothing needed for the calculation to account for the finite angular resolution of the beam. The absolute cross section is then fixed by the target thickness, number of incident ^{7}Be, the yield in each bin, and the solid angle. In both cases, the optical model calculations are compared to the data without additional normalization coefficients. Overall, the agreement between the measured absolute cross sections and the optical model predictions is excellent thus providing confidence that our normalization procedure is correct.

The ANC for ^{8}B $\rightarrow$ ^{7}Be + p was extracted based on the fit to the present data and the known ANCs [4,5] for the other vertices ^{10}B $\rightarrow$ ^{9}Be + p and ^{14}N $\rightarrow$ ^{13}C + p, following the procedure outlined above in our test case. Two ^{8}B orbitals, $1p_{1/2}$ and $1p_{3/2}$, contribute to the transfer reaction but the $1p_{3/2}$ dominates in both cases. Angular distributions for the (^{7}Be,^{8}B) reactions populating the ground states of ^{9}Be and ^{13}C are compared to DWBA calculations in Fig. 3.

The astrophysical S-factor for ^{7}Be(p,γ)^{8}B has been determined from the ANC which includes a 8% uncertainty for optical model parameters, a 11% uncertainty for experimental fits and normalization of the absolute cross section and the uncertainty in the ANC's for ^{10}B $\rightarrow$ ^{9}Be + p and ^{14}N $\rightarrow$ ^{13}C + p. The relative contribution of the two angular momentum couplings to the S-factor is straightforward to calculate and introduces a negligible additional uncertainty in our result [3]. The values that we find are $S_{17}(0)$ = 18.4 $\pm$ 2.5 eV b for the ^{10}B target, and 16.6 $\pm$ 1.9 eV b for the ^{14}N target. Both are consistent with each other and in good agreement with the recommended value [13] of 19^{+4}_{-2} eV b.

We concluded only last week a successful run for a similar measurement with radioactive ^{11}C on a melamine target. Proton transfer reaction ^{14}N(^{11}C,^{12}N)^{13}C was observed with good resolution and statistics using a beam of 10^6particle/sec. ^{11}C at 120 MeV. We intend to obtain data for the evaluation of the contribution of the direct capture in the reaction ^{11}C(p,γ)^{12}N of importance in the hot CNO cycle.

In conclusion, in a series of experiments we show that transfer reactions induced by radioactive beams can be successfully used to obtain accurate data for nuclear astrophysics. We start with proving that peripheral proton trans-

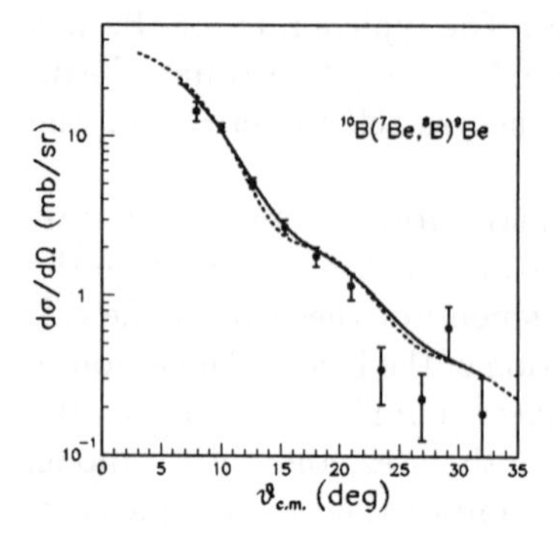

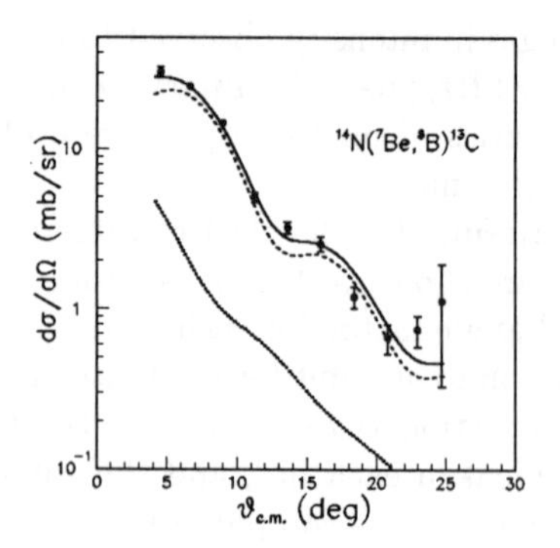

Figure 3. Angular distributions for ^{8}B populating the ground state of ^{9}Be from the ^{10}B target (left) and ^{13}C from the Melamine target (right). In both cases, the solid curve is smoothed over the angular acceptance of each bin.

fer reactions can be used to extract asymptotic normalization coefficients and show them to be a more precise and relevant quantity than the usual spectroscopic factors extracted in such cases. We describe a test case where, based on an ANC extracted from the transfer reaction $^{16}O(^3He,d)^{17}F$, we calculate S factors for the radiative capture $^{16}O(p,\gamma)^{17}F$ that compare very well with those measured directly. Finally we discuss the radioactive beam transfer reactions $^{10}B(^7Be,^8B)^9Be$ and $^{14}N(^7Be,^8B)^{13}C$ and astrophysical factor S_{17} extracted with the method above, including sources for uncertainties. A similar experiment with ^{11}C is in the data analysis phase.

Acknowledgments

One of us (FC) acknowledges the support of the Cyclotron Institute, Texas A&M University, for a stay in College Station, TX, during which a part of this work was done. This work was supported in part by the U. S. Department of Energy under Grant no DE-FG03-93ER40773 and by the Robert A. Welch Foundation.

References

1. *Opportunities in Nuclear Astrophysics, Origin of the Elements*, the white paper of the Town Meeting organized by the Division of Nuclear Physics of the American Physical Society, Notre Dame, June 7-8, 1999.

2. G. Baur and H. Rebel, J. Phys. G: Nucl. Part. Phys. 20, 1 (1994) and refs. therein; G. Baur, Phys. Lett. **B178**, 135 (1986); C. Spitaleri et al., Phys. Rev. C **60**, 055802 (1999).
3. H. M. Xu *et al.*, Phys. Rev. Lett. **73**, 2027 (1994); C. A. Gagliardi *et al.*, Nucl. Phys. **A588**, 327c (1995).
4. A. M. Mukhamedzhanov *et al.*, Phys. Rev. C **56**, 1302 (1997).
5. L. Trache *et al.*, Phys. Rev. C **58**, 271 (1998).
6. C.A. Gagliardi *et al.*, Phys. Rev. C **59**, 1149 (1999).
7. A. Azhari *et al.*, Phys. Rev. Lett. **82**, 3690 (1999).
8. A. Azhari *et al.*, Phys. Rev. C **60**, 055803 (1999).
9. L. Trache *et al.*, Phys. Rev. C **61**, 024612 (2000).
10. J.N. Bahcall, *Neutrino Astrophysics* (Cambridge University Press, Cambridge, 1989).
11. J. P. Jeukenne, A. Lejeune and C. Mahaux, Phys. Rev. C **16**, 80 (1977).
12. G. R. Satchler and W. G. Love, Phys. Rep. **55**, 183 (1979).
13. E.G. Adelberger *et al.*, Rev. Mod. Phys. Vol. **70(4)**, 1265 (1998).

THE ATLAS EXPERIMENT AT THE CERN LARGE HADRON COLLIDER

PETER JENNI
CERN, Geneva, Switzerland
E-mail: Peter.Jenni@cern.ch

The ATLAS project has entered its construction phase. The experiment will explore particle physics at the energy frontier for decades to come. The ATLAS detector is being built by a large world-wide collaboration with 35 countries, including an important team from the Horia Hulubei National Institute of Physics and Nuclear Engineering (IFIN-HH). I present the status of the project and the prospect for discovery physics, briefly reviewing in particular the Romanian participation in the project.

1 Introduction

The CERN Large Hadron Collider (LHC) opens a new frontier in particle physics. It will accelerate 2835 $\times$ 2835 bunches of 10^{11} protons each to an energy of 7 TeV (7×10^{12} eV) per beam, with a luminosity (intensity) of 10^{34} cm^{-2} s^{-1}. At a crossing rate of bunches of 40 MHz, the rate of proton collisions is about 10^{7}- 10^{9} Hz. Compared to the existing accelerators, LHC has 10 times higher collision energy and 100 times higher luminosity, which make it a "discovery machine", with a great potential for "new physics", expected and unexpected. This potential will be exploited by four large detectors which are at present under construction at LHC: two general experiments (ATLAS and CMS), and two dedicated experiments (LHC-b and ALICE).

The ATLAS detector is being built by a large world-wide collaboration of 147 institutions from 35 countries, including Romania. The first ideas and prestudies of LHC detectors were initiated in the period 1984-1989, the ATLAS Letter of Intent was submitted in 1992 and the Technical Proposal [1] in 1994. The general construction approval was given by the CERN Council in 1997. During 1996-1998 the Technical Design Reports (TDR) for systems were elaborated. At present the project has entered its construction phase, which will cover the period 1998-2004. In the interval 2003-2005 the detector will be installed on the LHC, the physics data taking being planned to start in 2005.

2 Physics motivation of ATLAS

ATLAS is a general purpose proton-proton experiment, which will investigate a new territory of physics, exploiting as much as possible the full discovery potential of LHC. The theoretical basis of the physics programme [2] is the Standard Model (SM) of strong, weak and electromagnetic interactions, which gives a satisfactory description of the structure of matter up to the highest energies accessible to current experiments. Despite its big successes, the SM has also some limitations. One is the large number of independent parameters (19), which suggests the existence of an underlying theory. Also, while the gauge theory part of the SM is well tested, there is no direct evidence either for, or against the simple Higgs mechanism for the electroweak symmetry breaking. Some difficulties can be overcome if new dynamics or new particles are present at the scale of 1 TeV. An appealing concept for such new physics is supersymmetry, which offers the only presently known mechanism for incorporating gravity into the quantum theory of particle interactions, solves the fine tuning problem and preserves the present successes of the SM, such as precision electroweak predictions.

In the first three years of low-luminosity at LHC (10^{33} cm^{-2}s^{-1}), the ATLAS experiment will function as a factory for QCD processes, heavy flavours and gauge-boson production, which will allow a large number of precision measurements, in particular on the top mass, masses and couplings of the gauge bosons and CP violating parameters in B-decays [2]. Although at LHC there is an experiment (LHC-b) dedicated to B-physics, ATLAS will be competitive in some channels and will play an important role in maximising the combined precision of B-physics measurements from LHC.

If the Higgs boson will not be discovered before LHC begins operation, the searches for it and its possible supersymmetric extensions, in particular in the Minimal Supersymmetric Standard Model scenario, will be a main focus of activity. The ATLAS detector would find the Higgs boson in the mass range 100 GeV to 1000 GeV (1 TeV), the bounds expected after LEP2 and FNAL Tevatron. The detector is also sensitive to new structures of the WW and ZZ interactions which must appear at the scale of 1 TeV, if the mass of the Higgs exceeds the value of 800 GeV. Therefore, the ATLAS experiment will be able to give a definite answer to the question of the electroweak symmetry breaking.

The ultimate structure of matter, in particular the question whether quarks are elementary or composite, will be investigated by searching signals of technicolor, excited quarks, leptoquarks, new gauge bosons, right-handed neutrinos and monopoles.

3 The ATLAS detector

The ATLAS detector design is a consequence of the main physics motivation of the ATLAS experiment. The detector optimization was guided by physics issues such as sensitivity to the largest possible Higgs mass range, to searches for heavy mass W- and Z-like objects, for supersymmetric particles, for compositness of fundamental fermions, as well as the investigation of CP violation in B-decays and detailed studies of the top quark.

The main requirements for the ATLAS detector are the following :

• very good electromagnetic calorimetry for electron and photon identification and measurements, complemented by full-coverage hadronic calorimetry for accurate jets and missing transverse energy measurements;

• high-precision muon momentum measurements;

• efficient tracking at high luminosity for high-p_T lepton momentum measurements, electron and photon identification, τ-lepton and heavy flavor identification, and full event reconstruction capability at lower luminosity;

• large acceptance in pseudorapidity with almost full azimuthal angle coverage everywhere;

• triggering and measurements of particles at low-p_T thresholds, providing high efficiencies for most physics processes of interest at LHC.

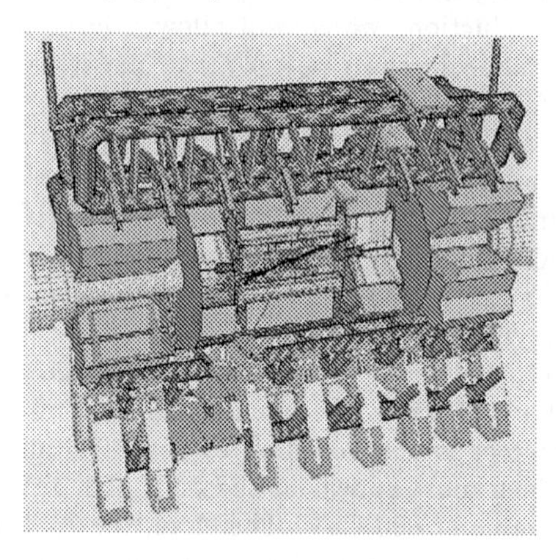

The overall concept of the ATLAS Detector, shown in the above figure, is based on four major components: the inner detector, the calorimeter system, the muon spectrometer and the magnetic system, supplemented by a complex trigger and data aquisition system.

The inner detector is contained within a cylinder of length 7m and a radius of 1.15m, in a solenoid magnetic field of 2T. Pattern recognition, momentum and vertex measurements, and electron identification are achieved with a combination of discrete high-resolution semiconductor pixel and strip detectors in the inner part of the tracking volume, and continuous straw-tube tracking detectors with transition radiation capability in its outer part.

The calorimeter system consists of electromagnetic and hadronic sampling calorimeters with performances corresponding to the particle characteristics whose energy deposition they are measuring. Highly granular liquid-argon (LAr) electromagnetic (EM) sampling calorimetry [3] with excellent performance in terms of energy and position resolution, covers the pseudorapidity range $|\eta| < 3.2$. In the end-caps, the LAr technology is also used for hadronic calorimeters, which share the cryostats with the EM end-caps. The same cryostats also house the special LAr forward calorimeters which extend the pseudorapidity coverage to $|\eta| < 4.9$. The central hadronic calorimetry is provided by a scintillator-tile calorimeter which is separated into a large barrel and two smaller extended barrel cylinders, one on each side of the the barrel [4]. The overall calorimeter system provides the very good jet and E_T^{miss} performance of the detector.

The calorimeter is surrounded by the muon spectrometer. The superconducting air-core toroid system, with a long barrel and two inserted end-cap magnets, generates a large magnetic field volume with strong bending power within a light and open structure. Multiple-scattering effects are thereby minimized, and excellent muon momentum resolution is achieved with three station of high-precision tracking chambers. The muon instrumentation also includes as a key component trigger chambers with very fast time response.

The interactions in the ATLAS detectors will create an enormous data-flow, which will be handled by the ATLAS trigger and data-acquisition (T/DAQ) system. It is based on three levels of online event selection. Each trigger level refines the decisions made at the previous level and, if necessary, applies additional selection criteria. Starting from an initial bunch-crossing rate of 40MHz, the rate of selected events must be reduced to ~100 Hz for permanent storage. While this requires an overall rejection factor of 10^7 against minimum bias events, excellent efficiency must be retained for the rare new physics processes.

4 Romanian participation in the ATLAS project

The Bucharest IFIN-HH is a member of the ATLAS collaboration since 1993. The Romanian group participates mainly in the systems of calorimetry and T/DAQ, and also in the physics programme.

4.1 The hadronic tile calorimeter

The design of the ATLAS hadronic barrel calorimeter was initated in the frame of the R&D 34 project. The Romanian group was among the first teams joining this project. The ATLAS hadronic barrel calorimeter is a sampling calorimeter based on steel absorber and plastic scintillator tiles read out by wavelength shifting fibers. The innovative geometry of this calorimeter consists in the orientation of the scintillator tiles in planes perpendicular to the colliding beams. For better sampling homogeneity the tiles are staggered in the radial directions.

The physics programme of ATLAS and the difficult experimental environment existing at LHC set stringent requirements on the hadronic barrel calorimeter. A granularity $\Delta\eta \times \Delta\phi = 0.1 \times 0.1$ is needed, mainly imposed by $W \rightarrow$ jet-jet decay at high-p_T. The longitudinal segmentation in three samplings is also driven by particle identification and by the possibility of achieving a better energy resolution via weighting in a non-compesanting calorimeter. A good energy resolution is crucial for jet-jet mass reconstruction as well as for missing p_T measurements for physical processes of interest. Thus well established performances are required for the energy resolution. Stringent linearity requirements are imposed by studies of quark compositness, where the jet energy scale has to be linear within 2% up to an energy of 4 TeV.

Several prototype modules of the scintillating-tile calorimeter have been constructed and extensively tested in pion, electron and muon beams at different energies from 10 GeV up to 400 GeV. The beam tests were performed using only prototypes of the scintillating-tile calorimeter (standalone tests) and in a combined mode with the electromagnetic liquid argon accordion calorimeter prototype in front, to reproduce as much as possible the ATLAS configuration. The standalone tests used firstly prototypes of reduced size (1m lenght along the beam axis), and then one barrel module or two extended barrel modules, in real size, with the 1m prototypes placed on the bottom and on the top of real size modules [5]. The test beam results obtained during 1993 - 1996 have played a major role in establishing the final design of the scintillating-tile calorimeter described in the TDR [4]. The good performances of the scintillating tile calorimeter have been demonstrated and it has been shown that these performances fulfill the requirments for ATLAS barrel hadronic calorimetry.

The combined tests have demonstrated the even better performances of the system of the two barrel calorimeters, electromagnetic and hadronic. A special attention has been devoted to determining the best algorithm for energy reconstruction, determining the weights by minimizing the energy resolution. Two combined tests have been carried out until now, in 1994 and in

1996. The results obtained in the 1996 tests will be soon published. They are confirming the good results obtained within the first combined tests [6].

The test beam data obtained in the standalone tests as well as in the combined tests have beem compared with different hadronic shower simulation codes (FLUKA, GHEISHA, CALOR) in the framework of the GEANT program. Until now none of the packages is fully adequate to describe all the experimental data.

The series production of the modules of the hadronic calorimeter has already began in 1999. During the first year the Romanian team has sent to the assembly site of the barrel part of the scintillating tile calorimeter 18 support girders, constructed by FORTPRES-Cluj, with financial contribution from CERN and IFIN-HH. Until March 2001 a total number of 65 support girders has to be fabricated.

4.2 ATLAS Data Acquisition and Event Filter Prototype -1 Project

The Romanian participation in the data aquisition started within the R&D 11 and R&D 13 projects, which investigated architectures for the Level2 Trigger and the DAQ system for test beams at LHC, respectively. In the present construction phase the Bucharest group participates in the design and implementation of a Data Acquisition (DAQ) and Event Filter (EF) prototype, based on the functional architecture described in the ATLAS Technical Proposal [1].

The prototype consists of a full "vertical" slice of the ATLAS Data Acquisition and Event Filter architecture, including all the hardware and software elements of the data flow, its control and monitoring, as well as all the elements of a complete on-line system [7]. The DAQ/EF system consists of 4 sub-systems: detector interface, data-flow, back-end DAQ and event filter. The Bucharest group is a member of DAQ/EF collaboration and is involved mainly in the data-flow and back-end activities.

The data-flow sub-system is designed to cover three main functions: the collection and buffering of the data from the detector, the merging of fragments into full events, and the interaction with the Event Filter sub-farm. The event building function is covered by the Event Builder block, the other functions of the data-flow system being covered by two modular building blocks: the read-out crate and the sub-farm crate. The Bucharest group participates in the activities related to the high level design, initial implementation and tests of the read-out crate. In particular test programs have been developed and measurements have been done in order to estimate the performance achieved with the actual implementation of the read-out buffers receiving data from the read-out link (S-Link).

The back-end software encompasses the software for configuring, controlling and monitoring the DAQ, but specifically excludes the management, processing or transportation of physics data. The Bucharest group participates to the evaluation of several software technologies [8] and to the high level design, implementation and testing of three components (information service, message reporting system and integrated graphical user interface) of the actual back-end model. The information service and the message reporting system were implemented as libraries using a communication package build on a free CORBA implementation. The integrated graphical user interface is written in Java and uses CORBA to interact with other components. Unit and integration tests (including functionality, error recovery, performance and scalability tests) have been performed for each component [9]. Work will continue to cover a wider spectrum of configurations and to achieve information about eventual limits and boundaries.

4.3 Studies in the ATLAS physics programme

The Romanian group participates in the ATLAS working groups on the search for exotic particles and on B-physics.

Theoretical models like grand unified theories, composite models, left-right symmetric models, technicolor models or superstring-inspired models predict the existence of new particles with masses around of 1 TeV, allowing thus the existence of new generations of fermions. A special interest is devoted to the search for experimental signatures of heavy leptons. Simulations with PHYTIA have been started for heavy lepton production via Z' decay into a pair of heavy charged leptons, with each heavy lepton decaying into a W boson and a massive neutrino.

Although the main focus of the ATLAS experiment is the physics beyond the SM, an important range of B-physics studies is planned. The B-physics measurements aim to test the SM by overconstraining the Cabibbo-Kobayashi-Maskawa (CKM) quark mixing matrix, indirect information about the physics beyond the SM being obtained if inconsistencies are found. Precise theoretical results leading to a better determination of the elements of the CKM matrix are therefore of much importance for understanding CP violation in the SM and detecting signals of new physics.

The Bucharest group performed theoretical studies on the semileptonic decays $\bar{B} \rightarrow D^{(*)}\ell\bar{\nu}$ using a combined method which exploits perturbative QCD, heavy quark effective theory (HQET) and analyticity properties [10]. The shape of the decay rates near zero recoil was predicted with a theoretical error of less than 2%, which improves the accuracy of the V_{cb} extraction from recoil spectrum measurements.

The Romanian physicists are investigating also the nonleptonic decays of B into light pseudoscalar mesons, in particular the decay $B^0 \to \pi^+\pi^-$. The time dependent asymmetry in this process is one of the benchmark measurement in ATLAS, leading to the determination of one angle of the CP-violating unitarity triangle. The accuracy of the determination depends in the crucial way on the additional complex phases generated by the final state strong interactions, whose magnitudes are still controversial. The problem was studied recently in a formalism [11] based on dispersion relations in external masses, combined with Regge theory and flavour SU(3) symmetry. Constraints on the penguin and tree amplitudes of the $B^0 \to \pi^+\pi^-$ decay have been derived, the results supporting the idea of parton-hadron duality. Research work in this direction will continue.

Acknowledgments

With great pleasure I thank Profs. G. Mateescu and D. Poenaru for having been such pleasant hosts during this symposium. I highly appreciate the many years of so competent and friendly collaboration with the IFIN-HH team. The help of Drs. I. Caprini, M. Caprini and S. Dita for these proceedings was simply invaluable.

References

1. ATLAS Technical Proposal, CERN/LHCC/94-43, LHCC/P2.
2. ATLAS Detector and Physics Performance TDR, LHCC 99-14/15.
3. Liquid Argon Calorimeter TDR, CERN/LHCC 96-41 (1996).
4. Tile Calorimeter TDR, CERN/LHCC 96-42 (1996).
5. R. Arsenescu et al, Tilecal Collaboration, ATLAS, CERN, *Nucl. Instr. and Meth. A* **349** 384 (1994).
6. Z. Ajaltouni et al, Tilecal Collaboration, ATLAS, CERN , *Nucl. Instr. and Meth. A* **387**, 333 (1997).
7. G.Ambrosini et al, *Comp. Phys. Comm.* **110**, 95 (1998).
8. D.Burckhart et al, *Comp. Phys. Comm.* **110**, 113 (1998).
9. A.Amorin et al, *IEEE Trans. Nucl. Sci.* **45**, 1978 (1998).
10. I.Caprini, L.Lellouch and M. Neubert, *Nucl. Phys. B* **530**, 153 (1998).
11. I. Caprini, L. Micu and C. Bourrely, *Phys. Rev. D* **60**, 074016 (1999).

THE DEAR EXPERIMENT AT DAΦNE

presented by Catalina Petrascu on behalf of the DEAR Collaboration:

M. AUGSBURGER[J], G. BEER[B], S. BIANCO[C], A. M. BRAGADIREANU[C,I], M. BREGANT[D], W. H. BREUNLICH[A], M. CARGNELLI[A], D. CHATELLARD[J], J. -P. EGGER[K], F. L. FABBRI[C], B. GARTNER[A], C. GUARALDO[C], R. S. HAYANO[F], M. ILIESCU[C,**], T. ISHIWATARI[E], T. M. ITO[C,L], M. IWASAKI[E], R. KING[A], P. KNOWLES[J], T. KOIKE[H], B. LAUSS[C,*], V. LUCHERINI[C], J. MARTON[A], E. MILOTTI[D], F. MULHAUSER[J], S. N. NAKAMURA[G], C. PETRASCU[C,**], T. PONTA[I], A. C. SANDERSON[B], L. A. SCHALLER[J], L. SCHELLENBERG[J], H. SCHNEUWLY[J], R. SEKI[L,M], D. TOMONO[E], T. YONEYAMA[E], E. ZAVATTINI[D], J. ZMESKAL[A]

[A] *Institute for Medium Energy Physics, Austrian Academy of Sciences, Boltzmanngasse 3, A-1090 Vienna, Austria*
[*] *on leave of absence from above*
[B] *Univ. of Victoria, Dept. of Physics and Astronomy, P.O. Box 3055 Victoria, BC V8W 3P6, Canada*
[C] *INFN - Laboratori Nazionali di Frascati, C.P. 13, Via E. Fermi 40, I-00044 Frascati , Italy*
[D] *Univ. degli Studi di Trieste, Dip. di Fisica and INFN Sezione di Trieste, Via A. Valerio 2, I-34127, Trieste, Italy*
[E] *Tokyo Institute of Technology, 2-12-1 Ookayama Meguro, Tokyo 152, Japan*
[F] *Univ. of Tokyo, Dept. of Physics, 7-3-1 Hongo, Bunkyo, Tokyo 113, Japan*
[G] *Inst. of Physical and Chemical Research (RIKEN), 2-1 Hirosawa, Wako, Saitama 351-01, Japan*
[H] *KEK, High Energy Accelerator Research Organization, Tanashi Campus 3-2-1 Midori, Tanashi, Tokyo 188, Japan*
[I] *Inst. of Physics and Nuclear Engineering "Horia Hulubei", Dept. of High Energy Physics, P.O. Box MG-6 R-76900 Magurele, Bucharest, Romania*
[**] *on leave of absence from above*
[J] *Univ. de Fribourg, Inst. de Physique, Bd. de Perolles, CH-1700 Fribourg, Switzerland*
[K] *Univ. de Neuchatel, Inst. de Physique, 1 rue A.-L. Breguet, CH-2000 Neuchatel, Switzerland*
[L] *W.K. Kellogg Radiation Laboratory, California Institute of Technology, Pasadena, CA 91125, USA*
[M] *Department of Physics and Astrophysics, California State University, Northridge, CA 91330, USA*

This paper gives an overwiew of the DEAR (DAΦNE Exotic Atom Research) experiment, which is fully ready to collect data at the new DAΦNE ϕ-factory at Laboratori Nazionali di Frascati dell'INFN. The objective of the DEAR experiment is to perform a precision measurement of the strong interaction shifts and widths of the K-series lines in kaonic hydrogen and the first observation of the same quantities in kaonic deuterium. The aim is to obtain a precise determination of the isospin-dependent kaon-nucleon scattering lengths what will represent a breakthrough in $\bar{K}N$ low-energy phenomenology and will allow to determine the kaon-nucleon sigma terms. The sigma terms give a direct measurement of chiral symmetry breaking and are connected to the strangeness content of the proton. First results on background measurements with the DEAR NTP setup installed on DAΦNE are reported.

1 Introduction

DEAR (DAΦNE Exotic Atom Research) [1]– one of the first experiments which will collect data at the new ϕ-factory DAΦNE [2] of the Laboratori Nazionali di Frascati dell'INFN, will observe X rays from kaonic hydrogen and kaonic deuterium, using the "K^- beam" from the decay of ϕs produced by DAΦNE, a cryogenic pressurized gaseous target and Charge-Coupled Devices (CCDs) as X ray detectors.

DEAR is aiming to perform a *precision* measurement of the K-series lines shift and width due to strong interaction for the kaonic hydrogen, while the kaonic deuterium will be measured for the first time. In this way, a precise determination of the isospin dependent $\bar{K}N$ scattering lengths will be obtained. The number of events which DEAR can collect in *one week* in the case of kaonic hydrogen at $10^{32}\text{cm}^{-2}\text{s}^{-1}$ machine luminosity will surpass by an order of magnitude the current world data set.

2 The DEAR scientific program

A kaonic atom is formed when a negative kaon entering a target loses its kinetic energy and is captured, replacing the electron, in an excited orbit. Three processes then compete in the deexcitation of the newly formed kaonic atom: dissociation of the surrounding molecules, external Auger transitions and radiative transitions.

When a kaon reaches low-n states with small angular momentum, it is absorbed through a strong interaction with the nucleus. This strong interaction is causing a shift in the energies of the low-lying levels from their purely electromagnetic values, whilst the finite lifetime of the state results in an increase of the observed level width. The shift ϵ and the width Γ of the $1s$ state of kaonic hydrogen are related in a fairly model-independent way to the real and imaginary parts of the complex S-wave scattering length:

$$\epsilon + \frac{i}{2}\Gamma = 2\alpha^3 \mu_{K^-p}^2 a_{K^-p} = (412\ eV\ fm^{-1}) \cdot a_{K^-p} \tag{1}$$

where α is the fine structure constant and μ_{K^-p} the reduced mass of the K^-p system. This expression is better known as the Deser-Trueman formula [3]. A similar relation applies for the case of kaonic deuterium and the corresponding scattering length a_{K^-d}:

$$\epsilon + \frac{i}{2}\Gamma = 2\alpha^3 \mu_{K^-d}^2 a_{K^-d} = (601\ eV\ fm^{-1}) \cdot a_{K^-d} \tag{2}$$

where μ_{K^-d} is the reduced mass of the K^-d system.

These observable scattering lengths (a_{K^-p} and a_{K^-d}) are related to the isospin dependent scattering lengths (a_0 and a_1):

$$a_{K^-p} = \frac{1}{2}(a_0 + a_1) \tag{3}$$

$$a_{K^-d} = 2(\frac{m_N + m_K}{m_N + m_K/2})a^{(0)} + C \tag{4}$$

where

$$a^{(0)} = \frac{1}{2}(a_{K^-p} + a_{K^-n}) = \frac{1}{4}(a_0 + 3a_1) \tag{5}$$

represents the lowest order impulse approximation and C contains higher order contributions, including three-body effects.

One order of magnitude improvement in the precision of the K^-p scattering length determination, with respect to the more accurate data presently available, the result of a recent experiment at KEK [4], is the expected performance of DEAR. This will represent a breakthrough in the low-energy $\bar{K}N$ phenomenology and make possible to discriminate different theoretical approaches and method of analysis. Moreover, an accurate determination of the K^-N isospin dependent scattering lengths will place strong constraints on the low energy K^-N dynamics, which in turn constrains the SU(3) description of chiral symmetry breaking. Crucial information about the nature of chiral symmetry breaking, and to what extent chiral symmetry is broken, is provided by the calculation of the meson-nucleon sigma terms [5]. The meson-nucleon sigma term is defined as the expectation value of a double commutator of the chiral symmetry breaking part of the strong-interaction Hamiltonian, H_{SB}:

$$\sigma_N^{ba} = i < p|[Q_b^5, [Q_a^5, H_{SB}]]|p > \tag{6}$$

with the proton state $|p>$ of momentum p, Q_a^5 and Q_b^5 representing the axial-vector charges.

The low energy theorem relates the sigma terms to the meson-nucleon scattering amplitude [5]. A phenomenological procedure starting from the experimental scattering amplitudes is then used to determine the sigma terms and, therefore, to measure chiral symmetry breaking [6]. According to an evaluation based on the uncertainities in the phenomenological procedure [7], the sigma terms can be extracted at a level of 20%, by combining the DEAR precision measurement at threshold with the bulk of most recent analyses of low energy $K^{\pm}N$ scattering data.

The sigma terms are important also as inputs for the determination of the strangeness content of the proton. The strangeness fraction is dependent on both kaon-nucleon and pion-nucleon sigma terms, but is more sensitive to the first one [8].

3 DEAR Experimental setup

The scientific program of DEAR will be performed in two steps: with an NTP (Normal Temperature and Pressure) target filled by nitrogen, and with a cryogenic target with hydrogen (deuterium).

The NTP target, presented in Fig. 1, consists of a pure nitrogen volume at room temperature equipped with four CCD-05. The purpose is to measure background, compare it with Monte Carlo calculations and tune the degrader thickness by optimizing the signal coming from kaonic nitrogen transitions: $7 \rightarrow 6$ ($\sim$ 4.6 keV) and $6 \rightarrow 5$ ($\sim$ 7.6 keV). The yield of these transitions being higher (by a factor of ten) than the one of kaonic hydrogen a faster feedback is then possible. Kaonic nitrogen X rays transitions have been never measured and will be *for the first time* measured by DEAR.

For the measurement of kaonic hydrogen (deuterium), a pressurized cryogenic target was built, to achieve a balance between high kaon stopping power in the targel cell and the loss of x-ray yield due to the Stark mixing effect. The target will operate at 25 K temperature and 3 bar of pressure. X-ray detectors are 16 CCD-22. According to cascade calculations [9] and to recent KEK results [4] the expected yield of K_α X rays per stopped kaon in the mentioned conditions is (1-3)%.

Further details regarding the NTP setup can be found in [10], while a description of the DEAR cryogenic setup can be found in [1].

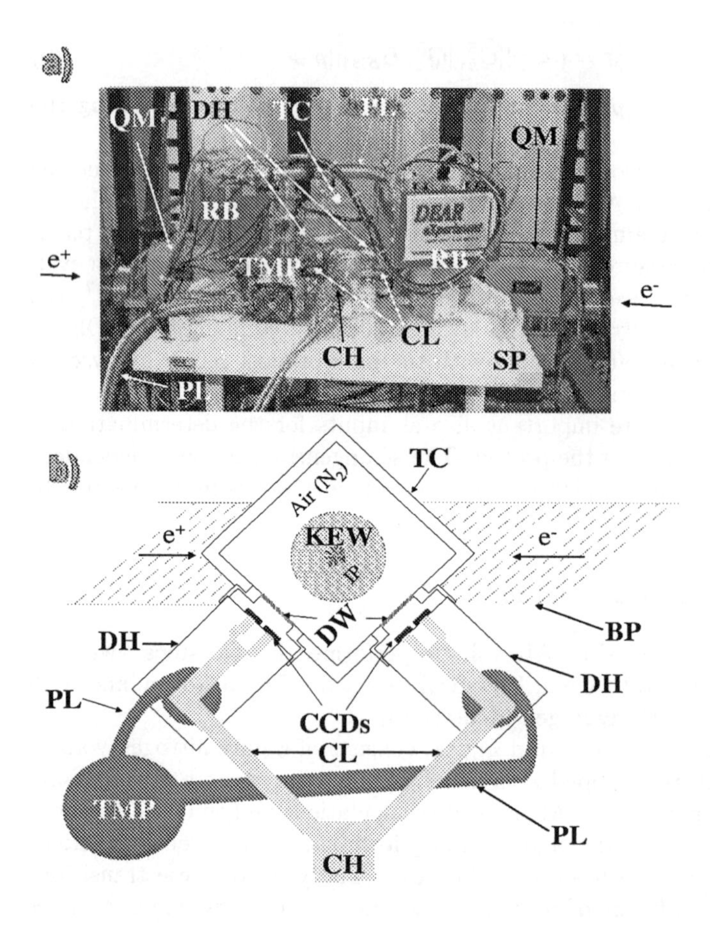

Figure 1. a) The NTP setup installed in the DAΦNE hall; b) Schematic drawing of the NTP setup: the displaied components are: RB = readout box; TMP = turbomolecular pump; CH = cooling head; Cl - cooling line; PL = pumping line; TC = target chamber; Dh = detector housing; QM - quadrupole magnet; BP = beam pipe; KEW = kaon entrance window; DW = detector window; IP = interaction point; SP = setup platform.

4 First background measurement on DAΦNE with the NTP setup

The NTP setup was installed on DAΦNE in February 1999.

At the date of this Symposium (9-10 December, 1999), *no e^+e^- collisions have been yet produced in the DEAR interaction region*, being the machine trying to optimize the luminosity in the other interaction region, where the KLOE solenoid is installed.

The background relevant for the DEAR experiment are low energy (≤ 20 keV) X rays and ionizing particles. A low-energy X ray hit gives a "single-pixel" event, a pixel with charge content above a selected noise threshold, surrounded by eight neghbour pixels displaying charge contents below threshold. Hits of more than one pixel are called "clusters", and are mainly caused by charged background particles.

Using the fully shielded DEAR NTP setup the following types of beams have been monitored in the period March to July 1999: a) only positron beams; b) only electron beams; c) electron and positron beams at the same time. Background data, in terms of *clusters*, were then compared with DEAR Monte Carlo simulation.

Fig. 2 shows, as an example, the number of clusters per half-CCD per minute observed with electron beams circulating in DAΦNE. The time dependence is given in correlation with the injection time of the beam into the accelerator ring. The time between two adjacent points represent the exposure time, which was 10 minutes. The time behaviour of the number of clusters follows the time behaviour of the injected electrons (reported in the upper part of the figure). The simulations agree very well with the measurements.

In the working conditions of the measurements (no collisions) and at the present status of the accelerator, the most important background comes from Touschek scattering [11] with a contribution from beam-gas interaction [12].

All measurements can be well reproduced by the DEAR Monte Carlo simulation using the standard beam parameters for the DEAR interaction region. The quantitative agreement between simulation and measurements is very convincing.

During this measurements the *shielding factor* could be as well measured and compared with the Monte Carlo value; it came out an experimental value of 3.6 ± 0.2, perfectly compatible with the calculated value of 3.8 ± 0.4. More details can be found in [10].

Since the goal of DEAR is to detect kaonic hydrogen X rays, the study of x-ray background is very important, as the final statistical accuracy attainable with DEAR will critically depend on the level of x-ray background. The observed energy spectra do not show any signal of unwanted structure. This proves the cleanness of the used materials.

Tests of the possibility of using fluorescence lines, for on-line energy calibration, from titanium and zirconium foils, placed in the target cell and

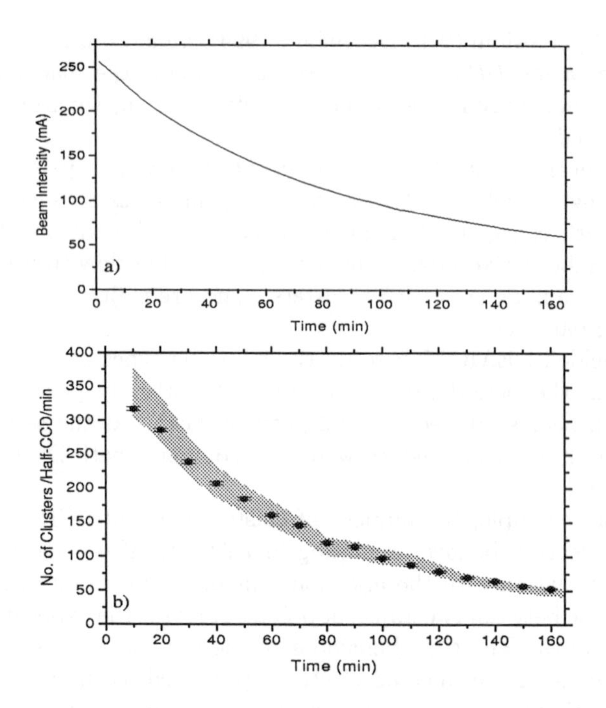

Figure 2. a) Electron beam intensity (mA) for the period 6:21 - 9:01 a.m. June 27, 1999 with only electrons circulating in the DAΦNE main ring; b) Observed number of clusters per half - CCD per minute (filled circles) during the corresponding beam period. The shaded area shows the results of the DEAR Monte Carlo simulation.

excited by the background particles, were also successfully performed.

5 Summary

The DEAR experiment will improve the precision in the measurement of the K^-p scattering length by a factor of ten and make the first measurement of the scattering length for the kaonic deuterium. A percent level measurement of the isospin dependent $\bar{K}N$ scattering length will represent a breakthrough in the low-energy $\bar{K}N$ phenomenology, implying the possibility of discriminating between many existent theoretical approaches and methods of analysis. Moreover, it will be possible the determination of the kaon-nucleon sigma terms, which give directly the degree of chiral symmetry breaking and allow to obtain an indication of the strangeness content of the proton.

At the time of writing this paper, the first stage of the DEAR scientific program is in progress. The NTP nitrogen target is installed on DAΦNE, has performed background measurements and the first kaons are eagerly awaited.

The cryogenic hydrogen target is being tested in the laboratory and is ready for installation at the beginning of 2000.

References

1. S. Bianco it et al., Rivista del Nuovo Cimento **22**, No. 11, 1 (1999).
2. G. Vignola, *DAΦNE: the first ϕ-Factory*, Proceedings of the *"5th European Particle Accelerator Conference"* (EPAC'96), Eds. S. Myres *et al.*, 1996, p.22.
3. S. Deser *et al.*, Phys. Rev. **96**, 774 (1954);
   T.L. Trueman, Nucl. Phys. **26**, 57 (1961);
   A. Deloff, Phys. Rev. **C13**, 730 (1976).
4. M. Iwasaki *et al.*, Phys. Rev. Lett. **78**, 3067 (1997);
   S.N. Nakamura *et al.*, Nucl. Inst. and Meth. **A408**, 438 (1998);
   T.M. Ito *et al.*, Phys. Rev. **C58**, 2366 (1998).
5. E. Reya, Rev. Mod. Phys **46**, 545 (1974);
   H. Pagels, Phys. Rep. **16**, 219 (1975).
6. E. Reya, Phys. Rev. **D7**, 3472 (1973).
7. B. Di Claudio *et al.*, Lett. Nuovo Cimento **26**, 555 (1979);
   A.D. Martin and G. Violini, Nuovo Cimento **30**, 105 (1981).
8. R.L.Jaffe and C.L. Korpa, Comments Nucl. Part. Phys. **17**, 163 (1987).
9. T. Koike, T. Harada and Y. Akaishi, Phys. Rev. **C53**, 79 (1996).
10. M. Augsburger *et al.* "First measurements at the DAΦNE ϕ-factory with the DEAR experimental setup", LNF-99/32(P) (1999), submitted to N.I.M.
11. S. Guiducci, *Background calculations for the DAΦNE experiments*, Proceedings of the *"5th European Particle Accelerator Conference"* (EPAC'96), Eds. S. Myres *et al.*, 1996, p.1365.
12. S. Guiducci and M.A. Iliescu, LNF-97/002 (IR) (1997).

THEORETICAL CONSTRAINTS ON THE SEMILEPTONIC AND NONLEPTONIC B DECAYS

IRINEL CAPRINI
National Institute of Physics and Nuclear Engineering, IFIN-HH
POB MG 6, Bucharest, R-76900 Romania
E-mail: caprini@theor1.theory.nipne.ro

The decays of the B meson are an important source of information about the elements of the Cabibbo-Kobayashi-Maskawa quark mixing matrix, which parametrize the mechanism of CP violation in the standard model. I briefly review recent results on the theoretical description of the semileptonic decays $\bar{B} \to D^{(*)}\ell\bar{\nu}$, which allow the accurate extraction of the element V_{cb}, and the nonleptonic B decays into light pseudoscalar mesons, which are of interest for the measurement of the CP-violating phases at present and future experiments.

1 Introduction

In the standard model of electroweak interactions, CP violation is described by a nonzero complex phase in the Cabibbo-Kobayashi-Maskawa (CKM) matrix, which parametrize the weak couplings of the charged gauge bosons $W^{\pm}$ to the quarks. In the standard parametrization [1], the only complex elements are V_{ub} and V_{td}, whose phases are denoted by $-\gamma$ and $-\beta$, respectively. Together with $\alpha = \pi - \beta - \gamma$, they are the angles of the so-called unitarity triangle [1].

The exclusive decays of the B meson are known as best candidates for the study of the phase structure of the CKM matrix. The advantage comes from the large variety of modes, which by interference produce detectable CP violating observables. Many present and future experiments, among which three experiments at the CERN future collider LHC (ATLAS, CMS and LHC-b), will measure with great accuracy a large number of observables related to B decays, including time-dependent asymmetries and direct CP violation. Various strategies have been proposed for extracting independently the angles of the unitarity triangle. By over-constraining this triangle one hopes to give a definite answer to the problem of whether CP violation originates in the standard model mixing, or requires new concepts.

On the theoretical side, the main difficulty in the description of B- decays is produced by the strong interactions. Both short distance and long distance effects contribute in the decays. While short distance effects are well described in perturbative QCD, the long distance physics is much poorly known, and more model dependent. The modern approaches use heavy quark effective theory (HQET), QCD sum rules, lattice calculations, quark models and dispersive techniques.

I shall first briefly review recent progress on the precise theoretical description of the semileptonic decays $\bar{B} \to D^{(*)}\ell\bar{\nu}$, using a combined method which exploits HQET and analyticity properties. Then I will consider the nonleptonic B decays into light pseudoscalar mesons, such as $B \to \pi\pi$ and $B \to \pi K$, discussing in more detail a recent dispersive approach, which yields constraints on the hadronic parameters of interest for the experimental measurements.

2 Semileptonic B decays into charmed mesons

The exclusive semileptonic B decays into charmed mesons were measured experimentally by several collaborations [2–6]. The decay amplitudes assume a factorized form, in which the weak leptonic and the hadronic currents appear in different matrix elements. Much progress in the description of the hadronic matrix elements of the hadronic currents $V^\mu = \bar{c}\gamma^\mu b$ and $A^\mu = \bar{c}\gamma^\mu\gamma^5 b$ was achieved by Isgur *et al.* [7], who noticed that in the weak transition of a heavy quark to another heavy quark, the degrees of freedom of the light quarks can be parametrized in an almost universal way (for a review of HQET, see Neubert [8]). The matrix elements $\langle D^{(*)}(v')|V^\mu|\bar{B}^{(*)}(v)\rangle$and $\langle D^{(*)}(v')|A^\mu|\bar{B}^{(*)}(v)\rangle$, where $B(B^*)$ and $D(D^*)$ denote the pseudoscalar (vector) mesons, involve a set of twenty form factors, which depend on the kinematical variable $w = v \cdot v'$, where v and v' are the meson velocities. In the heavy-quark limit, some form factors coincide with the universal Isgur - Wise function [7,8], the remaining ones being negligibly small. The differential rates of the decays $\bar{B} \to D^{(*)}\ell\bar{\nu}$ are expressed in terms of some of these hadronic form factors, the parameter $|V_{cb}|$ being obtained by the extrapolation of the differential decay rates to zero recoil. To reduce the uncertainties associated with this extrapolation, the precise shape of the semileptonic form factors near zero recoil is crucial. A detailed theoretical analysis [9,10] was performed to this end.

We used a dispersive technique [11], which is based on the QCD expression of the vacuum polarization tensor of the weak currents in the deep euclidian region, combined with dispersion relations and hadronic unitarity, generalized to all the ground state transitions [12,13]. The predictions of HQET with both short-distance and $1/m_Q$ corrections were taken into account near zero recoil. Defining two reference form factors, $V_1(w)$ for the vector, and $A_1(w)$ for the axial current respectively, we obtained simple one-parameter expressions, which describe these functions in the semileptonic region with an accuracy better than 2%. These parametrizations are [10]

$$
\begin{aligned}
V_1(w) &\approx V_1(1)\left[1 - 8\rho_1^2 z + (51.\rho_1^2 - 10.)z^2 - (252.\rho_1^2 - 84.)z^3\right], \\
A_1(w) &\approx A_1(1)\left[1 - 8\rho_{A_1}^2 z + (53.\rho_{A_1}^2 - 15.)z^2 - (231.\rho_{A_1}^2 - 91.)z^3\right].
\end{aligned} \tag{1}
$$

where $z = (\sqrt{w+1} - \sqrt{2})/(\sqrt{w+1} + \sqrt{2})$ is a conformal mapping variable, w the velocity transfer, and ρ_1^2 and $\rho_{A_1}^2$ are the slope parameters at zero recoil of the corresponding functions. The values of the form factors at zero recoil are $V_1(1) = 0.98 \pm 0.07$ and $A_1(1) = 0.91 \pm 0.03$, and the slopes are restricted to the intervals $-0.17 < \rho_1^2 < 1.51$ and $-0.14 < \rho_{A_1}^2 < 1.54$.

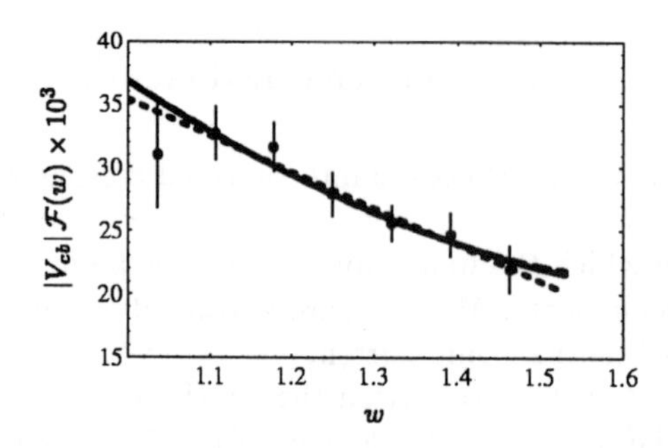

Figure 1. Experimental data for the product $|V_{cb}|\,\mathcal{F}(w)$ as a function of w, extracted from semileptonic $\bar{B} \to D^*\ell\,\bar{\nu}$ decays [2]. The solid curve correspunds to the parametrization of $\mathcal{F}(w)$ which follows from (1), the dashed one is a linear fit.

All the other form factors of definite spin and parity are obtained from (1) using heavy quark symmetry. In Figure 1 we indicate for illustration the behaviour of the form factor $\mathcal{F}(w)$ relevant for the decay rate of $\bar{B} \to D^*\ell\,\bar{\nu}$, together with the CLEO experimental values [2]. The theoretical shape deviates significantly from a simple linear fit of the experimental data and improves the accuracy of the determination of V_{cb}. We mention that the most recent world-averaged value [1] is $V_{cb} = 0.040 \pm 0.003$.

3 Nonleptonic B decays into light pseudoscalar mesons

The weak nonleptonic B decays into light pseudoscalar mesons are of much interest for the experimental determination of the CP violating phases in the standard model. As we mentioned above, a nontrivial theoretical problem related to these decays is the description of the final state strong interactions, which generate additional complex phases in the decay amplitudes, besides the CP violating phases of the CKM matrix. For instance, the determination of the angle α of the unitarity triangle from the time dependent asymmetry in the $B^0 \to \pi^+\pi^-$ decay requires the knowledge of the ratio and the strong phase difference between the tree and the penguin amplitudes describing the process.

The magnitude of the strong phases generated by the final state interactions is still controversial: arguments based on "color transparency" and calculations using heavy quark effective theory and perturbative QCD suggest that these phases are small [14], while the general features of high energy hadronic physics do not exclude quite appreciable strong phases [15].

Recently, we investigated the rescattering effects in the nonleptonic B-decays into light pseudoscalar mesons in a formalism [16] based on dispersion relations in terms of the external masses. We define the weak decay amplitude $A_{B\to P_1P_2} = A(m_B^2, m_1^2, m_2^2)$, where P_1, P_2 are pseudoscalar mesons. By applying the Lehmann-Symanzik-Zimmermann (LSZ) reduction formalism [17] to the final meson which does not contain the spectator quark, one can show that the weak amplitude can be analytically continued in the complex plane of the mass variable k_1^2, and satisfies a dispersion representation of the form

$$A(m_B^2, m_1^2, m_2^2) = A^{(0)}(m_B^2, m_1^2, m_2^2) + \frac{1}{\pi}\int_0^{(m_B-m_2)^2} \mathrm{d}z \frac{\mathrm{Disc}A(m_B^2, z, m_2^2)}{z - m_1^2 - i\epsilon}, \quad (2)$$

where $A^{(0)}(m_B^2, m_1^2, m_2^2)$ is the amplitude in the factorization limit and the discontinuity is

$$\mathrm{Disc}A(m_B^2, k_1^2, m_2^2) = \frac{1}{2\sqrt{2\omega_1}}\sum_n \delta(k_1 + k_2 - p_n)\langle P_2(k_2)|\eta_1|n\rangle\langle n|\mathcal{H}_w|B(p)\rangle\,. \quad (3)$$

The representation (2) allows one to recover the amplitude in the factorization approximation when the strong rescattering is switched-off, which is a reasonable consistency condition. As concerns the discontinuity (3), it describes the final state rescattering effects in the decay $B \to P_1P_2$, since in each term of the unitarity sum the amplitude of the weak decay of B into an intermediate state $|n\rangle$ is multiplied by the amplitude of the strong rescattering of $|n\rangle$ to the final state P_1P_2. Moreover, one can show that in the standard model the spectral function can be written as

$$\mathrm{Disc}\,A(m_B^2, z, m_2^2) = \sigma_1(z) + \sigma_2(z)\,\mathrm{e}^{i\gamma}\,, \quad (4)$$

where $\gamma = \mathrm{Arg}(V_{ub}^*)$ and the functions σ_1 and σ_2 are real. The proof is based on the transformation properties of the weak hamiltonian under space and time reversal, and on the completeness of the set of asymptotic states "in" and "out".

In the two-particle approximation, when the intermediate states are $|n\rangle = |P_3P_4\rangle$, the on shell weak decay amplitudes $A_{B\to P_3P_4}$ appearing in (3) are independent of the phase space integration variables and the dispersion variable z. Therefore, the dispersion representation (2) becomes an algebraic relation among on shell weak amplitudes [16]. Following a famous procedure of Goldberger and Treiman [18], it is convenient to write the complete set of hadronic states $|n\rangle$ entering the discontinuity Eq. (3) as a combination $1/2|n, in\rangle + 1/2|n, out\rangle$, since this procedure respects at all stages of approximation reality property of the discontinuity mentioned below (4), which is valid when the sum is not truncated [16]. Then the dispersion relation (2) can be written as

$$A_{B\to P_1P_2} = A^{(0)}_{B\to P_1P_2} + \frac{1}{2}\sum_{\{P_3P_4\}} \Gamma_{34;12}\bar{A}_{B\to P_3P_4} + \frac{1}{2}\sum_{\{P_3P_4\}} \overline{\Gamma}_{34;12}A_{B\to P_3P_4}\,. \tag{5}$$

In this relation $A^{(0)}_{B\to P_1P_2}$ is the amplitude in the factorization limit, $\bar{A}_{B\to P_3P_4}$ are obtained from $A_{B\to P_3P_4}$ by changing the sign of the strong phases, the coefficients $\Gamma_{34;12}$ are computed as dispersive integrals

$$\Gamma_{34;12} = \frac{1}{\pi}\int_0^{(m_B-m_2)^2} dz\,\frac{C_{34;12}(z)}{z - m_1^2 - i\epsilon}\,, \tag{6}$$

and $\overline{\Gamma}_{34;12}$ are defined as in (6), with the numerator $C_{34;12}$ replaced by $C^*_{34;12}$, where

$$C_{34;12}(z) = \frac{1}{2}\frac{1}{(2\pi)^2}\int \frac{d^3\mathbf{k}_3}{2\omega_3}\frac{d^3\mathbf{k}_4}{2\omega_4}\delta^{(4)}(p - k_3 - k_4)\mathcal{M}_{P_3P_4\to P_1P_2}(s,t)\,. \tag{7}$$

The strong amplitudes $\mathcal{M}_{P_3P_4;P_1P_2}(s,t)$ entering this expression are evaluated for an off-shell meson P_1 of mass squared equal to z, at the c.m. energy squared $s = m_B^2$, which is high enough to justify the application of Regge theory. A detailed calculation [16] takes into account both the t-channel trajectories describing the scattering at small angles, and the u-channel trajectories describing the scattering at large angles. Then, after performing the phase space integral in (7), the coefficients $C_{34;12}$ can be expressed as

$$C_{34;12}(z) = \sum_{\{V_t\}} \xi_{V_t}\gamma^{V_t}_{34;12}\,\kappa^{V_t}_{34;12}(z) + \sum_{\{V_u\}} \xi_{V_u}\gamma^{V_u}_{34;12}\,\kappa^{V_u}_{34;12}(z)\,, \tag{8}$$

Table 1. Values of the Regge residua of the rescattering amplitudes in $B^0 \to \pi^+\pi^-$: column II indicates the channel, III the Regge trajectories, IV the coupling given by $SU(3)$ Clebsch-Gordan coefficients, V an additional sign due to the definition of the meson states, and VI the signature factor ξ_V.

I	II	III	IV	V	VI
$\pi^+\pi^- \to \pi^+\pi^-$	t	ρ	γ_1^2	+	$i\sqrt{2}$
	t	f_8	γ_2^2	+	$-\sqrt{2}$
	u	exotic			
$\pi^0\pi^0 \to \pi^+\pi^-$	t	ρ	γ_1^2	−	$i\sqrt{2}$
	u	ρ	$-\gamma_1^2$	−	$i\sqrt{2}$
$K^0\bar{K}^0 \to \pi^+\pi^-$	t	exotic			
	u	K^*	$-1/2\gamma_1^2$	−	$i\sqrt{2}$
	u	K^{**}	$3/2\gamma_2^2$	−	$-\sqrt{2}$
$\eta_8\eta_8 \to \pi^+\pi^-$	t	A_2	γ_2^2	−	$-\sqrt{2}$
	u	A_2	γ_2^2	−	$-\sqrt{2}$
$\eta_1\eta_1 \to \pi^+\pi^-$	t	A_2	$5\gamma_2^2$	−	$-\sqrt{2}$
	u	A_2	$5\gamma_2^2$	−	$-\sqrt{2}$
$\eta_8\eta_1 \to \pi^+\pi^-$	t	A_2	$\sqrt{5}\gamma_2^2$	−	$-\sqrt{2}$
	u	A_2	$\sqrt{5}\gamma_2^2$	−	$-\sqrt{2}$

where the first (second) sum includes the contribution of the $t(u)$-channel trajectories. In Eq. (8), ξ_V is a numerical factor related to the signature of the trajectory, $\gamma_{34;12}^{V_t}$ denote the Regge residua, and $\kappa_{34;12}^{V_t}(z)$ are kinematical coefficients depending on the Regge trajectories [16].

We apply for illustration the dispersive formalism to the decay $B^0 \to \pi^+\pi^-$, taking as intermediate states in the dispersion relation (5) the pseudoscalar mesons $\pi^+\pi^-$, $\pi^0\pi^0$, K^+K^-, $K^0\bar{K}^0$, $\eta_8\eta_8$, $\eta_1\eta_1$ and $\eta_1\eta_8$, where η_8 (η_1) denote the octet (singlet) combination of the physical η and η'. Assuming $SU(3)$ flavor symmetry and keeping only the contribution of the dominant quark topologies discussed by Gronau *et al.* [19], the dispersion relation (5) becomes an algebraic equation involving the tree and the penguin amplitudes, A_T and A_P. The determination of the Regge residua $\gamma_{34;12}^{V}$ which appear in the expression (8) of the coefficients $C_{34;12}$ was described in detail in [16]. Using the optical theorem and the usual Regge parametrization of the total hadronic cross sections [1] we obtained the values $\gamma_P^2 = 25.6$, $\gamma_1^2 = 31.4$ and $\gamma_2^2 = 35.3$ for the residua of the Pomeron, the ρ and the f trajectories in the elastic $\pi^+\pi^-$ scattering, respectively. All the other residua are expressed in terms of the couplings γ_1^2 and γ_2^2 by flavour $SU(3)$ symmetry. For completeness we give their values in Table 1. The quantity $\xi_V\, \gamma_{34;12}^V$ appearing in (8) is obtained by taking the product of the values in the last three columns of Table 1.

With the Regge parameters given in Table 1, the dispersion relation (5) can be written as

$$\begin{aligned} e^{i\gamma} + Re^{i\delta}e^{-i\beta} &= \frac{e^{-i\delta_T}}{A_T}[A_T^{(0)}e^{i\gamma} + A_P^{(0)}e^{-i\beta}] \qquad (9) \\ &- \left[(0.010 + 1.273\, i) + (0.754 - 1.010\, i)\, e^{-2i\delta_T}\right] e^{i\gamma} \\ &+R\left[-(1.973 + 2.643\, i)\, e^{i\delta} - (1.787 - 1.995i)\, e^{-i\delta}e^{-2i\delta_T}\right] e^{-i\beta} . \end{aligned}$$

Here $A_T^{(0)}$ and $A_P^{(0)}$ are the amplitudes in the factorization approximation, $R = |A_P/A_T|$ and $\delta = \delta_P - \delta_T$, δ_T (δ_P) being the strong phase of A_T (A_P), respectively. The weak angles appear in the combination $\gamma + \beta = \pi - \alpha$. Solving the complex equation (9) for R and α, we derive their expressions as functions of δ_T and δ. In Figure 1 we represent R and α as functions of

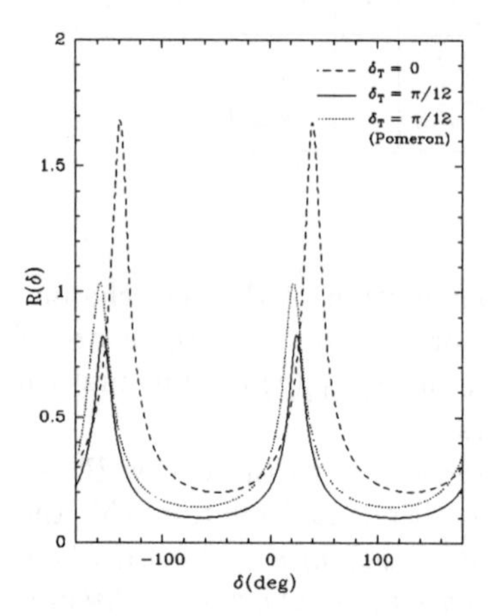

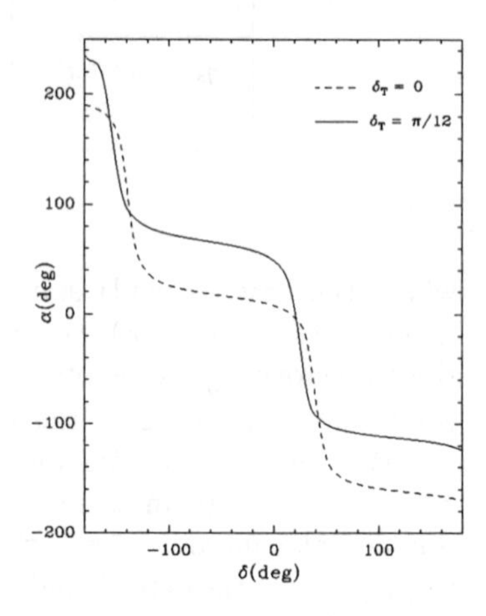

Figure 2. The ratio $R = |A_P/A_T|$ (*left*) and the weak phase α (*right*), as functions of the strong phase difference δ, solid curve $\delta_T = \pi/12$, dashed curve $\delta_T = 0$.

the phase difference δ, for two values of δ_T, using as input $A_P^{(0)}/A_T^{(0)} = 0.08$

and $A_T^{(0)}/A_T \approx 0.9$, suggested by Beneke *et al.* [14]. Values of the ratio R less than one, as expected from quark diagrams, are obtained for both $\delta_T = 0$ and $\delta_T = \pi/12$. The dominant contribution is given by the elastic channel, more precisely by the Pomeron, as is seen in Figure 1, where the dotted curve shows the ratio R for $\delta_T = \pi/12$, keeping only the contribution of the Pomeron in the Regge amplitudes.

The above results show that the dispersive formalism is not inconsistent with the treatment by Beneke *et al.*[14], based on factorization and perturbative QCD in the heavy quark limit, which predicts rather small strong phases. The dispersion representations in the external mass lead to nontrivial correlations among the hadronic parameters, which can be used as theoretical constraints in measurements of CP-violating observables. Work is in progress for improving the formalism, in particular by including the amplitudes of the quark diagrams neglected in the first approximation.

References

1. Particle Data Group, (C. Caso et al.), *Eur. Phys. J. C* **3**, 1 (1998).
2. CLEO Collaboration (B. Barish et al.), *Phys. Rev. D* **51**, 1014 (1995).
3. ARGUS Collaboration (H. Albrecht et al.), *Phys. Rep.* **276**, 223 (1996).
4. ALEPH Collaboration (D. Buskulic et al.), *Phys. Lett. B* **395**, 373 (1997).
5. DELPHI Collaboration (P. Abreu et al.), *Z. Phys. C* **71**, 539 (1996).
6. OPAL Collaboration (K. Ackerstaff et al.), *Phys. Lett. B* **395**, 128 (1997).
7. N. Isgur and M.B. Wise, *Phys. Lett. B* **232**, 113 (1989); **237**, 527 (1990).
8. M. Neubert, *Phys. Rep.* **245**, 259 (1994).
9. I. Caprini and M. Neubert, *Phys. Lett. B* **380**, 376 (1996).
10. I.Caprini, L.Lellouch and M. Neubert, *Nucl. Phys. B* **530** (1998) 153.
11. C. Bourrely, B. Machet and E. de Rafael, *Nucl. Phys. B* **189**, 157 (1981).
12. I. Caprini and C. Macesanu, *Phys. Rev. D* **54**, 5686 (1996).
13. C.G. Boyd, B. Grinstein and R.F. Lebed, *Phys. Rev. D* **56**, 6895 (1997).
14. M. Beneke, G. Buchalla, M. Neubert, and C.T. Sachrajda, *Phys. Rev. Lett.* **83**, 1914 (1999).
15. J.F. Donoghue et al, *Phys. Rev. Lett.* **77**, 2178 (1996).
16. I. Caprini, L. Micu, and C. Bourrely, *Phys. Rev. D* **60**, 074016 (1999); hep-ph/9910297, CPT-99/P.3893, submitted to *Phys.Rev. D.*
17. H. Lehmann, K. Symanzik, and W. Zimmermann, *Nuovo Cim.*, **1**, 205 (1956); **2**, 425 (1957).
18. M.L. Goldberger and S.B. Treiman, *Phys. Rev.* **111**, 354 (1958).
19. M. Gronau *et al*, *Phys. Rev. D* **50**, 4529 (1994).

FROM ELEMENTARY PARTICLES TO STARS

C. BESLIU AND AL. JIPA
Atomic and Nuclear Physics Department, Faculty of Physics, University of Bucharest, P.O. Box MG-11, R 76900, Bucharest-Magurele, ROMANIA
E-mail: besc@scut.fizica.unibuc.ro
E-mail: ajip@scut.fizica.unibuc.ro

An analysis of some connections between Particle Physics and Cosmology is presented in this work. The role of the Relativistic Nuclear Physics as a bridge between the two fields is stressed. The presence of the dibaryons in nuclear matter formed in nucleon-nucleon and nucleus-nucleus collisions at high energy is evidenced. Their influences on the quark-gluon plasma formation and its signature, as well as some connections with the Universe evolution after Big Bang are discussed.

1 Introduction

In 1948 Freier and coworkers discovered, in the cosmic rays, the first relativistic nuclei [1], and in 1974 a common international effort detected the charmons.

A new begin for the classical Nuclear Physics was signed by the Relativistic Nuclear Physics and Quantum Chromodynamics. This new age for the fundamental hadronic matter leads to strong connections with the Astrophysics and Cosmology. Nowadays any cosmological scenario asks as "ingredients" dark matter, neutrino oscillations, multiquark systems, chaos and nuclear reactions at high energies. This work is devoted to the path from elementary particles to stars. The aspects that will be discussed are related to the searches for non-strange dibaryons in neutron-proton and nucleus-nucleus collisions at high energies and to the consequences of the dibaryon existence for the behaviour of the nuclear matter and the neutron stars.

2 Signals of the dibaryonic resonances in neutron-proton collisions

It is well known that the fundamental group of the Quantum Chromodynamics, SU(3), predicts as mutiplets of colour the multiquark systems. Among them, the dibaryon consisting of 6-8 quarks (fundamental and excited states,

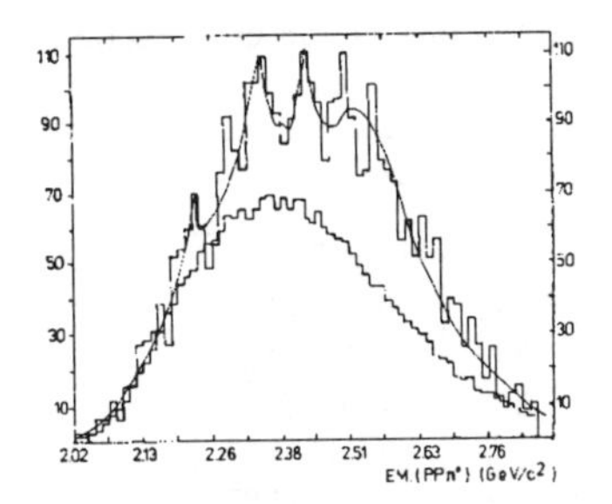

Figure 1. The integral ppπ^o effective mass distribution. The lower curve depicts the background contribution and the upper one the fitted curve.

respectively) represents the simple system. The analysis of the invariant mass spectra for NN, NNπ, NN2π was the common way to detect such a state, impossible to be construct as quantum numbers using the SU(3) group.

Our collaboration belonging to the Faculty of Physics, University of Bucharest performed, commonly with a group from JINR Dubna (Dr.Y.U.Trojan, Dr.V.I.Moroz, Dr.S.Ierusalimov), an experiment involving neutron-proton collisions. The neutron beam had high performances, namely: monochromatic energy of the neutron beam, with $\delta p/p \leq 3\%$; 7 neutron beam momenta, from 1.25 GeV/c up to 5.1 GeV/c. The detector was a 1m bubble chamber which permitted a dispersion in the invariant mass under 10 MeV/c^2 [2].

For the np reactions with 3 and 5 charged particles in the final state (for example: np$\rightarrow$ppπ^- (π^o), np$\rightarrow$np$\pi^+\pi^+\pi^-\pi^-$, respectively) the invariant mass spectra for different combinations NN, NNπ, NN2π indicated a number of peaks with the halfwidths less than 10 MeV/c^2. Fig.1 shows the ppπ^o invariant mass distribution for I=2 isospin states in the np$\rightarrow$pp$\pi^-\pi^o$ channel at 5.1 GeV/c [3]. The background curve contains, mainly, the peripheral and the baryon exchange contributions. Some Coulombian poles, the D(1232) and N* resonances are contained too. The fitted curve determines, as excited dibaryonic resonances, the signals with $\Gamma \prec 10$ MeV/c^2. The determined dibaryonic states are listed in Table I.

The same analysis identified a number of signals for isospin combinations with I=1 [4]. We point out here that for other incident momenta of the neutron beam (3.86 GeV/c, 4.36 GeV/c, respectively) the same signals were found. A number of 12 Breit-Wigner peaks belonging to the class of the non-strange dibaryonic resonances {DB$\rightarrow$NN (fundamental state), DB*$\rightarrow$NNπ, DB*$\rightarrow$NN2π (excited state with l=1,3)} were detected in this way.

Table 1. The masses, widths number of standard deviations and production cross sections for T=2 dibaryonic candidates.

Combination	Mass [MeV/c^2]	Γ [MeV/c^2]	NSD	$\sigma[\mu b]$
$pp\pi^+$	$2195 \pm 9 \pm 10$	37 ± 39	5.7	$16.01^{+17.29}_{-3.31}$
	$2316 \pm 12 \pm 18$	80 ± 15	4.6	$20.50^{+1.87}_{-0.41}$
	$2395 \pm 7 \pm 19$	32 ± 16	3.3	$14.72^{+3.34}_{-1.10}$
	$2447 \pm 4 \pm 21$	8 ± 2	3.2	$21.30^{+10.10}_{-3.2}$
$pp\pi^o$	$2234 \pm 8 \pm 12$	90 ± 17	4.9	$24.50^{+13.56}_{-5.4}$
	$2348 \pm 4 \pm 19$	53 ± 16	4.4	$34.30^{+13.90}_{-4.10}$
	$2428 \pm 5 \pm 21$	41 ± 16	3.7	$21.88^{+18.70}_{-2.01}$
	$2521 \pm 8 \pm 23$	101 ± 14	5.6	$40.40^{+18.48}_{-7.39}$
$pn\pi^+$	$2179 \pm 1 \pm 10$	6 ± 1	4.5	$21.27^{+12.08}_{-4.50}$
	$2389 \pm 1 \pm 19$	1 ± 5	4.2	$27.52^{+14.50}_{-1.10}$
	$2513 \pm 17 \pm 23$	1 ± 1	2.2	$46.80^{+12.05}_{-4.45}$
$pn\pi^-$	$2310 \pm 3 \pm 18$	23 ± 24	3.9	$24.30^{+12.60}_{-3.40}$
	$2476 \pm 6 \pm 22$	11 ± 4	3.1	$41.60^{+15.90}_{-2.40}$
	$2539 \pm 18 \pm 24$	1 ± 1	3.1	$46.80^{+21.30}_{-4.80}$
	$2620 \pm 8 \pm 26$	1 ± 1	2.2	$47.20^{+19.60}_{-5.10}$
	$2645 \pm 1 \pm 27$	2 ± 1	2.1	$32.13^{+12.57}_{-2.90}$

Our suppositions were compared with a simple model elaborated for such a multiquark system, namely: a diquark-four quarks cluster $[(q^2)_{3*} - (q^4)_3]$ interacting by colour magnetic forces inside a spherical bag. The final form led to a mass formula depending to the main quantum numbers: spin (S), isospin (I), orbital momentum (l), colour magnetic operator (F_c), namely:

$$M_o(R) = (4\pi/3)BR^2 - Z_oR + (N\alpha_n)/R + (\alpha_c M_{nn})/(R\Delta), \tag{1}$$

where R is the bag radius, B = 59.2 MeV$\cdot$Fm^{-2}, α_n = 403 MeV$\cdot$Fm, Z_o = 363 MeV$\cdot$Fm, α_c = 22, and

$$\Delta = (-1/3)N(6-N) + (1/3)S^2 + I^2 + F_c^2/2. \tag{2}$$

Here F_c is the Casimir operator for the magnetic colour forces, and R = $roN^{1/3}$, with N the number of quarks. Minimizing the bag radius the fundamental state is obtained.

For the excited dibaryons a Regge representation, own to the SU(3) hadronic group, was extended. The following relation was obtained:

$$M^*(l) = M_o^2 + l/\alpha, \tag{3}$$

Here, $1/\alpha = 1.1[(3f_c^2)]^{1/2}$.

The masses of the main dibaryonic resonances were compared with the predicted values defined for I, S, l quantum numbers. The comparison results are presented in the Table II. A satisfactory agreement in the limit of the experimental errors is observed. The order of magnitude for the half-widths ($\approx$10 MeV/c^2) were obtained too, using the overlap of the eigen functions for diquark-quarks clusters. Because this model uses only the SU(3) constants, the fits procedures are rejected and the confidence in our hypothesis increases.

3 The dibaryonic signals in the nucleus-nucleus reactions

The dibaryon presence in the so-called resonance matter could establish a way to easy understand the phase transitions in nuclear matter. Unfortunately, the classical methods based on the final decay identification are unrealistic and because that the experimental evidences for hadronic resonances are rare.

The study of the negative pion emission in He-A_T and C-A_T reactions at 4.5 A GeV/c performed with the SKM 200 Spectrometer from JINR Dubna, in a collaboration University of Bucharest-JINR Dubna, permitted the search for the possible signatures of the nonstrange dibaryons. Table III summaries the main features for 11 nucleus-nucleus collisions [4]. The relative small numbers of participant protons, Q, and participant nucleons, Q_N, favour the presence of resonances. Some resonances were identified [4]. The most present is the $\Delta(1232)$ resonance.

Figure 2 shows the attempt to identify the pp dibaryonic fundamental state (M_o = 1.96 GeV/c^2). The invariant mass spectrum was constructed inside of the maximum of the correlation function between $\Delta p = |p_1 - p_2|$ and $\Delta\theta = |\theta_1 - \theta_2|$, where p_1, p_2 and θ_1, θ_2 are the momenta and the angles, respectively, for the two protons. The He-Li collisions were used for Fig.2. The result is in agreement with the presence of such a state in nucleus-nucleus collision [5].

The second proof tried by us was based on the $\pi^-\pi^-$ invariant masses which carry the fingerprints of the excited dibaryons . Inspecting Fig.3, anomalous values in the low range of $M(\pi^-\pi^-)$ masses are observed. The contribution of the main dibaryonic nonstrange resonances detected as excited states with one or two π^- in the decay channels is the single possibility to explain it [5]. A complete fit evaluated at 20% the contribution of the nonstrange dibaryons for He-Li reaction. This percentage decreases to 4% for He-Pb reaction because the strong interaction with the spectator nucleons is noticeable. The two kind of evidences argued on the presence of the nonstrange dibaryons as products of the nucleus-nucleus reactions.

Table 2. The predicted mass values for dibaryonic resonances and their quantum numbers, in comparison with the experimental candidates.

		l=0			l=1	
(i:s)	M_{model}	M_{exp}	Contrib.	M_{model}	M_{exp}	Contrib.
(0:1)	1930	1952±8	7.56	2196	2195±19	0.003
		1948±8	5.06		2180±11	2.12
(1:0)	2008	1980±9	9.68	2246	2234±20	0.35
(1:1)	2047	2046±12	0.007	2301	2316±30	0.25
		2047±11	0.0			
(1:2,1,0)	2080	2083±14	0.05	2336	2348±23	0.27
(1:2)	2123	2126±16	0.04	2371	2389±20	0.81
(2,1,0:1)	2165	2146±17	1.25	2406	2395±26	0.18
(2,1,0:2,1,0)	2204	2230±24	1.17	2441	2447±25	0.06
		2225±27	0.60			
(2,1,0:3,2,1)	2283	2284±24	0.002	2511	2539±42	0.44
(3,2,1:2,1,0)	2240			2632		

		l=2	
(i:s)	M_{model}	M_{exp}	Contrib.
(0:1)	2432	2428±26	0.02
(1:0)	2496	2476±28	0.51
(1:1)	2528	2521±31	0.05
		2513±40	0.78
(1:2,1,0)	2560	2539±42	1.36
(1:2)	2592		
(2,1,0:1)	2624	2620±34	0.01
(2,1,0:2,1,0)	2656	2645±28	0.15
(2,1,0:3,2,1)	2720		
(3,2,1:2,1,0)	2849		

4 Possible consequences of the dibaryonic states in the high energies processes

a. Theoretical studies and experimental signals at high energies add to the phase diagrams of the hadronic matter new intermediate states between hadronic plasma and quark-gluon plasma. The dibaryonic component considered previously suggests the possibility to have a diquark plasma phase

Table 3. The average values of different general interesting physical quantities for 11 inelastic and central nucleus-nucleus collisions at 4.5 A GeV/c.

	He-Li	He-C	He-Ne	He-Al	He-Cu	He-Pb
N_{ev}	4026	1099	312	330	804	1048
$\sigma_{in}\ [mb]$	320±15	450±20	615±40	720±30	1150±50	2400±170
Q_n	4.0±1.0	5.8±0.6	7.2±0.6	7.8±0.6	12.5±1.1	24.9±2.5
$p_T\ [MeV/c]$	241±3	238±4	230±5	228 6	227±6	204±4
$T_\pi\ [MeV]$	69.0±2.0	71.0±3.0	76.0±3.3	77.0±4.1	78.0±4.0	94.0±3.0
N_{ev}	1658	927	221	573	522	930
$\sigma_{in}\ [mb]$	120±12	180±15			663±50	1840±160
Q_n	5.6±0.6	9.4±0.4	12.2±1.6	13.8±1.4	18.0±1.1	37.1±3.0
$p_T\ [MeV/c]$	198±3	195±5	189±5		186±6	167±4
$T_\pi\ [MeV]$	98.6±2.4	100.5±3.7	105.6±4.0		107.6±4.9	123.6±3.7

	C-C	C-Ne	C-Cu	C-Zr	C-Pb
N_{ev}	2033	2224	1170	883	784
$\sigma_{in}\ [mb]$	780±30	1040±60	1700±90	2025±120	3025±160
Q_n	8.4±0.4	10.2±0.8	19.6±1.7	22.0±1.8	38.2±2.0
$p_T\ [MeV/c]$	236±6	232±8	220±4		
$T_\pi\ [MeV]$	72.2±3.8	76.2±4.4	82.8±2.7		
N_{ev}	1663	2199	2154	650	2255
$\sigma_{in}\ [mb]$	38±6	98±8	356±50	530±50	1020±100
Q_n	15.6±0.6	17.8±1.8	42.8±2.2	49.3±2.0	87.0±3.2
$p_T\ [MeV/c]$	194±6	190±7	181±4		
$T_\pi\ [MeV]$	101.7±4.6	103.6±6.8	112.3±3.3		

between the hadronic and quark-gluon plasmas. This diquark plasma can be formed from the dissociation of the dibaryonic components. In our model, presented previously, the dibaryon contains a diquark string. Equating the thermodynamic phase relations a diquark phase could be possible. In Fig.4 the diquark plasma, situated between hadronic and quark-gluon plasmas, was introduced in the phase diagrams for different values estimated for the diquark mass. This hypothesis could drastically modify the passage towards the quark-gluon plasma phase, reducing or screening their specific signatures. New experimental data and new good statistics will decided on this hypothesis.

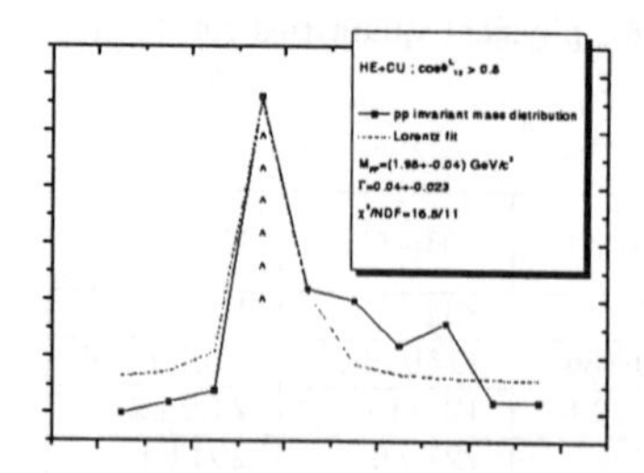

Figure 2. The pp effective mass distribution and the Lorentz fit for He+Cu interaction at 4.5 A GeV/c incident momentum. The cosinus between the two proton momenta in

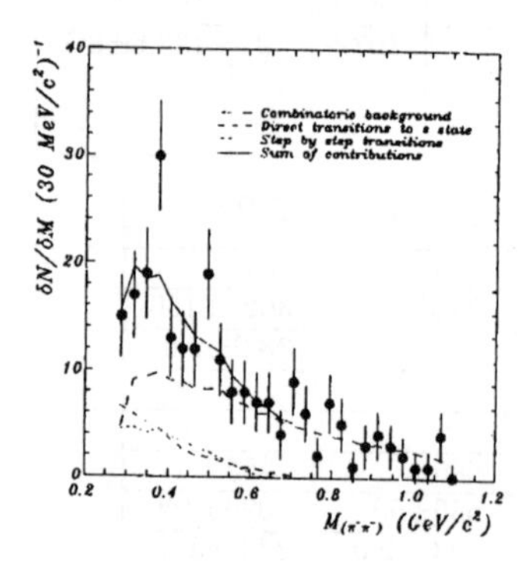

Figure 3. Contributions to the negative pion-negative pion effective masses.

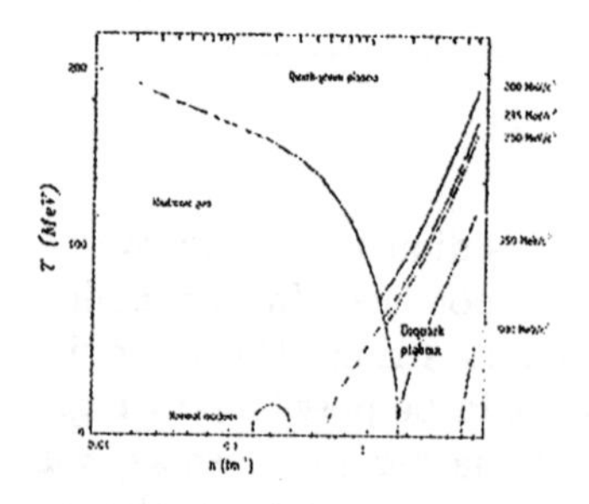

Figure 4. Phase diagram including the equilibrium condition between quark-gluon plasma and the diquark plasma for different quark masses values.

Table 4. The variation of the rest masses with the nuclear matter density.

$\rho\left(fm^{-3}\right)$	R(fm)	M(GeV/c^2)
0.17	1.276	2.061
0.30	1.245	2.106
0.90	0.908	2.924
1.40	0.700	3.815

b. The structure of the neutron stars is compatible with a hard core of dibaryons. A number of papers [6] and references quoted there pertinently include the strange dibaryons with heavy rest masses. The neutron-neutron dibaryons can be also a main component of the neutron star. The specific pressure, comparable with the value supposed in model, could synthetize a cover of n-n dibaryons.

Even the three masses detected by us can assume the gravitational stability: the dibaryon states determine the reduction of the degrees of freedom, N, with 2 units. The main consequence is the increase of the rest mass by the increase of the pressure and the decrease of the bag radius. The change of the rest masses with the density is evidenced by an example. For the fundamental state $M_o = 2.06$ GeV/c^2 the results included in the Table IV are obtained. The existence of the nonstrange dibaryons could define new meanings at the cosmological scale justifying the strong connection particles-nuclei-stars.

5 Final remarks

From the beginning the Relativistic Nuclear Physics has been defined as a bridge among Nuclear Physics, Particle Physics and Cosmology [1]. The relativistic and ultrarelativistic nuclear collisions can offer the experimental conditions for quark-gluon plasma formation and observation [7]. The quark-gluon plasma formation could give interesting information on different time moments after the "Big Bang" [[8], [9]]. To compare the physical processes that have been appeared after "Big Bang" with those related to the quark-gluon plasma formation in nucleus-nucleus collisions at high energies it is necessary to know the specific thermodynamic conditions for each stage, as well as the time evolution of the considered systems. Different experimental results on temperature, density and time evolution of the systems obtained in relativistic nuclear collisions [[10], [11]] or from cosmological observations [[9], [12]] were used to compare the physical processes after the "Big Bang" with those from the nuclear matter formed in relativistic nuclear collisions. For different time moments from the evolution of the two great global phenomena similar thermodynamic parameters were found [13].

Using different thermodynamic approaches, quark models and a geometric model [[14]- [18]] estimations of the mentioned thermodynamic parameters have been performed. For example, in nucleus-nucleus collisions at 4.5 A GeV/c temperatures between 95 MeV and 135 MeV and densities around $4\rho_o$ were obtained [[10], [16], [17]]. The corresponding time in the evolution after "Big Bang" is around 10^{-5} s, in good agreement with different cosmological scenarios [[8], [9], [12]]. Similar agreements are obtained for other collisions [[10], [16], [17]]. There is the possibility to obtain interesting and relevant experimental information on the Universe evolution using the experimental results obtained in relativistic nuclear collisions.

Acknowledgments

We are indebted to Radu Zaharia, Daniel Felea, and Ion-Sorin Zgura for their active contributions to the nuclei-nuclei reactions study.

References

1. D.K. Scott, .Part.Nucl.Phys. **IV**, 5 (1980).
2. A. Abdivaliev, C. Besliu, F. Cotorobai, S. Gruia, A.P. Ierusalimov, V.I. Moroz, A.V. Nikitin and Yu.A. Troyan, .Phys.B **99**, 445 (1975).
3. C. Besliu, L. Popa and V. Popa, .Phys.G:Nucl.Part.Phys **18**, 807 (1992).
4. Al. Jipa, C. Besliu, M. Iosif and R. Zaharia, Nuovo Cimento **112A**, 179 (1999).
5. C. Besliu, L. Popa, V. Popa and V. Topor-Pop, .Phys.G:Nucl.Part.Phys **19**, 1831 (1993).
6. J. de Pacheco, S. Stoica, F. Thevenin and N. J. Horvath, *Phys. Rev.* D **59**, 27303 (1999).
7. B. Müller, .Prog.Phys. **58**, 611 (1995).
8. S. Weinberg in *Primele trei minute ale Universului*, (Editura Politica, colection "Idei contemporane", Bucuresti, 1984).
9. H. Reeves, .Rep. **201**, 335 (1991).
10. Al. Jipa .Phys.G:Nucl.Part.Phys **22**, 231 (1996).
11. I.G. Bearden *et al* (NA44 Collaboration), .Rev.Lett. **78**, 2080 (1997).
12. H.G.Rebel in *Course of Astrophysics* - taught to the students of the Faculty of Physics Bucharest, october 1997.

13. C.Besliu *et al*, National Physics Conference, Constanta, Romania, 16-18.09.1998.
14. H. Meyer-Ortmans, .Mod.Phys. **68**, 473 (1996).
15. D.H.E. Gross, .Rep **279**, 119 (1997).
16. C.Besliu, Al.Jipa, .J.Phys. **33**, 1011 (1992); Nuovo Cimento **A106**, 317 (1993).
17. C.Besliu *et al*, .J.Phys. **43**, 489 (1998); International Nuclear Physics Conference, Paris, France, 24-28.08.1998.
18. L.McLerran, Preprint FERMILAB, FERMILAB-Conf 84/101-T(1984)

KILLING-YANO TENSORS ON MANIFOLDS ADMITTING "HIDDEN" SYMMETRIES

MIHAI VISINESCU

Department of Theoretical Physics, Horia Hulubei Institute of Physics and Nuclear Engineering. P.O.Box. MG-6, Magurele, Bucharest, Romania
E-mail: mvisin@theor1.theory.nipne.ro

Relationships among the existence of Stäckel-Killing and Killing-Yano tensors are investigated for four dimensional Riemannian manifolds. The general results are applied to the extended Euclidean Taub-NUT space.

1 Introduction

It is well known that space-time isometries give rise to constants of motion along geodesics. However it is worth to mention that not all conserved quantities arose from isometries of the manifolds and associated Killing vector fields. Other integrals of motion are related to "hidden" symmetries of the manifold, which manifest themselves as tensor of valence $n > 1$, satisfying a generalized Killing condition: [1]

$$K_{(\mu_1 \dots \mu_r ; \mu_{r+1})} = 0. \tag{1}$$

They are usually referred to as Stäckel-Killing tensors. The conserved quantities along geodesics are homogeneous functions in momentum p_μ of degree r, and which commute with the Hamiltonian

$$H = \frac{1}{2} g^{\mu\nu} p_\mu p_\nu \tag{2}$$

in the sense of Poisson brackets.

An illustration of the existence of extra conserved quantities is provided by the Taub-NUT geometry which is involved in many modern studies in physics. The Kaluza-Klein monopole was obtained by embedding the Taub-NUT gravitational instanton into five-dimensional Kaluza-Klein theory. On the other hand, in the long-distance limit, neglecting radiation, the relative motion of two monopoles is described by the geodesics of this space. [2,3] For the geodesic motions in the Taub-NUT space, there is a conserved vector, analogous to the Runge-Lenz vector of the Kepler type problem, whose existence is rather surprising in view of the complexity of the equations of motion. [4,5,6,7] The Runge-Lenz vector is quadratic in four-velocities and its components are Stäckel-Killing tensors and they can be expressed as symmetrized products of the Killing-Yano tensors. [8,5,9,10,11]

The Killing-Yano tensors play an important role in the models for relativistic spin one half particles involving anticommuting vectorial degrees of freedom, usually called the spinning particles. [12] In the spinning case the generalized Killing equations are more involved and new procedures have been conceived. [12,13,11] In particular, if the Killing tensors can be written in terms of Killing-Yano tensors (and that is the case of the Taub-NUT space), the generalized Killing equations can be solved explicitly in a simple, closed form.

In the last time, Iwai and Katayama [14,15,16,17] extended the Taub-NUT metric so that it still admits a Kepler-type symmetry. This class of metrics, of course, includes the original Taub-NUT metric.

In what follows we shall investigate if the Stäckel-Killing tensors involved in the conserved Runge-Lenz vector of the extended Taub-NUT metrics can be also expressed in terms of Killing-Yano tensors.

The relationship between Killing tensors and Killing-Yano tensors has been investigated to the purpose of the Lorentzian geometry used in general relativity. [18,19] In the next section we re-examine the conditions that a Killing tensor of valence 2 be the contracted product of a Killing-Yano tensor of valence 2 with itself. The procedure is quite simple and devoted to the Riemannian geometry appropriate to Euclidean Taub-NUT metrics.

In Section 3 we show that in general the Killing tensors involved in the Runge-Lenz vector cannot be expressed as a product of Killing-Yano tensors. The only exception is the original Taub-NUT metric.

Our comments and concluding remarks are presented in Section 4.

2 The relationship between Killing tensors and Killing-Yano tensors

Suppose we are given a 4−dimensional Riemannian manifold M and a metric $g_{\mu\nu}(x)$ on M in local coordinates x^μ. The distance ds between two infinitesimally nearby points x^μ and $x^\mu + dx^\mu$ is given by

$$ds^2 = g_{\mu\nu}(x)dx^\mu dx^\nu \tag{3}$$

where the $g_{\mu\nu}(x)$ are the components of a symmetric covariant second-rank tensor. We decompose the metric into vierbeins $e^a{}_\mu$

$$g_{\mu\nu} = \eta_{ab}e^a{}_\mu e^b{}_\nu \tag{4}$$

where $\eta_{ab} = \delta_{ab}, a, b = 0, 1, 2, 3$.

Let Λ^2 be the space of two-forms $\Lambda^2 := \Lambda^2 T^*(\mathbb{R}^4 - \{0\})$. We remind that a tensor $f_{\mu_1 \dots \mu_r}$ is called a Killing-Yano tensor of valence r if it is totally antisymmetric and it satisfies the equation

$$f_{\mu_1 \ldots (\mu_r;\lambda)} = 0. \tag{5}$$

Let us define self-dual and anti-self dual bases for Λ^2 using the vierbein one-forms e^a: [20]

$$basis\ of \quad \Lambda^2_\pm = \begin{cases} \lambda^1_\pm = e^0 \wedge e^1 \pm e^2 \wedge e^3 \\ \lambda^2_\pm = e^0 \wedge e^2 \pm e^3 \wedge e^1, \\ \lambda^3_\pm = e^0 \wedge e^3 \pm e^1 \wedge e^2 \end{cases} \qquad *\lambda^i_\pm = \pm\lambda^i_\pm \quad . \tag{6}$$

Let Y be a Killing-Yano tensor of valence 2 and $*Y$ its dual. The symmetric combination of Y and $*Y$ is a self-dual two-form

$$Y + *Y = \sum_{1=1,2,3} y_i \lambda^i_+ \tag{7}$$

while their difference is an anti-self-dual two-form.

$$Y - *Y = \sum_{1=1,2,3} z_i \lambda^i_- . \tag{8}$$

Let us suppose that a Stäckel-Killing tensor $K_{\mu\nu}$ can be written as the contracted product of a Killing-Yano tensor $Y_{\mu\nu}$ with itself:

$$K_{\mu\nu} = Y_{\mu\lambda} \cdot Y^\lambda{}_\nu = (Y^2)_{\mu\nu} \quad , \mu, \nu = 0,1,2,3. \tag{9}$$

An explicit evaluation shows that :

$$K + \frac{1}{16}\left[\sum_i (y_i^2 - z_i^2)\right]^2 K^{-1} + \frac{1}{2}\sum_i (y_i^2 + z_i^2) \cdot \mathbf{1} = 0. \tag{10}$$

On the other hand the Killing tensor K is symmetric and it can be diagonalized with the aid of an orthogonal matrix. Its eigenvalues satisfy an equation of degree two:

$$\lambda_\alpha{}^2 + \frac{1}{2}\sum_i (y_i^2 + z_i^2)\lambda_\alpha + \frac{1}{16}\left[\sum_i (y_i^2 - z_i^2)\right]^2 = 0 \tag{11}$$

with at most two distinct roots.

In conclusion a Stäckel-Killing tensor K which can be written as the square of a Killing-Yano tensor has at the most two distinct eigenvalues.

3 Generalized Taub-NUT metrics

For a special choice of coordinates the generalized Euclidean Taub-NUT metric considered by Iwai and Katayama [14,15,16,17] takes the form:

$$ds_G^2 = f(r)[dr^2 + r^2 d\theta^2 + r^2 \sin^2\theta \, d\varphi^2] + g(r)[d\chi + \cos\theta \, d\varphi]^2. \tag{12}$$

where $r > 0$ is the radial coordinate of $\mathbb{R}^4 - \{0\}$, the angle variables $(\theta, \varphi, \chi), (0 \leq \theta < \pi, 0 \leq \varphi < 2\pi, 0 \leq \chi < 4\pi)$ parameterize the unit sphere S^3, and $f(r)$ and $g(r)$ are arbitrary functions of r.

Spaces with a metric of the form above have an isometry group $SU(2) \times U(1)$. The four Killing vectors are

$$D_A = R_A^\mu \, \partial_\mu, \qquad A = 0, 1, 2, 3. \tag{13}$$

D_0 which generates the $U(1)$ of χ translations, commutes with the other Killing vectors. In turn the remaining three vectors, corresponding to the invariance of the metric (12) under spatial rotations ($A = 1, 2, 3$), obey an $SU(2)$ algebra.

Let us consider geodesic flows of the generalized Taub-NUT metric which has the Lagrangian L on the tangent bundle $T(\mathbb{R}^4 - \{0\})$. Since χ is a cyclic variable

$$q = g(r)(\dot{\theta} + \cos\theta \dot{\varphi}) \tag{14}$$

is a conserved quantity. This is known in the literature as the "relative electric charge".

Taking into account this cyclic variable, the dynamical system for the geodesic flow on $T(\mathbb{R}^4 - \{0\})$ can be reduced to a system on $T(\mathbb{R}^3 - \{0\})$. The reduced system admits manifest rotational invariance, and hence has a conserved angular momentum:

$$\vec{J} = \vec{r} \times \vec{p} + q \frac{\vec{r}}{r}. \tag{15}$$

where $\vec{r}$ denotes the three-vector $\vec{r} = (r, \theta, \varphi)$ and $\vec{p} = f(r)\dot{\vec{r}}$ is the mechanical momentum.

If $f(r)$ and $g(r)$ are taken to be

$$f(r) = \frac{4m + r}{r} \quad , \quad g(r) = \frac{16m^2 r}{4m + r} \tag{16}$$

the metric ds_G^2 becomes the original Euclidean Taub-NUT metric. The Taub-NUT geometry also possesses four Killing-Yano tensors of valence 2. [5] The first three are

$$f_i = 8m(d\chi + \cos\theta d\varphi) \wedge dx_i - \epsilon_{ijk}(1 + \frac{4m}{r})dx_j \wedge dx_k,$$
$$D_\mu f^\nu_{i\lambda} = 0, \quad i = 1,2,3. \tag{17}$$

They are covariantly constant (with vanishing field strength), mutually anticommuting and square the minus unity Thus they are complex structures realizing the quaternion algebra. Indeed, the Taub-NUT manifold defined by (12) and (16) is hyper-Kähler. The fourth Killing-Yano tensor is

$$f_Y = 8m(d\chi + \cos\theta d\varphi) \wedge dr$$
$$+4r(r+2m)(1+\frac{r}{4m}) \sin\theta d\theta \wedge d\varphi \tag{18}$$

and has only one non-vanishing component of the field strength

$$f_{Y\,r\theta;\varphi} = 2(1+\frac{r}{4m})r\sin\theta. \tag{19}$$

In the original Taub-NUT case there is a conserved vector analogous to the Runge-Lenz vector of the Kepler-type problem:

$$\vec{K} = \frac{1}{2}\vec{K}_{\mu\nu}\,\dot{x}^\mu\dot{x}^\nu = \vec{p}\times\vec{j} + \left(\frac{q^2}{4m} - 4mE\right)\frac{\vec{r}}{r} \tag{20}$$

where the conserved energy E, from eq. (2), is

$$E = \frac{\vec{p}^{\,2}}{2f(r)} + \frac{q^2}{2g(r)}. \tag{21}$$

The components $K_{i\mu\nu}$ involved with the Runge-Lenz type vector (20) are Killing tensors and they can be expressed as symmetrized products of the Killing-Yano tensors f_i (17) and f_Y (18): [10,11]

$$K_{i\mu\nu} - \frac{1}{8m}(R_{0\mu}R_{i\nu} + R_{0\nu}R_{i\mu}) = m\left(f_{Y\mu\lambda}f_i{}^\lambda{}_\nu + f_{Y\nu\lambda}f_i{}^\lambda{}_\mu\right). \tag{22}$$

Returning to the generalized Taub-NUT metric, on the analogy of eq.(20), Iwai and Katayama [14,15,16,17] assumed that in addition to the angular momentum vector there exist a conserved vector $\vec{R}$ of the following form:

$$\vec{R} = \vec{p}\times\vec{J} + \kappa\frac{\vec{r}}{r} \tag{23}$$

with an unknown constant κ.

It was found that the metric (12) still admits a Kepler type symmetry (23) if the functions $f(r)$ and $g(r)$ take, respectively, the form

$$f(r) = \frac{a+br}{r} \quad , \quad g(r) = \frac{ar+br^2}{1+cr+dr^2} \tag{24}$$

where a, b, c, d are constants. The constant κ involved in the Runge-Lenz vector (23) is

$$\kappa = -aE + \frac{1}{2}cq^2 \tag{25}$$

where E is the energy(21). If $ab > 0$ and $c^2 - 4d < 0$ or $c > 0, d > 0$, no singularity of the metric appears in $\mathbb{R}^4 - \{0\}$. On the other hand, if $ab < 0$ a manifest singularity appears at $r = -a/b$. [15]

Our task is to investigate if the components of the Runge-Lenz vector (23) can be the contracted product of Killing-Yano tensors of valence 2. Taking into account eq.(22) from the original Taub-NUT case, in general, a component K_i of the Runge-Lenz vector could not be expressed directly as a symmetrized product of Killing-Yano tensors. The components K_i can be combined with other trivial Stäckel-Killing tensors formed from symmetrized pairs of Killing vectors to get the appropriate tensor which has to be decomposed in products of Killing-Yano tensors.

In order to use the results from the previous section, we shall write the symmetrized product of two different Killing-Yano tensors f' and f'' as a the contracted product of $f' + f''$ with itself, extracting adequately the contribution of f'^2 and f''^2. We note that the contracted products of f'^2 and f''^2 are not new conserved quantities, being expressed in terms of the scalar conserved quantities $E, q, \vec{J}^2$.

Therefore we are looking for a general linear combination between a component K_i of the Runge-Lenz vector and symmetrized pairs of Killing vectors which has at the most two distinct eigenvalues. A long and detailed evaluation shows that the components K_i only in the combination (22) with Killing vectors, just like in the original Taub-NUT metric, could be written as the contracted product of Killing-Yano tensors provided that

$$d = \frac{c^2}{4}. \tag{26}$$

Hence the constants involved in the functions f, g are constrained, restricting accordingly their expressions.

Finally the condition stated for a Stäckel-Killing tensor to be written as the square of a skew symmetric tensor in the form (9) must be supplemented with (5) which defines a Killing-Yano tensor. To verify this last condition we shall use the Newman-Penrose formalism for Euclidean signature. [21] We introduce a tetrad which will be given as an isotropic complex dyad defined by the vectors l, m together with their complex conjugates subject to the normalization conditions

$$l_\mu \bar{l}^\mu = 1, \quad m_\mu \bar{m}^\mu = 1 \tag{27}$$

with all others vanishing. Therefore the one-forms e^a will be given by

$$e^a = e^a{}_\nu dx^\nu = \{\bar{l}, l, \bar{m}, m\} \tag{28}$$

and the metric is expressed in the form

$$ds^2 = l \otimes \bar{l} + \bar{l} \otimes l + m \otimes \bar{m} + \bar{m} \otimes m. \tag{29}$$

For a Stäckel-Killing tensor K with two distinct eigenvalues one can choose the tetrad in such that

$$K_{ij} = 2\lambda_1^2 l_{(i}\bar{l}_{j)} + 2\lambda_2^2 m_{(i}\bar{m}_{j)}. \tag{30}$$

In this case, the form of the would-be Killing-Yano tensors are

$$f_{ij} = 2\lambda_1 l_{[i}\bar{l}_{j]} + 2\lambda_2 m_{[i}\bar{m}_{j]}. \tag{31}$$

Again, a standard evaluation shows that the above quantity is a Killing-Yano tensor only if

$$c = \frac{2b}{a}. \tag{32}$$

With this constraint, together with (26), the extended metric (12) coincides, up to a constant factor, with the original Taub-NUT metric on setting $4m = a/b$.

4 Concluding remarks

The aim of this paper is to prove that only the original Taub-NUT space admits Killing-Yano tensors and any extension of it, in spite of the existence of Stäckel-Killing tensors, has not this property.

The extended Taub-NUT metrics are not Ricci flat and, consequently, not hyper-Kähler. On the other hand the existence of the Killing-Yano tensors f_i is correlated with the hyper-Kähler, self-dual structure of the metric.

A "hidden" symmetry is encapsulated in a Stäckel-Killing tensor of valence $r > 1$. The generalized Killing equations on spinning spaces including a Stäckel-Killing tensor is more involved. Assuming that the Stäckel-Killing tensors can be written as symmetrized products of pairs of Killing-Yano tensors, the evaluation of the spin corrections is feasible. [12,10,11,13] In fact the antisymmetric Killing-Yano tensors are the natural geometric objects which could be contracted with even products of the Grassmann variables.

Therefore, if the Killing-Yano tensors are missing, it is not evident how to compute the spin corrections to the conserved quantities corresponding to "hidden" symmetries or even if these corrections exist.

Summing up, we believe that the relation between the Stäckel-Killing and Killing-Yano tensors could be fruitful and that it should be pursued beyond the results we have presented.

References

1. G.W. Gibbons and C.A.R. Herdeiro, *hep-th/9906098.*
2. N.S. Manton, *Phys. Lett.* B **110**, 54 (1985); *id,* **B154**, 397 (1985); *id,* (E) **B157**, 475 (1985).
3. M.F. Atiyah and N. Hitchin, *Phys. Lett.* A **107**, 21 (1985).
4. G.W. Gibbons and N.S. Manton, *Nucl. Phys.* B **274**, 183 (1986).
5. G.W. Gibbons and P.J. Ruback, *Phys. Lett.* B **188**, 226 (1987); *Commun. Math. Phys.* **115**, 267 (1988).
6. L.Gy. Feher and P.A. Horvathy, *Phys. Lett.* B **182**, 183 (1987): *id,* (E) **B188**, 512 (1987).
7. B. Cordani, L.Gy. Feher and P.A. Horvathy, *Phys. Lett.* B **201**, 481 (1988).
8. K. Yano, *Ann. Math.* **55**, 328 (1952).
9. J.W.van Holten, *Phys. Lett.* B **342**, 47 (1995).
10. D. Vaman and M. Visinescu, *Phys. Rev.* D **57**, 3790 (1998).
11. D. Vaman and M. Visinescu, *Fortschr. Phys.* **47**, 493 (1999).
12. G.W. Gibbons, R.H. Rietdijk and J.W. van Holten, *Nucl. Phys.* B **404**, 42 (1993).
13. J.W.van Holten, *gr-qc/9910035.*
14. T. Iwai and N. Katayama, *J. Geom. Phys.* **12**, 55 (1993).
15. T. Iwai and N. Katayama, *J. Phys. A: Math. Gen.* **27**, 3179 (1994).
16. T. Iwai and N. Katayama, *J. Math. Phys.* **35**, 2914 (1994).
17. Y. Miyake, *Osaka J. Math.* **32**, 659 (1995).
18. C.D. Collinson, *Int. J. Theor. Phys.* **15**, 311 (1976).
19. W. Dietz and R. Rüdinger, *Proc. R. Soc. Lond.* A **375**, 361 (1981).
20. T. Eguchi, P.B. Gilkey and A.J. Hanson, *Phys. Rep.* **66**, 213 (1980).
21. A.N. Aliev and Y. Nutku, *Class. Quant. Grav.* **16**, 189 (1999); *gr-qc/9805006.*

$1/N$-EXPANSION AND FLUCTUATIONS IN ISOTROPIC LATTICE SYSTEMS

N. ANGELESCU
Department of Theoretical Physics, IFIN-HH
E-mail: nangel@theor1.theory.nipne.ro

We consider classical and quantum n-vector lattice models with short-range isotropic interactions, suited for the description of ferromagnets and ferroelectric anharmonic crystals. The approximation method, named $1/n$-expansion, consisting in deriving complete asymptotic series of the equilibrium state around its $n = \infty$ limit is elaborated with full mathematical rigour. The Feynman-Kac formula allows to view the quantum model as a classical model for Brownian paths. In both cases, the components are decoupled in the $n = \infty$ limit, i.e. the limit state is an infinite product of Gaussian distributions with self-consistently defined covariance operator on the corresponding space (known as the spherical-, and Hartree-Fock-, approximations, respectively), and all higher order corrections to the correlations ivolving a finite number of components can be systematically calculated in terms of one-component Gaussian moments. However, for observables involving all components, different order corrections may add up with nontrivial effects. For instance, the joint distribution of the local fluctuations is asymptotically close to a Gaussian distribution *different*, in the ordered phase, from the spherical or Hartree-Fock approximation. In particular, the large distance behaviour of their correlations is drastically changed, leading to different degree of abnormality of the global fluctuations at the critical point.

1 INTRODUCTION

I am honoured by the invitation to contribute this anniversary session. The Institute in Magurele, whatever its name during history, has been my affiliation for 35 years. However, as I belong to a small group of people working in condensed matter theory and statistical physics, somewhat singular among so many nuclear physicists and engineers, and often overlooked in spite of its achievements, the invitation came up as a surprise.

This session is taking place in hard times for the Institute, meaning that the society we live in seems uninterested in developing, and is certainly unwilling to support, science. Let me say that we, theoretical physicists, are in a better position than experimentalists, as almost all we have to loose from lack of financing is our salaries, which, as you well know, mean little money. I find remarkable that, in spite of the unfriendly environment, the Institute survives, meaning that there are still people interested in, and able to read and produce, good physics. On this occasion, and having also in mind the time left to my retirement, I would like to wish the Institute to survive for at least another decade!

***

I want to review here a piece of work done with M.Bundaru and G.Costache [1], [2], and recently extended to the quantum case in collaboration with A.Verbeure (K.U.Leuven) and V.A.Zagrebnov (C.P.T. Marseille) [3], [4]. Some 30 years ago, we managed to "sell" a mathematical work on the spectral properties of the linear Boltzmann operator as nuclear reactor theory. Now, as I grew older and wiser, I shall not even try to make our work look as "Advances in Nuclear Physics". I prefer thinking that our nuclear physicists started to keep an open eye to satistical physics, as, in my opinion, it is a right thing to do.

The talk is devoted to a method of approximation for the equilibrium properties of isotropic n-vector lattice systems. The latter are composed of identical subsystems, each living at a site x of the cubic lattice $\mathbf{Z}^d$ and described by a position vector $\vec{q}_x \in \mathbf{R}^n$. The subsystems are interacting among themselves and with the lattice site via potentials invariant to a simultaneous rotation of their position vectors. In fact, the main feature of interest here is that the components of the vectors $\vec{q}_x = (q_x^\alpha)_{\alpha=1,\ldots,n}$ enter in a symmetric way, i.e. that the potential energy associated to a configuration is invariant under permutations of the axes of $\mathbf{R}^n$. Typical examples are:

a) *the classical Heisenberg model* used for describing magnetic ordering in crystals, where $\vec{q}_x$ is a vector of fixed length in $\mathbf{R}^3$ having the meaning of spin orientation,

b) *the quantum anharmonic crystal model* used to describe structural phase transitions in ferroelectrics, where $\vec{q}_x \in \mathbf{R}^3$ denotes the displacement of an ion from its equilibrium position x.

As with usual perturbation theory in Quantum Mechanics or ε-expansion in Renormalisation Group, one includes the given model in a larger class of models indexed by some parameter, in which an exactly solvable limit is known, and calculates corrections as series expansions around that limit. *In the case at hand, the parameter is the dimension n of the vector $\vec{q}_x$, the limit is $n = \infty$, and the corrections are given as a $1/n$-expansion.*

The solvability of the limit is assisted by very general theorems: de Finetti's theorem in the classical case, adapted by Störmer to the quantum framework. Namely, the equilibrium states of the model at every finite n are symmetric under permutations of components, and so are their limit points as $n \to \infty$; the theorems referred to above say that a symmetric state on infinite sequences is a superposition of product states. Simpler stated, this means that *components decouple in the limit*, so one is reduced to solving a one-component problem, i.e. *a scalar continuous-spin model, with an effec-*

tive, self-consistently defined, Hamiltonian. In the models considered by us the effective one-component model turns out to be Gaussian, what allows one, after having solved the selfconsistency equations, to derive closed expressions for the limit state and its corrections.

Having described the general framework, let me give a few hints on the history and on our contribution to the subject.

a) The idea to look at the n-vector models and their $n \to \infty$ limit originates to E. Stanley [5], who identified the limit to be the *spherical model* invented, with quite different justification, by Berlin and Kac in 1952 [6]. In fact, this is rigorously true only for translation invariant situations [7], otherwise the limit is a generalized spherical model [8], in which the spherical constraint is imposed (in the mean) at every lattice site. The nice feature of the spherical model, which explains its extensive study [9], is that it has, unlike usual mean-field models, phase diagram and critical behaviour non-trivially dependent on the lattice dimension and also, in its generalized version, it can give account on finite size effects.. Formal $1/n$-expansion has been initiated by K.Abe [10] in the translation invariant case. The first proof of the limit at the level of states and of the asymptotic nature of the $1/n$-expansion is contained in our 1979 paper [1], followed in 1980 by an independent derivation (however, with techniques valid only in the translation invariant case) by Kupiainen [11]. Further properties, such as Borel summability, have been derived by J.Fröhlich et al in 1983 [12]. There are many interesting physical applications, but I shall confine myself to remind a few papers of our group devoted to phase separation [13], correlation inequalities [1] and decay of correlations [2] in isotropic ferromagnets.

b) In the quantum case, the limit scalar model is a *Hartree-Fock approximation* [14], i.e. the harmonic crystal with effective strength at each site, having the free energy closest to that of the original anharmonic one [15]. Quantum effects play an important role in such models: the structural phase transition is completely suppressed for small ion masses due to large quantum fluctuations, in both the finite and infinite n cases [16], [15]. The $1/n$-expansion, proved by reducing the quantum case to a classical one via functional integration and adapting to it the methods of [1], is contained in [3]. The fluctuations of the components of the displacement field $\vec{q}_x$ related with large n, are shown to have different asymptotic distribution in the exact (finite n) and and limit (Hartree-Fock) equilibrium states. This leads to significant qualitative differences between the exact critical behaviour and that predicted by the Hartree-Fock approximation [4]. This phenomenon bears the main physical message of our work, namely that the Hartree-Fock approximation is too rough at the level of fluctuations and care is needed in taking over its predictions.

2 $1/n$-EXPANSION

We restrict, for definiteness, our presentation to the quantum anharmonic crystal with quartic anharmonic term depending solely on the magnitude of the ion-displacement from the equilibrium position. The Hamiltonian

$$H_{N,n} = T_{N,n} + V_{N,n}(\vec{q}),$$

where $T_{N,n} = \sum_{\alpha=1}^{n} \sum_{x\in\Lambda_N} \frac{1}{2m}(p_x^\alpha)^2$ is the kinetic energy and the potential energy

$$V_{N,n}(\vec{q}) = \sum_{\alpha=1}^{n} \{ \sum_{x\in\Lambda_N} [-hq_x^\alpha + (a/2)(q_x^\alpha)^2 + (b/2)(\vec{q}_x^{\,2}/n)^2] + \tfrac{1}{2} \sum_{x,y\in\Lambda_N;\, |x-y|=1} (q_x^\alpha - q_y^\alpha)^2 \} \tag{1}$$

contains an external electric field h along the main diagonal (such that equivalence of the components is not affected), a harmonic two-body interaction between nearest neighbour ions and the anharmonic (quartic) binding potential to the equilibrium position. Λ_N is a cube of side N of the cubic lattice $\mathbf{Z}^d$. Remark that the only coupling between different components enters through the anharmonic term $(b/2)(\vec{q}_x^{\,2}/n)^2$.

I shall now describe the decoupling which settles in as $n \to \infty$. To this aim, let us substract and add free parameters c_x from $\vec{q}_x^{\,2}/n$ in the quartic term, expand the square and isolate the "renormalized" quadratic terms in $V_{N,n}(\vec{q})$ (which do not couple different components):

$$V_{N,n}(q) = \sum_{\alpha=1}^{n} V_N^{\text{harmonic}}(q^\alpha; c) + nb/2 \sum_{x\in\Lambda_N} (\frac{\vec{q}_x^{\,2}}{n} - c_x)^2 \tag{2}$$

Neglecting the last term in Eq.(2), one obtains a system made of n independent copies of the scalar harmonic model of Hamiltonian

$$\tilde{H}_N(c) = \sum_{x\in\Lambda_N} \frac{1}{2m}(p_x)^2 + V_N^{\text{harmonic}}(q; c). \tag{3}$$

By choosing c_x equal to the expectation of $\vec{q}_x^{\,2}/n$ in the equilibrium state of the latter model, i.e. as the solution of the equation system

$$c_x = \langle q_x^2 \rangle_{\tilde{H}_N(c)}, \; x \in \Lambda_N, \tag{4}$$

the quadratic term is "minimized" to a sum of squares of the local fluctuation operators

$$F^{(n)}(q_x^2) = \frac{1}{\sqrt{n}} \sum_{\alpha=1}^{n} \left[(q_x^\alpha)^2 - \left\langle (q_x^\alpha)^2 \right\rangle_{\tilde{H}_N(c)} \right] . \tag{5}$$

The Hartree-Fock approximation corresponds to the infinite sequence of independent harmonic systems defined by Eqs. (3),(4).

Our approach to the asymptotic expansion of the equilibrium state $\langle \cdot \rangle_{N,n}$ of $H_{N,n}$ around the Hartree-Fock state is based on a steepest descents method adapted to handle quantum systems. As an illustration, consider the partition function $Z_{N.n} = tr \exp(-\beta H_{N,n})$ and its Hartree-Fock approximation $\tilde{Z}_{N,1}^n$, where $\tilde{Z}_{N,1} = tr \exp\left(-\beta \tilde{H}_N(c)\right)$. In order to compare them, let us use the Feynman-Kac representation to express $Z_{N,n}$ as an integral over a Brownian bridge process $\omega(t)$:

$$\begin{aligned} Z_{N,n} &= \int d\omega \exp\left[-\tfrac{\beta}{m} \int_0^1 V_{N,n}(\omega(\tfrac{\beta}{m}t))dt \right] \\ &= \tilde{Z}_{N,1}^n \int d\mu(\omega) \prod_x \exp\left[-\tfrac{\beta b}{m} \int_0^1 F^{(n)}(\omega_x(\tfrac{\beta}{m}t)^2)^2 dt \right], \end{aligned}$$

where $d\omega$ denotes the Brownian bridge distribution and where we included the contribution of V^{harmonic} in the functional probability measure $d\mu$, which now corresponds to an "oscillator bridge process". The exponent is a squared L_2-norm, and can be linearized using a Gaussian identity like

$$e^{-\frac{1}{2}\|A_x\|^2} = \int e^{i(A_x, \xi_x)} dG(\xi_x),$$

where $A_x(t) = \sqrt{\frac{2\beta b}{m}} F^{(n)}(\omega_x(\frac{\beta}{m}t)^2)$ and the "conjugate" variable $\xi_x(t)$ is a white noise generalized process of distribution dG. In this way the components are no longer coupled under the integral over the processes $\xi_x(\cdot)$ and the integrals over $\omega_x^\alpha(\cdot)$ can be performed, yielding an expression

$$Z_{N,n} = \tilde{Z}_{N,1}^n \int \exp n f\left(n^{-1/2}\xi\right) dG(\xi),$$

suited for applying steepest descents. Here, the function f satisfies $f(0) = 0$, $\partial f(0) = 0$ in view of the selfconsistency Eq.(4), hence the exponent converges to a quadratic form, i.e. provides a new Gaussian measure, and higher order terms in the series expansion can be calculated as moments of the latter. This linearization trick is the core of the proof (some work is of course needed to settle all convergence details [3]), and in fact it proves quite useful in other circumstances as well (e.g. in obtaining Bender-Wu formula for the ground state of the anharmonic oscillator).

In precise terms, the following theorem holds:

Let $\langle\cdot\rangle_n$ be the canonical Gibbs state for $H_{N,n}$, Eq.(1). Then for all $\lambda_x^\alpha, \mu_x^\alpha \in \mathbf{C}, x \in \Lambda_N$, $\alpha = 1, \ldots, n$, the characteristic function (which completely determines $\langle\cdot\rangle_n$)

$$E(\lambda,\mu)_n \equiv \left\langle e^{i\ \sum_{\alpha=1}^N (\lambda^\alpha, q^\alpha)} e^{i\ \sum_{\beta=1}^N (\mu^\beta, p^\beta)} \right\rangle_n \tag{6}$$

has a complete asymptotic series in $\frac{1}{n}$ and the leading term is given by

$$E^{HF}(\lambda,\mu) \equiv \lim_{n\to\infty} E(\lambda,\mu)_n = \prod_{\alpha=1}^N \left\langle e^{i(\lambda^\alpha, q)} e^{i(\mu^\alpha, p)} \right\rangle_{\tilde{H}_N(c)}. \tag{7}$$

3 THE LOCAL FLUCTUATION FIELD, CLUSTERING PROPERTIES

In a system with a large number of equivalent subsystems there are two kinds of macroscopic variables of interest for the thermodynamics: *the extensive observables*, which are obtained from sums of copies of a given observable of one subsystem, and *the fluctuations.* In our case, we have n equivalent subsystems associated with the different components α.

For defining an extensive observable, one has to look at limits of means,e.g. $n^{-1} \sum_{\alpha=1}^{\infty} q^\alpha$, i.e. to prove a law of large numbers. As a rule, these limit exist and are c-numbers, equal to the expectation in the limit state. For instance, $n^{-1} \sum_{\alpha=1}^{\infty} q_x^\alpha$ converges to the Hartree-Fock expectation of the displacement $\langle q_x \rangle_{\tilde{H}_N(c)} = \tilde{m}$, which is the order parameter of the model. We have the following phase diagram: in the $(1/\beta, 1/m)$ plane, there exists a critical line isolating the origin, above which we have normal phase ($\tilde{m} = 0$) and below which we have ferroelectric phase ($\tilde{m} > 0$). Remark that for small ion mass (i.e. strong quanticity) the system is normal at all temperatures, while for large mass it has a transition to a ferro-phase at low temperature.

The fluctuations measure the error distribution from the mean and are used to estimate the confidence one can have in the limit above. In the particular case of the ion displacements, we have to look at the observables, which we call "local displacement fluctuations",

$$F^{(n)}(q_x) = \frac{1}{\sqrt{n}} \sum_{\alpha=1}^{n} (q_x^\alpha - \tilde{m}), x \in \mathbf{Z}^d, \tag{8}$$

in the large n limit. $F^{(n)}(q_x)$ are to be looked at as a random field, with joint distribution given by the equilibrium state $\langle\cdot\rangle_n$ of the model (1). Our main result concerns the limit in distribution, $\{F^{(\infty)}(q_x)\,;x\in\mathbf{Z}^d\}$ of this random field as $n\to\infty$ and in the thermodynamical limit, $N\to\infty$.

Let us remark that one has at his disposal the Hartree-Fock approximation $\langle\cdot\rangle^{HF}$ of the state $\langle\cdot\rangle_n$, to get an approximate joint distribution for the variables $F^{(n)}(q_x)$. This will however define other random fields, denoted $\tilde{F}^{(n)}(q_x)$, $\tilde{F}^{(\infty)}(q_x)$, respectively, in order to avoid confusion. Our point is that *the two fields have different limits, due to the higher order* $1/n$ *corrections.* For example, in view of the permutation symmetry and of the convergence of the states, one has for the dispersions

$$\begin{aligned}\lim_{n\to\infty}\left\langle F^{(n)}(q_x)^2\right\rangle_n &= \langle q_x^2\rangle_{\tilde{H}_N(c)}-\tilde{m}^2+\lim_{n\to\infty}(n-1)\left\langle q_x^\alpha q_x^\beta\right\rangle_n,\ \text{any}\,\alpha\neq\beta\\ \lim_{n\to\infty}\left\langle F^{(n)}(q_x)^2\right\rangle^{HF} &= \langle q_x^2\rangle_{\tilde{H}_N(c)}-\tilde{m}^2\end{aligned}\tag{9}$$

Global displacement fluctuations and their Hartree-Fock counterparts are defined in terms of local ones by summing over the sites in cubes growing to infinity and dividing by a suitable power of the number of sites such that a nontrivial random variable is obtained in the limit:

$$\begin{aligned}F^{(\delta)} &= \lim_{N\to\infty}N^{-d(1+\delta)/2}\sum_{x\in\Lambda_N}F^{(\infty)}(q_x)\,,\\ \tilde{F}^{(\delta)} &= \lim_{N\to\infty}N^{-d(1+\tilde{\delta})/2}\sum_{x\in\Lambda_N}\tilde{F}^{(\infty)}(q_x)\,,\end{aligned}\tag{10}$$

with $\delta=0$ meaning normal, and $\delta>0$ ($\delta<0$) abnormally large (small), fluctuations.

The result is the following:

In the ferroelectric phase, $\{F^{(\infty)}(q_x)\,;x\in\mathbf{Z}^d\}$ *and* $\left\{\tilde{F}^{(\infty)}(q_x)\,;x\in\mathbf{Z}^d\right\}$ *are massless Gaussian field of covariance matrix having the decay at infinity*

$$\begin{aligned}\langle F^{(\infty)}(q_x)\,F^{(\infty)}(q_y)\rangle &\sim \tfrac{1}{\tilde{m}^2}\,|x-y|^{2(d-2)}\,.\\ \left\langle\tilde{F}^{(\infty)}(q_x)\,\tilde{F}^{(\infty)}(q_y)\right\rangle &\sim |x-y|^{d-2}\,.\end{aligned}\tag{11}$$

The global fluctuations are Gaussian variables for the following values of the exponents

$$\delta=\begin{cases}1/6, & \text{if}\,d=3\\ 0+, & \text{if}\,d=4\\ 0, & \text{if}\,d>4\end{cases};\ \tilde{\delta}=\frac{1}{d},\forall d\geq 3.\tag{12}$$

In the normal phase both fields coincide and are massive (i.e. have exponential decay), and the global fluctuations are normal.

Hence, the "exact" displacement fluctuations are less singular than their Hartree-Fock approximations In particular, in contradistinction with the Hartree-Fock result, they are normal in the ferro-phase for d larger than 4 (where the mean-field behaviour is expected to set in). Scaling laws are valid when approaching the critical line from the ferro-phase, which allow one to introduce a notion of diverging correlation length, in spite of the power law of the correlation decay.

The classical analogue of this result [2] can also be expressed as a difference in the clustering properties of the longitudinal and transversal spin-spin correlations. The longitudinal correlations share properties of the scalar spin models, such as Ising, and for $d > 4$, imply finite longitudinal susceptibility in the ferromagnetic phase, while the transverse susceptibility is infinite in all dimensions.

References

1. N.Angelescu, M.Bundaru, G.Costache, *J.Phys.A:Math.Gen.* **12**, 2457 (1979)
2. N.Angelescu, M.Bundaru, G.Costache, *J.Phys.A:Math.Gen.* **32**, 4571 (1999)
3. N.Angelescu, A.Verbeure, V.A.Zagrebnov, *Comm.Math.Phys.* **205**, 81 (1999)
4. N. Angelescu, A. Verbeure, V. A. Zagrebnov, *J.Statistical Phys.* (submitted)
5. H.E.Stanley, *Phys.Rev.***176**, 718 (1968)
6. T.H.Berlin, M.Kac, *Phys.Rev.***86**, 821 (1952)
7. M.Kac, C.J.Thompson, *Physica Norvegica* **5**, 163 (1971)
8. H.J.F.Knops, *J.Math.Phys.* **14**, 1918 (1973)
9. G.S.Joyce in *Phase Transitions and Critical Phenomena*, vol.2, Eds. C.Domb and M.S.Green (Academic Press,London,1972)
10. R.Abe, *Progr.Theor.Phys.* **49**, 113 and 442 (1973)
11. A.J.Kupiainen, *Comm.Math.Phys.* **73**, 273 (1980)
12. J.Fröhlich, A.Mardin, V.Rivasseau, *Comm.Math.Phys.* **86**, 87 (1982)
13. N.Angelescu, M.Bundaru, G.Costache, *J.Phys.A:Math.Gen.* **16**, 2479 (1983)
14. T.Schneider, H.Beck, E.Stoll, *Phys.Rev.B* **13**, 1123 (1976)
15. A.Verbeure, V.A.Zagrebnov, *J.Stat.Phys.* **69**, 329 (1992)
16. A.Verbeure, V.A.Zagrebnov, *J.Phys.A:Math.Gen.* **28**, 5415 (1995)

CURRENT STATUS OF HEAVY ION PHYSICS

K. PAECH, S. A. BASS[*], M. BELKACEM, M. BLEICHER[†], J. BRACHMANN, L. BRAVINA, D. DIETRICH, A. DUMITRU[‡], C. ERNST, L. GERLAND, M. HOFMANN, L. NEISE, M. REITER, S. SCHERER, S. SOFF, C. SPIELES, H. WEBER, E. ZABRODIN, D. ZSCHIESCHE, J.A. MARUHN, H. STÖCKER, W. GREINER

Institut für Theoretische Physik, Johann Wolfgang Goethe-Universität, Robert Mayer-Str. 8–10, D-60054 Frankfurt am Main, Germany
E-mail: paech@th.physik.uni-frankfurt.de

Nonequilibrium models (UrQMD and three-fluid hydrodynamics) are used to discuss the uniqueness of often proposed experimental signatures for quark matter formation in relativistic heavy ion collisions. It is demonstrated that these two models – although they do treat the most interesting early phase of the collisions quite differently (coherent color fields with virtual particles vs. thermalizing QGP) – both yield a reasonable agreement with a large variety of the available heavy ion data. Hadron/hyperon yields, including J/Ψ meson production/suppression, dileptons, and directed flow are investigated. We emphasize the need for systematic future measurements to search for simultaneous irregularities in the excitation functions of several observables in order to come close to pinning the properties of hot, dense QCD matter from data.

1 Introduction

In the last few years researchers at Brookhaven and CERN have succeeded to measure a wide spectrum of observables with heavy ion beams, $Au + Au$ and $Pb + Pb$. While these programs continue to measure with greater precision the beam energy-, nuclear size-, and centrality dependence of those observables, it is important to recognize the major milestones passed thusfar in that work. The experiments have conclusively demonstrated the existence of strong nuclear A dependence of, among others, J/ψ and ψ' meson production and suppression, strangeness enhancement, hadronic resonance production, stopping and directed collective transverse and longitudinal flow of baryons and mesons – in and out of the impact plane, both at AGS and SPS energies –, and dilepton-enhancement below and above the ρ meson mass. These observations support that a novel form of "resonance matter" at high energy- and

[*]PRESENT ADDRESS: DEPARTMENT OF PHYSICS, MICHIGAN STATE UNIVERSITY, EAST LANSING, USA

[†]PRESENT ADDRESS: LAWRENCE BERKELEY LABORATORY, BERKLEY, USA

[‡]PRESENT ADDRESS: PHYSICS DEPARTMENT, COLUMBIA UNIVERSITY, NEW YORK, USA

baryon density has been created in nuclear collisions. The global multiplicity and transverse energy measurements prove that substantially more entropy is produced in $A + A$ collisions at the SPS than simple superposition of $A \times pp$ would imply. Multiple initial and final state interactions play a critical role in all observables. The high midrapidity baryon density (stopping) and the observed collective transverse and directed flow patterns constitute one of the strongest evidence for the existence of an extended period ($\Delta\tau \approx 10$ fm/c) of high pressure and strong final state interactions. The enhanced ψ' suppression in $S + U$ relative to $p + A$ also attests to this fact. The anomalous low mass dilepton enhancement shows that substantial in-medium modifications of multiple collision dynamics exists, probably related to in-medium collisional broadening of vector mesons. The non-saturation of the strangeness (and anti-strangeness) production shows that novel non-equilibrium production processes arise in these reactions. Finally, the centrality dependence of J/ψ absorption in $Pb + Pb$ collisions presents further hints towards the nonequilibrium nature of such reactions. Is there evidence for the long sought-after quark-gluon plasma that thusfar has only existed as a binary array of predictions inside teraflop computers?

As we will discuss, it is too early to tell. Theoretically there are still too many "scenarios" and idealizations to provide a satisfactory answer. Recent results from microscopic transport models as well as macroscopic hydrodynamical calculations differ significantly from predictions of simple thermal models, e. g. in the flow pattern. Still, these nonequilibrium models provide reasonable predictions for the experimental data. We may therefore be forced to rethink our concept of what constitutes the deconfined phase in ultrarelativistic heavy-ion collisions. Most probably it is not a blob of thermalized quarks and gluons. Hence, a quark-gluon plasma can only be the source of *differences* to the predictions of these models for hadron ratios, the J/Ψ meson production, dilepton yields, or the excitation function of transverse flow. And there are experimental gaps such as the lack of intermediate mass $A \approx 100$ data and the limited number of beam energies studied thusfar, in particular between the AGS and SPS. In the future, the field is at the doorstep of the next milestone: $A + A$ at $\sqrt{s} = 30 - 200$ AGeV due to begin at RHIC/BNL hopefully in 2000.

2 Nonequilibrium models

In the present survey we employ two sharply distinct nonequilibrium models for relativistic heavy ion collisions, namely the macroscopic 3-fluid hydrodynamical model[1] and the Ultra-relativistic Quantum Molecular Dynamical

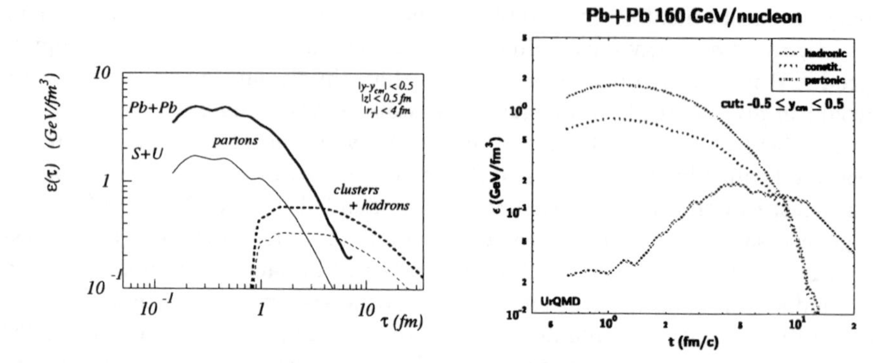

Figure 1. (Left) Parton cascade (VNI) and (Right) UrQMD results for the time evolution of energy density in central $Pb + Pb$ reactions at 160 AGeV. At an early stage, most of the energy is contained in the partonic degrees of freedom (VNI) or in constituents (UrQMD).

model, UrQMD[2]. The first model assumes that a projectile- and a target fluid interpenetrate upon impact of the two nuclei, creating a third fluid via new source terms in the continuity equations for energy- and momentum flux. Those source terms are taken from energy- and rapidity loss measurements in high energy *pp*-collisions. The equation of state (EoS) of this model assumes equilibrium only in each fluid separately and allows for a first order phase transition to a quark gluon plasma in fluid 1, 2 or 3, if the energy density in the fluid under consideration exceeds the critical value for two phase coexistence. Pure QGP can also be formed in every fluid separately, if the energy density in that fluid exceeds the maximum energy density for the mixed phase. The UrQMD-model, on the other hand, assumes an independent evolution of hadrons, strings, and constituent quarks and diquarks in a nonequilibrium multiparticle system. The collision terms in this system of coupled Boltzmann (partial differential-/integral-) equations are taken from experimental data, where available, and otherwise from additive quark model and string phenomenology.

Fig. 1 demonstrates the role of partonic degrees of freedom in heavy ion reactions at the SPS. It shows the time evolution of the energy density ϵ in central $Pb + Pb$ reactions at 160 AGeV as obtained within *a)* the parton cascade approach VNI[3], *b)* the UrQMD model[4]. It can be seen that in both models and at early times of the collision, a large fraction of the energy density is contained in partonic degrees of freedom (VNI) or to nearly equal parts in constituent diquarks and quarks from the strings and in virtual hadrons. This (virtual) "partonic" phase in $Pb+Pb$ reactions at 160 AGeV is, however, not to be identified with an equilibrated QGP. Note that the absolute values differ by a factor 2 in the two models and depend heavily on the rapidity cuts imposed to discriminate between virtual free streaming and interacting matter.

3 Yields of Hadronic Probes

Let us now discuss the results obtained from hadronic probes, such as observed production of J/Ψ mesons and particle ratios. Observed hadrons include feeding by the decay of resonances.

3.1 J/ψ suppression

Debye screening of heavy charmonium mesons in an equilibrated quark-gluon plasma may reduce the range of the attractive force between heavy quarks and antiquarks[5]. Mott transitions then dissolve particular bound states, one by one. NA38 found evidence of charmonium suppression in light ion reactions. Then also in $p+A$ such suppression was observed. New preliminary $Pb+Pb$ data of NA50 show "anomalous" suppression.

One of the main problems in the interpretation of the observed suppression as a signal for deconfinement is that non-equilibrium dynamical sources of charmonium suppression have also been clearly discovered in $p+A$ reactions, where the formation of an equilibrated quark-gluon plasma is not expected. A recent development is the calculation of the hard contributions to the charmonium- and bottonium-nucleon cross sections based on the QCD factorization theorem and the non-relativistic quarkonium model[8]. Including non-perturbative contributions, the calculated $p+A$ cross section agrees well with the data. Whereas these descriptions of nuclear absorption can account for the $p+A$ observation, the corrections needed for an extrapolation to $A+A$ reactions are, however, not yet under theoretical control.

Purely hadronic dissociation scenarios have been suggested[9] which could account for J/ψ and ψ' suppression without invoking the concept of deconfinement ("comover models"). Suppression in excess to that due to preformation and nuclear absorption is ascribed in such models to interactions of the charmonium mesons with "comoving", but probably off-equilibrium, mesons and baryons, which are produced copiously in nuclear collisions. Fig. 2 shows an UrQMD calculation which employs a microscopic free streaming simulation for J/ψ production and a microscopic transport calculation for nuclear and comover dynamics as well as for rescattering[7]. The dissociation cross

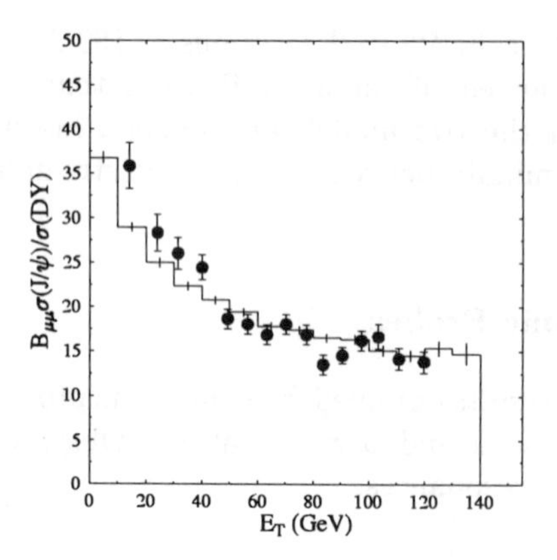

Figure 2. The ratio of J/ψ to Drell-Yan production as a function of E_T for $Pb + Pb$ at 160 GeV. The experimental data are from Ref. [6], the histogram ia a UrQMD calculation [7]. No scaling factor has been applied to the x-axis for either the calculations or the data.

sections are calculated using the QCD factorization theorem[8], feeding from ψ' and χ states is taken into account, and the $c\bar{c}$ dissociation cross sections increase linearly with time during the formation of the charmonium state. Taking into account the non-equilibrium "comovers" ($\sigma_{meson} \approx 2/3\sigma_{nucleon}$), the agreement between theory and data is reasonable (Fig. 2). New, unpublished data agree better with the model predictions, but the high and low E_T regions remain to be studied carefully in the experiment. At present, no ab initio calculation does predict sudden changes in the suppression. In fact, from three-fluid calculations, even with QGP phase included, only a moderate change of the average and local energy density with bombarding energy is predicted. This seems to strongly speak against drastic threshold effects in the charmonium production.

The strong dependence of these results on details, such as the treatment of the formation time, quantum effects on energy dependent formation length or the time dependent dissociation cross section, remain to be studied further before definite statements can be made with regard to the nature of the J/ψ suppression.

Hence, the theoretical debate on the interpretation of the pattern of charmonium suppression discovered by NA38/NA50 at the SPS is far from settled.

It is not clear whether the suppression is the smoking gun of nonequilibrium dynamics or deconfinement. It is not likely to be due to simple Debye screening.

The major goal of further theoretical work is not to continue to try to rule out more "conventional" explanations, but to give positive proof of additional suppression by QCD-calculations which actually *predict* the E_T-dependence of the conjectured signature. Consistency tests and a detailed simultaneous analysis of all other measured observables are needed, if at least the same standards as for the present calculations are to be hold up.

3.2 Particle ratios

The study of particle ratios has recently attended great interest at the AGS[10,11,12] and at the SPS[13].

Assuming a thermalized system with the same density, temperature and linear radial and longitudinal flow velocity profiles for all hadrons and fragments, the system decouples as a whole at some break-up time, and the particles are emitted instantaneously from the whole volume of the thermal source. A complete loss of memory results, due to thermalization – the emitted particles carry no information about the evolution of the source. A two parameter fit (μ_q, T, μ_s is fixed by strangeness conservation) to the hadronic freeze-out data describes the experimental results well, if feeding from Δ's etc. and the proper Hagedorn volume correction is included[13]. Does this compatibility with a thermal source proof volume emission from a globally equilibrated source?

The ideal gas thermal fit to experimental data for hadron ratios in $S+Au$ collisions at 200 AGeV gives values for the parameters T and $\mu_{\rm B}$ which can be used as input for a $SU(3)$ chiral mean-field model[15] extended to finite temperatures[16]. Feeding from the decay of higher resonances is included. One finds that in such a model (which selfconsistently contains a chiral phase transition at $T \approx 150$ MeV) the ideal gas model values $T = 160$ MeV and $\mu_B = 170$ MeV lead to strong deviations from the experimental data. Only the Ω/Ξ^--ratio is in a good agreement, in contrast to the ideal gas model. Hence, the system can not be close to the chiral phase transition – the T and μ values extracted from the free thermal model cannot be identified with the real temperature and chemical potential of the system!

The chiral mean-field model does reproduce the data compiled in[17] for relative abundances in $Pb+Pb$ collisions at 160 AGeV (Fig. 3, left) for $T =$ 125 MeV and $\mu_B = 180$, much lower than the thermal model results[18,17] ($T = 160 - 175$ MeV, $\mu_B = 200 - 270$ MeV).

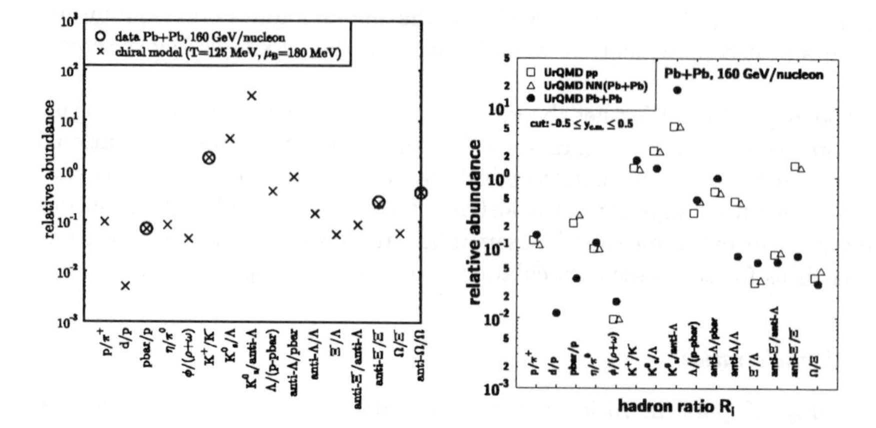

Figure 3. (Left) Fit of hadron ratios from the chiral model to preliminary data from $Pb+Pb$ collisions at SPS. The obtained values of T and μ allow the prediction of further ratios. T and μ are much lower than thermal model results from the ideal hadron gas . (Right) UrQMD prediction for hadron ratios in $Pb+Pb$ collisions at midrapidity (full circles), compared to a superposition of pp, pn and nn reactions with the isospin weight of the $Pb+Pb$ system (open triangles), i.e. a first collision approach.

The microscopic UrQMD transport model is in good agreement with the measured hadron ratios of the system $S+Au$ at CERN/SPS[19]. A thermal model fit to the calculated ratios yields a temperature of $T = 145$ MeV and a chemical potential of $\mu_B = 165$ MeV. However, these ratios exhibit a strong rapidity dependence. Thus, thermal model fits to data may be distorted due to different acceptances for the individual ratios.

Hadron ratios for the system $Pb+Pb$ are predicted by UrQMD and can be fitted by a thermal model with $T = 140$ MeV and $\mu_B = 210$ MeV (Fig. 3, right). Analyzing the results of non-equilibrium transport model calculations by an equilibrium model may, however, be not meaningful.

There is a problem in the definition of equilibrium in itself: Do heavy ion collisions ever reach a thermalized system? Or are there transient steady states off equilibrium[14] Due to the rapid dynamics of the system, the assumption of detailed balance is not fulfilled in the initial stage. This drives the system into a steady state far from equilibrium, but stationary in time. This steady state is easily visible in an enhanced production of light mesons, as compared to thermal models.

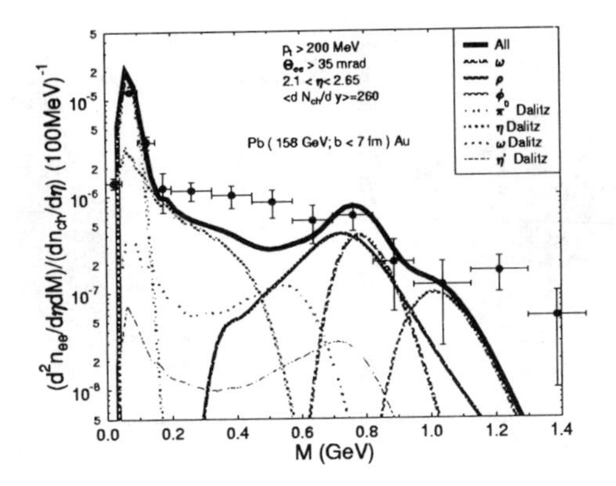

Figure 4. Microscopic calculation of the dilepton production in the kinematic acceptance region of the CERES detector for $Pb + Au$ collisions at 158 GeV. No in-medium effects are taken into account. Plotted data points are taken at CERES in '95.

4 Dilepton production

Beside results from hadronic probes, electromagnetic radiation – and in particular dileptons – offer an unique probe from the hot and dense reaction zone: here, hadronic matter is almost transparent. The observed enhancement of the dilepton yield at intermediate invariant masses ($M_{e^+e^-} > 0.3$ GeV) received great interest: it was prematurely thought that the lowering of vector meson masses is required by chiral symmetry restoration (see e.g.[21] for a review).

Fig. 4 shows a microscopic UrQMD calculation of the dilepton production in the kinematic acceptance region of the CERES detector for $Pb + Au$ collisions at 158 GeV. This is compared with the '95 CERES data[23]. Aside from the difference at $M \approx 0.4$ GeV there is a strong enhancement at higher invariant masses. It is expected that this discrepancy at $m > 1$ GeV could be filled up by direct dilepton production in meson-meson collisions[24] as well as by the mechanism of secondary Drell-Yan pair production proposed in[25].

5 Collective Flow and the softening of the EoS

The in-plane flow has been proposed as a measure of the "softening" of the EoS [26], therefore we investigate the excitation function of directed in-plane flow. A three-fluid model with dynamical unification of kinetically equilli-

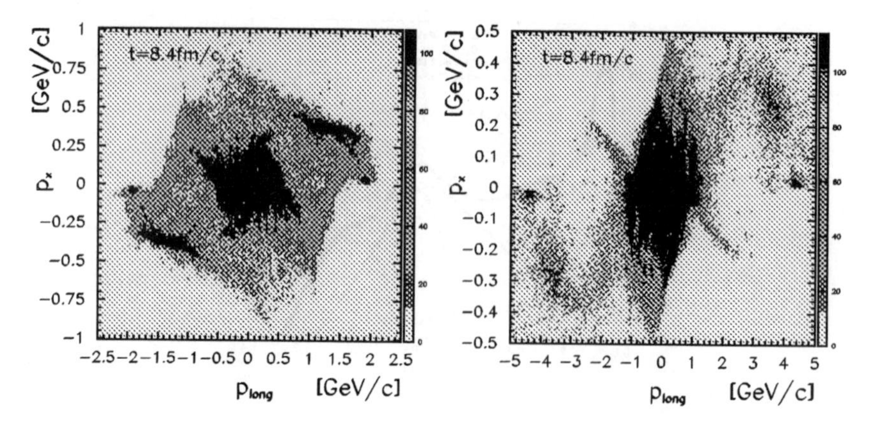

Figure 5. (Left) Net-baryon density in momentum space, $Pb + Pb$ (8 AGeV) at $b = 3$ fm and (Right) for $Pb + Pb$ (40 AGeV)at $b = 3$ fm.

brated fluid elements is applied [1].

Fig. 5 shows the time-like component of the net baryon current in momentum space ($p_x - p_{long}$ plane) for a $Pb + Pb$ collsion with 8 GeV and 40 GeV. One clearly observes the directed in-plane flow (before the collision, there is no matter at $p_x \neq 0$). However, there is almost no momentum of baryons in the upper left or bottom right quadrants, where $p_x \cdot p_{long} < 0$ (except for two "jets"). This is due to the fact that the central region is in the phase coexistence region at the depicted time, with a rather small average isentropic speed of sound. Therefore, isentropic expansion of the highly excited matter is inhibited. Note the central region is also in the phase coexistance region but due to a high baryon density $\langle c_s \rangle$ is not small.

Integrating up the collective momentum in x-direction at given rapidity, and dividing by the net baryon number in that rapidity bin, we obtain the so-called directed in-plane flow per nucleon.

Its excitation function (Fig. 6) shows a local minimum at 8 AGeV and rises until a maximum around 40 AGeV is reached. Fig. 6 shows the excitation function of directed flow calculated in the three-fluid model in comparison to that obtained in a one-fluid calculation. Due to non-equilibrium effects in the early stage of the reaction, which delay the build-up of transverse pressure[28], the flow shifts to higher bombarding energies. While measurements of flow at AGS[27] have found a deacrease of directed flow with increasing bombarding energy, a minimum has so far not been observed.

Its slope at midrapidity, $d(p_x/N)/dy$, is shown in Fig. 6 as a function of

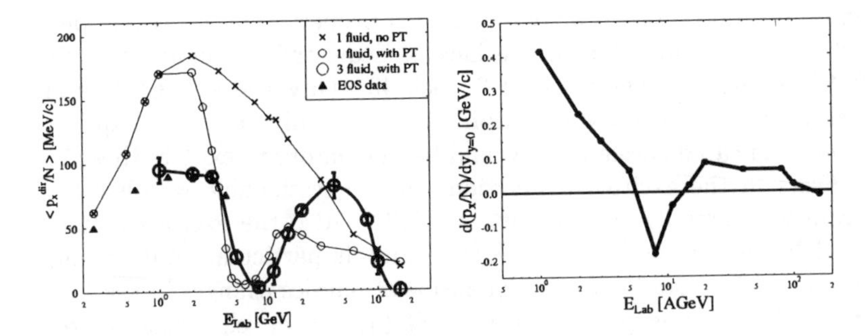

Figure 6. (Left) Excitation function of transverse flow as obtained from three fluid hydrodynamics with a first order phase transition and (Right) the slope of the directed in-plane momentum per nucleon at midrapidity.

beam energy. We find a steady decrease of $d(p_x/N)/dy$ up to about top BNL-AGS energy, where the flow around midrapidity even becomes negative due to prefered expansion towards $p_x \cdot p_{long} < 0$. At higher energy, $E_{Lab} \simeq 40A$ GeV, the isentropic speed of sound becomes small and we encounter the expansion pattern depicted in Fig. 5: flow towards $p_x \cdot p_{long} < 0$ can not build up ! Consequently, $d(p_x/N)/dy$ increases rapidly towards $E_{Lab} = 20 - 40A$ GeV, decreasing again at even higher energy because of the more forward-backward peaked kinematics which is unfavorable for directed flow.

Thus, the $Pb + Pb$ collsions (40 GeV) runs performed recently at the CERN-SPS may provide a crucial test of the picture of a quasi-adiabatic first-order hadronization phase transition at small isentropic velocity of sound.

Outlook

The latest data of CERN/SPS on flow, electro-magnetic probes, strange particle yields (most importantly multistrange (anti-)hyperons) and heavy quarkonia will be interesting to follow closely. Simple energy densities estimated from rapidity distributions and temperatures extracted from particle spectra indicate that initial conditions could be near or just above the domain of deconfinement and chiral symmetry restoration. Still the quest for an *unambiguous* signature remains open.

Directed flow has been discovered – now a flow excitation function, filling the gap between 10 AGeV (AGS) and 160 AGeV (SPS), will be extremely interesting: look for the softening of the QCD equation of state in the coexistence region. The investigation of the physics of high baryon density (e.g. partial restoration of chiral symmetry via properties of vector mesons) is presently not accessible due to the lack of dedicated accelerators in the 10 – 200 AGeV regime. The planned 8 week 40 AGeV run at CERN is an absolute necessity into this new direction, but it can only be a first step.

However, dedicated accelerators would be mandatory to explore these intriguing effects in the excitation function. It is questionable whether this key program will actually get support at CERN. Also the excitation function of particle yield ratios ($\pi/p, d/p, K/\pi ...$) and, in particular, multistrange (anti-)hyperon yields, can be a sensitive probe of physics changes in the EoS. The search for novel, unexpected forms of $SU(3)$ matter, e.g. *hypermatter*, *strangelets* or even *charmlets* is intriguing. Such exotic QCD multi-meson and multi-baryon configurations would extend the present periodic table of elements into hitherto unexplored dimensions. A strong experimental effort should continue in that direction.

Experiments and data on ultra-relativistic collisions are essential in order to motivate, guide, and constrain theoretical developments. They provide the only terrestrial probes of non-perturbative aspects of QCD and its dynamical vacuum. The understanding of confinement and chiral symmetry remains one of the key questions at the beginning of the next millennium.

Acknowledgments

This work was supported by DFG, GSI, BMBF, Graduiertenkolleg Theoretische und Experimentelle Schwerionenphysik, the A. v. Humboldt Foundation, and the J. Buchmann Foundation.

References

1. J. Brachmann, A. Dumitru, J.A. Maruhn, H. Stöcker, W. Greiner, D.H. Rischke, Nucl. Phys. **A619** (1997) 391.
   J. Brachmann et al., Phys. Rev. **C61** (2000); nucl-th/9912014
2. S.A. Bass, M. Belkacem, M. Bleicher, M. Brandstetter, L/ Bravina, C. Ernst, L. Gerland, M. Hofmann, S. Hofmann, J. Konopka, G. Mao, L. Neise, S. Soff, C. Spieles, H. Weber, L.A. Winckelmann, H. Stöcker, W. Greiner, C. Hartnack, J. Aichelin, N. Amelin, Prog. Part. Nucl. Phys. **41** (1998) 225.
3. K. Geiger, Nucl. Phys. **A638** (1998) 551c.
4. H. Weber et al., to be published
5. T. Matsui, H. Satz, Phys. Lett. **178B** (1986) 416.

6. A. Romana (NA50 Collab.), in *Proceedings of the XXXIIIrd Rencontres de Moriond*, March 1998, Les Arcs, France.
7. C. Spieles, R. Vogt, L. Gerland, S. A. Bass, M. Bleicher, L. Frankfurt, M. Strikman,H. Stöcker, W. Greiner, hep-ph/9810486
8. L. Gerland, L. Frankfurt, M. Strikman, H. Stöcker and W. Greiner, Phys. Rev. Lett. **81** (1998) 762.
9. D. Neubauer, K. Sailer, B. Müller, H. Stöcker and W. Greiner, Mod. Phys. Lett. **A4** (1989) 1627.
   S. Gavin, R. Vogt, Nucl. Phys. **B345** (1990) 104.
10. H. Stöcker, A.A. Ogloblin, W. Greiner, LBL-12971 (1981).
11. J. Cleymanns, H. Satz, Z. Phys. **C57** (1993) 135.
12. E. Schnedermann, J. Sollfrank, U. Heinz, Phys. Rev. **C48** (1993) 2462.
13. P. Braun-Munzinger, J. Stachel, J.P. Wessels, N. Xu, Phys. Lett. **B344** (1995) 43.
    P. Braun-Munzinger, J. Stachel, J.P. Wessels, N. Xu, Phys. Lett. **B365** (1996) 1.
14. L. Bravina, et al. , J. Phys. **G25** (1999) 351
15. P. Papazoglou, D. Zschiesche, S. Schramm, J. Schaffner-Bielich, H. Stöcker, W. Greiner, nucl-th/9806087 , Phys. Rev. **C59** (1999) 411.
16. D. Zschiesche et al., to be published
17. G. D. Yen, M. I. Gorenstein, Phys. Rev. **C59** (1999) 2788.
18. P. Braun-Munzinger, J. Stachel, Nucl. Phys. **A638** (1998) 3.
19. S. Bass, M. Belkacem, M. Brandstetter, M. Bleicher, L. Gerland, J. Konopka, L. Neise, C. Spieles, S. Soff, H. Weber, H. Stöcker, W. Greiner, Phys. Rev. Lett. **81** (1998) 4092.
20. E. Andersen et al., Phys. Lett. **B433** (1998) 209.
21. V. Koch, Int. Jour. Mod. Phys. **E6** (1997) 203.
22. W. Cassing, E. L. Bratkovskaya, R. Rapp, and J. Wambach, Phys. Rev. **C57** (1998) 916
23. G. Agakishiev et al., Phys. Lett. **B402** (1998) 405.
24. G. Q. Li and C. Gale, Phys. Rev. **C58** (1998) 2914.
25. C. Spieles et al., Eur. Phys. J. **C5** (1998) 349
26. D. H. Rischke, Y. Pürsün, J.A. Maruhn, H. Stöcker, W. Greiner, Heavy Ion Physics **1** (1995) 309.
27. H. Liu et al. (E895 Collaboration), Nucl. Phys. **A638**, 451c (1998)
28. H. Sorge, Phys. Rev. Lett. **78**, 2309 (1997)

DYNAMICAL FEATURES IN NUCLEUS-NUCLEUS COLLISIONS STUDIED WITH INDRA

M.F. RIVET[1], F. BOCAGE[2], B. BORDERIE[1], J. COLIN[2], P. LAUTESSE[3], J. LUKASIK[1]*, A.M. MASKAY[3], P. PAWŁOWSKI[1], E. PLAGNOL[1], G. AUGER[4], CH.O. BACRI[1], N. BELLAIZE[3], R. BOUGAULT[2], B. BOURIQUET[4], R. BROU[2], P. BUCHET[5], J.L. CHARVET[5], A. CHBIHI[4], D. CUSSOL[2], R. DAYRAS[5], N. DE CESARE[7], A. DEMEYER[3], D. DORÉ[5], D. DURAND[2], J.D. FRANKLAND[4], E. GALICHET[1,8], E. GENOUIN-DUHAMEL[2], E. GERLIC[3], S. HUDAN[4], D. GUINET[3], F. LAVAUD[1], J.L. LAVILLE[4], J.F. LECOLLEY[2], C. LEDUC[3], R. LEGRAIN[5], N. LE NEINDRE[2], O. LOPEZ[2], M. LOUVEL[2], L. NALPAS[5], J. NORMAND[2], M. PÂRLOG[6], J. PÉTER[2], E. ROSATO[7], F. SAINT-LAURENT[4]†, J.C. STECKMEYER[2], M. STERN[3], G. TĂBĂCARU[6], B. TAMAIN[2], L. TASSAN-GOT[1], O. TIREL[4], E. VIENT[2], M. VIGILANTE[7], C. VOLANT[5], J.P. WIELECZKO[4]

(INDRA COLLABORATION)

1 *Institut de Physique Nucléaire, IN2P3-CNRS, F-91406 Orsay Cedex, France.*

2 *LPC, IN2P3-CNRS, ISMRA et Université, F-14050 Caen Cedex, France.*

3 *Institut de Physique Nucléaire, IN2P3-CNRS et Université, F-69622 Villeurbanne Cedex, France.*

4 *GANIL, CEA et IN2P3-CNRS, B.P. 5027, F-14076 Caen Cedex, France.*

5 *DAPNIA/SPhN, CEA/Saclay, F-91191 Gif sur Yvette Cedex, France.*

6 *National Institute for Physics and Nuclear Engineering, RO-76900 Bucharest-Măgurele, Romania.*

7 *Dipartimento di Scienze Fisiche e Sezione INFN, Universitį Napoli 'Federico II", I-80126 Napoli, Italy.*

8 *Conservatoire National des Arts et Métiers, F-75141 Paris cedex 03.*

Dynamical emissions observed in heavy-ion collisions around the Fermi energy are reviewed. Their main interest lies in their connection with the in-medium nucleon-nucleon cross section (two-body nuclear viscosity, stopping of nuclear matter and transparency). Among non-equilibrium emissions, direct cluster production, and aligned fission are detailed. Fusion cross sections for a light system are reported.

1 Introduction

The ensemble of nuclear reactions occurring between two heavy ions colliding at energies between 20 and 100 MeV/u can be well studied thanks to

*permanent address: Institute of Nuclear Physics, ul. Radzikowskiego 152, 31-342 Kraków, Poland.

†present address: DRFC/STEP, CEA/Cadarache, F-13018 Saint-Paul-lez-Durance cedex, France.

4π arrays, among which INDRA [1] is one of the most powerful. In this energy domain, nucleon-nucleon collisions are less and less Pauli blocked and thus progressively overcome mean fields effects. Non-equilibrium (or dynamical) emissions become important, and their separation from emissions from thermalised objects should allow to extract fundamental information such as relaxation times, nuclear viscosity, in-medium σ_{nn} ...The classification as equilibrated or non-equilibrated emission is not obvious, as the intervening time scales are very similar: reaction times vary from 40 fm/c (peripheral collisions at 100 MeV/u) to 250 fm/c (central collisions at 30 MeV/u); the characteristic energy relaxation time is $\tau_{th} \sim$ 20-30 fm/c, independent of the incident masses and energy [2]; the life-time of a nucleus heated at T=5 MeV is around 50 fm/c [3]. Moreover the reaction times are long enough to allow the growth of fluctuations, leading to shape, volume, surface instabilities, either in the bulk of the system or in the overlap zone of the incident partners.

The dominant binary character of the collisions is now well established, most of them ending up with two hot (projectile-like PLF and target-like TLF) fragments in the exit channel [4]. In most of the reactions, one observes, besides particles which seems to originate from evaporation by these fragments, the emission of nucleons and light clusters with a parallel velocity intermediate between those of the two main fragments, and perpendicular velocities which can reach high values. Obviously these products are emitted in the early stages of the collisions, but many different processes can be invoked, such as direct emissions of nucleons or clusters, neck rupture, aligned fission ...Fusion processes vanish above 40-50 MeV/u. For light systems, phenomena similar to fusion-evaporation are still observed, while for heavy systems fusion-multifragmentation occurs. In this presentation, after some global quantitative estimates of the dynamical emissions we will focus on two well characterised processes among those giving rise to these products, namely direct emission of nucleons followed by coalescence, and aligned break-up of the projectile-like fragment. Then fusion cross sections for a light system will be discussed in the light of the stopping power of nuclear matter in dynamical simulations.

2 Global estimate and composition of dynamical emissions

If one defines dynamical emissions as all products but those coming from the isotropic evaporation of *PLF/TLF having reached thermodynamical and shape equilibrium*, then a global estimate of the mass lost in dynamical emissions can be tempted. Because of detection threshold effects one generally works on the PLF side only. Isotropic PLF evaporation is quantified and subtracted from

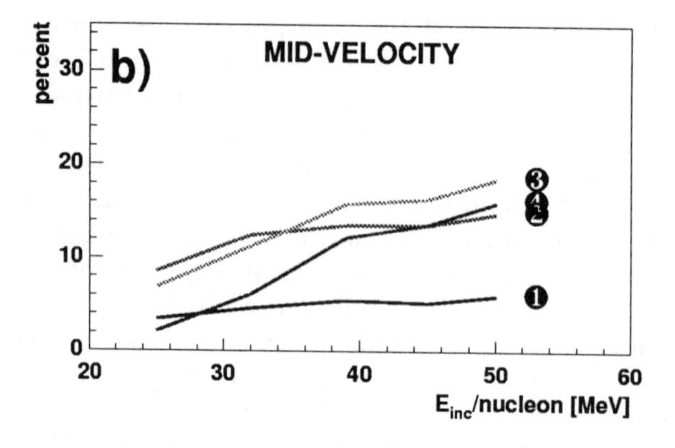

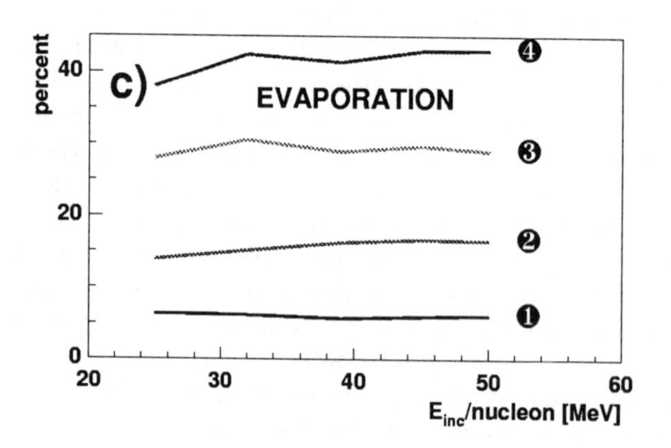

Figure 1. Xe+Sn system: Percentage of the forward detected charge emitted through dynamical (b) and evaporation (c) processes, as a function of incident energy. The numbers in black circles refer to an estimated (experimental) impact parameter, from peripheral (1) to mid-central (4). The heaviest (PLF) residue is not included, neither in b) nor in c). Extracted from [5].

the measured products, and the remaining part is attributed to dynamical emissions. The results, which strongly depend on the estimated PLF velocity (see refs [5,6] for more details) are shown in fig. 1 for the ^{129}Xe+^{nat}Sn system between 25 and 50 MeV/u. Dynamical emission becomes sizeable around 20 MeV/u, and significantly increases with energy, and with the centrality of the collision (curves labelled 1 to 4 correspond to average impact parameters from 9.2 to 4.7 fm); Conversely the evaporated charge increases with the centrality of the collisions, but appears to be independent of the energy *at a given impact parameter*: this could be interpreted as a saturation of the thermal excitation energy (*total, but not per nucleon*). For a lighter system, ^{36}Ar+^{58}Ni, the same evolution with centrality is observed, but there is no dependence on the energy between 52 and 95 MeV/u [6]. In this case the amount of charge attributed to dynamical emissions corresponds to the geometrical overlap of the incident nuclei, at least for impact parameters larger than $b_{max}/2$.

Dynamical and statistical emissions so isolated strongly differ in their compositions. While protons and α particles dominate in evaporation, the other H, He isotopes and the light fragments mostly arise from other processes. Neutron rich fragments are more abundant in dynamical emissions [7], in a heavy system with a large neutron excess such as Xe+Sn, as well as in a light system with isospin close to 1, ^{36}Ar+^{58}Ni. It was suggested that this could result from the coalescence of small symmetric clusters, favored in the low density intermediate region, with the neutron rich gas simultaneously present [8,9]. A variant would be to view a stretched neck zone as a molecular cluster structure, such structures being stabilized by an excess of neutrons [10].

The equilibrium non-equilibrium separation so realised is very crude. Besides imposing drastic restrictions on what is qualified as evaporation (for instance deformed systems, density anisotropy of the semi-spaces backward and forward of the PLF due to the close proximity of the target, are not considered), it mixes in the "dynamical emissions" products whose origin and time of emission may be rather different. In the next two sections, we will focus on high energy particles, probably emitted very early, and on the aligned fission, which should be a process occurring later in the collision course.

3 High energy light charged particles: Coalescence of direct nucleons

The origins of high energy nucleons and light clusters in intermediate energy heavy ion collisions can be manyfold. It was proposed that they come from the fast ejection of nucleons or clusters preformed inside the nucleus [11], or from projectile break-up [12]. In the INDRA experiments, it was observed that protons and light clusters of high transverse energy are focused in the reaction plane [13]. This suggests angular momentum effects on nucleons ejected after a single collision; direct light clusters would then be formed through the coalescence of such nucleons. Coalescence naturally occurs in AMD simulations of heavy ion collisions, where it becomes more important than mean field effects for cluster formation when the incident energy increases [14]. Coalescence has been included in BUU calculations [15], leading to reasonable reproduction of some experimental light charged particles (lcp) spectra [16]. A revival of another approach was used in the INDRA collaboration. Direct nucleon densities in the momentum space given by an intranuclear cascade code (ISABEL [17], with $\sigma_{nn} = \sigma_{nn}^{free}$) were used as input of a coalescence model, instead of the measured nucleon spectra as previously done [18]. The calculation was intentionally limited to the cascade stage,without subsequent

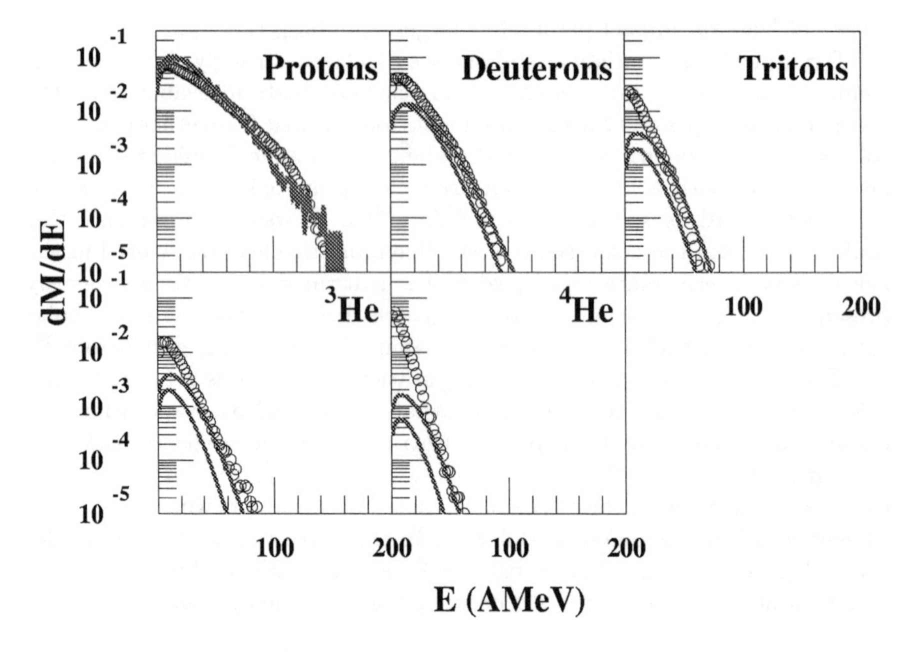

Figure 2. ^{36}Ar+^{58}Ni at 95 MeV/u: Comparison of the experimental (points) and calculated (lines) spectra of lcp emitted between 60 and 120°c.m. The double line for the calculation reflects uncertainties on the coalescence radius p_0. The impact parameter bin is 5-6 fm, and there is no normalisation factor. Extracted from [19].

evaporation. Comparison was done with the experimental spectra of lcp measured in the range 60-120°c.m. in the reaction ^{36}Ar+^{58}Ni at 95 MeV/u [19]; in this angular range, high energy particles appear more separated from the lcp evaporated by the PLF and TLF. Fig. 2 shows, for mid-peripheral collisions, that such a process quantitatively accounts for the higher energy part of the measured spectra (without any normalisation factor); the fraction of the lcp multiplicity which may arise from coalescence decreases when the particle mass increases (inherent to coalescence), and with decreasing centrality, due to smaller direct nucleon multiplicities.

4 Fission and aligned break-up of projectile-like fragments

For heavy projectiles ($Z_p > 40$) the binary break-up in two large fragments ("fission ") of the PLF, was studied in great details by the Nautilus and IN-

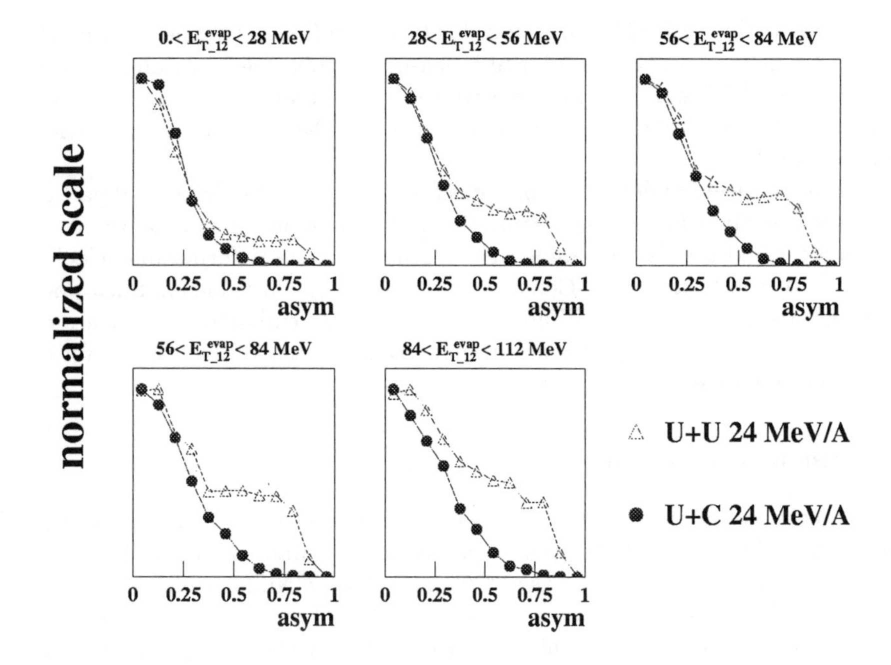

Figure 3. Fission fragment mass distributions (expressed vs asymmetry $\eta = (Z_1 - Z_2)/(Z_1 + Z_2)$) of a uranium projectile-like fragment after collisions with a U target (triangles) or a C target (points). The distributions are arbitrarily normalised at $\eta = 0$. Increasing $E_{T_1 2}^{evap}$ scale the impact parameter, from peripheral to mid-central collisions. Extracted from [20].

DRA collaborations [20], as it was in ref [21]. For a given projectile, "fission" mass distributions show striking differences depending on the bombarded target. Figure 3 illustrates this effect. A 24 MeV/u uranium colliding with a C target presents a "standard" fission distribution, where symmetric break-up dominates. When the target is a U nucleus, another component appears in addition to the previous one: it corresponds to an asymmetric break-up, and its relative importance increases with the centrality of the collision. The angular distributions of the two phenomena are also very different: while the scission axis is randomly oriented with respect to the recoil direction of the PLF, as expected, for standard fission, the asymmetric fission is strongly aligned on the PLF-TLF separation axis, and the smaller fragment is between the larger partner and the TLF. The relative velocity of the fission fragments is much higher for aligned break-up. Aligned break-up was observed for different projectiles (Xe, Gd, Ta, Pb, U). Its relative weight in binary break-ups increases for decreasing fissility of the PLF, so as to become the dominant phenomenon for Xe-like PLF.

The privileged fission direction signs the memory of the dynamical part of the reaction, the PLF emerging from the nuclear reaction with a deformation already beyond the fission saddle point. Then an extra deformation velocity adds up to the "standard" (Coulomb + thermal) fission velocity. Such phenomena are important contributors to "dynamical" emissions. They are of primary interest in bringing information on the viscosity of nuclear matter in the intermediate energy domain.

5 Fusion cross sections

Incomplete fusion cross sections for light systems were found to vanish beyond 30-40 MeV/u [22,23]. Similar data have been obtained with INDRA for different systems, ^{36}Ar+KCl, ^{36}Ar+^{58}Ni, and ^{58}Ni+^{58}Ni, for which the fusion excitation function [24] is shown in fig. 4. At 32 MeV/u the fusion cross section has already dropped to 4-5% of the reaction cross section, and it represents less than 1% beyond 40 MeV/u. The occurrence of fusion is linked to the stopping power of nuclear matter, and thus depends on in-medium σ_{nn}. It is therefore a strong experimental constraint to test dynamical simulations. Recently Landau-Vlasov simulations were performed in order to determine the threshold energy E_{th} defined as the energy at which, for a given impact parameter, the system fuses below E_{th} and remains binary above [25]. The mean field was implemented through a Gogny force with K_∞=228 MeV, and the residual interaction was simulated with the free, energy and isospin dependent σ_{nn} arbitrarily scaled with a constant factor which was varied from 0. to 1.5. The Ni+Ni excitation function of fig. 4 puts, in a sharp cut-off picture, the threshold energy for b=2 fm at 37 MeV/u; from the simulations it would require a very high σ_{nn}, around 1.5 times σ_{nn}^{free}. Similar high σ_{nn} were required to reproduce the angular and velocity distributions of fragments produced in dissipative Ar+Ag collisions at 27 MeV/u *with the same simulation* [26]. The enhancement of σ_{nn} was suggested in calculations with the Brueckner G-matrix [27,28], and we feel that it is striking to need the same enhancement of σ_{nn} to reproduce data obtained on different exit channels with different experimental devices. For fusion data however, one should keep in mind that a sharp cut-off picture is unrealistic, and semi-classical simulations including fluctuations predict a non-negligible percentage of fusion events for

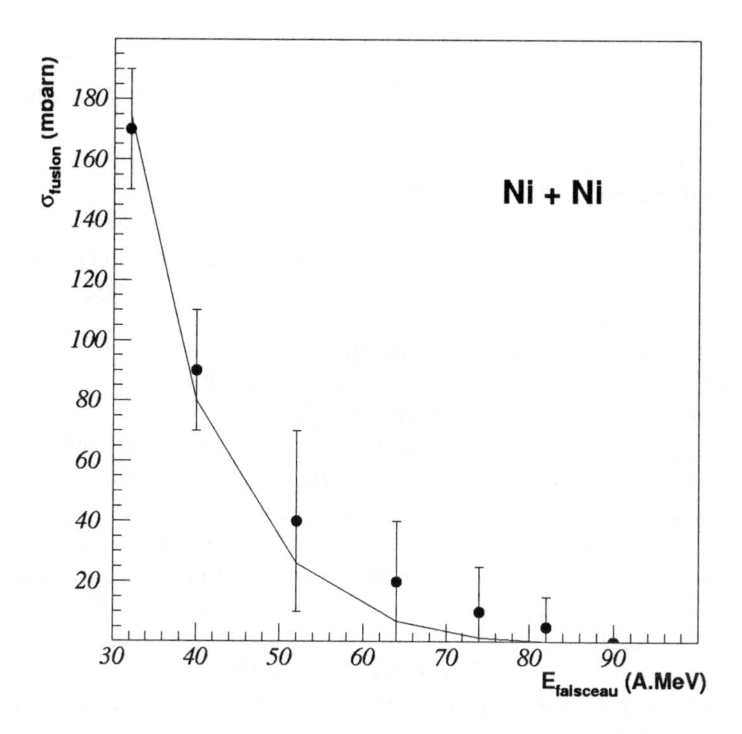

Figure 4. Fusion excitation function for the $^{58}Ni+^{58}Ni$ system. The points are experimental data and the line shows the prediction of the event generator SIMON. The abscissa axis is the lab. energy per nucleon. Extracted from [24].

rather large impact parameters [29].

6 Summary

In this paper we have touched on some dynamical features of heavy ion collisions which clearly stand out. Dynamical emissions may occur within different time scales, but can all be linked to the stopping (viscosity) of nuclear matter.The analyses presented here tempted to separate dynamical and statistical emissions. Such separations may be avoided if one only search to constrain the ingredients of transport models with experimental results, and approaches in this sense are underway [30].

References

1. J. Pouthas *et al.*, (INDRA coll.), *Nucl. Instrum. Methods* A **357**, 418 (1995).
2. B. Borderie *et al.*, *Z. Phys.* A **357**, 7 (1997).
3. B. Borderie, *Ann. Phys. Fr.* **17**, 349 (1992).
4. V. Métivier *et al.*, (INDRA coll.), *Nucl. Phys.* A , (in press).
5. E. Plagnol *et al.*, (INDRA coll.) *Phys. Rev.* C **61**, 014606 (1999)
6. T. Lefort *et al.*, (INDRA coll.), *Nucl. Phys.* A , (in press).
7. F.Dempsey *et al.*,PRC **54**, 1710 (1996).
8. L.G. Sobotka *et al.*,PRC **55**, 2109 (1997).
9. Ph. Chomaz and F. Gulminelli, PLB **447**, 221 (1999).
10. W. Von Oertzen, *Z. Phys.* A **357**, 355 (1997).
11. Chinmay Basu and Sudip Ghosg, *Phys. Rev.* C **56**, 3248 (1997).
12. H. Fuchs and K. Möhring, *Rep. Prog. Phys.* **57**, 231 (1994).
13. R. Dayras *et al.*, (INDRA coll.), unpublished data.
14. A. Ono, in *Proceedings of the 7th Conference on Clustering Aspects of Nuclear Structure and Dynamics* Rab, Croatia, 1999, (World Scientific).
15. P. Danielewicz *et al.*, *Nucl. Phys.* A **553**, 712 (1991) and *Phys. Rev.* C **46**, 2002 (1992).
16. D. Prindle *et al.*, *Phys. Rev.* C **57**, 1305 (1998).
17. Y. Yariv and Z. Fraenkel, *Phys. Rev.* C **20**, 2227 (1979) and *Phys. Rev.* C **24**, 488 (1981).
18. T.C. Awes *et al.*, *Phys. Rev.* C **24**, 89 (1981) and C **25**,2361 (1982).
19. P. Pawłowski *et al.*, (INDRA coll.), to be published.
20. F. Bocage *et al.*, (INDRA and Nautilus coll.), subm. to *Nucl. Phys.* A.
21. A. Stefanini *et al.*, *Z. Phys.* A **351**, 167 (1995).
22. A. Fahli *et al.*, *Phys. Rev.* C **34**, 161 (1986)
23. P. Box *et al.*, *Phys. Rev.* C **50**, 934 (1993)
24. A.M. Maskay, *thèse*, Université Lyon-1, LYCEN - T 9969.
25. Z. Basrak and P. Eudes in *Proceedings of the XXXVII International Winter Meeting on Nuclear Physics*, Bormio, 1999, edited by I. Iori (University of Milan Press, Milan, 1999).
26. F. Haddad *et al.*, *Z. Phys.* A **354**, 321 (1996)
27. A. Bohnet *et al.*, *Nucl. Phys.* A **494**, 349 (1989).
28. J. Cugnon *et al.*, *Phys. Rev.* C **35**, 861 (1987)
29. M.Colonna *et al.*, *Nucl. Phys.* A **642**, 449 (1998).
30. J.F. Lecolley *et al.*, (INDRA coll.), *Nucl. Instrum. Methods* A , (in press).

THERMODYNAMICAL FEATURES OF NUCLEAR SOURCES STUDIED WITH INDRA

B. BORDERIE[1], CH.O. BACRI[1], R. BOUGAULT[2], A. CHBIHI[3], PH. CHOMAZ[3], M. COLONNA[4], J.D. FRANKLAND[1,3], A. GUARNERA[1], F. GULMINELLI[2], N. LE NEINDRE[2], M. PÂRLOG[5], M.F. RIVET[1], S. SALOU[3], G. TĂBĂCARU[1,5], L. TASSAN-GOT[1], J.P. WIELECZKO[3] G. AUGER[3], N. BELLAIZE[2], F. BOCAGE[2], B. BOURIQUET[3], R. BROU[2], P. BUCHET[6], J.L. CHARVET[6],, J. COLIN[2], D. CUSSOL[2], R. DAYRAS[6], N. DE CESARE[7], A. DEMEYER[8], D. DORÉ[6], D. DURAND[2], E. GALICHET[1,9], E. GENOUIN-DUHAMEL[2], E. GERLIC[8], S. HUDAN[3], D. GUINET[8], F. LAVAUD[1], P. LAUTESSE[8], J.L. LAVILLE[3], J.F. LECOLLEY[2], C. LEDUC[3], R. LEGRAIN[6], O. LOPEZ[2], M. LOUVEL[2], A.M. MASKAY[8], L. NALPAS[6], J. NORMAND[2], J. PÉTER[2], E. PLAGNOL[1], E. ROSATO[7], F. SAINT-LAURENT[3]*, J.C. STECKMEYER[2], B. TAMAIN[2], O. TIREL[3], E. VIENT[2], M. VIGILANTE[7], C. VOLANT[6],
(INDRA COLLABORATION)

1 *Institut de Physique Nucléaire, IN2P3-CNRS, F-91406 Orsay Cedex, France.*
2 *LPC, IN2P3-CNRS, ISMRA et Université, F-14050 Caen Cedex, France.*
3 *GANIL, CEA et IN2P3-CNRS, B.P. 5027, F-14076 Caen Cedex, France.*
4 *Lab. Nat. del Sud, Via S. Sofia 44, I-95123 Catania, Italy.*
5 *National Institute for Physics and Nuclear Engineering, RO-76900 Bucharest-Măgurele, Romania.*
6 *DAPNIA/SPhN, CEA/Saclay, F-91191 Gif sur Yvette Cedex, France.*
7 *Dipartimento di Scienze Fisiche e Sezione INFN, Universitį Napoli 'Federico II", I-80126 Napoli, Italy.*
8 *Institut de Physique Nucléaire, IN2P3-CNRS et Université, F-69622 Villeurbanne Cedex, France.*
9 *Conservatoire National des Arts et Métiers, F-75141 Paris cedex 03.*

Vaporized and multifragmenting sources produced in heavy ion collisions at intermediate energies have been detected with the multidetector INDRA. The properties of such highly excited nuclear sources which undergo a simultaneous disassembly into particles (vaporization) are discussed in terms of thermal and chemical equilibrium at freeze-out. For multifragmenting sources which undergo a simultaneous disassembly into particles and fragments, properties of fragments are compared to different models assuming statistical fragment emission or spinodal instabilities. Experimental charge correlations for fragments show a weak but non ambiguous signal of the enhancement of events with nearly equal size fragments. Such a signal can be interpreted as a trace of spinodal instabilities.

---

*present address: DRFC/STEP, CEA/Cadarache, F-13018 Saint-Paul-lez-Durance cedex, France.

1 Introduction

The decay of highly excited nuclear systems through a simultaneous disassembly is, at present time, a subject of great interest in nucleus-nucleus collisions at intermediate energies. Although this process has been observed for many years, its experimental knowledge in the Fermi energy domain was strongly improved only recently with the advent of powerful 4π devices such as INDRA [1]. Well defined systems or subsystems which undergo vaporization or multifragmentation can be thus carefully selected.

To describe the simultaneous disassembly of highly excited nuclear sources produced in nucleus-nucleus collisions, many models presuppose that when they disintegrate these sources have achieved partial or complete thermodynamical equilibrium [2,3,4,5,6,7,8]. This hypothesis is essential if one wants to describe the sources and understand the generated mechanisms for disassembly as well as the properties by means of macroscopic variables such as temperature and related pressure and density. However such an assumption is *a priori* not obvious if we bear in mind the shortening of all time scales with respect to those involved at low and moderate excitation energies. The very fundamental questions of the degree of equilibrium reached before disassembly and of the mechanism responsible for disassembly can be only answered by confronting these models to complete data on the deexcitation of well identified nuclear sources.

2 Thermal and chemical equilibrium for vaporizing sources

Let us first report on a comparison of the properties of vaporized quasi-projectiles (QP), produced in binary dissipative collisions between ^{36}Ar and ^{58}Ni nuclei at 95 MeV/u incident energy [9], with a quantum statistical model describing a gas of fermions and bosons in thermal and chemical equilibrium which includes a final state volume interaction (van der Waals-like behavior) and side feeding [10,11].

These sources are very interesting because they represent an extreme deexcitation mode for very hot pieces of nuclear matter, close to the expectation of a supercritical nuclear gas [12,13]. In the model, to cover the experimental range in excitation energy per nucleon of the source (ε^*), the temperature had to be varied from 10 to 25 MeV. Isospin (N/Z) was fixed to 1, which is very close to the N/Z of the system. Finally the freeze-out density has been fixed to $\rho = \rho_0/3$, in order to reproduce the experimental ratio between the proton and alpha yields at ε^*=18.5 MeV.

Fig 1 shows the evolution of the chemical composition as a function of

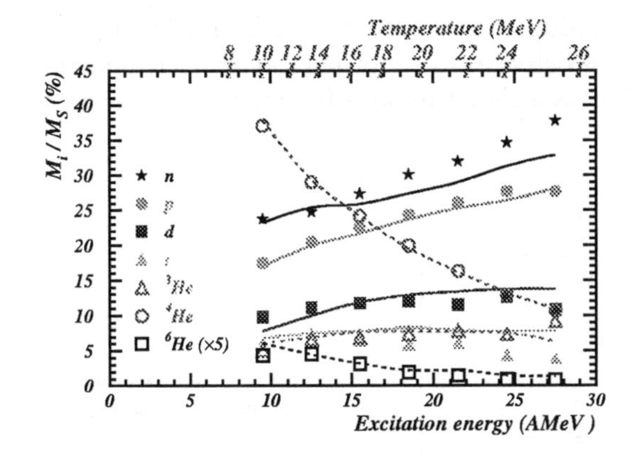

Figure 1. Composition of the QP as a function of its excitation energy. Symbols are for data while the lines (dashed for He isotopes) are the results of the model. The temperature values used in the model are also given. (from [11]).

excitation energy of the source; we observe an excellent agreement between data and model. Average kinetic energies of the different particles as well as second moments of the chemical composition are also well reproduced over the whole excitation energy range [11].

The scenario producing such sources may be the following: after an interaction time of around 40 fm/c (estimated from the time needed by the two nuclei to pass through each other), partners of collisions are on the way to thermalization. Then, due to the very high excitation energies involved, the partners start to dissociate and reach a density around $\rho = \rho_0/3$ within about 30 fm/c (see for example ref. [14,15]). Finally at low density and at very high temperatures in the supercritical region of the phase diagram [12,13], the emission properties of vaporized sources at freeze-out are fixed by thermodynamical equilibrium .

3 Experimental evidence for bulk effect in multifragmentation

Using the INDRA detector, multifragmenting fused systems have been carefully selected for two reactions leading to the same available excitation energy per nucleon ($\sim 7MeV$) : $^{129}Xe + ^{nat}Sn$ at 32 MeV/u and $^{155}Gd + ^{nat}U$ at 36 MeV/u. The selection was performed by examining the evolution of fragment kinematics as a function of the dissipated energy and loss of memory of the entrance channel [16,17].

Fig 2 shows that, for the two fused systems, we observe the same Z distribution for fragments while, as a consequence, the fragment multiplicities scale as the size of the total systems. This independence of the Z distribution, experimentally observed for the first time [16], can be considered as a strong evidence for a bulk effect to produce fragments. It can be related to bulk instabilities in the liquid-gas coexistence region of nuclear matter (spinodal region) or perhaps simply taken as a signature of a full exploration of phase space for such heavy systems.

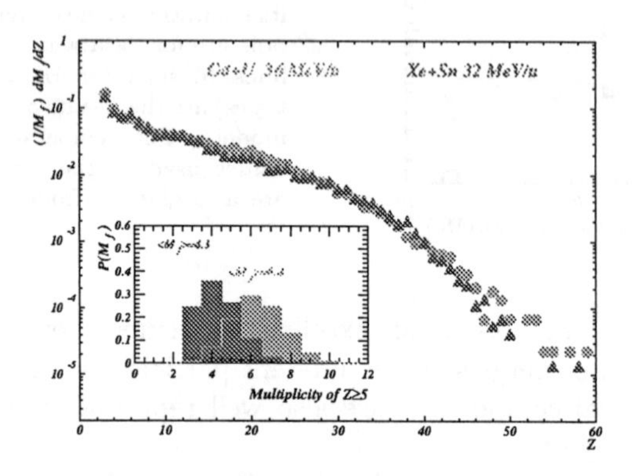

Figure 2. Experimental fragment multiplicity distributions and differential charge multiplicity distributions for the 32 MeV/u Xe+Sn (black histogram and triangles) and 36 MeV/u Gd+U (grey histogram and circles).

4 Comparison to models

Many theories have been developed to explain multifragmentation (see for example ref. [18] for a general review of models). Among the models some are related to statistical approaches whereas others, more ambitious, try to describe the dynamical evolution of systems, from the beginning of the collision between two nuclei to the fragment formation. In the latter, theoretical scenarios to be compared with experimental data are simulated (with approximations) via molecular dynamics, eventually including a stochastic implementation [19] or momentum fluctuations [20], or stochastic mean field approaches which account for the dynamics of the phase transition. It is this last type of approach that we shall first use for a comparison with our experimental observables. In simulations spinodal decomposition of hot and dilute nuclear systems are mimicked and, relative to the standard nuclear Boltzmann treatment, a powerful approximate tool is provided by the Brownian One-Body (BOB) dynamics [21,8]. The BOB dynamics introduces a noise by means of a brownian force in the mean field whenever the local conditions correspond to spinodal instability. Thus the magnitude of the force is adjusted to produce the same growth rate as the full Boltzmann-Langevin theory for the most unstable modes in nuclear matter prepared at the corresponding density and temperature [22].

In reference [16], it was shown that the multiplicity and charge distributions for both systems were correctly reproduced, but that the calculated fragment kinetic energies were too low. Since then two major improvements have been implemented in the simulations. Firstly, in deterministic or stochastic simulations, the collision term is assumed to be local in both space and time; this simplification is no more justified when fast unstable modes are present as in the spinodal region. Then quantal fluctuations connected with collisional memory effects are now taken into account with the determinant result of about doubling the overall amplitude of fluctuations [23]. Secondly the previous simulation method consisted in simply injecting the correct (classical) fluctuations when the system enters the spinodal region (Stochastic Initialisation Method) [24]. This method led to a damping of the fluctuations. The fragment formation time was thus artificially (and incorrectly) increased, leading to smaller kinetic energies due to the decrease with time of the radial expansion of the system. Then a better simulation method (BOB) was proposed [21,8] which takes into account the time dependence of the fluctuation source. This method is analogous to the Langevin treatment of Brownian motion.

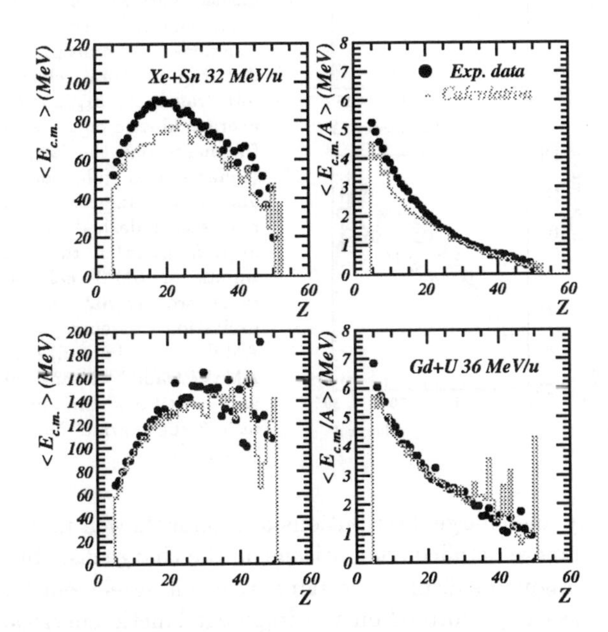

Figure 3. Experimental (points) and simulated (histograms) average fragment kinetic energies. Histograms are the calculated values when the thermal kinetic energy is added; indeed the introduced fluctuations act mostly on the configuration space but they do not provide a good description of fluctuations in momentum space. The results are for total energies on the left, and in MeV/u on the right.

The simulation (for head-on collisions, as fluctuations calculated for infinite matter can be rather safely extrapolated only to systems having a spherical symmetry) thus proceeds as follows [17]: the dynamics is followed through a BNV/BOB simulation, up to the time where the fragments are formed and well separated. The fluctuating force starts acting at the time of maximum compression (25-30% higher than normal density) when local thermal equilibrium is fulfilled. Slightly after entering the spinodal region, the systems are in thermal equilibrium at low density (0.4 the normal density), with a temperature of 4 MeV. The radial velocity at the surface is rather large ~0.1c. When the fragments are well separated, they still bear an average excitation energy of ~3.2 MeV/u. Then the de-excitation of the hot primary fragments, whose mass accounts for 80% of the total system mass is followed through the evaporation part of the SIMON code. Finally the results are filtered to take into account the experimental set-up.

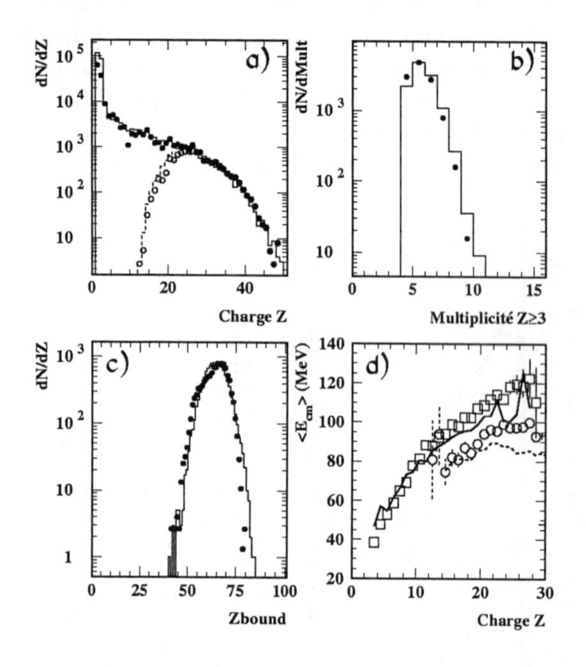

Figure 4. Experimental (points, open square and open points) and SMM simulations (histograms and full and dashed curves): a) fragment charge distrib., b) fragment multiplicity distrib., c) total charge into fragments Z_{bound}, d) average kinetic energies of fragments but the largest (squares and full curve) and of the largest (open circles and dashed curve) as a function of their Z. Inputs in SMM calculations are A=202, Z=85, excitation energy=5.0 AMeV (thermal)+0.6 AMeV (radial expansion) and partitions are fixed at 1/3 the normal density (from [27]).

Calculated multiplicity and charge distributions of fragments well match the experimental ones and more detailed comparisons on the charge distributions of the three largest fragments display the same excellent agreement [17]. Finally the most crucial test is performed on the fragment kinetic energies. Figure 3 shows that the simulations now correctly reproduce the fragment energies for the Gd+U system. For Xe+Sn, the calculated energies fall within 20% of the measured values, which is satisfactory if one remembers that there were no adjustable parameters in the simulation.

A very good agreement with experimental data is also observed when they are compared to the statistical model SMM [2,25,26,27,28]. Figure 4 shows this comparison for the Xe+Sn system at 32 MeV/u. For Gd+U the thermal part of the excitation energy is also fixed at 5.0 AMeV to well reproduce experimental data. However in this case the dynamical phase of the reaction is ignored and parameters such as the mass and charge of the multifragmenting system, its excitation energy, its volume (or density) and the added radial expansion have to be backtraced to the experimental data.

5 Charge correlation of fragments and spinodal decomposition

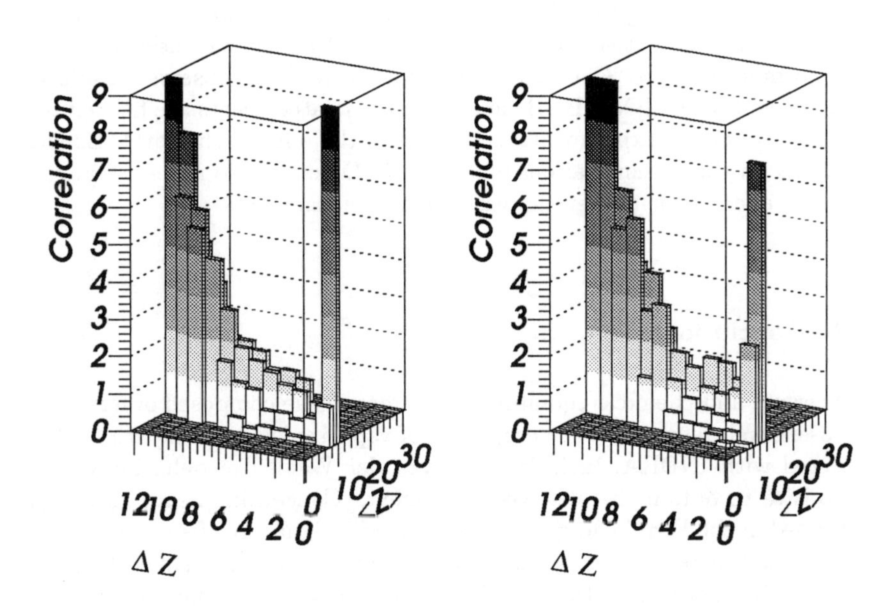

Figure 5. Fragment charge correlations for the reaction ^{129}Xe+^{nat}Sn at 32 MeV/u: Comparison between experiment (left) and BOB calculations (right) for fragment multiplicity equal to 6 (from [29]).

To put ultimate constraints on models, we can and should also compare fragment correlations in events (fragment size [29] and fragment reduced velocities [25]). In particular, for fragment size correlations, if spinodal instabilities occur the most unstable modes present in the spinodal region are predicted to favor "primitive" partitions of equal size fragments (Z:10-15) [24]. But this simple picture is blurred by several effects: the beating of different modes, the coalescence of the "primitive " fragments and the finite size of the system and consequently experimental Z distributions (see fig 2) do not present any visible enhancement around Z=10-15. Then how to search for a possible very weak "fossil" signature of spinodal decomposition? A few years ago a new method called higher order charge correlations was proposed in [30]. To search for very weak signals, all fragments in one event (average fragment charge Z and the standard deviation per event ΔZ) are used to build the charge correlation for each fragment multiplicity. Due to statistics in experiments this method could be only applied on the Xe+Sn system [29]. A signal is observed for the different multiplicities from 3 to 6. An example of the observed experimental correlations is shown in figure 5 (left part). Note that if we enlarge our data sample to the very dissipative collisions (dominated by binary collisions) the signal is no more present. Concerning the models we again observe an impressive agreement of the BNV/BOB simulation (simulated events are filtered to take into account the experimental set-up) with the data (right part). Conversely results on charge correlations built with SMM events do not show any signal for an enhancement of events with equal size fragments.

6 Conclusions

Thermal and chemical equilibrium for a gas phase composed of fermions and bosons well reproduces the properties of vaporized sources experimentally selected with INDRA. Inclusion of a van der Waals-like behavior was found decisive to obtain the observed agreement. Concerning heavy fused sytems formed in the Fermi energy domain, a "fossil" signature of spinodal decomposition as the mechanism responsible for multifragmentation of such systems is probably observed for the first time. It consists in an enhancement of events with equal size fragments. A full dynamical model including also the dynamics of spinodal instabilities reproduces impressively all the experimental observables. However, except the weak "fossil" signal and the precise description at freeze-out (velocity correlations for fragments) [25], statistical models, when adding a radial expansion, reproduce also very well the experimental observables. This fact seems to be a strong indication that dynamical instabilities responsible for mutifragmentation lead to an exploration of practically all the phase space.

References

1. J. Pouthas *et al.*, (INDRA coll.), *Nucl. Instrum. Methods* A **357**, 418 (1995).
2. J. Bondorf et al., *Phys. Rep.* **257**, 133 (1995) and references therein.
3. D.H.E. Gross, *Rep. Prog. Phys.* **53**, 605 (1990) and references therein.
4. A. Z. Mekjian, *Phys. Rev.* C **17**, 1051 (1978).
5. H. Stöcker and W. Greiner, *Phys. Rep.* **5**, 277 (1986), J. Konopka et al., *Phys. Rev.* C **50**, 2085 (1994).
6. W. A. Friedman, *Phys. Rev.* C **42**, 667 (1990).
7. B. Borderie, *Ann. Phys. Fr.* **17**, 349 (1992), P. Abgrall et al., *Phys. Rev.* C **49**, 1040 (1994).
8. A. Guarnera et al., *Phys. Lett.* B **403**, 191 (1997).
9. M.F. Rivet et al., (INDRA coll.) *Phys. Lett.* B **388**, 219 (1996).
10. F. Gulminelli and D. Durand, *Nucl. Phys.* A **615**, 117 (1997).
11. B. Borderie *et al.*, (INDRA coll.), *Eur. Phys. J.* A **6**, 197 (1999), *Phys. Lett.* B **388**, 224 (1996).
12. H. R. Jaqaman et al., *Phys. Rev.* C **29**, 2067 (1984).
13. J. N. De et al., *Phys. Rev.* C **55**, 1641 (1997).
14. D. Vautherin et al., *Phys. Lett.* B **191**, 6 (1987).
15. L. Vinet et al., *Nucl. Phys.* A **468**, 321 (1987).
16. M.F. Rivet et al., (INDRA collaboration), *Phys. Lett.* B **430**, 217 (1998).
17. J.D. Frankland, thèse, Université Paris XI Orsay, 1998, IPNO-T-98-06.
18. L. G. Moretto and G. J. Wozniak, *Ann. Rev. of Nuclear and Particle Science* **43**, 379 (1993) and references therein.
19. A.Ono an H. Horiuchi, *Phys. Rev.* C **53**, 2958 (1996).
20. Y. Sugawa and H. Horiuchi, *Phys. Rev.* C **60**, 064607-1 (1999) and references therein.
21. Ph. Chomaz et al., *Phys. Rev. Lett.* **73**, 3512 (1994).
22. Ph. Chomaz, *Ann. Phys. Fr.* **21**, 669 (1996).
23. S. Ayik and J. Randrup, *Phys. Rev.* C **50**, 2947 (1994).
24. A. Guarnera et al., *Phys. Lett.* B **373**, 267 (1996).
25. S. Salou, *thèse*, Université de Caen, GANIL T 97 06.
26. R. Bougault et al., (INDRA coll.) in *Proc. of the XXXV Int. Winter Meeting on Nuclear Physics*, Bormio, Italy (1997), ed I. Iori, Ricerca scientifica ed educazione permanente, page 251.
27. N. Le Neindre, *thèse*, Université de Caen, LPCC T 99 02.
28. Ch. O. Bacri, private communication
29. G. Tăbăcaru, *thèse*, Université Paris XI Orsay, in preparation.
30. L. G. Moretto et al.,*Phys. Rev. Lett.* **77**, 2634 (1996).

SPECIAL ASPECTS OF THE BARYONIC FLOW IN RELATIVISTIC HEAVY ION COLLISIONS

M. PETROVICI
National Institute for Physics and Nuclear Engineering,
P.O. - MG6, 76900 Bucharest, ROMANIA
E-mail: mpetro@ifin.nipne.ro

FOPI COLLABORATION

Specific experimental results and qualitative and quantitative comparisons with model predictions relative to the azimuthal and polar angle dependence of the flow phenomena in semi-central and respectively central heavy ion collisions are presented.

1 Introduction

After about three decades of theoretical and experimental studies of relativistic heavy ion collisions the question of equation of state (EoS) of nuclear matter produced in such collisions does not have yet a definite answer. Dynamical effects, local thermal and chemical equilibration, momentum dependent interaction and in medium effects turned out to play an important role. Complete and accurate experimental information turned out to be mandatory for understanding these effects using realistic force in transport microscopic calculations. Based on this force one could calculate the equation of state of nuclear matter. The results presented here are based on a series of experiments on symmetric heavy ion collisions at energies between 90 and 400 A·MeV using a complete configuration of the FOPI setup. Details on the experimental configuration and analysis technique can be found elsewhere [1,2,3,4,6,7].

Section 2 is dedicated to the experimental results on transverse momentum and reaction product dependence of the incident energy where a transition from in-plane to out-of-plane azimuthal enhancement takes place. Qualitative explanations of the observed trends based on a semi-analytical hydrodynamical model are presented. The influence of the baryonic and N/Z content of the spectator matter on the mean kinetic energy and yield azimuthal distributions, respectively are analysed in Section 3. In Section 4 is presented the polar angle- and A_{part}-dependence of the azimuthally symmetric correlated motion in highly central collisions. Conclusions are presented in Section 5.

2 Transition energy - transverse momentum and reaction product dependence

Detailed studies of the incident energy at which the azimuthal distributions change their character from an in-plane, rotational-like to an out-of-plane, squeeze-out pattern have been done in the last period [9,10,11,4,12,13,6,5,18]. For a given centrality range, transverse momentum and mass of the analysed reaction products, the squeeze-out ratio - R_N=(1-a_2)/(1+a_2), (a_2 being the second coefficient of a Fourier expansion in azimuth), is represented as a function of incident energy (E_{inc}). Squeeze-out signal corresponds to $R_N > 1$ while in-plane enhancement of the azimuthal distribution is characterized by $R_N < 1$. The incident energy corresponding to $R_N = 1$ value is defined as transition energy - E_{tran}. Fig.1 shows the experimental E_{tran} values as a function of scaled transverse momentum $p_t^{(0)} = (p_t/A)/(p_P^{cm}/A_P)$ for different reaction products for a centrality corresponding to 4-6 fm impact parameter interval. A continuous decrease of the E_{tran} value as a function of $p_t^{(0)}$ is evidenced for all analysed particles and the difference in E_{tran} values for different particles for a given $p_t^{(0)}$ is decreasing towards larger values of $p_t^{(0)}$. For a rotating emitting source one would expect a larger in-plane alignment for heavier fragments [10]. This effect alone can not explain the mass dependence of E_{tran}. Therefore a dynamical effect has to be considered besides the pure geometrical one of shadowing. For this centrality the number of participating nucleons is about 200.

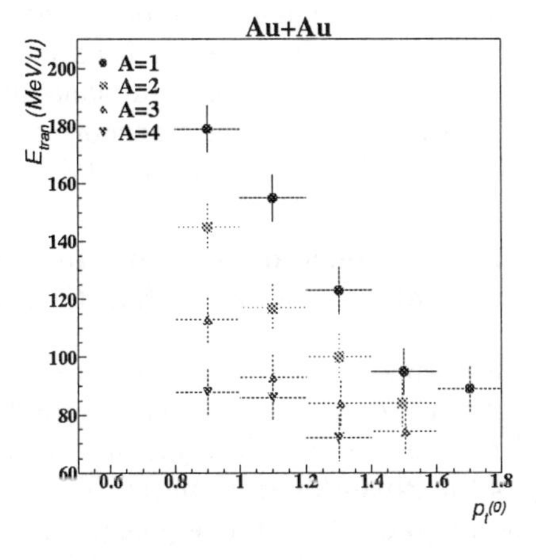

Figure 1. E_{tran} values as a function of $p_t^{(0)}$ for CM3 centrality and A=1,2,3 and 4.

For such a fireball we calculated the expansion dynamics using a semi-analitical hybrid model [14]. A two dimensional $p_t^{(0)}$ versus break-up time of the yield distribution for p, d, α and Li fragments can be followed in Fig.2 At large $p_t^{(0)}$ the contribution comes from the same range in the break-up time for all particles, corresponding to large expansion velocities. As far as the squeeze-out signal is the largest at high $p_t^{(0)}$, it follows that at this break-up times the shadowing reaches the maximum value. At lower values of $p_t^{(0)}$, the expansion zones are less localized specially for light fragments due to larger contribution of the thermal velocities relative to the collective ones, the shadowing being less effective on the in-plane yields. One could observe also that heavier fragments at lower $p_t^{(0)}$ are preferentially produced later in the expansion process relative to the light ones.

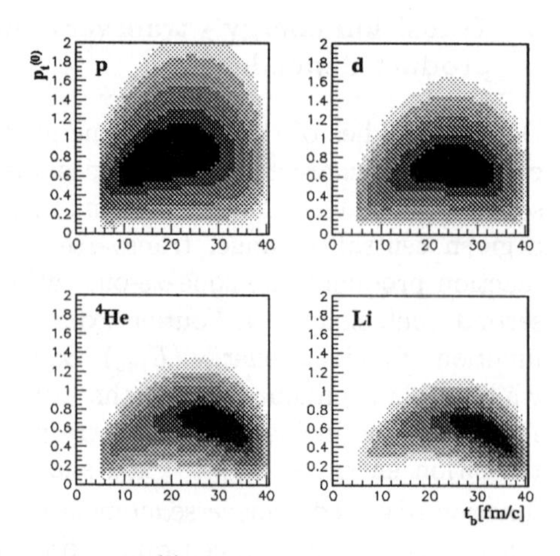

Figure 2. $p_t^{(0)}$ - break-up time distribution for A=1,2,4 and 7 reaction products

During the expansion process the rotational-like motion of the source is attenuated by the faster emitted particles, the slower ones being exposed at weaker rotational effect - in-plane alignement. Detailed comparison with predictions of microscopic transport codes is on the way, the main difficulty coming from the low statistics of IMFs produced by them.

3 The influence of the baryonic and N/Z spectator matter content on the azimuthal distributions

Incident energy excitation functions of the azimuthal distributions of flow, temperature or mean kinetic energy for different particles, different centralities and mass of the colliding systems are of real interest for deeper understanding of the expansion mechanism. They could be used to discriminate between pressure gradient or simple preequilibrium or shadowing contributions to the observed azimuthal anisotropies [16,17,18]. In this contribution we will concentrate on the influence of the baryonic and N/Z content of the spectator matter on the azimuthal distributions which seems to be sensitive observables to the in medium nucleon-nucleon interaction and its isospin dependence.

$< E_{kin}^{cm} >$ azimuthal distributions for different reaction products follow a $< E_{kin}^{cm} > = E_0$ - ΔE·cos2Φ behaviour. $E_{max}=E_0$ + ΔE and $E_{min}=E_0$ - ΔE corresponding to out-of-plane and respectively in-plane mean kinetic energies and 2·ΔE values for Au + Au, Xe + CsI, Ni + Ni, CM3 centrality (4-6 fm), at 250 A·MeV, Z=2 fragments, as a function of A_{part} (the number of

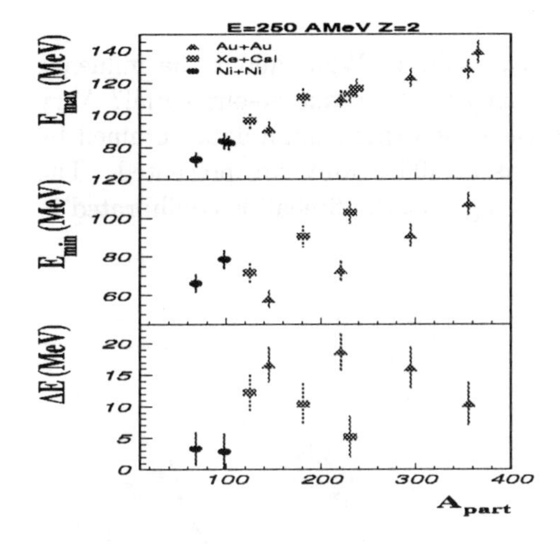

Figure 3. E_{max}, E_{min} and 2·ΔE as a function of A_{part}

the participaing nucleons estimated using straight trajectories) can be followed ig Fig.3. The out-of-plane mean kinetic energy E_{max} shows almost a linear dependence as a function of A_{part} with minor difference between different systems (aspect ratio of the overlaping zone) while a large differences in E_{min} between the three systems is observed. Simple geometrical considerations (see Fig.4,5) show that for A_{part}=200 nucleons, while the aspect ratio of Au+Au and Xe+CsI differ by 20%, the thickness of the spectator matter in the reaction plane varies by almost a factor of three. As far as small changes in the aspect ratio do not show large effects in the mean kinetic energy in the region not hindered by the spectators and the difference in the transit time is of the order of 1 fm/c, the conclusion is that the thickness of the spectator matter is the main reason

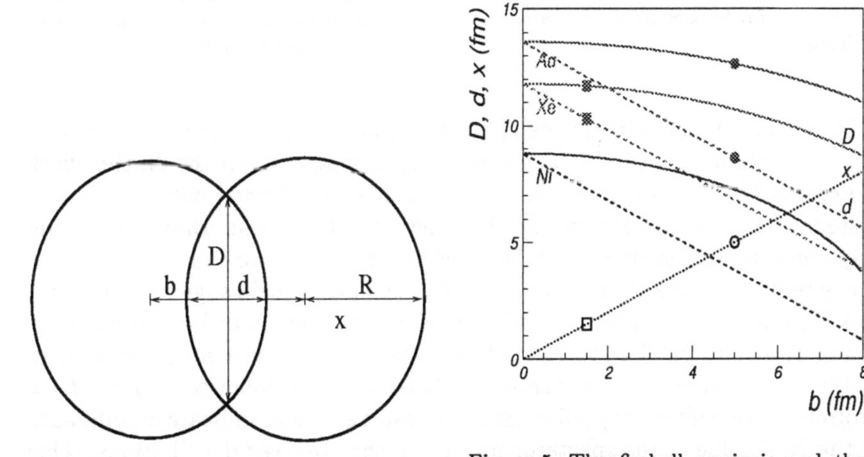

Figure 4. Collision geometry projected on a plane perpendicular to the collisions axis

Figure 5. The fireball semiaxis and the in-plane spectator thickness for the projection presented in Fig.4

for the observed difference in the E_{min} values. What about the influence of the N/Z content of the spectator matter on the squeeze-out signal? Very preliminary results of such a study using the experimental data obtained by us in Ru + Zr and Zr + Ru collisions at 400 A·MeV are presented. The situation is schetched in Fig.6. Supposing that the fireball is equilibrated in

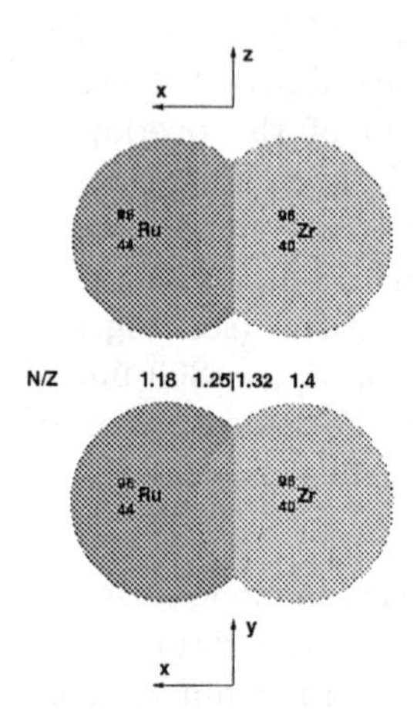

Figure 6. Collision geometry projected on the reaction plane - up and perpendicular to the collisions axis - down. N/Z values corresponding to different zones for 5 fm impact parameter are indicated.

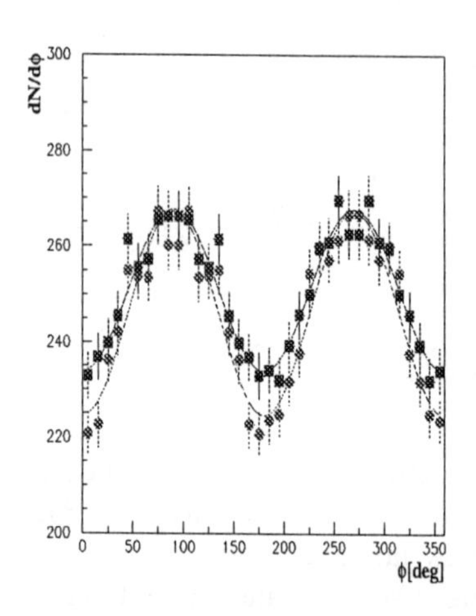

Figure 7. Azimuthal proton distributions for Ru + Zr (dots) and Zr + Ru (squares) measured between 0-90 and 270-360 deg and reflected relative to 90 and 270, respectively

N/Z we have to do with an expanding object in the presence of spectator matter asymmetric in N/Z. As far as the asymmetry is very small, the effect is expected to be also small. For this reason we performed our studies in a system of coordinates rotated by the sidewards flow angle. Breaking in this way the azimuthal symmetry of the experimental device we had to symmetrise the experimental distributions obtained in the azimuthal range 0-90 deg and 270-360 deg relative to 90 deg and 270 deg respectively in order to obtain the complete azimuthal distribution. The results for 4-6 fm centrality and a range in the scaled transverse momentum of 0.8-1.2 can be followed in Fig. 7 for protons. Although quite preliminary, the result seems to indicate an influence of the N/Z value of the spectator matter on the observed distributions. This could become a sensitive probe for studying the isospin dependence of the in-medium nucleon-nucleon interaction used in microscopic transport codes.

4 Polar angle- and A_{part}- dependence of the correlated motion in heavy ion central collisions

Highly central collisions (1% cross section) have been slected using $E_{rat} = \sum_i E_{\perp,i} / \sum_i E_{\parallel,i}$ observable, where i reffers to all particles identified by our detector. The experimental average kinetic energies represented as a function of mass of the corresponding reaction products show a linear dependence. The extracted slopes from such representations are presented in Fig.8 as a function of incident energy for the three symmetric systems and for two polar regions, $\theta_{cm} \in$ (25°-45°) (with a condition in azimuth of $\phi \in$ (90°(270°) ± 45°)) and $\theta_{cm} \in$ (80°-100°).

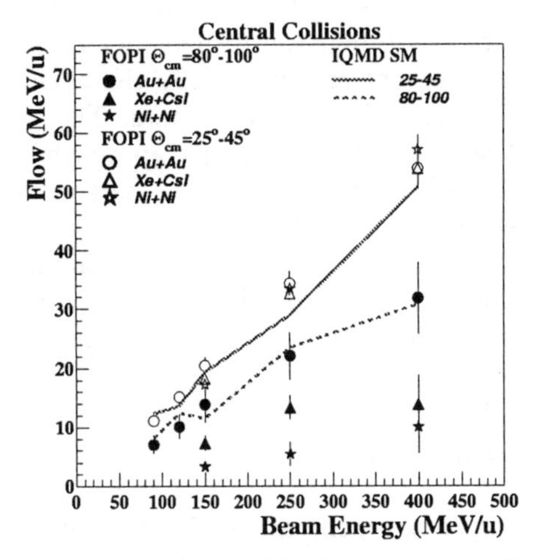

Figure 8. Excitation function for the correlated motion for Au+Au, Xe+CsI and Ni+Ni in two polar regions and comparison with the IQMD-SM model predictions

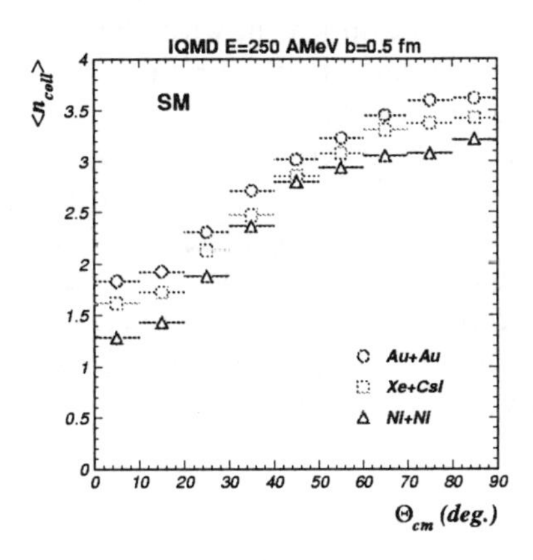

Figure 9. The average number of collisions of each nucleon as function of c.m. polar angle, predicted by IQMD model.

One could easily follow the difference between the extracted slopes, representing the amount of the correlated motion per nucleon, in the forward and transversal direction relative to the collision axis. While in the forward region ($\theta_{cm} \in$ (25°-45°)) no dependence of the extracted slopes on the mass of the colliding systems is observed, in the transversal direction ($\theta_{cm} \in$ (80°-100°)) the dependence on the baryonic content of the fireball is quite clear.

The continuous and dashed lines represent the results of IQMD - SM[15] estimates of the corresponding slopes for Au + Au in the corresponding polar regions. The agreement with the experimental data is quite satisfactory. IQMD simulations show a dependence of the average number of collisions per nucleon as a function of polar angle, see Fig. 9. In the average two times more colisions are suffered by the nucleons emitted in the transversal direction relative to the forward ones. The experimental trends corroborated with the results of the microscopic approach bring to the conclusion that the higher values of the slopes and their independency on the baryonic content of the fireball in the forward direction is mainly due to transparency effect. A better estimate of the amount of expansion and its dependence as a function of the participating number of nucleons is visible in the transverse direction relative to the collision axis where the contamination from other effects has the lowest value.

5 Conclusions

Using a new generation of data collected with the complete FOPI experimental device we performed a systematic study of the azimuthal- and polar angle- dependence of the flow phenomena in semi-central and respectively central heavy ion collisions. Clear transverse momentum and reaction product dependence of the E_{tran} value has been evidenced in the experimental data. Qulitative explanations of these trends could be understood in the frame of a semianalitical hybrid model which predicts a breakup which, with elapsing time, starts at the outer shells and evolves to inner ones. Using three symmetric systems and selecting the geometry such to have the same baryonic content of the fireball, it was clearly evidenced the role of the spectator matter for the observed kinetic energy azimuthal distributions. Preliminary results on the azimuthal distributions for Ru+Zr and Zr+Ru indicate an influence of the N/Z value of the spectator matter on the squeeze out phenomena. A clear polar angle- and A_{part}-dependence of the correlated motion in highly central collisions was evidenced. Soft EoS in a microscopic transport model (IQMD) explains better the trends observed in highly central collisions.

Acknowledgments

This work has been supported in part by the German BMBF under contract RUM-005-95, Deutsche Forschungsgemeinschaft (DFG) under projects 436 RUM-113/10/0 and contract B7/1999 financed by the Romanian Agency for Science, Technology and Inovation.

References

1. J. Ritman, FOPI Collaboration, *Nucl. Phys. B (Proc. Suppl.)* **44** (1995) 708
2. D. Pelte, FOPI Collaboration, *Z. Phys. A* **357** (1997) 215
3. M. Petrovici, FOPI Collaboration, *Heavy Ion Physics at Low, Intermediate and Relativistic Energies using 4π Detectors*, ed. M. Petrovici et al., World Scientific, 1997, p. 216
4. A. Andronic, FOPI Collaboration, *Heavy Ion Physics at Low, Intermediate and Relativistic Energies using 4π Detectors*, ed. M. Petrovici et al., World Scientific, 1997, p. 209
5. A. Andronic, PhD Thesis - Bucharest, 1998
6. A. Andronic, FOPI Collaboration, will be published
7. F. Rami, FOPI Collaboration, *Phys. Rev. Lett.*, accepted for publication
8. W.K. Wilson et al., *Phys. Rev. C* **41** (1990) R1881
9. A. Buţă, FOPI Collaboration, GSI Scientific Report 1994, 95-1 (1995) 57
10. W.K. Wilson et al., *Phys. Rev. C* **51** (1995) 3136
11. M.B. Tsang et al., *Phys. Rev. C* **53** (1996) 1959
12. N. Bastid, FOPI Collaboration, *Nucl. Phys. A* **622** (1997) 573
13. P. Crochet, FOPI Collaboration, *Nucl. Phys. A* **624** (1997) 755
14. M. Petrovici, FOPI Collaboration, *Phys. Rev. Lett.* **74** (1995) 5001
15. C. Hartnack et al., *Phys. Lett. B* **336** (1994) 131, *Mod. Phys. Lett. A* **9** (1994) 1151
16. S. Wang et al., *Phys. Rev. Lett.* **76** (1996) 3911
17. A. Andronic, FOPI Collaboration, *International School of Nuclear Physics" September 17-25, Erice, Italy, 1998*
18. M. Petrovici, FOPI Collaboration, 7th *International Conference on Clustering Aspects of Nuclear Structure and Dynamics, June 14-19, 1999, Rab, Croatia*, will be published by World Scientific

FORMATION AND DECAY OF HOT NUCLEI IN $^{40}CA + ^{40}CA$ AT 35 MEV/NUCLEON

R.PŁANETA, W.GAWLIKOWICZ, K.GROTOWSKI, J.BRZYCHCZYK, P.HACHAJ, S.MICEK, P.PAWŁOWSKI, Z.SOSIN, A.WIELOCH

M.Smoluchowski Institute of Physics, Jagellonian University, Reymonta 4, 30-059 Cracow, Poland

A.J.COLE, P.DÉSESQUELLES, A.CHABANE, M.CHARVET, A.GIORNI, D.HEUER, A.LLÉRES, J.B.VIANO

Institut des Sciences Nucléaires de Grenoble, IN2P3-CNRS/ Université, Joseph Fourier 53, Avenue des Martyrs, F-38026 Grenoble Cedex, France

D.BENCHEKROUN, E.BISQUER, B.CHEYNIS, A.DEMEYER, E.GERLIC, D.GUINET, P.LAUTESSE, L.LEBRETON, M.STERN, L.VAGNERON

Institut de Physique Nucléaire de Lyon, IN2P3-CNRS/ Université Claude Bernard 43, Boulevard du 11 Novembre 1918, F-69622 Villeurbanne Cedex, France

Properties of multifragmentation of "hot sources" produced in the $^{40}Ca + {}^{40}Ca$ reaction have been studied at a beam energy 35 MeV/nucleon. Two signatures of prompt multifragmentation which make use of special features of particle emission from the "freeze out volume" together with an analysis of the reduced relative velocity between pairs of intermediate mass fragments indicate the presence of a transition from the sequential decay to prompt multifragmentation at an excitation energy of about 3 MeV/nucleon.

1 Introduction

Phase transitions in finite systems are, at present time, a subject of great interest. Prompt multifragmentation of highly excited nuclei is of particular interest since it may yield information concerning the liquid-gas phase transition in nuclear matter [1]. It may be induced by a nuclear collision transferring a system into a spinodal region of instability [2]. Multifragmentation may also appear as a natural extrapolation of the evaporation mechanism characteristic of nuclear decay at low excitation energies. The essential question concerns the typical time interval between successive emissions. In the limit of very short times nuclear multifragmentation can be considered as prompt multifragmentation (PM), whereas the opposite limit is usually referred to as a sequential or a binary sequential decay (BSD).

Atomic nuclei, at low excitation energies ($\cong 2MeV/nucleon$), decay by emission of neutrons and light charged particles. In this scenario particle emission is such a rare event that a chain of subsequent emissions may be assumed to take place from a corresponding sequence of equilibrated parent nuclei. With increasing energy the emission of heavier particles, intermediate mass fragments (IMF, $Z > 2$) competes with light particle emission. Assumption of step by step equilibration of parent nuclei may break down with further increase of the excitation energy and reduction of the decay time scales because the parent nucleus may not have time to equilibrate between successive emissions [3]. Emitted particles may not have time to leave the vicinity of the parent and consequently the presence of previously emitted fragments may influence the decay process.

For excitation energy high enough to transfer a nuclear system in the spinodal region below the critical point ($T < T_{cr}$), the system should break into fragments due to instability of nuclear matter. In this case one can postulate the existence of a set of fragments, enclosed in some finite spatial region (the "freeze-out volume"), which interact via the inter-fragment repulsive Coulomb force only [4]. Although such a process is considered to be prompt it should not be treated as simply a short-time limit of the sequential decay.

The experimental search for differences between the BSD and PM processes is usually based on dynamic correlations between IMF's emitted from the excited PLF [5]. IMF's from the PM are localized closer in space and in time as compared to IMF's from the BSD. Consequently, at small values of the relative velocity the number of coincidences is smaller for the PM than for the BSD. In the above method the main difference between BSD and PM is due to the decay time.

In this work we make use of two novel signatures of prompt multifragmentation which instead of the decay time, exploit special features of particle emission from the "freeze out volume". They are: (i) the shape of the distribution of squared momentum of the heaviest emitted fragment, p_1^2, [6,7], (ii) the focusing of fragments by the Coulomb field of the decaying system [8].

2 The $^{40}Ca + ^{40}Ca$ experiment and data

Experiment was performed at the Grenoble SARA facility using the upgraded multi-detector system AMPHORA [9,10]. For the symmetric $^{40}Ca + ^{40}Ca$ system most fragments are detected in the forward part of detector, where the granularity of AMPHORA provides a good angular resolution for particle-particle correlation measurements. Experimental details are described in [11,12].

2.1 The PLF hot source

At 35 MeV/nucleon the dynamics of peripheral $^{40}Ca+^{40}Ca$ collisions exhibits a predominantly binary mechanism characterized by strong energy dissipation [11,12,13]. Thus, in the exit channel, we should observe two excited and decaying "sources": a target-like and a projectile-like fragment (TLF and PLF). In our experiment because of kinematics as well as due to detection energy thresholds only reconstruction of the primary PLF was possible. For primary PLF's fragments charge, mass and excitation energy have been determined [13,14].

Angular distributions of fragments measured in the center of mass of the decaying PLF indicate the forward backward symmetry for the IMF emission, as expected for a thermalized source [13].

The excitation energy distribution of the primary PLF is presented in Fig.1 (upper part). As one can see, deep inelastic collisions produce PLF nuclei with an excitation energies extending from zero up to over 10 MeV/nucleon. Such a span of excitation energy should provide a possibility to trace a transition from BSD to PM.

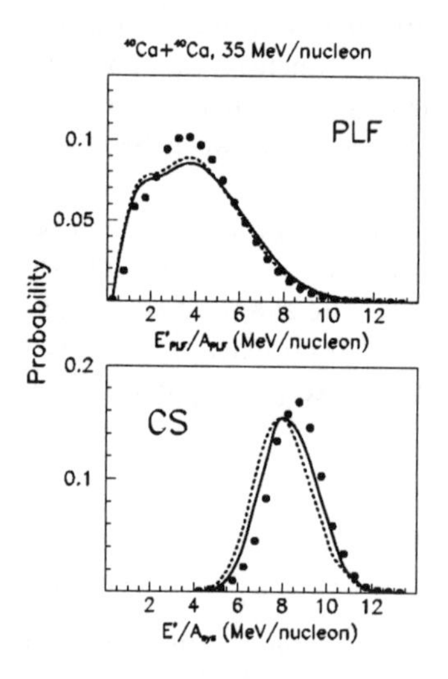

Figure 1. The PLF and CS excitation energy distributions (black dots) together with model predictions (PM - solid line; BSD - broken line).

2.2 The CS hot source

For more central collisions a composite system (CS) is formed as a result of incomplete fusion, with a very low cross section (several milibarns). To select events belonging to more central collisions we use a cut in the coplanarity (C) sphericity (S) plane [15]. Parameters C and S are related to the shape of event in the linear momentum space. The cut is defined by:

$$C < 0.7(S - 0.3). \qquad (1)$$

In model simulations (see Sec.3) we can tag events produced according to incomplete fusion or DIC scenario and estimate their relative contributions. With the restriction (1) and additional conditions: total detected charge $Z_{total} > 34$ and the LAB angle of the heaviest fragment $\Theta_1 < 20$ degrees, about 60 percent of events come from the incomplete fusion.

The excitation energy distribution of the primary "hot CS source", reconstructed under the above conditions is presented in Fig.1 (lower part). The CS excitation energy is quite high (about 8.5 MeV/nucleon) and width of its distribution is about 3 MeV/nucleon.

3 Simulations

In order to study different decay scenarios of "hot sources" one has to use special gating techniques and particle correlation methods. To check them one needs a model which reproduces some necessary details of the reaction picture. We use a computer code elaborated by Sosin [16]. It belongs to a family of models [17] which are based on the Randrup assumption [18] that for higher collision energies, energy dissipation proceeds mainly through stochastic transfer of nucleons between colliding ions.

The computer code of Sosin includes competition between mean field effects and effects of nucleon-nucleon interactions in the overlap zone of colliding nuclei. Activated nucleons are transfered to PLF or TLF. They may also escape form the system or form a cluster. In the limit of smaller angular momenta (more central collisions) such clustering process may result in formation of a composite system.

An excited PLF, TLF or CS emits particles. In the model the deexcitation process is simulated by the GEMINI statistical code [19] which treats the cooling process as a sequence of binary decays.

As an option, the code permits decay of an excited PLF (TLF) or CS by prompt multifragmentation. In this work we follow a suggestion of López and Randrup [4] applied in our previous work [8]. For a given system with some initial angular momentum L, and excitation energy E* (obtained from the Sosin code) we use a partition provided by the GEMINI code as an initial distribution of fragments inside the "freeze- out" volume. Initial fragments are randomly positioned by a Monte Carlo subroutine [8,20] inside a spherical region of space such that no two fragments overlap.

Fragments of the initial configuration are accelerated in the mutual Coulomb field along proper trajectories, which are integrated numerically [21]. Predictions of the code are filtered by a software replica of the AMPHORA detector system [22]. For comparison with experimental data the same reconstruction procedure of a PLF or CS has been applied to the model predictions as for the data (see lines in Fig.1).

4 Decay characteristics of hot PLF's

To study the PLF decay characteristics one can use a conventional method based on correlations between IMF's emitted from the excited PLF. Here we use a $1 + R(v_{red})$ correlation function for pairs of IMF's moving apart with reduced relative velocity, v_{red} [5].

The reduced velocity correlation functions measured for pairs of IMF's with $3 \leq Z \leq 8$ are shown in Fig.2a for different bins of the excitation energy of the primary PLF. The "Coulomb hole" seen at small values of v_{red} clearly broadens for higher PLF excitation energies. The sequential binary scenario explains the experimental data at low excitations only, below 3 MeV/nucleon. At higher excitations one has to use a correlation function calculated according to the PM scenario.

As an alternative, instead of the 1 + R correlation method we can examine the distribution of the squared momentum of the heaviest fragment. For this signature fragment momenta should be measured in the PLF center of mass system whose location is determined, event by event, in the PLF reconstruction procedure [13]. To avoid the influence of the TLF we take these heaviest fragments only which are emitted at an angle smaller than 90 degrees in the coordinate system oriented by the running PLF.

The measured distributions of the squared momentum p_1^2 of the heaviest fragment are displayed in Fig.2b. They become distinctly broader for PLF excitation energies higher than 3 MeV/nucleon, in agreement with the PM model prediction.

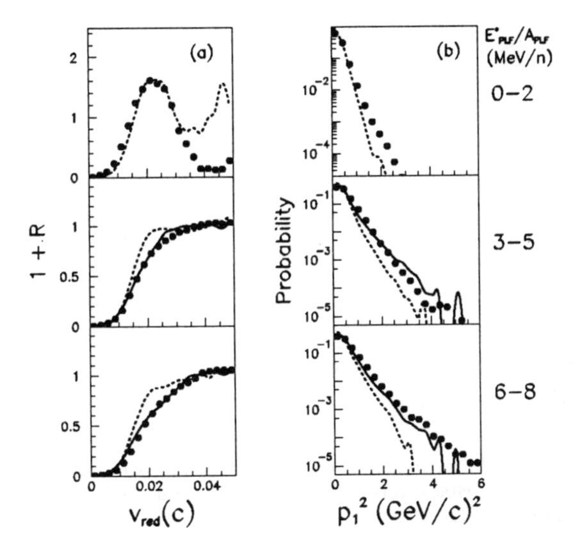

Figure 2. Reduced velocity correlation functions for IMF's with $3 \leq Z \leq 8$ (a), and distributions of p_1^2 (b) for different PLF excitation energy bins. Black dots indicate experimental data. Model predictions: solid line - PM; broken line - BSD.

5 Decay characteristics of hot CS's

Fig.3a displays the 1+R correlation function for events selected by conditions described in Sec. 2.2. It shows a broad "Coulomb hole" in agreement with the PM model prediction but too broad for the BSD decay scenario.

The p_1^2 distribution (Fig.3b) is well predicted by the PM reaction scenario and is in disagreement with the BSD curve.

The Coulomb focusing effect is observed in the IMF velocity distribution, $d\sigma(\Theta_v)/d\Omega$ displayed in a reference frame defined by the relative velocity, $\vec{v}_1 - \vec{v}_2$, of the two heaviest fragments. Here Θ_v is an angle between the IMF velocity and the $\vec{v}_1 - \vec{v}_2$ vector. As expected the two heaviest fragments generate a strong Coulomb field, focusing the velocities of IMF's around $\Theta_v = 90$ degrees (see Fig.3c). The experimental points agree with the PM prediction. For the sequential binary decay the $d\sigma(\Theta_v)/d\Omega$ distribution is distinctly more flat but not isotropic because of momentum conservation and due to very short time intervals between consecutive emissions, which create a dependence of subsequent decays.

The agreement of the measured Z distribution with the model prediction is good, for both BSD and PM reaction scenarios (see Fig.3d).

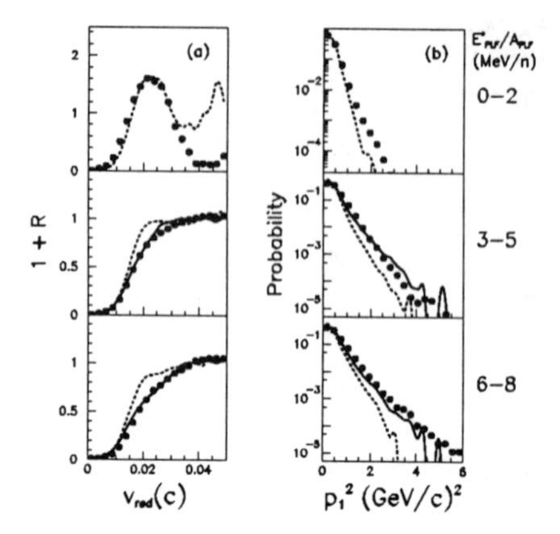

Figure 3. Composite system: (a) 1+R correlation function; (b) p_1^2 distribution; (c) Coulomb focusing; (d) secondary charge distribution. Black dots indicate experimental data. Model predictions: solid line - PM; broken line - BSD.

6 Summary and conclusions

Multifragmentation of excited nuclei of the $^{40}Ca + ^{40}Ca$ reaction at E_{LAB} = 35 MeV/nucleon has been studied using the multidetector system AMPHORA. Using special gating and reconstruction procedures we could observe projectile-like fragments with different degrees of excitation, and also highly excited systems from incomplete fusion. These "hot sources" possess features of thermalized systems. To investigate their decay characteristics we have used the conventional reduced velocity correlation method and also two signatures based on the distribution of the squared momentum of the heaviest fragment, and on the Coulomb focusing effect, respectively.

For the PLF, both methods, the reduced velocity correlation, and the p_1^2 distribution, support the binary sequential decay scenario below 3 MeV/nucleon excitation energy and prompt multifragmentation for higher excitations. For CS which has about twice the PLF electric charge the Coulomb focusing effect could be also observed. In that case all three signatures indicate prompt multifragmentation of the hot system.

Consistency of all these observations show that both the p_1^2 distribution and the Coulomb focusing effect can be used as signatures of prompt multifragmentation.

At high excitations the PM signal (the difference between the BSD and PM prediction) is for the new signatures quite strong. This may be used as an argument that prompt multifragmentation should not be treated simply as a short time limit of the binary sequential decay.

This work was supported by the Polish-French(IN_2P_3) agreement, the Committee of Scientific Research of Poland (KBN Grant No. $2P03B13914$) and by the M.Skłodowska-Curie Fund ($MEN/DOE - 97 - 318$). Calculations for this work were performed using facilities of the Cracow Academic Computer Center CYFRONET.

References

1. H.Jaquaman et al., Phys.Rev. **C27** (1983) 2782.
2. See e.g. J.D.Gunton and M.Droz, "Introduction to the Theory of Metastable and Unstable States", Lecture Notes in Physics 183, Springer-Ferlag Berlin Heidelberg New York Tokyo.
3. B.Borderie, Ann.Phys.Fr. 17, 349 (1992)
4. J.A.Lopez and J.Randrup, Nucl.Phys. **A491**,477 (1989).
5. R.Trockel et al., Phys.Rev.Lett. **59**, 2844 (1987).
6. D.Heuer et al., Nucl.Phys. **A583**, 537 (1995).
7. K.Grotowski, Acta Phys.Pol. **B24**, 285 (1993).
8. W.Gawlikowicz and K.Grotowski, Acta Phys. Pol. **22**, 885 (1991); Nucl.Phys. **A551**, 73 (1993).
9. D.Drain et al., Nucl.Inst.Meth. **A281**, 528 (1989).
10. T.Barczyk et al., Nucl.Inst.Meth. **A364**, 311 (1995).
11. P.Pawłowski et al., Phys.Rev. **C54**, R10 (1996).
12. P.Pawłowski et al., Z.Phys. **A357**, 342 (1997).
13. R.Płaneta et al., preprint, Cracow IF UJ, ZFGM-99-01.
14. J.Péter et al., Nucl.Phys. **A593**, 95 (1995).
15. J.Cugnon et al.,Phys.Lett. **109B**, 167 (1982); M.Gyulassy et al., Phys.Lett. **110B**, 185 (1982); G.Fai and J.Randrup, Nucl.Phys. **A404**, 551 (1983).
16. Z.Sosin, preprint, Cracow IF UJ, ZFGM-99-03.
17. B.G.Harvey, Nucl. Phys. **A444**, 498 (1985); A.J.Cole, Z.Phys. **A322**, 315 (1985); Phys. Rev. **C35**, 117 (1987); L.Tassan-Got, and C.Stéphan, Nucl.Phys. **A524**, 121 (1991); D.Durand, Nucl.Phys. **A541**, 266 (1992); Z.Sosin et al., Acta Physica Polonica **B25**, 1601 (1994).
18. J.Randrup, Nucl.Phys. **A307**, 490 (1979).
19. R.J.Charity et al., Nucl.Phys. **A483**, 371 (1988).
20. W.Gawlikowicz, Ph.D.Thesis, Cracow 1994.
21. W.Gawlikowicz, Acta Phys. Pol. **28**, 7 (1997).
22. See the SIR code, D.Heuer et al., unpublished.

STUDY OF SPALLATION RESIDUES OF GOLD AT 0.8 GEV/N IN REVERSE KINEMATICS

L. TASSAN-GOT, B. MUSTAPHA, F. REJMUND, C. STÉPHAN, M. BERNAS, J. TAIEB
IPN-IN2P3, F-91406 Orsay - France

P. ARMBRUSTER, K.H. SCHMIDT, T. ENQVIST, W. WLAZLO
GSI Planckstrasse 1, D-64291 Darmstadt - Germany

J. BENLLIURE
University of Santiago de Compostela, E-15706 Santiago de Compostela - Spain

S. LERAY, A. BOUDARD, C. VOLANT, R. LEGRAIN
DAPNIA/SPhN CEA/Saclay, F-91191 Gif/Yvette - France

J.P. DUFOUR, S. CZAJKOWSKI, M. PRAVIKOFF
CENBG IN2P3, F-33175 Gradignan - France

Spallation residue cross-sections of gold have been measured in reverse kinematics. For the first time isotopic distributions have been obtained for such a heavy nucleus, at a level of accuracy of 10 %. When compared to model calculations the results appear to be very sensitive to the distribution of excitation energy in the fast step of the reaction. This allows an insight on the mechanism of the energy and momentum deposition.

1 Introduction

Spallation reactions have been studied for at least forty years. The expected possibility of using such reactions to generate high neutron fluxes either for accelerator-driven systems or for more fundamental purposes, gave rise to a renewal of spallation studies both in experimental and theoretical fields. Indeed in prospect of such applications a better accuracy is needed in the quantitative description of these reactions, particularly on the characteristics of particle emission, and production of radioactive species.

Regarding the latter the reverse kinematical method, consisting here in bombarding a proton target with a high energy gold projectile, is valuable and reveals several advantages : an intrinsic good accuracy of residue cross-sections at the level of 10 %, the measurement of full isotopic distributions independently of any radioactivity, which allows the detection of short-lived and stable isotopes as well, and a sensitive comparison to model calculations.

In addition the method delivers a measurement of the momentum transfer imparted to the spallation fragments.

Following this line we have started an experimental program of cross-section measurements at GSI where the synchrotron SIS can deliver high quality beams of very heavy nuclei in the GeV per nucleon range. We report here on the results of the first experiment which involved a 0.8 GeV/n gold beam impnging on a liquid hydrogen target.

2 Experimental setup

We just recall here the main aspects of the experiment, a more detailed description can be found elsewhere [1].

In order minimize the error related to the difference method, especially for large mass losses, the proton target was made of cryogenic liquid hydrogen sealed by thin titanium foils. The reaction probability of incident gold ions was close to 10 %.

The spallation products were collected at 0° by the Fragment Recoil Separator (FRS) [7] which allows a full identification of isotopes. This was achieved by tracking of fragments for magnetic rigidity and angle determination, by energy loss measurement in a MUSIC ionization chamber [8] for Z determination, and time of flight information delivered by plastic scintillators for mass identification. A thick degrader (3.5 g/cm^2 of aluminum) was added to give a redundant Z determination through the imparted loss in magnetic rigidity.

The target thickness, related to the window swelling, was accurately measured at 2.5 % by comparison of the energy loss of particles at the center and the edge of the target, the thickness at the edge being determined by mechanical construction. The beam intensity was monitored by a secondary electron emission monitor calibrated by the counting rate in a ionization chamber. The current in this chamber was used as an intermediate variable making the bridge between the high intensity domain measured by the monitor and the low one covered by particle counting in the ionization chamber.

The FRS accepted the full angular distribution of evaporation residues, while the momentum acceptance was only 2 %, needing usually several field settings to cover the whole distribution.

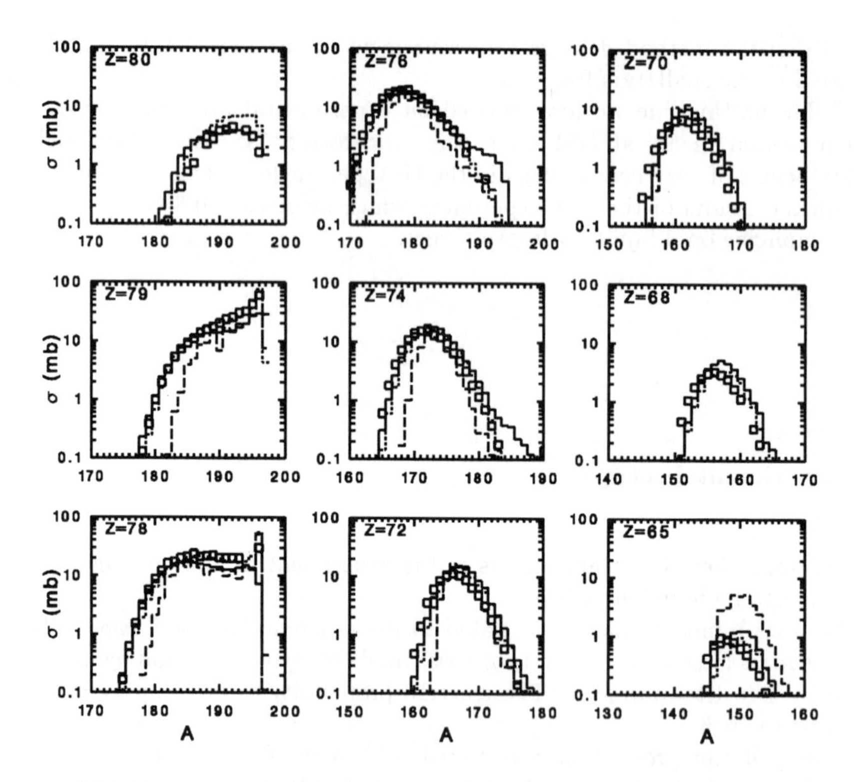

Figure 1. Isotopic distribution of selected elements. Open squares are for experimental data, while lines represent different model calculations : solid line : INCL [2]+ABLA [4], dotted line : ISABEL [3]+ABLA, dashed line : LAHET [5].

3 Results and discussion

3.1 Isotopic distributions

Figure 1 shows the isotopic distributions associated to selected elements. Open squares represent the experimental distributions which have been obtained for the first time for such a heavy nucleus. They exhibit an asymmetric bell-shape which is more and more deformed in the vicinity of the incident projectile, retaining to a large extent a memory of the fragment origin.

The different curves indicate the predictions of model calculations based on the two-step model : a fast step described by an intranuclear cascade calculation (INC), and a longer phase simulating the statistical decay of the hot

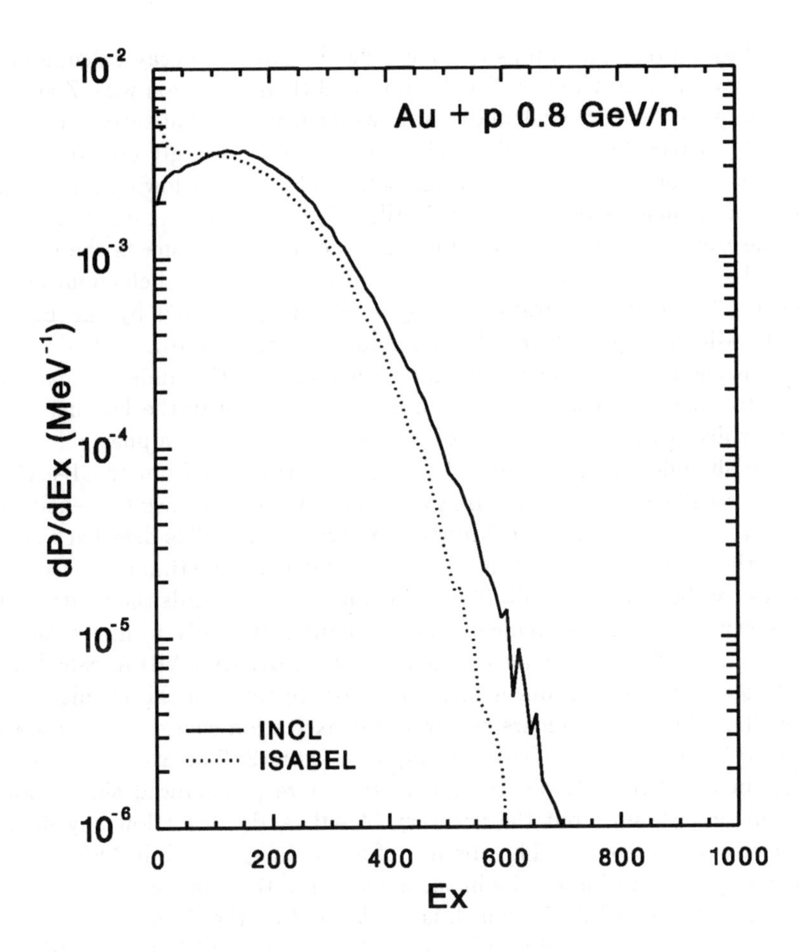

Figure 2. Excitation energy spectrum generated in the intranuclear cascade step. The solid line corresponds to INCL [2], while the dotted line is for ISABEL [3].

pre-fragment formed in the fast step. The solid curve corresponds to the Liège code [2], hereafter called INCL, followed by the ABLA statistical decay code [4] developed at GSI. The dotted line is for the combination of the intranuclear calculation ISABEL [3] with ABLA. Finally the dashed line shows the prediction of the LAHET code [5] based on the Bertini-type intranuclear cascade followed by a preequilibrium step preceding the Dresner decay calculation.

The overall behaviour of distributions and their evolution with Z are well reproduced by the three calculations, however from a quantitative viewpoint significant departures are visible. The LAHET code strongly underestimates the cross-sections of proton-rich fragments, and also to a low extent the isotopes and elements close to the projectile. Conversely for light elements, for example Z=65, the predicted yield is too large. In these respects the LAHET code is less accurate than the two other calculations. This behaviour can be understood from the excitation energy spectrum generated by the Bertini-cascade, which extends to much higher values than ISABEL or INCL. This feature is clearly ruled out by the data. Furthermore the underestimation of the proton-rich tails could come from the barriers used in the Dresner calculation, which favour the emission of composite particles as alphas.

The shoulder in the predicted isotopic distribution from INCL+ABLA of Z=76 results from some pre-fragments of low, or even negative excitation energy, remaining unchanged through the decay stage. This drawback is specific of the version 2 of INCL here used. A slight underestimation of heavier isotopes can be noticed for Z=79 and 78, and a shift towards the neutron-rich side is visible for lighter elements, corresponding probably to non-optimized barriers, this effect being also present in the ISABEL+ABLA calculation. The latter however reproduces nicely all cross-sections in the vicinity of the projectile. This can be understood from the excitation energy spectrum associated to the INC step, which is displayed in figure 2. The average excitation energy in ISABEL is lower, leading to the more pronounced shift towards the neutron-rich side, but the accuracy of ABLA doesn't allow any definite conclusion on this point. The absence of a low energy peak in INCL excitation energy distribution results in the depletion of the cross-section of heavier isotopes of Z=79 and Z=78, which is invalidated by the data.

It must be emphasized that both INCL+ABLA and ISABEL+ABLA are able to predict within a factor 2 the production of mercury corresponding to a charge exchange channel in the fast step.

3.2 Recoil momentum

The magnetic spectrometer allows an accurate determination of the longitudinal momentum of each detected fragment. For each isotope the momentum spectrum is well represented by a gaussian shape, and when transformed to the gold projectile rest frame, this gaussian shape can be characterized by its mean value and width. Instead of these quantities, following Morrissey's definitions [6], we have plotted in figure 3 $P'_{\|}$ which is tightly related to the mean

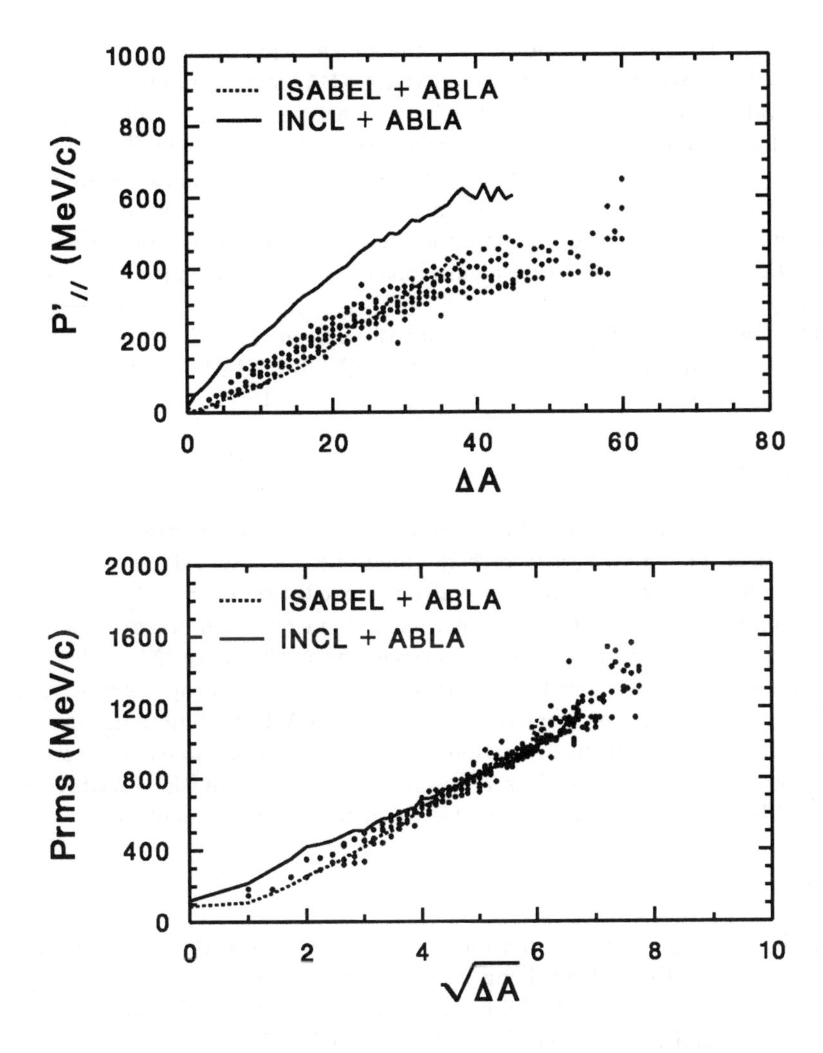

Figure 3. Upper part : $P'_{\parallel}$ versus the mass loss in the reaction. For the same mass loss different points correspond to different elements. Lower part : P_{rms} versus the square root of the mass loss.

momentum, and P_{rms} which is derived from the longitudinal width, assuming isotropy : P_{rms} is taken as the longitudinal width multiplied by $\sqrt{3}$.

Figure 3 (upper plot) shows that the mean recoil momentum is fairly reproduced by ISABEL+ABLA while INCL+ABLA overestimates the experimental values by a factor of 2. This features are not changed when another decay calculation is applied, and they reflect the behaviour of the INC calculation. The apparent scattering of data points is essentially due to a systematic dependency of $P'_{||}$ with Z : the mean momentum increases with Z. In the figure, only the Z-averaged prediction is shown for the calculations. However it must be emphasized that the calculations quantitatively reproduce this dependency.

The lower part of figure 3 shows that the agreement of both calculations with the experimental widths is satisfactory, INCL+ABLA slightly overestimating the data at low mass loss.

4 Conclusion

We have presented the accurate measurement of spallation cross-sections of gold obtained at 0.8 GeV/n in reverse kinematics. These data severely constrain theoretical calculations, especially the excitation energy deposition in the intranuclear cascade step. In this respect the excitation energy spectrum generated by the Bertini cascade is too large, and in the Liège INC code a low energy peak is missing. This last feature could be cured by properly describing the diffuseness of nuclear density. The ISABEL+ABLA combination seems provide the most accurate prediction of isotopic cross-sections. Concerning the recoil properties, the widths are well predicted by ISABEL+ABLA and INCL+ABLA but INCL overestimates the mean recoil momentum.

References

1. L. Tassan-Got *et al*, International Conference on the Physics of Nuclear Science and Technology, Long Island, New York, october 1998, p 1334
2. J. Cugnon, C. Volant, S. Vuillier, *Nucl. Phys.* A **620**, 475 (1997).
3. Y. Yariv and Z. Fraenkel, *Phys. Rev.* C **20**, 2227 (1979).
4. J. Benlliure, A. Grewe, M. de Jong, K.-H. Schmidt, S. Zhdanov, *Nucl. Phys.* A **628**, 458 (1998)
5. R.E. Prael and H. Lichtenstein, User Guide to LCS : the LAHET Code System, Los Alamos Natinal Laboratory report **LA-UR-89-3014**, 1989.
6. D.J. Morrissey, *Phys. Rev.* C **39**, 460 (1989)
7. H. Geissel *et al*, *Nucl. Instrum. Methods* B **70**, 286 (1992)
8. M. Pfützner *et al*, *Nucl. Instrum. Methods* B **86**, 213 (1994)

ISOSPIN AND F-SPIN CHANGING $M1$ TRANSITIONS IN NUCLEI

P. VON BRENTANO[1], A. DEWALD[1], C.FRANSEN[1], C. FRIEßNER[1], R. V. JOLOS[1,2], A. F. LISETSKIY[1], N. PIETRALLA[1,3], I. SCHNEIDER[1], A. SCHMIDT[1]

[1] *Institut für Kernphysik, Universität zu Köln, 50937 Köln, Germany*
E-mail: brentano@ikp.uni-koeln.de

[2] *Bogoliubov Theoretical Laboratory, Joint Institute for Nuclear Research, 141980 Dubna, Russia*
E-mail: jolos@thsun1.jinr.ru

[3] *Wright Nuclear Structure Laboratory, Yale University, New Haven, Connecticut 06520-8124, USA*
E-mail: pietrall@galileo.physics.yale.edu

Isovector ΔT=1 and ΔF=1 M1 transitions between low-lying states in odd-odd $N = Z$ nuclei and in the even-even nucleus ^{94}Mo, respectively are discussed. The data on low spin states in the odd-odd nuclei ^{54}Co and ^{46}V investigated with the ^{54}Fe(p, nγ)^{54}Co and ^{46}Ti(p, nγ)^{46}V fusion reactions at the FN-TANDEM accelerator in Cologne are reported. The proton-neutron states of *quasideuteron* character are identified. The identification of the mixed-symmetry $J^\pi = 1^+_{sc}, 2^+_{ms}$ and 3^+_{ms} states in the nucleus ^{94}Mo is reported. The predictions of the interacting boson model for the basic mixed-symmetry $J^\pi = 1^+_{sc}, 2^+_{ms}, 3^+_{ms}$ states agree with the observation. The similar physics of ΔT=1 and ΔF=1 M1 transitions is stressed.

1 Introduction

Enhanced magnetic dipole ($M1$) γ transitions between low-lying states in atomic nuclei are of great interest [1]. Due to the dominantly isovector character of the $M1$ transition operator, $M1$ transitions are very efficient tools to explore the nucleonic interaction in the proton-neutron channel and to test the symmetries in nuclei related to this interaction. According to the fermionic shell model and to the interacting boson model (IBM-2) an enhanced $M1$ strength is a general feature of a proton-neutron (pn) degree of freedom. The pn symmetry of nuclei is quantified by the isospin quantum number T [3,4] and the pn symmetry of low-lying collective states in heavy nuclei is quantified in the IBM-2 approach by the F-spin quantum number [2]. F-spin is the isospin for the elementary proton and neutron bosons. The shell model predicts strong isovector $M1$ transitions between low-lying states with isospin quantum numbers $T = 0$ and $T = 1$ in odd-odd $N = Z$ nuclei. Likewise the IBM-2 predicts enhanced $M1$ transitions between states with F-spin quantum number F_{max} and $F_{max} - 1$. The states with $F \neq F_{max}$ are not fully symmetric with re-

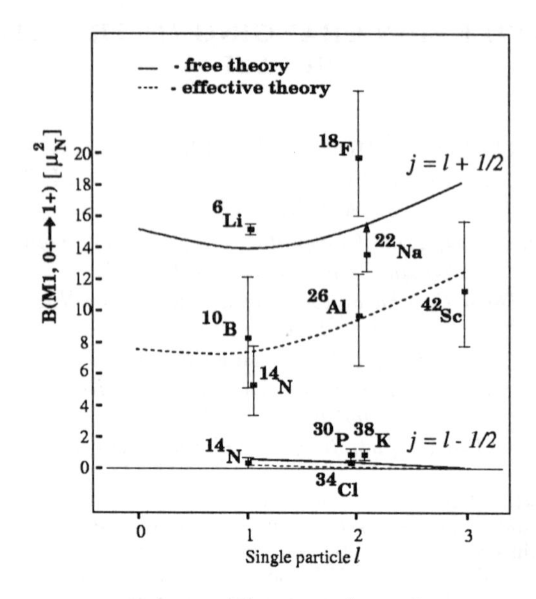

Figure 1. Experimental and calculated $\sum_i$ B(M1;$0_1^+ \rightarrow 1_i^+$) values for odd-odd $N = Z$ nuclei as a function of the appropriate single particle orbital angular momentum l of the unpaired valence nucleons. The value for ^{22}Na represents a lower limit. $B(M1)$ values measured in odd-odd $N = Z$ nuclei with *quasideuteron* character, i.e., $j = l + 1/2$, belong to the largest $M1$ transitions observed in nuclei. Results for the quasideuteron concept are shown with lines. Full lines correspond to the usage of free spin g factors and the dashed lines correspond to an effective theory with quenched g factors. The experimental values lie between or in the vicinity of the two lines for both $j = l + 1/2$ and $j = l - 1/2$ cases supporting the quasideuteron concept. From A.F. Lisetskiy, *et al.* [13].

spect to the pn degree of freedom and are called mixed-symmetry (MS) states. Therefore, enhanced $M1$ transitions between low-lying states of nuclei on the $N = Z$ line provide sensitive tests for the isospin symmetry while enhanced $M1$ transitions between collective nuclear states test the F-spin symmetry of the IBM-2.

Much work (see [5–13] and references therein) has recently been carried out for the investigation and understanding of the $N = Z$ nuclear structure. The competition between $T = 0$ and $T = 1$ channels in the isospin degree of freedom can be studied by γ-spectroscopy of bound states in odd-odd $N = Z$ nuclei, where the lowest states with total isospin quantum numbers $T = 0$ and $T = 1$ are almost degenerate. Very strong isovector $M1$ transitions (see Fig. 1) occur between them [13]. These are among the largest $M1$ transitions known in nuclei. In the present work we focus on the low-energy structure of the odd-odd $N = Z$ nuclei ^{54}Co and ^{46}V. Our experimental data on low-spin states in ^{54}Co [12] are considered as a confirmation of the quasideuteron scheme. Less pronounced manifestation of quasideuteron configurations is found to take place in the nucleus ^{46}V [11].

Related important issue is the identification of mixed-symmetry collective states in heavy nuclei. We report on the recent identification [19,20] of the $J^{\pi} = 1^{+}_{sc}, 2^{+}_{ms}$ and 3^{+}_{ms} states in ^{94}Mo in experiments done at the FN tandem accelerator in Cologne and at the dynamitron accelerator in Stuttgart. These states were identified from measured M1 strengths and represent the discovery of a MS one- and two-phonon structure [21]. We discuss the decays of the observed MS states including the first measurement of a transition rate between MS states.

2 Recent experimental data for the $N = Z$ nuclei ^{54}Co and ^{46}V

2.1 *Quasideuteron states in ^{54}Co*

In very recent work [12] in Cologne we have investigated the low-spin structure of the odd-odd $N = Z$ nucleus ^{54}Co up to an excitation energy of 4 MeV. From the $\gamma\gamma$-coincidence relations a low spin level scheme of ^{54}Co was constructed, a part of which is displayed in Fig. 2.

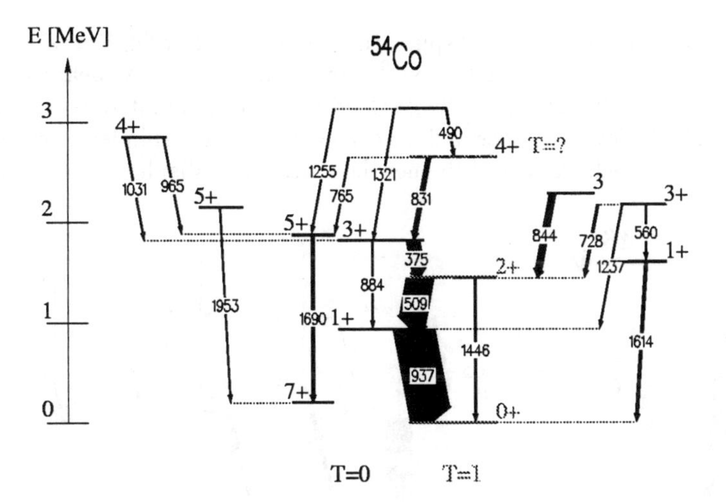

Figure 2. A part of the low-spin level scheme of ^{54}Co observed in the ^{54}Fe(p, nγ)^{54}Co reaction at 13 MeV beam energy. From I. Schneider , *et al.* [12].

There are also some recent medium spin data for ^{54}Co from Rudolph *et al.* [7,8]. Together with our low spin data there is now a consistent and extensive level scheme available for ^{54}Co. We have compared the rich data on γ-transitions, branching ratios and multipole mixing ratios to shell model calculations (for details see [12]). We obtain good agreement for the branching ratios of transitions between states with rather pure $\pi(f_{7/2}^{-1}) \times \nu(f_{7/2}^{-1})$ quasideuteron configurations. From our calculations we can note the following general properties of the quasideuteron states:
1) The isoscalar $\Delta T = 0$, $\Delta J = 2$ transitions have large $B(E2)$ values. We note further that the calculated $B(E2)$ values for the $2_1^+ \to 0_1^+$ and $3_1^+ \to 1_1^+$ transitions are almost equal.
2) The isovector $\Delta T = 1$, $\Delta J = 1$ transitions between $(\pi f_{7/2}^{-1} \times \nu f_{7/2}^{-1})_{J,T}$ states have large $B(M1)$ values and small $B(E2)$ values. The shell model predicts $B(M1; 0^+ \to 1^+) = 12\ \mu_N^2$ for ^{54}Co. This $M1$ strength is comparable to the known strong quasideuteron transitions in other odd-odd $N = Z$ nuclei (see Fig. 1).

From these and other arguments we tentatively identify the 1_1^+, 3_1^+, 5_1^+, 7_1^+, T=0 and 0_1^+, 2_1^+,T=1, (4_1^+,T=1) levels as the quasideuteron states in ^{54}Co.

2.2 Low-lying states in ^{46}V

Besides ^{54}Co we studied the low-spin structure of the odd-odd $N = Z$ nucleus ^{46}V in Cologne [11]. The ^{46}Ti(p, nγ)^{46}V reaction at 15 MeV beam energy, provided by the Cologne FN-TANDEM accelerator, populated levels up to 3.2 MeV in ^{46}V. From the observed $\gamma\gamma$-coincidences a low spin level scheme of ^{46}V was constructed. A part of the new level scheme is displayed in Fig. 3. In

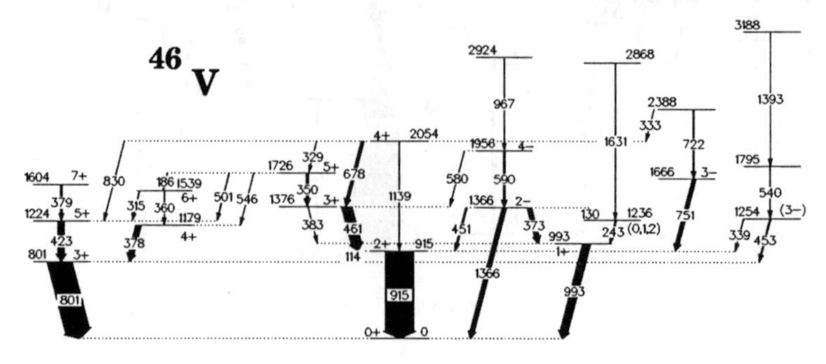

Figure 3. A part of the low spin level scheme of ^{46}V observed in the ^{46}Ti (p, nγ) ^{46}V reaction at 15 MeV beam energy.

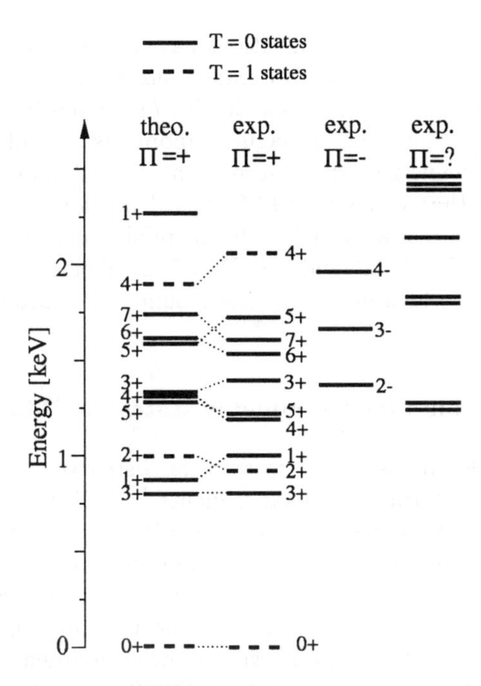

Figure 4. Low-spin level scheme of ^{46}V compared to a shell model calculation with the KB3 force in the full pf-shell. The calculations were done by T.Otsuka and Y. Utsuno and published in [11]. A one-to-one correspondence exists between the theory and the observed positive parity states below 2 MeV.

total seven new spin assignments and five new parity assignments were made. Together with new high spin results from Legnaro [9] and Copenhagen [10] there exists now a consistent and rather extensive level scheme for ^{46}V.

The data are compared to shell model calculations for the positive parity states of ^{46}V in the full pf-shell without truncation which were done by T. Otsuka and Y.Utsuno from the Tokyo group [11]. The KB3 parametrization [14] of the residual interaction was used.

The calculated excitation energies for the $T = 0$ and $T = 1$ levels below 3 MeV are compared to the data in Fig. 4. The measured ordering of the lowest excited levels is theoretically well reproduced. Alternative shell model calculations for ^{46}V have been performed by the Legnaro-Madrid-Strasbourg-Aarhus group and are reported by Lenzi *et al.* [9]. Also experimental branching ratios and $E2/M1$ multipole mixing ratios δ are well reproduced [11] by the shell model. In particular, $\Delta J = 1$ $\Delta T = 1$ isovector transitions are of dominant

$M1$ character and the $\Delta T = 0$ transitions have a stronger $E2$ character.

To understand the isovector $M1$ transitions in the quasideuteron scheme we have applied the relation between the $B(M1)$ values for two nucleon configurations and the experimental magnetic moments μ_π and μ_ν of the ground states $J^\pi = j^\pi$ in the neighboring odd-proton and odd-neutron nuclei, respectively (see [13]). In that way we could find evidence for the $1_1^+, 3_2^+, 5_2^+, T = 0$ and $0_1^+, 2_1^+, 4_2^+, T = 1$ states in ^{46}V being predominantly $(7/2_\pi^-) \times (7/2_\nu^-)$ states, where the $J_\rho^\pi = 7/2_\pi^-$ and the $J_\rho^\pi = 7/2_\nu^-$ denote effective single particle configurations, which determine the ground states properties of ^{45}V and ^{45}Ti, respectively.

3 Proton-neutron mixed-symmetry states in ^{94}Mo

In the early 1980s, Richter and co-workers discovered the MS $J^\pi = 1_{sc}^+$ state in electron scattering (e, e') experiments [15] in Darmstadt. This discovery was supported by photon scattering (γ, γ') experiments [16] in Stuttgart. Subsequent (e, e') [1] and extensive (γ, γ') experiments of the Stuttgart-Köln-Darmstadt collaboration (see [17,18,21] and references therein) accumulated knowledge about the 1^+ scissors mode. The information about other MS states from absolute B(M1) values is sparser or even absent and therefore the assignments are often much less sure (see Ref.[22] from [19]).

Recently [19] the 2^+ MS and the 1^+ MS states in ^{94}Mo were identified using a powerful combination of γ-singles photon scattering experiment on ^{94}Mo and $\gamma\gamma$-coincidence measurements of transitions following the β-decay of ^{94}Tc to ^{94}Mo. The part of the spectrum of γ rays following the β decay spectrum is shown in Fig. 5. From this new combination of techniques we obtain a new richness of information on absolute M1 and E2 transition strengths from MS states. The discovery of a 3^+ MS state in ^{94}Mo using the ^{91}Zr(α, n) fusion evaporation reaction was very recently reported, too [20].

From the lifetime information together with the branching ratios and $E2/M1$ multipole mixing ratios, the $E2$ and $M1$ transition strengths were obtained. The measured reduced $M1$ matrix elements of the $(1_1^+ \rightarrow 0_1^+, 2_2^+)$, $(2_3^+ \rightarrow 2_1^+)$ and $(3_2^+ \rightarrow 4_1^+, 2_2^+)$ transitions are of the order of 1 nuclear magneton. In accordance with the IBM-2 such large values for $M1$ matrix elements indicate the mixed symmetry character of the $1_1^+, 2_3^+$ and 3_2^+ states, to which the F-spin value $F = F_{max} - 1$ can be assigned. The measured transition rates for the $1_1^+ \rightarrow 2_3^+$ and $3_2^+ \rightarrow 2_3^+$ MS $\rightarrow$ MS transitions agree with collective $E2$ transitions and are comparable with the $2_1^+ \rightarrow 0_1^+$ $E2$ transition strengths. This fact supports the collective nature of the $1_1^+, 2_3^+$ and 3_2^+ states of ^{94}Mo and the multi-phonon interpretation [23] for MS states

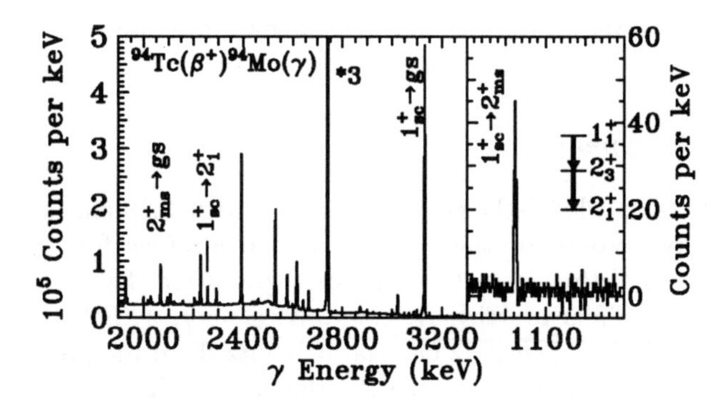

Figure 5. Left: part of the observed spectrum of γ rays following the β-decay of the $J^\pi = (2^+)$ isomer of ^{94}Tc populated in the ^{94}Mo(p,n) reaction. Right: part of the $\gamma\gamma$-coincidens spectrum gated with the $2^+_3 \to 2^+_1$ transition.

of ^{94}Mo. We conclude, that the 2^+_1 state is a collective, pn-symmetric one-phonon excitation of the ground state and the $2^+_2, 4^+_1$ states are two-phonon excitations. The 2^+_3 state is interpreted as the collective mixed-symmetry one-quadrupole phonon excitation of the ground state and acts as building block for the 1^+_1 and 3^+_2 states, which are considered as symmetric–mixed-symmetry two-phonon states. Schematic IBM-2 predictions for low-lying symmetric and mixed-symmetry states in ^{94}Mo agree resonably well with the observations (see [19,20]).

4 Summary

Odd-odd $N = Z$ nuclei are the only nuclei in which nearly degenerate $T = 0$ and $T = 1$ excitations exist at low excitation energy. This allows a γ spectroscopy of the strong $\Delta T = 1$ $M1$ transitions between bound states which are among the largest observed $M1$ transitions in atomic nuclei. We reported here on experiments on the low spin level schemes of ^{54}Co and ^{46}V nuclei done at the FN-TANDEM accelerator in Cologne. We identified $T = 0$ and $T = 1$ states in these nuclei and discussed their properties. Basing on the comparison of the observations to theory the $\Delta T = 1$ M1 transitions and $\Delta T = 0$ E2 transitions are found to be strong.

The identification of $1^+_1, 2^+_3$ and 3^+_2 mixed-symmetry states in ^{94}Mo was discussed. The $\Delta F = 1$ transitions are observed as the strong M1 transitions and $\Delta F = 0$ as collective E2 transitions. The analogy with strong $\Delta T = 1$ M1 transitions and $\Delta T = 0$ transitions in odd-odd N=Z nuclei suppose very similar physics of M1 excitations in collective even-even and in odd-odd N=Z nuclei.

5 Acknowledgments

We gratefully acknowledge the good cooperation with T.Otsuka and Y.Utsuno from the Tokyo group for the shell model calculations for ^{46}V and with U.Kneissl and H.H.Pitz from the Stuttgart group for the $\gamma\gamma'$-experiments. We thank R. S. Chakrawarthy, A. Gade, A. Gelberg,J. Eberth, K. Jessen, H. Klein, U.Kneissl, S. Lenzi, T. Otsuka, H.H.Pitz, D. Rudolph, V.Werner for valuable discussions. One of us (R.V.J.) thanks the Universität zu Köln for a Georg Simon Ohm guest professorship. This work was partly supported by the DFG under Contracts no. Br 799/9-1 and Pi 393/1-1.

References

1. A. Richter, *Prog.Part.Nucl.Phys.* **34**, 261 (1995).
2. T.Otsuka *et al., Nucl. Phys.* A **309**, 1 (1978).
3. W. Heisenberg, *Z.Phys.* **77**, 1 (1932).
4. E.P. Wigner, *Phys. Rev.* **51**, 106 (1937).
5. C.E. Svensson, *et al., Phys. Rev.* C **58**, R2621 (1998).
6. S. Skoda, *et al., Phys. Rev.* C **58**, R5 (1998).
7. D. Rudolph, *et al., Phys. Rev. Lett.* **76**, 376 (1996).
8. D. Rudolph,*et al., Nucl. Phys.* A **630**, 417c (1998).
9. S.M. Lenzi, *et al., Phys.Rev.* C **60**, 021303 (1999).
10. C.D. O'Leary, *et al., Phys. Lett.* B **459**, 73 (1999).
11. C. Frießner, *et al., Phys. Rev.* C **60**, 011304 (1999).
12. I. Schneider, *et al.*, to be published in *Phys. Rev.* C **61**, (2000).
13. A.F. Lisetskiy *et al., Phys. Rev.* C **60**, 064310 (1999).
14. A. Poves and A. Zuker, *Phys. Rep.* **70**, 235 (1981).
15. D.Bohle *et al., Phys. Lett.* B **137**, 27 (1984).
16. U.E.P. Berg *et al., Phys. Lett.* B **149**, 59 (1984).
17. U. Kneissl *et al., Prog.Part.Nucl.Phys.* **37**, 349 (1996).
18. N. Pietralla *et al., Phys. Rev.* C **58**, 184 (1998).
19. N. Pietralla *et al., Phys. Rev. Lett.* **83**, 1303 (1999).
20. N. Pietralla *et al.*, submitted to *Phys. Rev. Lett.*
21. N. Pietralla *et al.*, in "CGS10", Santa Fe 1999 (World Scientific) in press.
22. T. Otsuka *et al., Phys. Rev.* C **50**, R1768 (1994).
23. N. Pietralla *et al., Phys. Rev.* C **57**, 150 (1998) and references therein.

OCTUPOLE CORRELATIONS IN ^{143,145}BA AND ^{147}PR

J.H. HAMILTON[1], A.V. RAMAYYA[1], J.K.HWANG[1,2], S.J. ZHU[1,2,3] E.F. JONES[1], P. GORE[1], AND GANDS95 COLLABORATIONS

[1] *Physics Department, Vanderbilt University, Nashville, TN 37235, USA*
[2] *Joint Institute for Heavy Ion Research, Oak Ridge, TN 37831, USA*
[3] *Physics Department, Tsinghua University, Beijing 100084, People's Republic of China*

High spin states in neutron-rich odd-Z ^{143,145}Ba nuclei have been investigated from the study of prompt γ-rays in the spontaneous fission of ^{252}Cf. Alternating parity bands are identified for the first time in ^{145}Ba and extended in ^{143}Ba. A new side band with equal, constant dynamic and kinetic moments of inertia equal to the rigid body value, as found in superdeformed bands, is discovered in ^{145}Ba. Enhanced E1 transitions between the negative- and positive-parity bands in these nuclei give evidence for strong octupole deformation in ^{143}Ba and in ^{145}Ba. These collective bands show competition and co-existence between symmetric and asymmetric shapes in ^{145}Ba. The first evidence is found for crossing M1 and E1 transitions between the s=+i and s=-i doublets in ^{143}Ba.
Neutron-rich ^{147}Pr also was studied in the spontaneous fission of ^{252}Cf. Possible parity doublets observed in ^{147}Pr with N=88 indicates that neutron-rich $^{147}_{59}$Pr$_{88}$ nucleus exhibits strong octupole correlations like those observed in the $^{146}_{58}$Ce$_{88}$ core.

1 Introduction

Theoretical calculations in the deformed shell model suggested the existence of an island of stable octupole deformed nuclei around Z=56 and N=88 [1-2]. Leander et al. [3] predicted that the odd-N ^{145}Ba is a good candidate for octupole deformation. Searches for octupole deformation in ^{145}Ba, including β-decay work [4] and spontaneous fission studies found some collective bands [5-7] but no evidence for octupole deformation. The first evidence for octupole deformation in this region was reported in ^{144,146}Ba [8], ^{146}Ce [9], and then in ^{148}Nd [10]. The first evidence for octupole deformation in an odd-A system in this region was discovered in ^{143}Ba [5,11]. Both s=±i parity doublets were then reported in ^{143}Ba [6] and confirmed in our work [12]. Evidence for octupole correlations and deformation is also observed in ^{139}Xe, ^{140,141,142}Ba and ^{144}Ce [5,7,11-14]. The odd-Z ^{145,147}La also are reported to have strong octupole correlations [15] with the evidence significantly extended in our work [16]. Thus, an island of stable octupole deformation around Z=56, N=88 is established. However, evidence for octupole deformation was not found in ^{145}Ba as predicted to occur [3].

Here we report the first evidence of octupole deformation in ^{145}Ba and expanded level structures in ^{143}Ba. A surprising new type of band structure with equal kinetic and dynamic moments of inertia and equal to the rigid body is found in ^{145}Ba. These properties are characteristic of the superdeformed bands first observed in ^{152}Dy [17]. Also we find the first evidence for crossing transitions between the s=±i parity doublets in ^{143}Ba. These transitions test the purity of these doublets.

The strong octupole correlations are observed in $^{139-141}$Xe (Z=54) [13], $^{142-146}$Ba (Z=56) [5,18], ^{145,147}La (Z=57) [15,16] and ^{144,146}Ce (Z=58) [11,19] isotopes but not in Cs (Z=55) [20]. Therefore a search for the possible octupole correlations in Pr(Z=59) may be useful for mapping out the systematics in this region. If the octupole correlations are present, parity doublet bands should be observed. Since strong octupole correlations have been established in ^{146}Ce, the ^{147}Pr nucleus with N=88 is the best candidate where parity doublet bands can be observed. These two nuclei are very similar to N=88 ^{144}Ba and ^{145}La where clear octupole correlations are observed. In the present work, we investigated the level scheme of ^{147}Pr to look for the similarities between ^{147}Pr and ^{145}La where coexistence of a strong coupled ground rotational band and parity doublets were observed[15,16]. Additional evidence for octupole correlation in ^{147}Pr was found.

The experimental details can be seen at Ref. [7,18,21].

2 Octupole correlations in ^{143,145}Ba

The new level schemes for ^{143}Ba and ^{145}Ba are shown in Figs. 1 and 2 based on $]gamma - \gamma - \gamma$ coincidence data. The bands connected by stretched E2 γ-transitions inside the band are numbered. All previously reported transitions in [5, 6, and 11] were confirmed. Many new transitions and levels are observed. In ^{143}Ba, a new level at 2425.9 keV was added to band (3). New E1 crossover transitions, 207.2, 418.2, and 160.9 keV, between bands (3) and (4), along with a tentative one at 274.7 keV were found. Other new transitions of (E1) 389.7 and (M1) 846.1 keV between the s=+i band (3) and s=-i bands (2) and (1) and M1 transitions of 458.7, 596.9, 706.5, and 727.2 keV and E1 transitions of 428, 571.7, and tentatively 717.3 keV between the s=+i band (4) and the s=-i bands (2) and (1), respectively, were also found.

In ^{145}Ba, two new collective bands, (1) with 671.1, 972.1, 1384.2, 1889.9, and 2429 keV levels and (5) with 1463.1, 1813.3, 2235.4, 2726.1, 3290.1, 3922.7, and 4624.5 keV levels were discovered. Two sets of intertwined, crossing transitions of 393.5, 330.3, 126.5, 285.6, 256.2, and 249.5 keV between bands (1) and (2), and of 595.8, 572.1, 515.8, 442.1, (197.9), and 366.1 keV between bands (4) and (5) were also observed and assigned as E1 based on systematics and the measured total conversion coefficient of the 126.5 keV transition (0.08(20)) which agrees with an E1 value (0.11) but cannot definitely exclude an M1 value (0.53). This ICC was extracted from the 301.0 - 505.7 keV double gated spectrum by comparing the γ-ray intensities of the intermediate 126.5 and 285.6 keV cascade transitions. A new level is added to bands (2), (3), and (4). Three new E1 side transitions, 23.5 and 231.6 keV between bands (2) and (4) and 282.6keV between bands (3) and (4) are also found. A total of 28 new transitions and 15 new levels were found in ^{145}Ba.

Spins and parities (J^{π}) for each band in ^{143}Ba have been assigned in previous reports [5,6] based on systematics and some angular correlation as well as internal conversion coefficient measurements. From Fig. 1, one can see that in ^{143}Ba, two sets of opposite parity bands, bands (1) and (2) and bands (3) and (4), each set with intertwined, strongly enhanced E1 crossing transitions, form structures characteristic of octupole deformation with the simplex quantum number s=-i and s=+i, respectively. Here the first evidence for M1 and E1 transitions between the s=+i and s=-i structures is reported. These new data in ^{143}Ba provide an important test of the purity of the simplex quantum numbers s=$\pm$i.

In ^{145}Ba, the spins and parities of bands (2), (3), and (4) also have been assigned based on systematics and some angular correlation as well as internal conversion measurements [6]. Based on systematic comparison with neighboring nuclei, ^{143}Ba, ^{144}Ba, and ^{146}Ba, and intertwined strong crossing transitions between bands (1) and (2) and bands (4) and (5) with B(E1)/B(E2) ratios similar to ^{143}Ba, the J^{π} of the 671.1 keV head of band (1) was assigned as ($11/2^+$), and J^{π} of the 1463.1 keV head of band (5) was assigned as ($19/2^-$). Bands (1) and (2) with $\Delta I = 2$ transitions in each band and intertwined E1 transitions between the bands form a typical octupole deformation structure similar to that in ^{143,144,146}Ba with simplex quantum number s=-i. Bands (4) and (5) with similar structural characteristics to bands (1) and (2) form the octupole deformation structure with s=+i. Thus, these data indicate the two sets of parity doublets with s=$\pm$i expected for octupole deformation at higher spins. The ground bands (2) and (3) at lower spins, linked by M1 transitions in ^{145}Ba, form a strong coupled collective structure with signature splitting. Above the 463.3 keV ($11/2^-$) level, the transition intensities are very weak as the levels become non-yrast. This strong-coupled collective structure represents a well-deformed symmetric rotor shape in ^{145}Ba and also is observed in ^{145}La [16]. It probably originates mainly from Coriolis-mixed $\nu h_{9/2}$ and $\nu f_{7/2}$ orbitals.

The B(E1)/B(E2) values in ^{143}Ba and ^{145}Ba in our investigation are

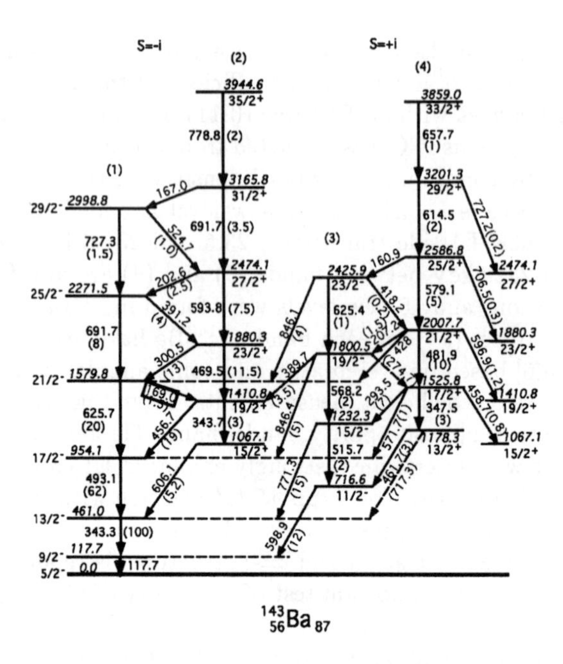

Figure 1. Level scheme of ^{143}Ba

listed in Table 1. The similarity of these data support the ^{145}Ba spin and parity assignments. For ^{143}Ba, the error weighted average values are $0.64(4)\times10^{-6}$ fm^{-2} for s=-i and $0.33(5)\times10^{-6}$ fm^{-2} for s=+i, respectively. For ^{145}Ba, the same average values are $0.55(6)\times10^{-6}$ fm^{-2} for s=-i and $0.36(5)\times10^{-6}$ fm^{-2} for s=+i. These compare favorably with the average in ^{144}Ba of $0.36(2)\times10^{-6}$ fm^{-2}. These data indicate that the octupole correlations are very strong leading to stable octupole deformation in ^{143,145}Ba.

Band (4) in ^{145}Ba based on the $13/2^+$ level becomes the yrast band and has the strongest transition intensities. It most probably originates from a $\nu i_{13/2}$ single particle orbital coupling. The fact that the average B(E1)/B(E2) value is less than in the s=-i band may indicate that the $i_{13/2}$ neutron single orbital coupling reduces the octupole correlations.

For the $i_{13/2}$ band (4), the J_1 is very large at low rotational frequency ($\hbar\omega$) and smoothly reduces as $\hbar\omega$ increases, but J_2 smoothly increases as $\hbar\omega$ increases. The J_1 and J_2 of band (5) are very large and are essentially constant and equal with increasing rotational frequency. Quite surprisingly, they essentially have a rigid body moment of inertia. This is the first such band

Table 1. B(E1)/B(E2) ratios in ^{143,145}Ba

$I_i^\pi \rightarrow I_f^\pi$	B(E1)/B(E2) ($10^{-6}fm^{-2}$)	$I_i^\pi \rightarrow I_f^\pi$	B(E1)/B(E2) ($10^{-6}fm^{-2}$)
	^{143}Ba, s=-i, Bands (1) and (2)		
$19/2^+ \rightarrow 15/2^+$	0.25(7)	$25/2^- \rightarrow 21/2^-$	1.0(1)
$19/2^+ \rightarrow 17/2^-$		$25/2^- \rightarrow 23/2^+$	
$21/2^- \rightarrow 17/2^-$	1.2(2)	$27/2^+ \rightarrow 23/2^+$	2.3(6)
$21/2^- \rightarrow 19/2^+$		$27/2^+ \rightarrow 25/2^-$	
$23/2^+ \rightarrow 19/2^+$	0.73(6)	$29/2^- \rightarrow 25/2^-$	0.7(3)
$23/2^+ \rightarrow 21/2^-$		$29/2^- \rightarrow 27/2^+$	
	^{143}Ba, s=+i, Bands (3) and (4)		
$17/2^+ \rightarrow 13/2^+$	0.36(6)	$23/2^- \rightarrow 19/2^-$	0.20(15)
$17/2^+ \rightarrow 15/2^-$		$23/2^- \rightarrow 21/2^+$	
$21/2^+ \rightarrow 17/2^+$	0.3(1)		
$21/2^+ \rightarrow 19/2^-$			
	^{145}Ba, s=-i, Bands (1) and (2)		
$15/2^+ \rightarrow 11/2^+$	0.45(10)	$19/2^+ \rightarrow 15/2^+$	0.63(12)
$15/2^+ \rightarrow 13/2^-$		$19/2^+ \rightarrow 17/2^-$	
$17/2^- \rightarrow 13/2^-$	0.8(3)	$21/2^- \rightarrow 17/2^-$	0.59(12)
$17/2^- \rightarrow 15/2^+$		$21/2^- \rightarrow 19/2^+$	
	^{145}Ba, s=+i, Bands (4) and (5)		
$27/2^- \rightarrow 23/2^-$	0.35(7)	$35/2^- \rightarrow 31/2^-$	0.50(15)
$27/2^- \rightarrow 25/2^+$		$35/2^- \rightarrow 33/2^+$	
$31/2^- \rightarrow 27/2^-$	0.34(7)		
$31/2^- \rightarrow 29/2^+$			

observed in neutron rich nuclei. This could indicate a pairing-free rotational band and if so is the first such example observed in this region. On the other hand, it may be related to the same phenomenon that occurs in superdeformed bands in the light Hg - Pb region where octupole deformation plays a role in the SD bands. Indeed the large, constant, and essentially equal J_1 and J_2 for band (5) are very similar to the large, constant, and essentially equal J_1 and J_2 in the first superdeformed band in ^{152}Dy [17]. The origin of this rigid body band in ^{145}Ba is not clear but definitely offers a new challenge for theory.

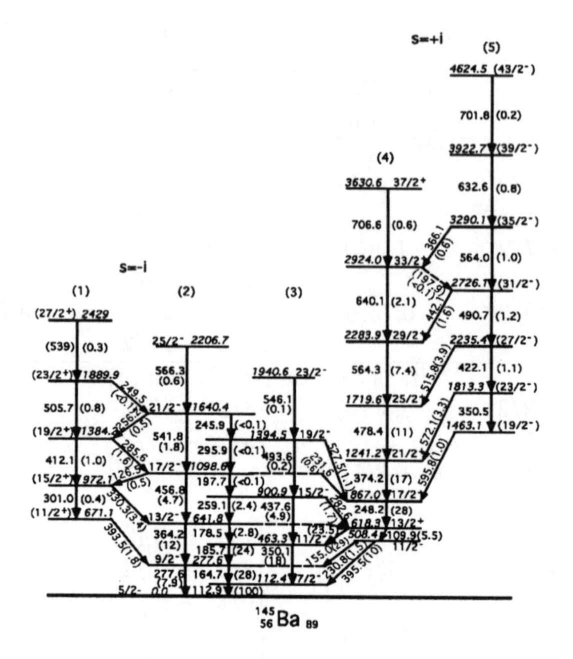

Figure 2. Level scheme of ^{145}Ba

3 Octupole correlations in ^{147}Pr

The new γ transitions in Pr were discovered by gating on the γ transitions in the partner Y fragments. The 58.5 keV transition (58.2, 58.0 and 57.7 keV in Ref. [22,23]) and 103.0 keV transition (104.5 keV in Ref. [24]) in ^{149}Pr were identified from the ^{149}Ce β decay. Also, the 220.3 keV transition (220.0 keV in Ref. [22]) in ^{149}Pr was identified in the reaction ^{150}Nd(d,^{3}He)^{149}Pr. The positions of 103.0 keV and 220.3 keV transitions were not known in the level scheme of ^{149}Pr [22,24]. A 95.5 keV transition in ^{151}Pr (96.8 keV in Ref. [25]) was discovered in studies of β decay from ^{151}Ce. We believe that the small differences between the present and previous energies for the low energy transitions such as 58.5, 95.5 and 103.0 keV transitions are related to differences in the energy calibration. Gating on these transitions in the Pr isotopes and some known transitions in the Y isotopes, one sees numbers of new γ transitions belonging to the Pr isotopes [26].

The level scheme of ^{147}Pr as discovered in this work is shown in Fig. 3. All the levels in ^{147}Pr are identified by assigning the transitions between the

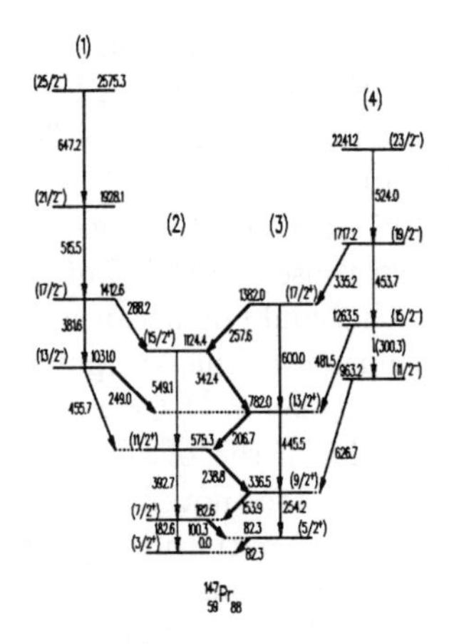

Figure 3. Level scheme of ^{147}Pr. All the transitions are new.

lowest levels of the rotational bands on the basis of their yields relative to the partner Y isotopes. This conclusion came from the comparison of them to some transitions of ^{149,151}Pr [26] in the coincidence spectra with double gates on the γ transitions in ^{99,101}Y. The very close similarity of the observed bands to those in ^{145}La with N=88 supports the assignment of these bands to ^{147}Pr. A similar rotational band was not observed in ^{146}La with N=89 [27].

The strong coupled ground rotational band in ^{145}La has a configuration of 5/2[313] based on the $1g_{7/2}$ proton orbital. But the ground state in ^{147}Pr has spin and parity of $3/2^+$ assigned on the basis of the β decay and the 3/2[411] orbital of the $2d_{5/2}$ proton is near the Fermi surface of ^{147}Pr. Therefore the 3/2[411] configuration is assinged to the discovered rotational band in ^{147}Pr. The spins and parities of the bands -1 and -4 in ^{147}Pr are tentatively assigned on the basis of the similarity to the negative parity bands in ^{145}La. The negative parity bands -1 and -4 in Fig. 3, have a large signature splitting without any linking transitions between them. This means that bands -1 and -2, and bands -3 and -4 can be parity doublets at the higher spin. The B(E1)/B(E2) ratio of $0.33\times10^{-6}\text{fm}^{-2}$ for the 288.2(E1) and 381.6(E2) keV transitions is extracted from the coincidence spectrum gated on the 515.6 and 647.2 keV transitions in ^{147}Pr. Also, the B(E1)/B(E2) ratio of $0.60\times10^{-6}\text{fm}^{-2}$ for 335.2(E1) and 453.7(E2) keV transitions is extracted from the coincidence spectrum gated on the 445.5 and 153.9 keV transitions in ^{147}Pr because the 257.6 keV transition is much weaker than the 600.0 keV transition. The B(E1)/B(E2) ratios for ^{146}Ce are 0.70×10^{-6}fm^{-2} (379.5(E1)-367.9(E2) keV transitions), 2.09×10^{-6}fm^{-2} (185.5(E1)-565.2(E2) keV transitions), 0.87×10^{-6}fm^{-2} (282.6(E1)-468.1(E2) keV transitions), and 0.78×10^{-6}fm^{-2} (332.5(E1)-614.5(E2) keV transitions). The B(E1)/B(E2) ratios in ^{147}Pr are smaller than those in ^{146}Ce but still show definite enhancement of the E1 transitions. This could be because of the $h_{11/2}$ proton blocking the proton $d_{5/2}$-$h_{11/2}$ contribution to the octupole correlation.

The parity doublets are not seen in ^{147}Ce with N=89. This indicates that the additional odd neutron blocks the octupole correlations in ^{147}Ce with N=89 but the additional proton does not change the degree of octupole correlation effects in ^{147}Pr. This level pattern in N=88 ^{147}Pr is very like that in N=88 ^{145}La where octupole correlations are observed [15,16]. The existence of the parity doublet bands in ^{147}Pr related to the octupole correlations is consistent with its core nucleus, ^{146}Ce, which shows octupole correlations. The ^{147}La nucleus with N=90 exhibits the parity doublets in the high spin region similar to the weak octupole correlations in ^{146}Ba. However, octupole correlation strength in ^{147}La is weakened by the decoupling effect of the $h_{11/2}$ proton from the core nucleus, ^{146}Ba. Thus the additional proton does not change the degree of octupole correlations in ^{145}La and ^{147}Pr with N=88.

4 Conclusion

In summary, new high spin states in ^{143}Ba and ^{145}Ba have been investigated. Stable octupole deformation or at least strong octupole correlations are observed in these nuclei. These new data confirm the long-standing theoretical prediction [3] of stable octupole deformation in ^{145}Ba. A new band with rigid body moments of inertia in ^{145}Ba may be the first example in neutron rich nuclei of a pairing-free structure or of a type of superdeformed band. This new structure offers a challenge for theory. The strong-coupling ground band and octupole deformation structures in ^{145}Ba show competition and coexistence between symmetric and asymmetric shapes. The first evidence for crossing M1 and E1 transitions between the s=±i doublets in ^{143}Ba was obtained.

Possible parity doublets observed in ^{147}Pr with N=88 indicate that neutron-rich $^{147}_{59}$Pr$_{88}$ exhibits strong octupole correlations like those observed in its core $^{146}_{58}$Ce$_{88}$.

Acknowledgments

The work at Vanderbilt University is supported in part by the U.S. Department of Energy under Grants No. DE-FG05-88ER40407. The Joint Institute for Heavy Ion Research is supported by its members, University of Tennessee, Vanderbilt University, and the U.S. Department of Energy.

References

1. W. Nazarewicz et al., Nucl. Phys., A429, 269 (1984).
2. W. Nazarewicz and P. Olanders, Nucl. Phys., A441, 420 (1985).
3. G.A. Leander et al., Phys. Lett., B152, 284 (1985).
4. J.D. Robertson et al., Phys. Rev. C 34, 1012 (1986).
5. S.J. Zhu et al., Phys. Lett., B357, 273 (1995).
6. M.A. Jones et al., Nucl. Phys., A 605, 133 (1996).
7. J.H. Hamilton et al., Prog. Part. Nucl. Phys., 38, 273 (1997).
8. W.R. Phillips et al., Phys. Rev. Lett., 57, 3257 (1986).
9. W.R. Phillips et al., Phys. Lett., B212, 402 (1988).
10. R.C. Ibbotson et al., Phys. Rev. Lett., 71, 1990 (1993).
11. J.H. Hamilton et al., Prog. Part. Nucl. Phys., 35, 635 (1995).
12. S.J. Zhu et al., Chin. Phys. Lett., 14, 569 (1997).
13. S.J. Zhu et al., J. Phys., G23, L77 (1997).
14. W. Urban et al., Nucl. Phys., A 613, 107 (1997).
15. W. Urban et al., Phys. Rev., C 54, 945 (1996).
16. S.J. Zhu et al., Phys. Rev. C 59, 1316 (1999).
17. P.J. Twin et al., Phys. Rev. Lett., 57, 811 (1986).
18. S.J. Zhu et al., Phys. Rev. **C60**, 051304 (1999).
19. J.H. Hamilton et al., **Proc. of Int. Conf. on Nuclear Structure 98**, ed. by C. Baktash, AIP Conf. Proc. **481**, (1999) p. 473.
20. T. Rzaca-Urban et al., Phys. Lett. **B348**, 336 (1995).
21. D.C. Radford, Nucl. Instrum. Methods Phys. Res. **A361**, 297 (1995).
22. R.B. Firestone and V.S. Shirley, **Table of Isotopes**, 8th ed. (John Wiley and Sons, Inc., New York, 1996).
23. J.A. Szucs, M.W. Johns and B. Singh, Nucl. Data Sheets, **46**, 1 (1985).
24. B. Pfeiffer et al., J. Phys.(Paris) **38**, 9 (1977).
25. C.M. Class, ORO-1316-168, p. D4 (1974).
26. J.K. Hwang et al., to be submitted to Phys. Rev. C (1999).
27. J.K. Hwang et al., Phys. Rev. **C58**, 3252 (1998).

NUCLEAR STRUCTURE CALCULATIONS FOR DOUBLE-BETA DECAY

S. STOICA

Department of Theoretical Physics, Horia Hulubei Institute for Physics and Nuclear Engineering, P.O. Box MG-6, 76900-Bucharest, Romania
E-mail: stoica@ifin.nipne.ro

We give a critical review of the QRPA-based methods used in the computation of the nuclear matrix elements for the $\beta\beta$ decay, stressing on second-QRPA and renormalized-QRPA approaches which go beyond the quasi boson approximation.

1 Introduction

Nuclear double-beta ($\beta\beta$) decay may proceed via several theoretical scenarios, but only the two-netrino (2ν) decay mode, allowed by the Standard Model (SM) of the electroweak interactions, has been up to now observed experimentally. However, theories more general than the SM, predict that the $\beta\beta$ decay may also occur without emission of any neutrino (0ν). If this would happen, several fundamental questions related to the nature and mass of the neutrino, existence of right-handed components in the weak interaction, lepton number conservation, validity of the SUSY theories, etc. will find quantitative answers. This is why the $0\nu\beta\beta$ decay is at present a very searched process. The parameters connected to the physics beyond the SM, which appear in the theory of the $\beta\beta$ decay, depend decisively on the relevant nuclear matrix elements (ME) which have to be evaluated theoretically. The models for calculating these ME are mainly based of two different approaches: the Shell Model (ShM) and the proton-neutron quasi random phase approximation (pnQRPA). Since the nuclei which undergo a $\beta\beta$ decay are generally rather far from the closed shells, there are the QRPA-based methods which are widely employed in the literature for their computation. The pnQRPA method, adapted from the standard QRPA for treating charge-exchanging nuclear processes, has been the first used for $\beta\beta$ decay calculations. One of its most important achievements was the success in explaining the suppression mechanism of the $2\nu\beta\beta$ decay ME [1-3]. However, this method faces with the problem of a strong dependence of the ME on the renormalization of the particle-particle component of the residual interaction. To overcome this problem several further developments of this method have been advanced during the recent past.

In this paper we will give a critical review of two of the most used higher-order approaches for treating the nuclear charge-changing proceses: i) the second-QRPA and ii) the renormalized-QRPA (RQRPA). The second-QRPA and RQRPA methods are described in section 2. The section 3 is devoted to a critical comparison between these approaches and the case of ^{76}Ge is presented as an example. Conclusions are presented in section 4.

2 Second-QRPA formalism

The first approach going beyond pnQRPA has been developed in ref. [5] and further, applied in ref. [6]. In this approach the extension of the pnQRPA was done using a boson expansion of both the phonon operators and transition $\beta^{\pm}$ operators and retaining the next order in this expansion beyond the quasi-boson approximation. Also, this method allowed, for the first time, the computation of $\beta\beta$ decay rates to excited final states. In the version when only two boson states contributions are taken into account in the improved wave function, one calls this method second-QRPA. To include higher-order corrections to the pnQRPA and restoring partially the Pauli principle the two quasiparticle and the quasiparticle-density dipole operators are expanded in a Beliaev- Zelevinski series:

$$A^{\dagger}_{1\mu}(pn) = \sum_{k} \left(A^{(1,0)}_{k_1} \Gamma^{+}_{1\mu}(k) + A^{(0,1)}_{k_1} \tilde{\Gamma}^{+}_{1\mu}(k) \right) \qquad 2.1$$

$$B^{\dagger}_{1\mu}(pn) = \sum_{k_1 k_2} \left(B^{(2,0)}_{k_1 k_2}(pn)[\Gamma^{\dagger}_{1}(k_1)\Gamma^{\dagger}_{2}(k_2)]_{1\mu} + B^{(0,2)}_{k_1 k_2}(pn)[\Gamma_{1}(k_1)\Gamma_{2}(k_2)]_{1\mu} \right) \qquad 2.2$$

where

$$B^{\dagger}_{1\mu}(pn) = \sum_{m_k, m_l} C^{JM}_{j_p m_p j_n m_n} a^{\dagger}_{j_p m_p} a_{j_n m_n}$$

$$\tilde{B}_{1\mu}(pn) = (-)^{J-M} B_{1\mu}(pn) \qquad 2.3$$

The boson expansion coefficients $A^{(1,0)}$, $A^{(1,0)}$, $B^{(2,0)}$, $B^{(0,2)}$ are determined so that the equations (2.1)-(2.2) are also valid for the corresponding ME in the boson basis.

Further, the transition $\beta^{\pm}$ operators in the quasiparticle representation can be expressed in terms of the dipole operators $A_{1\mu}$ and $B_{1\mu}$:

$$\beta^{-}_{\mu}(k) = \theta_k A^{\dagger}_{1\mu}(k) + \bar{\theta}_k \tilde{A}_{1\mu} + \eta_k B^{\dagger}_{1\mu}(k) + \bar{\eta}_k \tilde{B}_{1\mu}$$

$$\beta^+_\mu(k) = -\left(\bar{\theta}_k A^\dagger_{1\mu}(k) + \theta_k \tilde{A}_{1\mu} + \bar{\eta}_k B^\dagger_{1\mu}(k) + \eta_k \tilde{B}_{1\mu}\right) \qquad 2.4$$

where

$$\theta_k = \frac{\hat{j}_p}{\sqrt{3}}\langle j_p||\sigma||j_n\rangle U_p V_n;\ \bar{\theta}_k = \frac{\hat{j}_p}{\sqrt{3}}\langle j_p||\sigma||j_n\rangle U_n V_p;\ \hat{j} = \sqrt{2j+1}$$

$$\eta_k = \frac{\hat{j}_p}{\sqrt{3}}\langle j_p||\sigma||j_n\rangle U_p U_n;\ \bar{\eta}_k = \frac{\hat{j}_p}{\sqrt{3}}\langle j_p||\sigma||j_n\rangle V_p V_N \qquad 2.5$$

Using the boson expansions (2.1)-(2.2), one also gets expressions of the transition operators beyond the quasiboson approximation. Thus, in the second-QRPA method, the higher-order corrections to the pnQRPA are introduced not only in the RPA wave funtions (by improving the phonon operator with additional correlations), but also in the expressions of the $\beta^\pm$ operators, and the procedure is now more consistent.

3 RQRPA formalism

The RQRPA method [7], [8] is based on the idea of replacing the uncorrelated QRPA ground state (g.s.) by a correlated g.s., in the calculation of the expectation value of the commutator of the two bifermion operators involved in the derivation of the QRPA equations. The expectation values of the number operator in the QRPA correlated g.s. are introduced in the quasi-boson commutators of the pair operators and this leads to a renormalization of the QRPA forward- and backward-going amplitudes. Within the RQRPA a stabilization of the ME against g_{pp} and a shift of the RPA break-down point towards larger (un-physical) values of this constant are also observed.

In the RQRPA method the A, $A^\dagger$ operators are renormalized:

$$\bar{A}_{\mu\mu'}(k,l,J,M) = D^{-1/2}_{\mu k\nu k' J^\pi}\ A_{\mu\mu'}(k,l,J,M) \qquad 2.6$$

where the $D_{\mu k\nu k'}$ matrices are defined as follows:

$$D_{\mu k\nu k' J^\pi} = \mathcal{N}(k\mu, l\nu)N(k'\mu', l'\nu')\left(\delta_{\mu\mu'}\delta_{\nu\nu'}\delta_{kk'}\delta_{ll'} - \delta_{\mu\nu'}\delta_{\nu\mu'}\delta_{lk'}\delta_{kl'}(-)^{j_k+j_l-J}\right)$$

$$\left[1 - j_l^{-1}\langle 0^+_{RPA}|[a^\dagger_{\nu l}a_{\nu l'}]_{00}|0^+_{RPA}\rangle - j_k^{-1}\langle 0^+_{RPA}|[a^\dagger_{\mu k}a_{\mu k'}]_{00}|0{+}_{RPA}\rangle\right] \qquad 2.7$$

One observes that by this renormalization one goes beyond the QBA by taking into account the next terms in the commutator relations of the A, $A^\dagger$ operators which are just, essentially, the proton and neutron number operators. It is worth mentioning that they are taken into account within RQRPA only by their averages on the RPA g.s.. The renormalization of the $A, A^{\dagger}$ operators is further caried on the RPA amplitudes, on the $\mathcal{A}, \mathcal{B}$ matrices and on the RPA phonon operator also obtaining a renormalization of them:

$$\bar{X}^m = D^{1/2} X^m \; ; \; \bar{Y}^m = D^{1/2} Y^m \; ; \; \bar{\mathcal{A}}^m = D^{-1/2} \mathcal{A} D^{-1/2} \; ; \; \bar{\mathcal{B}}^m = D^{-1/2} \mathcal{B} D^{-1/2} \quad 2.8$$

$$\Gamma^{m+}_{JM^{\pi}} = \sum_{k,l,\mu \leq \mu'} \left[\bar{X}^m_{\mu\mu'}(k,l,J^{\pi}) \bar{A}^{\dagger}_{\mu\mu'}(k,l,J,M) + \bar{Y}^m_{\mu\mu'}(k,l,J^{\pi}) \tilde{\bar{A}}_{\mu\mu'}(k,l,J,M) \right] \quad 2.9$$

To calculate $\bar{\mathcal{A}}$ and $\bar{\mathcal{B}}$ we need to determine the renormalization matrices D. This is done by solving a system of non-linear equations for them by an iterative numerical procedure. As input values one can use their expressions in which the averages of the number operators are replaced by the back-forwarded amplitudes obtained as initial solutions of the QRPA equation.

In the QRPA-type methods, before starting the RPA procedure, we need the BCS occupation amplitudes and the quasiparticle energies, in order to get the image of the RPA operators in the quasiparticle representation. This is done by solving the HFB equations, which may include, in the general case, both like- and unlike-nucleon pairing. When one includes only like-nucleon pairing in these equations, the QRPA procedure described above was called RQRPA, while when both types of the pairing interaction are included it was called full-RQRPA.

4 Critical view on RQRPA and second-QRPA methods

4.1 ME calculations

As an example we have performed calculations of the nuclear ME involved in the $2\nu\beta\beta$ decay mode of ^{76}Ge using the pnQRPA, second-QRPA, RQRPA and full-RQRPA methods. For the s.p. basis we used two choices. We included: i) the 12 levels belonging to the full sd, pf and sdg shells and considering thus ^{16}O as core and ii) the 9 levels belonging to the full pf and sdg shells and considering thus ^{40}Ca as core.

In Fig. 1 we displayed the $M^{2\nu}_{GT}$ matrix elements function of g_{pp} calculated with pnQRPA and second-QRPA methods. The two curves for each method represent the calculations performed with the two different s.p. basis. On the figure it is also drawn the line representing the ME value corresponding to the latest experimental $2\nu\beta\beta$ decay half-life of the ^{76}Ge, obtained by the Heidelberg-Moscow experiment: $T^{2\nu}_{1/2} = 1.55 \times 10^{21}$.

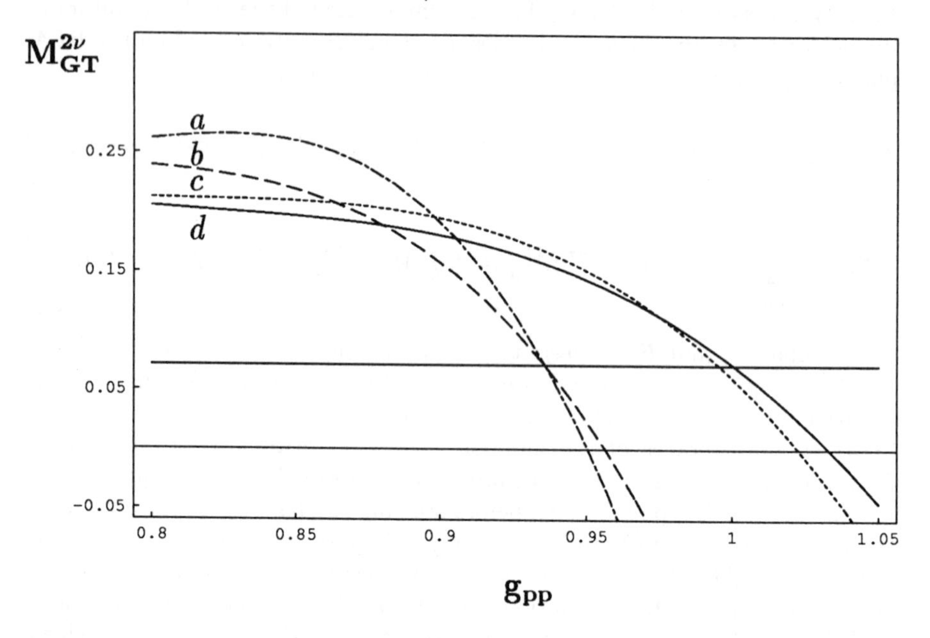

Figure 1. $a = pnQRPA - 9$,$b = pnQRPA - 12$,$c = second - QRPA - 9$, $d = second - QRPA - 12$

As it was already observed in previous calculations [5], the point where QRPA breaks down is pushed to higher values of g_{pp} in the framework of the second-QRPA method as compared with the pnQRPA. The calculation also shows that, within these two methods the values of the ME do not depend significantly on the size of the s.p. space, especially in the region around the experimental value. The values of g_{pp} which fit the best this experimental value are: 0.94 for both calculations performed with pnQRPA and 0.99 and 1.01 for the calculation with 12 and 9 levels, respectively performed with second-QRPA.

In Fig. 2 are displayed the same ME but calculated with RQRPA and full-RQRPA methods, Contrary to the previous calculations, in this case the difference between the results obtained with different choices of the s.p. space is rather large. At the value of g_{pp} where $M^{2\nu}_{GT}$ function cross the line representing the experimental value, the values for the ME, obtained with the same method, differ each-other up to $50 - 75\%$ when the two different choices of the s.p. basis are used. The values of g_{pp} for the best fits with the experimental value for the ME are: 0.977; 0.982 in the case of the RQRPA and 0.975; 1.012 in the case of the full-RQRPA for calculations including 9 and 12 levels in the s.p. basis, respectively. The different behavior of the calculation, performed with RQRPA and full-RQRPA on one side, and with pnQRPA and second-QRPA one the other side, in connection to the choices of the s.p. basis, reflects the sensitivity of the former methods in computing the $M_{GT}^{2\nu}$ ME.

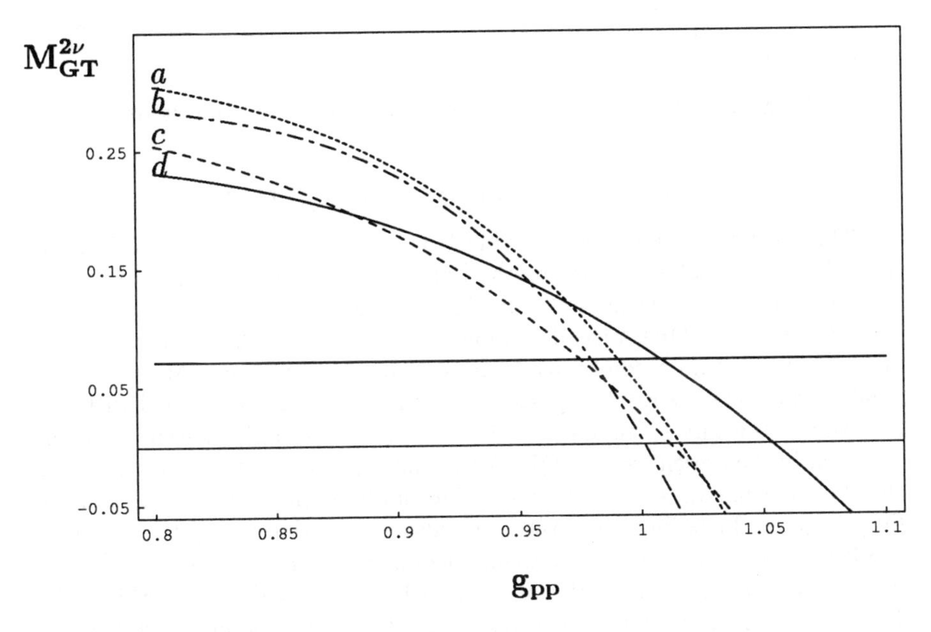

Figure 2. $a = fullRQRPA - 12, b = fullRQRPA - 9, c = RQRPA - 9$, $d = RQRPA - 12$

One possible source of this sensitivity might be in the origin of the numerical computation. Indeed, the double-iteration procedure, one for solving the full-RQRPA equation, the other for solving the non-liniar system of equations for the renormalization factors $D^{\mu k,\nu l;J^{\pi}}$, for all the multipolarities, is very time consuming which might affect the precison of the calculation, especially if the size of the s.p. model space is large.

On the other hand, there are some theoretical arguments which could explain the different results for the ME obtained with RQRPA-like methods as compared to the other two. We will discuss a bit later, after we shall refer to the ISM.

	pnQRPA		RQRPA		full-RQRPA		HQRPA	
^{76}Ge	0.23	0.26	20.06	21.34	17.68	17.29	2.7	3.7
^{76}Se	0.41	0.48	19.66	19.94	17.12	16.91	3.13	4.18

Table 1. The numbers represent the deviations (in percents) from the ISR calculated within the specified methods. The first (second) numbers in colons represent the calculation with a s.p. basis containing 12 (9) levels, respectively.

4.2 Ikeda sum rule

The ISM was also checked out in the framework of the four methods.

$$S_- - S_+ = \Sigma_m |\langle 0^+_{gs}||\sigma\tau_+||1^+_m\rangle|^2 - \Sigma_m |\langle 0^+_{gs}||\sigma\tau_+||1^+_m\rangle|^2 \qquad 3.1$$

The results are presented in Table 1, where the percentages of deviation from the correct value for each method and choice of the basis are given.

One can see that within pnQRPA ISM are very well fulfilled, while within RQRPA and full-RQRPA the deviations are between $17 - 21\%$. One can also observe that within second-QRPA the deviations from the ISM are rather small, confirming the result reported earlier in ref. [6], but performed with the inclusion of the only 1^+ channel and 9 levels in the s.p. basis. We should mention that in the present second-QRPA calculation we did not take into account the three boson states which may introduce undesirable spurious states in the QRPA space. Including such states one also gets deviations up to 17% from the ISR. A method to eliminate them adequately has not yet been developed up to now. One reason for the different behavior related to the ISM of the RQRPA and second-QRPA could be the existence of some inconsistencies of the RQRPA related to way of partial restoration of the Pauli principle. Indeed, as we already mentioned in section 2, within RQRPA method the Pauli principle is partially restored for the operators A, $A^\dagger$, by taking into account the averages of the quasiparticle-number operators in their commutator relations. However, it is no justification to neglect them in the $B, B^\dagger$ operator commutation relations. However, this is done within the second-QRPA and moreover, in this method the next higher-order corrections beyond pnQRPA are also introduced in the $\beta^\pm$ operators. The effect of the additional terms corroborated with a larger boson space (in the second-QRPA the boson space enlarges from one to two boson states) is a positive contribution to the ISM. On the other it is known that RQRPA underestimates ISR, so this could be one possible explanation why the ISR is better fulfilled within second-QRPA. Another shortcoming of the RQRPA method is a lack of consistency between

BCS and QRPA levels. While in the BCS still one assumes the g.s. to be the quasparticle vacuum at the level of RQRPA we are deling with the non-vanishing quasiparticle content of the g.s. due to the additional scattering terms taken into account in the commutation relations, [11].

5 Conclusions

We have performed a calculation of the $2\nu\beta\beta$ decay ME for the case of ^{76}Ge with the pnQRPA, second-QRPA, RQRPA and full-RQRPA methds, using two different choices of the s.p. basis. We can summarized the main results as follows:

i) for the $M_{GT}^{2\nu}$ we got a significant dependence of the results on the size of the s.p. basis, in the case of the RQRPA and full-RQRPA methods, while the results obtained with pnQRPA and second-QRPA do not display such a dependence.

ii) we also check the ISM within the four methods and found, it is fulfilled with good approximation within the second-QRPA method, while within the RQRPA and full-RQRPA methods the deviations are up to 21%. This result is not much dependent on the the size of the s.p. basis used. This result, besides the numerical arguments mentioned above, could also be explained by theoretical arguments related to the way that the partial restoration of the Pauli principle is done within the RQRPA. This restoration is made in the commutator relations of the operators A, $A^\dagger$ by taking into account the averages of the quasiparticle-number operators in their commutator relations. However, it is no justification to neglect them in the $B, B^\dagger$ operator commutation relations. This is done within the second-QRPA and moreover, in this method the next higher-order corrections beyond pnQRPA are also taken into account for the $\beta^\pm$ operators. The additional terms give a positive contribution to the ISM, while as it is known RQRPA underestimates ISR.

References

1. P. Vogel and M. R. Zirnbauer, Phys. Rev. Lett. **57** (1986) 3148.
2. J. Suhonen, A. Faessler, T. Taigel and T. Tomoda, Phys. Lett. **B 202** (1988) 174.
3. A. Staudt, T.T.S. Kuo and H.V. Klapdor-Kleingrothaus, Phys. Lett. **B 242** (1990) 17.
4. A. Staudt, K. Muto and H.V. Klapdor-Kleingrothaus, Europhys. Lett. **13** (1990) 31.
5. A. A. Raduta, A. Faessler, S. Stoica and W. A. Kaminski,

Phys. Lett. **B254** (1991) 7; A.A. Raduta, A. Faessler and S. Stoica, Nucl. Phys. **A 534** (1991) 149.
6. S. Stoica, Phys. Lett. **B 350** (1995) 152.
7. J. Toivanen and J. Suhonen, Phys. Rev. Lett. **75** (1995) 410; Phys. Rev. **C 55** (1997) 2314.
8. J. Schwieger, F. Simcovic and A. Faessler, Nucl. Phys. **A 600** (1996) 179.
9. O. Civitarese, P.O. Hess and J.G. Hirsch, P.O. Hess, Phys. Lett. **B412** (1997) 1; J.G. Hirsch, P.O.Hess and O. Civitarese, Phys. Rev. **C 54** (1996) 1976.
10. J. Suhonen and O. Civitarese, Phys. Rep. **300** (1998) 123.
11. A. Bobyk, W.A. Kaminski and P. Zareba, Eur. Phys. J. **A 5** (1999) 385.

QUADRUPOLE DEFORMATIONS OF NEUTRON-DRIP-LINE NUCLEI STUDIED WITHIN THE SKYRME HARTREE-FOCK-BOGOLIUBOV APPROACH

M.V. STOITSOV,[1,2] J. DOBACZEWSKI,[3] P. RING,[2] AND S. PITTEL[4]

[1] *Institute of Nuclear Research and Nuclear Energy, Bulgarian Academy of Sciences, Sofia-1784, Bulgaria*

[2] *Physik Department, Technische Universität, München, D-85748 Garching, Germany*

[3] *Institute of Theoretical Physics, Warsaw University, Hoża 69, PL-00-681 Warsaw, Poland*

[4] *Bartol Research Institute, University of Delaware, Newark, Delaware 19716*

We introduce a local-scaling point transformation to allow for modifying the asymptotic properties of the deformed three-dimensional Cartesian harmonic oscillator wave functions. The resulting single-particle bases are very well suited for solving the Hartree-Fock-Bogoliubov equations for deformed drip-line nuclei. We then present results of self-consistent calculations performed for the Mg isotopes and for light nuclei located near the two-neutron drip line. The results suggest that for all even-even elements with Z=10–18 the most weakly-bound nucleus has an oblate ground-state shape.

1 Introduction

In contrast to stable nuclei within or near the valley of beta stability, a proper theoretical description of weakly-bound systems requires a very careful treatment of the asymptotic part of the nucleonic density. This is particularly true in the description of pairing correlations near the neutron drip line, for which the correct asymptotic properties of quasiparticle wave functions and of one-particle and pairing densities is essential. In the framework of the mean-field approach, the best way to achieve such a description is to use the Hartree-Fock-Bogoliubov (HFB) theory in coordinate-space representation [5,6,7].

Such an approach presents serious difficulties, however, when applied to deformed nuclei. On the one hand, for finite-range interactions the technical and numerical problems arising when a two-dimensional mesh of spatial points is used are so involved that reliable self-consistent calculations in coordinate space should not be expected soon. On the other hand, for zero-range interactions existing approaches [8,9] are able to include only a fairly limited pairing phase space. The main complication in solving the HFB equations in coordinate space is that the HFB spectrum is unbounded from below, so that methods based on a variational search for eigenstates cannot be easily implemented. Because of this and other difficulties, one has to look for alternative solutions.

In principle, such an alternative solution is well known in the form of the configurational representation. In this approach, the system of partial differential HFB equations are solved by expanding the nucleon quasiparticle wave functions in an appropriate complete set of single-particle wave functions. In many applications, an expansion of the HFB wave function in a large harmonic oscillator (HO) basis of spherical or axial symmetry provides a satisfactory level of accuracy. For nuclei at the drip lines, however, expansion in an oscillator basis converges much too slowly to describe the physics of continuum states [7], which play a critical role in the description of weakly-bound systems. Oscillator expansions produce wave functions that decrease too steeply in the asymptotic region at large distances from the center of the nucleus. As a result, the calculated densities, especially in the pairing channel, are too small in the outer region and do not reflect correctly the pairing correlations of such nuclei.

In two recent works [10,11], a new transformed harmonic oscillator (THO) basis, based on a unitary transformation of the spherical HO basis, was discussed. This new basis derives from the standard oscillator basis by a local-scaling point coordinate transformation [12,13,14], with the precise form dictated by the desired asymptotic behavior of the densities. The transformation preserves many useful properties of the HO wave functions. Using the new basis, characteristics of weakly-bound orbitals for a square-well potential were analyzed and the ground-state properties of some spherical nuclei were calculated in the framework of the energy density functional approach[11]. It was demonstrated in [10] that configurational calculations using the THO basis present a promising alternative to algorithms that are being developed for coordinate-space solution of the HFB equations.

In the present work, we develop the THO basis for use in HFB equations of axially-deformed weakly-bound nuclei. As specific applications, we repeat previous calculations performed for the chain of Mg isotopes [8], but for different effective interactions, and then report a preliminary study of light, neutron-rich nuclei near the drip line.

2 Results

2.1 Drip-line-to-drip-line calculations in Mg

The chain of even-Z magnesium isotopes has been the subject of numerous recent theoretical analyses. The extreme interest in this isotopic chain is motivated by recent measurements in ^{32}Mg [48,49,50], which show a larger-then-expected quadrupole collectivity. Based on the relativistic and non-relativistic mean-field approaches and on shell-model calculations (see Ref. [51] for a review), it is now well documented that shape coexistence and configuration mixing occur in this N=20 nucleus. Moreover, recent advances in radioactive-ion-beam technology allow mass measurements of even heavier isotopes [52], giving hope that the neutron-drip line can be experimentally reached in the Z=12 chain [53].

In this section, we present results of an investigation of the deformation properties of the even-even Mg isotopes from the proton-drip-line to the neutron-drip-line. Our results are complementary to those of recent Skyrme+HFB calculations [8], in which the imaginary-time evolution method of finding eigenstates of the mean-field Hamiltonian was combined with a diagonalization of the HFB Hamiltonian within a relatively small set of these eigenstates. In that study, a complete set of results was given only for the SIII force and density-dependent pairing was used. Here, we present a complete set of results for the SLy4 force with a density-independent (volume) pairing interaction.

In Fig. 1, we plot the total HFB energies per nucleon E/A, the neutron chemical potentials λ_n, the neutron and proton deformation parameters, β_n and β_p, the neutron, proton, and total quadrupole moments, Q_n, Q_p, and Q_t, the average neutron and proton pairing gaps, $\widetilde{\Delta}_n$ and $\widetilde{\Delta}_p$ and the pairing energies E^n_{pair} and E^p_{pair} for the magnesium isotopes as functions of the mass number A. Ground-state values are shown by full symbols connected by lines, while the isolated open symbols correspond to secondary minima of the deformation-energy curves. In the top panel of Fig. 2, we compare the results for the two-neutron separation energies S_{2n} (open symbols) with those for the related quantity $-2\lambda_n$ (full symbols), and in the bottom panel we show the neutron and proton rms radii.

The lightest Mg isotope predicted by these calculations to be bound against two-proton decay is ^{20}Mg. The heaviest bound against two-neutron decay, on the basis of having a positive two-neutron separation energy, is ^{40}Mg. On this basis, the position of the two-neutron drip line obtained within the HFB+SLy4 approach is identical to that obtained in the finite-range droplet model [54], relativistic mean field (RMF) [55], and HFB+SIII [8] calculations. The RMF approach with the NL-SH effective interaction [56] predicts the two-neutron drip line at ^{42}Mg, and the relativistic Hartree-Bogoliubov (HB) approach with the NL3 effective interaction [57] predicts it at or beyond ^{44}Mg.

On the other hand, from Fig. 2, we see that both ^{42}Mg and ^{44}Mg, though having negative values of S_{2n}, have (small) negative values of the Fermi energy, λ_n.These nuclei, both of which exhibit oblate shapes, are bound against neutron emission. We will return to this point later.

The most deformed nucleus of the isotope chain is ^{24}Mg with almost the same neutron and proton deformations. At the other end of the chain, due to a large excess of neutrons over protons, significant differences exist between the proton and neutron quadrupole moments. The onset of large deformation in ^{36}Mg causes a decrease of the neutron chemical potential λ_n with respect to its value in ^{34}Mg. This gives an additional binding of ^{36}Mg, and correspondly to an increase and decrease of the two-neutron separation energies S_{2n} in ^{36}Mg and ^{38}Mg, respectively (see Fig. 2). In experiment [52], these changes are less pronounced and arrive two mass units earlier, giving rise to a small and large decrease of S_{2n} in ^{34}Mg and ^{36}Mg, respectively.

Concerning the ground-state deformation properties (full symbols connected by lines in Fig. 1), the proton drip-line nucleus ^{20}Mg displays a well defined spherical minimum (N=8 is a magic number). Then, there is a competition between prolate (^{22,24}Mg and ^{36,38,40}Mg) and oblate (^{26,30}Mg) deformations, while ^{28,32}Mg are spherical. The last two localized isotopes (with negative Fermi energies), ^{42,44}Mg, display oblate deformations. Secondary minima of the deformation energy curves (isolated symbols) exist for isotopes ^{22,24,26}Mg and ^{36,38,40}Mg.

Non-zero proton pairing correlations are present at all spherical or oblate minima. However, at these shapes, tangible neutron pairing exist only in ^{22,24}Mg and ^{34,36,38}Mg. Moreover, for all nuclei with prolate ground-state shapes, i.e., in ^{22,24}Mg and ^{36,38,40}Mg, both proton and neutron correlations are small or vanish altogether. These results are at variance with the Gogny-pairing HB calculations of Ref. [57], where non-zero pairing exists in all the heavy Mg isotopes. Also, in Ref. [8], stronger pairing correlations were obtained for the zero-range density-dependent pairing force. However, in that study, the strength parameters were not adjusted to odd-even mass staggering but rather taken from high-spin calculations of superdeformed bands. Our results suggest that the pure HFB-pairing approach is not necessarily the best way to treat pairing correlations in the Mg isotopes, and approximate or exact particle-number projection should probably be employed.

The bottom panel of Fig. 2 shows the neutron and proton rms radii, r_n and r_p. At the proton drip-line, the neutron rms radii are smaller than the proton rms radii, and then they increase with increasing neutron number. Around ^{24,26}Mg, r_n becomes almost equal to r_p, and for nuclei close to the neutron drip-line, r_n takes significantly larger values than r_p. The increase of r_n is fairly linear, similarly as in Refs. [8,56,57], and gives no hint of an existence of unusually larger neutron distributions at the neutron drip line (see also the discussion in Refs. [58,59]).

2.2 Neutron-drip-line calculations

Having at our disposal a viable method for performing deformed HFB calculations up to the drip lines, we have performed a systematic study of the equilibrium properties of the neutron-rich nuclei in all even-Z isotopic chains with proton numbers from Z=2–18. In this way, we have explored the neighborhood of the neutron-drip line for all neutron numbers from N=6–40.

We first performed spherical HFB+SLy4 calculations in coordinate space, using the methods and the code developed in Ref. [6]. We used volume delta pairing, with a coupling constant V_0=-218.5 MeV fm^3, adjusted as in Ref. [43]. This value is very close to the one used in our deformed THO code, suggesting that the effective pairing phase spaces used in the two approaches are very similar to one another.

From the spherical calculations, we obtained that the heaviest even isotopes, for which the Fermi energies are negative are: ^{8}He, ^{12}B, ^{22}C, ^{28}O, ^{30}Ne, ^{44}Mg, ^{46}Si, ^{50}S, ^{58}Ar. We used these spherical results as a starting point for our deformed calculations.

Next, within the deformed THO formalism, we found that the heaviest isotopes with negative Fermi energies are: ^{8}He, ^{12}B, ^{22}C, ^{28}O, ^{36}Ne, ^{44}Mg, ^{46}Si, ^{52}S, ^{58}Ar. The results obtained for these nuclei are summarized in Fig. 3. By comparing the deformed results to the spherical results, we see that the position of the last bound nucleus is influenced by deformation only in ^{36}Ne and ^{52}S. Volume pairing correlations are very weak in these nuclei; indeed, in all but the one case of ^{36}Ne, neutron pairing vanishes in the last bound nucleus of an isotopic chain. This suggests the necessity of using a surface pairing force here. Such a conclusion is supported by the fact that HB calculations [57], carried out with a Gogny pairing force, give sizable neutron pairing correlations in this region (Note that surface pairing and Gogny pairing produce quite similar distributions [7] over the single-particle states.)

Since neutron pairing vanishes in ^{12}B, our result is identical to that of Ref. [30], namely that the SLy4 force does not produce ^{14}B as bound, in disagreement with experiment [60]. Similarly, neither pairing nor deformation effects are present in the calculated ^{28}O nucleus, and hence this nucleus remains bound. On the other hand, the SLy4 force correctly describes ^{8}He [60] and ^{22}C as the last bound nuclei of their respective isotope chains [61].

3 Conclusions

In this paper, we have applied a local-scaling point transformation to the deformed three-dimensional Cartesian harmonic oscillator wave functions so as to allow for a modification of their unphysical asymptotic properties. In this way, we have obtained single-particle bases that remain infinite, discrete, and complete, but for which the wave functions have the asymptotic properties that are required by the canonical bases of Hartree-Fock-Bogoliubov theory. These bases preserve all the simplicity of the original harmonic-oscillator wave functions, and at the same time are amenable to very efficient numerical methods, such as Gauss-integration quadratures.

As a first application of this new methodology, we have carried out HFB calculations using the SLy4 Skyrme force and a density-independent (volume) pairing force. The calculations were performed for the chain of even-Z Mg isotopes and for the light even-Z nuclei located near the two-neutron drip line. We have presented results for binding energies, quadrupole moments, and for the pairing properties of these nuclei.

Perhaps the most interesting outcome of our calculations is that nuclei that are formally beyond the two-neutron drip line, i.e., those with negative two-neutron separation energies, may have tangible half lives, provided (i) that they have localized ground states (negative Fermi energies), and (ii) their ground-state configurations are significantly different than those of the (daughter) nuclei with two less neutrons. According to our calculations, precisely such a situation occurs in the chains of isotopes with Z=10, 12, 14, 16 and 18. In these chains, the prolate configuration becomes unbound before (i.e., for a smaller neutron number) than the oblate configuration. That change in the ground state structure leads to negative two-neutron separation energies and thus to the exotic conditions given above.

Acknowledgments

This work has been supported in part by the Bulgarian National Foundation for Scientific Research under project Φ-809, by the Polish Committee for Scientific Research (KBN) under Contract No. 2 P03B 040 14, by a computational grant from the Interdisciplinary Centre for Mathematical and Computational Modeling (ICM) of the Warsaw University, and by the National Science Foundation under grant #s PHY-9600445, INT-9722810 and PHY-9970749.

References

1. E. Roeckl, Rep. Prog. Phys. **55**, 1661 (1992).
2. A. Mueller and B. Sherril, Annu. Rev. Nucl. Part. Sci. **43**, 529 (1993).
3. P.-G. Hansen, Nucl. Phys. **A553**, 89c (1993).
4. J. Dobaczewski and W. Nazarewicz, Phil. Trans. R. Soc. Lond. A **356**, 2007 (1998).
5. A. Bulgac, Preprint FT-194-1980, Central Institute of Physics, Bucharest, 1980, nucl-th/9907088.
6. J. Dobaczewski, H. Flocard, and J. Treiner, Nucl. Phys. **A422**, 103 (1984).
7. J. Dobaczewski, W. Nazarewicz, T.R. Werner, J.-F. Berger, C.R. Chinn, and J. Dechargé, Phys. Rev. **C53**, 2809 (1996).
8. J. Terasaki, H. Flocard, P.-H. Heenen, and P. Bonche, Nucl. Phys. **A621**, 706 (1997).
9. N. Tajima, XVII RCNP International Symposium on *Innovative Computational Methods in Nuclear Many-Body Problems*, eds. H. Horiuchi *et al.* (World Scientific, Singapore, 1998) p. 343.
10. M.V. Stoitsov, P. Ring, D. Vretenar, and G.A. Lalazissis, Phys. Rev. **C58**, 2086 (1998).
11. M.V. Stoitsov, W. Nazarewicz, and S. Pittel, Phys. Rev. **C58**, 2092 (1998).
12. I.Zh. Petkov and M.V. Stoitsov, Compt. Rend. Bulg. Acad. Sci. **34**, 1651 (1981); Theor. Math. Phys. **55**, 584 (1983); Sov. J. Nucl. Phys. **37**, 692 (1983).
13. M.V. Stoitsov and I.Zh. Petkov, Ann. Phys. (NY) **184**, 121 (1988).
14. I.Zh. Petkov and M.V. Stoitsov, *Nuclear Density Functional Theory*, Oxford Studies in Physics, (Clarendon Press, Oxford, 1991).
15. J. Dobaczewski and J. Dudek, Comp. Phys. Commun. **102**, 166 (1997); **102**, 183 (1997).
16. D. Vautherin, Phys. Rev. **C7**, 296 (1973).
17. M. Abramowitz and I.A. Stegun, *Handbook of Mathematical Functions* (Dover, New York, 1970).
18. Y.K. Gambhir, P. Ring, and A. Thimet, Ann. Phys. (N.Y.) **198**, 132 (1990).

19. Y.M. Engel, D.M. Brink, K. Goeke, S.J. Krieger, and D. Vautherin, Nucl. Phys. **A249**, 215 (1975).
20. P. Ring and P. Schuck, *The Nuclear Many-Body Problem* (Springer-Verlag, Berlin, 1980).
21. V.E. Oberacker and A.S. Umar, Proc. Int. Symp. on *Perspectives in Nuclear Physics*, (World Scientific, Singapore, 1999), Report nucl-th/9905010.
22. J. Terasaki, P.-H. Heenen, H. Flocard, and P. Bonche, Nucl. Phys. **A600**, 371 (1996).
23. D. Gogny, Nucl. Phys. **A237**, 399 (1975).
24. M. Girod and B. Grammaticos, Phys. Rev. **C27**, 2317 (1983).
25. J.L. Egido, H.-J. Mang, and P. Ring, Nucl. Phys. **A334**, 1 (1980).
26. J.L. Egido, J. Lessing, V. Martin, and L.M. Robledo, Nucl. Phys. **A594**, 70 (1995).
27. A.V. Afanasjev, J. König, and P. Ring, Nucl. Phys. **A608**, 107 (1996).
28. P. Ring, Prog. Part. Nucl. Phys. **37**, 193 (1996).
29. E. Chabanat, P. Bonche, P. Haensel, J. Meyer, and F. Schaeffer, Nucl. Phys. **A635**, 231 (1998).
30. X. Li and P.-H. Heenen, Phys. Rev. **C54**, 1617 (1996).
31. S. Ćwiok, J. Dobaczewski, P.-H. Heenen, P. Magierski, and W. Nazarewicz, Nucl. Phys. **A611**, 211 (1996).
32. K. Rutz, M. Bender, T. Buervenich, T. Schilling, P.-G. Reinhard, J. A. Maruhn, and W. Greiner, Phys. Rev. **C56**, 238 (1997).
33. P.H. Heenen, J. Dobaczewski, W. Nazarewicz, P. Bonche, and T.L. Khoo, Phys. Rev. **C57**, 1719 (1998).
34. F. Naulin, J.Y. Zhang, H. Flocard, D. Vautherin, P.H. Heenen, and P. Bonche, Phys. Lett. **429B**, 15 (1998).
35. W. Satuła, J. Dobaczewski, and W. Nazarewicz, Phys. Rev. Lett. **81**, 3599 (1998).
36. T. Bürvenich, K. Rutz, M. Bender, P.-G. Reinhard, J.A. Maruhn, and W. Greiner, Eur. Phys. J. **A3**, 139 (1998).
37. D. Rudolph, C. Baktash, J. Dobaczewski, W. Nazarewicz, W. Satuła, M.J. Brinkman, M. Devlin, H.-Q. Jin, D.R. LaFosse, L.L. Riedinger, D.G. Sarantites, and C.-H. Yu, Phys. Rev. Lett. **80**, 3018 (1998).
38. S. Bouneau, F. Azaiez, J. Duprat, I. Deloncle, M.G. Porquet, A. Astier, M. Bergstrom, C. Bourgeois, L. Ducroux, B.J.P. Gall, M. Kaci, Y. Le Coz, M. Meyer, E.S. Paul, N. Redon, M.A. Riley, H. Sergolle, J.F. Sharpey-Schafer, J. Timar, A.N. Wilson, and R. Wyss, Eur. Phys. Jour. **A2**, 245 (1998).

39. S. Bouneau, F. Azaiez, J. Duprat, I. Deloncle, M.G. Porquet, A. Astier, M. Bergstrom, C. Bourgeois, L. Ducroux, B.J.P. Gall, M. Kaci, Y. Le Coz, M. Meyer, E.S. Paul, N. Redon, M.A. Riley, H. Sergolle, J.F. Sharpey-Schafer, J. Timar, A.N. Wilson, R. Wyss, and P.-H. Heenen, Phys. Rev. **C58**, 3260 (1998).
40. D. Rudolph, C. Baktash, M.J. Brinkman, E. Caurier, D.J. Dean, M. Devlin, J. Dobaczewski, P.-H. Heenen, H.-Q. Jin, D.R. LaFosse, W. Nazarewicz, F. Nowacki, A. Poves, L.L. Riedinger, D.G. Sarantites, W. Satuła, and C.-H. Yu, Phys. Rev. Lett. **82**, 3763 (1999).
41. C. Rigollet, P. Bonche, F. Flocard, and P.-H. Heenen, Phys. Rev. **C59**, 3120 (1999).
42. K. Bennaceur, J. Dobaczewski, and M. Płoszajczak, Phys. Rev. **C60**, 034308 (1999).
43. J. Dobaczewski, W. Nazarewicz, and T.R. Werner, Physica Scripta **T56**, 15 (1995).
44. O. Tarasov, R. Allatt, J.C. Angelique, R. Anne, C. Borcea, Z. Dlouhy, C. Donzaud, S. Grevy, D. Guillemaud-Mueller, M. Lewitowicz, S. Lukyanov, A.C. Mueller, F. Nowacki, Yu. Oganessian, N.A. Orr, A.N. Ostrowski, R.D. Page, Yu. Penionzhkevich, F. Pougheon, A. Reed, M.G. Saint-Laurent, W. Schwab, E. Sokol, O. Sorlin, W. Trinder, and J.S. Winfield, Phys. Lett. **409B**, 64 (1997).
45. H. Sakurai, S.M. Lukyanov, M. Notani, N. Aoi, D. Beaumel, N. Fukuda, M. Hirai, E. Ideguchi, N. Imai, M. Ishihara, H. Iwasaki, T. Kubo, K. Kusaka, H. Kumagai, T. Nakamura, H. Ogawa, Yu.E. Penionzhkevich, T. Teranishi, Y.X. Watanabe, K. Yoneda, and A. Yoshida, Phys. Lett. **448B**, 180 (1999).
46. A.T. Kruppa, P.-H. Heenen, H. Flocard, and R.J. Liotta, Phys. Rev. Lett. **79**, 2217 (1997).
47. E. Caurier, F. Nowacki, A. Poves, and J. Retamosa, Phys. Rev. **C58**, 2033 (1998).
48. T. Motobayashi, Y. Ikeda, Y. Ando, K. Ieki, M. Inoue, N. Iwasa, T. Kikuchi, M. Kurokawa, S. Moriya, S. Ogawa, H. Murakami, S. Shimoura, Y. Yanagisawa, T. Nakamura, Y. Watanabe, M. Ishihara, T. Teranishi, H. Okuno, and R.F. Casten, Phys. Lett. B **346**, 9 (1995).
49. D. Habs, O. Kester, G. Bollen, L. Liljeby, K.G. Rensfelt, D. Schwalm, R. von Hahn, G. Walter, P. Van Duppen, and the REX-ISOLDE Collaboration, Nucl. Phys. **A616**, 29c (1997).
50. F. Azaiez *et al.*, Int. Conf. on Nuclear Structure, Gatlinburg 1998; to be published.

51. P.-G. Reinhard, D.J. Dean, W. Nazarewicz, J. Dobaczewski, J.A. Maruhn, and M.R. Strayer, Phys. Rev. **C60**, 014316 (1999).
52. F. Sarazin, H. Savajols, W. Mittig, P. Roussel-Chomaz, G. Auger, D. Baiborodin, A.V. Belozyorov, C. Borcea, Z. Dlouhy, A. Gillibert, A.S. Lalleman, M. Lewitowicz, S.M. Lukyanov, F. Nowacki, F. de Oliveira, N. Orr, Y.E. Penionzhkevich, Z. Ren, D. Ridikas, H. Sakuraï, O. Tarasov, and A. de Vismes, Proc. of the XXXVII Int. Winter Meeting on Nuclear Physics, ed. I. Iori (*Università degli Studi di Milano*, Milano, 1999) Suppl. No. 114.
53. I. Tanihata, J. Phys. G **24**, 1311 (1998).
54. P. Möller, J.R. Nix, W.D. Myers, and W.J. Swiatecki, Atom. Data and Nucl. Data Tables **59**, 185 (1995).
55. Z. Ren, Z.Y. Zhu, Y.H. Cai, and G. Xu, Phys. Lett. B **380**, 241 (1996).
56. G.A. Lalazissis, A.R. Farhan, and M.M. Sharma, Nucl. Phys. **A628**, 221 (1998).
57. G.A. Lalazissis, D. Vretenar, P. Ring, M. Stoitsov, and L.M. Robledo, Phys. Rev. **C60**, 014310 (1999).
58. J. Dobaczewski, Acta Phys. Pol. **B30**, 1647 (1999).
59. S. Mizutori, G. Lalazissis, J. Dobaczewski, W. Nazarewicz, and P.-G. Reinhard, to be published.
60. G. Audi and A.H. Wapstra, Nucl. Phys. **A565**, 1 (1993); Nucl. Phys. **A565**, 66 (1993).
61. K. Yoneda, N. Aoi, H. Iwasaki, H. Sakurai, H. Ogawa, T. Nakamura, W.-D. Schmidt-Ott, M. Schaefer, M. Notani, N. Fukuda, E. Ideguchi, T. Kishida, S.S. Yamamoto, and M. Ishihara, J. Phys. **G24**, 1395 (19 .

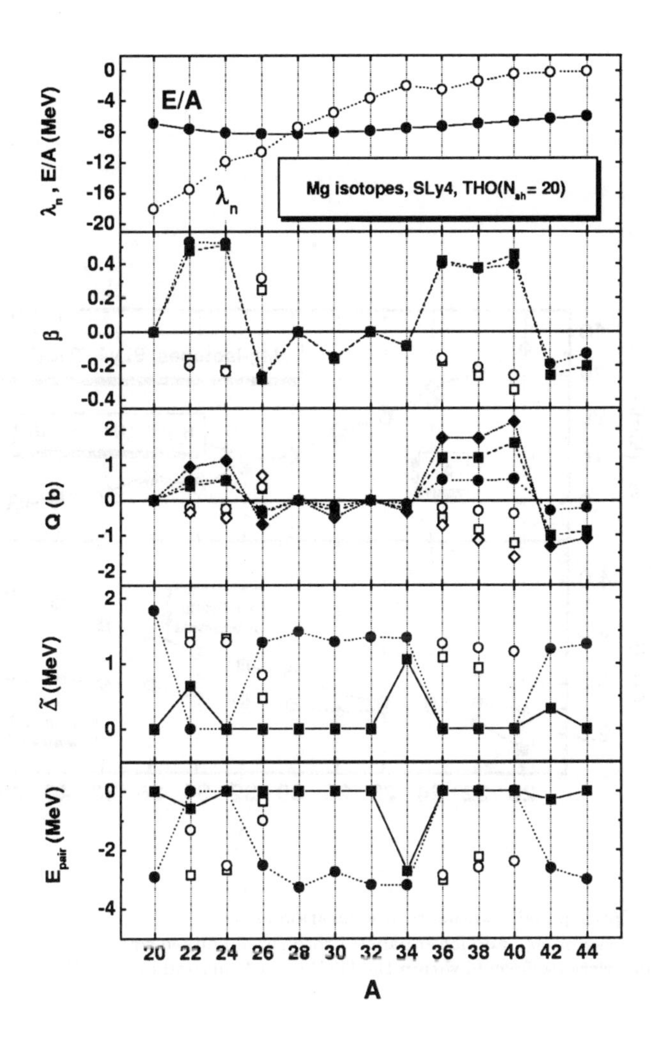

Figure 1. Neutron Fermi energies λ_n, energies per particle E/A, deformations β, quadrupole moments Q, pairing gaps $\widetilde{\Delta}$, and pairing energies E_{pair} calculated for the Mg isotopes within the HFB+SLy4 method in the THO basis (N_{sh}=20), as functions of the mass number A. Apart from the upper panel, circles, squares, and diamonds pertain to proton, neutron, and total results, respectively. Closed symbols connected with lines denote values for the absolute minima in the deformation-energy curve (axial shapes are assumed), while open symbols pertain to secondary minima.

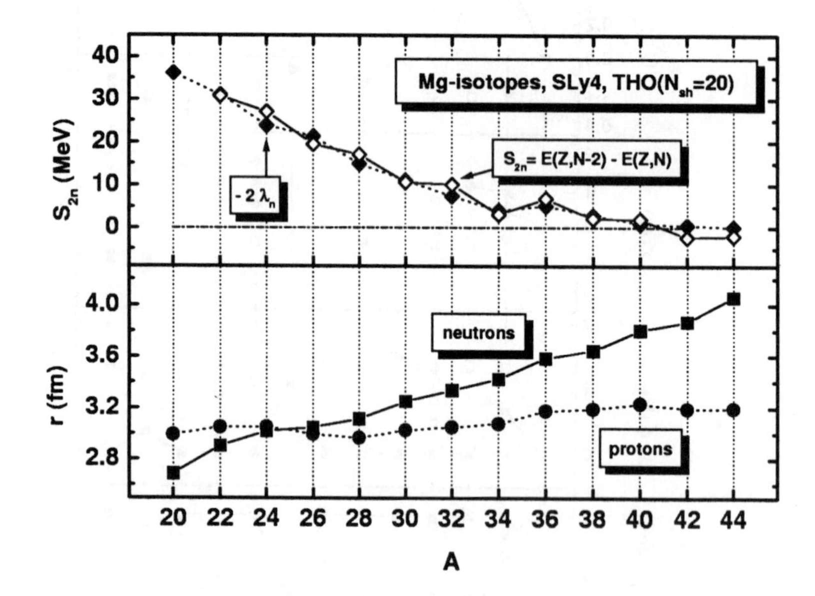

Figure 2. Upper panel: two-neutron separation energies S_{2n} (open symbols) compared to $-2\lambda_n$ (closed symbols), and lower panel: proton and neutron rms radii. Calculations for the Mg isotopes were performed within the HFB+SLy4 method in the THO basis for N_{sh}=20.

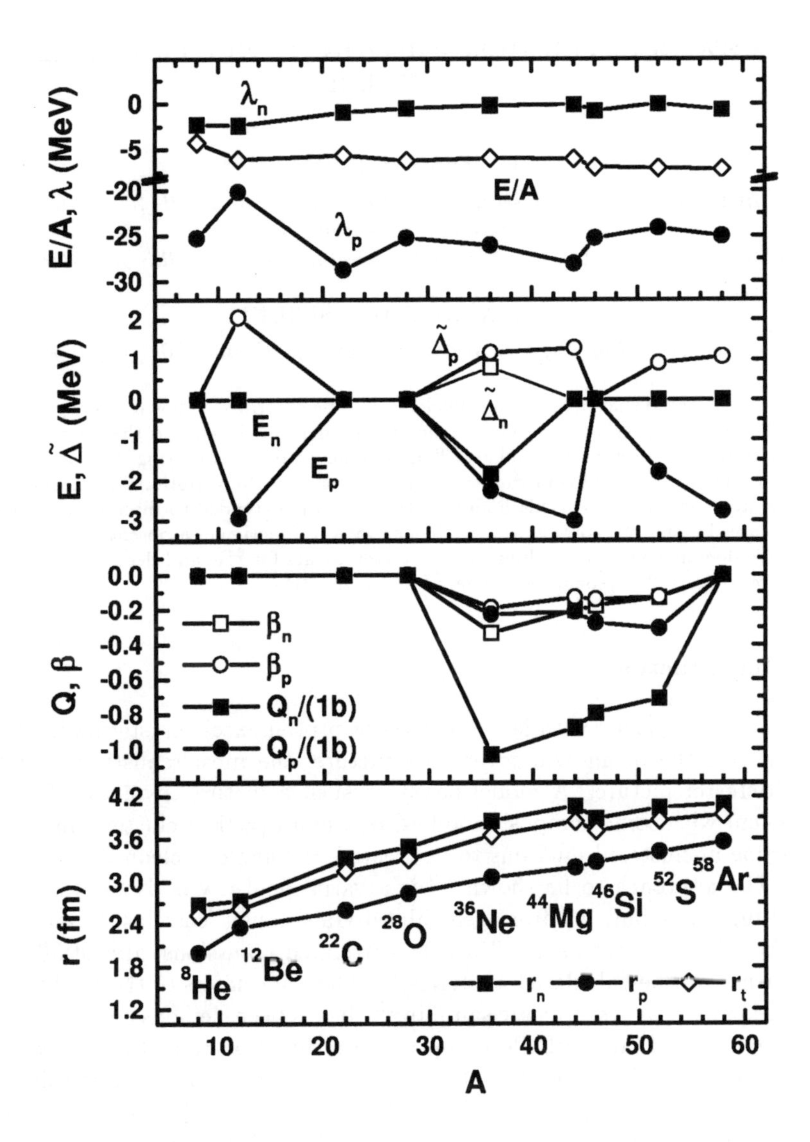

Figure 3. Neutron Fermi energies λ_n, energies per particle E/A, pairing gaps $\widetilde{\Delta}$, pairing energies E, deformations β, quadrupole moments Q, and rms radii r calculated for the neutron-drip-line nuclei (indicated in the lower panel) within the HFB+SLy4 method in the THO basis, as functions of the mass number A. Circles, squares, and diamonds pertain to proton, neutron, and total results, respectively.

FEW-CLUSTER DESCRIPTION OF NEUTRON-HALO NUCLEI

R. G. LOVAS[1,2], K. VARGA[1,3]

[1] *Institute of Nuclear Research, Debrecen, P. O. Box 51, H–4001, Hungary*

[2] *E-mail: rgl@atomki.hu*

[3] *Physics Division, Argonne National Laboratory, Argonne, Illinois 60439, USA*

K. ARAI AND Y. SUZUKI

Department of Physics, Niigata University, Niigata 950-2181, Japan

It is proposed that neutron-halo nuclei should be described as few-cluster systems. The scheme presented is microscopic at the nucleonic level, and the intercluster motion is treated variationalally. This is achieved by constructing the trial function from correlated Gaussian terms and performing the variational procedure in a stochastic manner. This bound-state method is extended to discrete unbound states by analytic continuation of the energy as a function of a potential strength. We demonstrate the goodness of the cluster anzatz for ^{6}He and describe the controversial $1/2^+$ resonances of ^{5}He, ^{5}Li, ^{9}Be and ^{9}B.

1 Introduction

Since in light nuclei with large neutron or proton excess clustering may dominate over the mean-field-governed structure, the most realistic picture is a multicluster picture. A model based on such a picture involves a few-body problem with some composite bodies, α, t and h(=^{3}He) clusters, in addition to some unclustered nucleons to be treated as "single-nucleon clusters".

The approach to be shown [1,2] is a variational few-body approach, with trial functions built up from generalized Gaussians ("correlated Gaussians") of the relative coordinates. With the correlated Gaussians, almost all matrix elements can be calculated analytically. The parameters of the trial function are generated by a stochastic sampling [3]. It has been demonstrated by numerical examples that, by building up the trial function term by term with the stochastic selection procedure, the solution converges to the exact solution of the relative-motion problem. Through the example of ^{6}He it will be shown that this is a viable approach to the description of light exotic nuclei [4].

Neutron-halo nuclei usually have just one bound state, and if one removes a neutron from a two-neutron halo, the other neutron becomes usually unbound. Thus to make the description reasonably complete, one has to cope with unbound states as well. We show how this may be done through examples for ^{5}He, ^{5}Li, ^{9}Be and ^{9}B [5].

2 Theory

The approach to be presented is based on a variational method with a trial function built up from generalized Gaussians called correlated Gaussians. A trial function has a suitable functional form if a wave function of any conceivable shape can be approximated with relatively few parameters. A radial wave function $F_l(r)$ is most conveniently expressed as $F_l(r) \approx \sum_k c_k g_{a_k}(r)$, where $g_a(r)$ is a function containing a continuous parameter a. A nodeless harmonic-oscillator (h.o.) function $g_a(r) = r^l \exp(-\frac{1}{2}ar^2)$ is qualified for this role because any wave function with angular momentum lm can be approximated, to any desired accuracy, by a linear combination of such functions: $F_{lm}(\mathbf{r}) \approx \sum_k c_k e^{-\frac{1}{2}a_k r^2} \mathcal{Y}_{lm}(\mathbf{r})$, with $\mathcal{Y}_{lm}(\mathbf{r}) = r^l Y_{lm}(\hat{\mathbf{r}})$. For an A-particle problem, one should introduce a set of intrinsic Jacobi coordinates $\mathbf{x} \equiv \{\mathbf{x}_1, \ldots, \mathbf{x}_{A-1}\}$. The expansion may then take the form

$$F_{LM_L}(\mathbf{x}) \approx \sum_{\{\{k\}\{l\}\}} c_{\{k\}\{l\}} \exp\left(-\tfrac{1}{2}\sum_{i=1}^{A-1} a_{\{k\}i}\mathbf{x}_i^2\right) \theta_{\{l\}LM_L}(\mathbf{x}), \qquad (1)$$

where $\{k\} \equiv \{k_1, \ldots, k_{A-1}\}$ label the different sets of product Gaussians, and $\theta_{\{l\}LM_L}(\mathbf{x})$ may be imagined as a vector-coupled product of the solid spherical harmonics for the individual Jacobi vectors; then the label $\{l\}$ may denote a set of orbital momentum quantum numbers:

$$\theta_{\{l\}LM_L}(\mathbf{x}) = \left[\mathcal{Y}_{l_1}(\mathbf{x}_1)\ldots\mathcal{Y}_{l_{A-1}}(\mathbf{x}_{A-1})\right]_{LM_L}.$$

The sum in eq. (1) may contain far fewer terms than if there were $A-1$ uncorrelated particles, owing just to the correlations that blur the mean-field picture.

To conform to some asymptotic correlations, one may choose other sets of relative coordinates. E.g., for four identical particles there are two topologically different sets of independent relative vectors: those which form a K pattern and those which form an H, with one particle sitting at the end of each arm. E.g., an asymptotic partitioning into two pairs of particles is described most economically with the H-shaped set of vectors. To have good asymptotics everywhere, it is economical to include terms with all possible relative vectors. Such terms will overlap with each other substantially, but, since even the terms with the same relative vectors are non-orthogonal, this does not pose extra difficulties.

This scheme becomes simpler by two straightforward generalizations. First, all possible relative coordinate systems, and even their mixtures, may be contained in a single sum as in eq. (1) if we replace the diagonal quadratic form $\sum_{i=1}^{A-1} a_{\{k\}i}\mathbf{x}_i^2$ in the exponent by a general positive-definite symmetric quadratic form, $\sum_i \mathbf{x}_i \cdot \left(\sum_j A_{\{k\}ij}\mathbf{x}_j\right)$. That is so because the relative coordinates transform into each other, and, *a fortiori*, into the Jacobi set $\mathbf{x}$ by orthogonal transformations. The function $\exp\left(-\frac{1}{2}\sum_{ij}\mathbf{x}_i \cdot A_{\{k\}ij}\mathbf{x}_j\right)\theta_{\{l\}LM_L}(\mathbf{x})$ is called a correlated Gaussian. The second generalization is that we allow $\theta_{\{l\}LM_L}(\mathbf{x})$ not to carry definite values of the intermediate angular momenta. Then, for states of parity $(-1)^L$, the function $\theta_{\{l\}LM_L}(\mathbf{x})$ may be chosen as

$$\theta_{\{l\}LM_L}(\mathbf{x}) = v_{\{l\}}^{2K}\mathcal{Y}_{LM_L}(\mathbf{v}_{\{l\}}), \qquad \text{with } \mathbf{v}_{\{l\}} = \sum_{i=1}^{A-1} u_{\{l\}i}\mathbf{x}_i.$$

The case of $(-1)^{L+1}$ is also simple. With the eigenfunctions of the intrinsic degrees of freedom (χ for the spin and $\mathcal{X}$ isospin) and the antisymmetrizer $\mathcal{A}$ included, the trial function will become $\Psi = \sum_{i=1}^{\mathcal{K}} c_i\psi_i$, where

$$\begin{aligned}\psi_i &\equiv \psi_{\{k_i\}\{l_i\}(L_iS_i)JMTM_T}(\mathbf{x}) \\ &= \mathcal{A}\left\{e^{-\frac{1}{2}\sum_{ij}\mathbf{x}_i\cdot A_{\{k\}ij}\mathbf{x}_j}\left[\theta_{\{l_i\}L}(\mathbf{x})\chi_{S_i}\right]_{JM}\mathcal{X}_{TM_T}\right\}.\end{aligned}$$

The matrix elements are calculated analytically by a generating function technique. The correlated Gaussians are generated by an integral transform of Slater determinants of shifted Gaussian single-particle (s.p.) states. The shifted Gaussians generate a wave function of A s.p. 'clusters' if all Gaussian centres are kept different, but they generate states with composite clusters whose internal states are 0s h.o. shell-model states, if some centres are constrained to coincide. Hence are they used in cluster models.

The trial function must have many terms, and each term contains a number of parameters: the coefficient c_i, the independent elements of $A_{\{k\}ij}$ and the numbers $u_{\{l_i\}j}$. Since, however, the basis elements are singled out from a continuous set of largely overlapping functions, there are infinitely many almost equally good parameter sets, and it would be useless to find the very best among them even if it were feasible. Good parameters can be obtained, however, by properly varying the linear parameters only, while the non-linear parameters are sampled stochastically term by term [6] and a new term is admitted only if it passes a test. The admittance criterion may then be loosened gradually, whereby rapid convergence to the same energy value can be achieved. This trial and error procedure can be regarded as an approximate optimization, and hence the term 'stochastic variation'. The approach can be best tested for structureless particles. The numerical values obtained agree with the exact model values whenever those are known. With some effort, an accuracy that satisfies the standard of atomic physics can be attained.

In particular, the tail of the wave function can be reproduced so well that Efimov states [2] can be produced in three-body systems. With present-day supercomputers, the approach is feasible up to seven-body systems.

3 Three-body models for neutron-halo nuclei

The two-neutron-halo nuclei, like ^{6}He, ^{11}Li and ^{14}Be, are mostly viewed as core+n+n systems. Doubt is cast on the validity of this model by the fact that the forces that describe the two-cluster subsystems correctly tend to underbind the three-cluster system. It gives an even stronger warning that, for even the simplest system, ^{6}He, an explicit inclusion of the only alternative clusterization (t+t) on top of core+n+n was found to increase the binding substantially [7]. To understand such "core distortion" effects, extensive calculations have been performed for ^{6}He.

In Table 1 the results of four models are summarized. The first is a pure α+n+n model, in which the the α-particle is described as a combination of 0s shell-model configurations of different extensions, such that the distorting effect of the two neutrons on α is assumed to be simple breathing excitations. In the second model the t+t clusterization is also explicitly included. In the third model, which is called the extended three-cluster model, the core is composed of a three-nucleon cluster, which is allowed to breathe, and a fourth nucleon. In the fourth model both extra effects are taken into account. The same effective interaction, the central plus spin-orbit type Minnesota force [8], is used. The force has a free parameter (a mixing parameter) with which the binding energy ε with respect to the α+n+n threshold can be set to its exact

Table 1. ^{6}He binding energies E, α+n+n separation energies ε (in MeV) and point nucleon (N), proton (p) and neutron (n) rms radii (in fm). The symbol $*$ signifies that the marked cluster is allowed to undergo breathing distortion.

Model	E	ε	Radii		
			N	p	n
α^*+n+n	−25.91	−0.32			
$\{\alpha^*$+n+n;t+t$\}$	−26.56	−0.96	2.42	1.81	2.68
$\left\{\binom{t^*+p}{h^*+n}+n+n\right\}$	−27.75	−1.20	2.34	1.75	2.59
$\left\{\binom{t^*+p}{h^*+n}+n+n;t+t\right\}$	−27.84	−1.29	2.33	1.76	2.57
Experiment	−29.271	−0.975	2.33	1.72	2.59
			2.48	2.21	2.61

value, thus the ε values are only informative relative to each other.

It is found that both types of breakup of the α-cluster deepen the binding energy much more than a breathing excitation, but the effect is less dramatic in the α+n+n separation energy. However, the effect of t+t on top of the extended model is minute. This shows that the inclusion of a t+t clusterization opens up the same segment of the state space as the allowance for the cluster breakups α=t+p and α=h+n. This implies that the large contribution of the t+t component does not invalidate the core+n+n model. It just shows that the core, both its density and and distortability, should be described more realistically. Since the α-particle is the most rigid core conceivable, one can expect larger core effects for other neutron halo nuclei. Thus, these results emphasize the importance of treating the core in a realistic microscopic model without invalidating the core+n+n picture of halo nuclei.

4 Description of unbound states

Since most neutron-halo nuclei have just one bound state or none, one has to find a way of generalizing the approach to unbound states. The treatment of various discrete states is unified by recognizing that all belong to poles of the S-matrix S considered to be a function of the complex energy or, rather, of the corresponding complex wave number k. The technique that has proved most powerful is based on the analytic continuation of the pole wave number $k(u)$ as a function of a strength parameter of the interaction, u. The bound-state poles lie on the positive imaginary axis, the virtual state poles are situated on the negative imaginary axis, and the resonance poles occur pairwise on the two lower quadrants of the complex k-plane, symmetrically with respect to the imaginary axis.

According to the analytic continuation technique [9], one changes the strength parameter u such that the unbound state to be described becomes bound. One performs a set of consistent bound-state calculations by varying u in the bound-state region, and extrapolates the results to the physical value of the strength parameter. For the extrapolation, it is useful to apply the Padé approximation technique. The only extra input that one has to add is the behaviour of the pole trajectory around the bifurcation point, $-i\bar{\kappa}$, where the pole departs from the imaginary axis:

$$k(u) \sim -i\bar{\kappa} + i\sqrt{u-\bar{u}} \qquad \text{for} \quad u \approx \bar{u},$$

where $\bar{u}$ is defined by $k(\bar{u}) = -i\bar{\kappa}$. The Padé approximant can absorb this bahaviour by choosing the Padé variable x as $x = \sqrt{u-\bar{u}}$. For two-body problems with non-zero angular momenta or with a Coulomb barrier, it is *a*

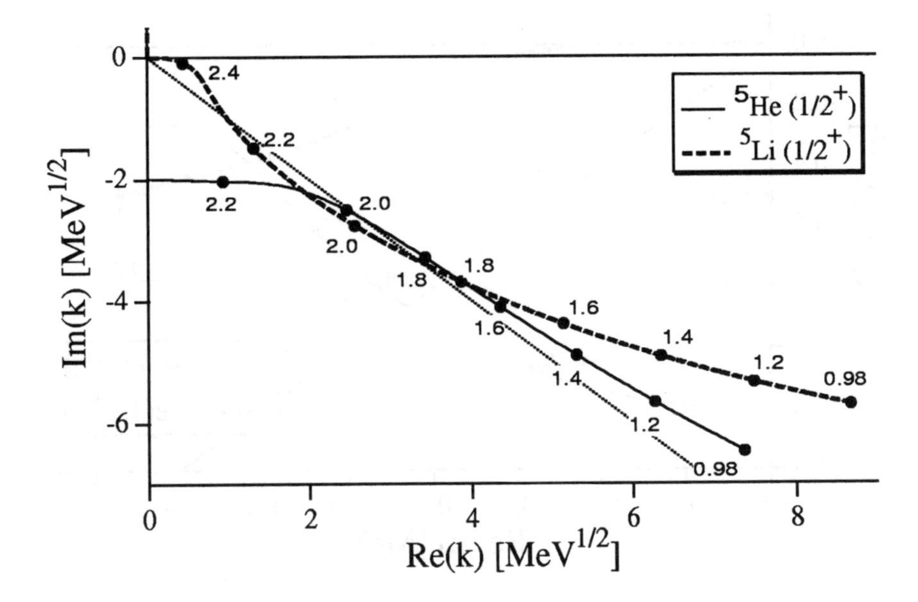

Figure 1. The trajectories of the $1/2^+$ states in ^{5}He and ^{5}Li as functions of the u parameter in the fourth quadrant of the complex k-plane. Here k is identified with $E^{1/2}$. The dotted line marks Im k=−Re k.

priori known that $\bar{\kappa} = 0$. This means that there is no virtual-state section of the pole trajectory. A finite barrier makes the s-wave case behave very similarly, and if a three- or more-body channel opens up, the pole bahaves as in a two-body problem with a barrier. For s-wave problems without barriers (or with small barriers), $\bar{\kappa} > 0$, and its value can be determined by trial and error. What should be used to determine $\bar{\kappa}$ is that the Padé approximant is most stable just at the correct $\bar{\kappa}$ value.

Two pairs of controversial unbound states are shown for illustration: the $1/2^+$ poles of ^{5}He as well as of ^{5}Li and the $1/2^+$ poles of ^{9}Be as well as of ^{9}B.

The shell-model calculations insistently produce a low-lying $1/2^+$ state for ^{5}He, which is interpreted as a resonance, and R-matrix analyses seem to corroborate this. But the shell model ignores the unbound nature of the state, and the empirical evidence is not strong enough. Fig. 1 shows a section of the pole trajectory obtained by analytic continuation. The physical value of u is 0.98, which belongs to energy E=12 MeV and width Γ=190 MeV for ^{5}He and

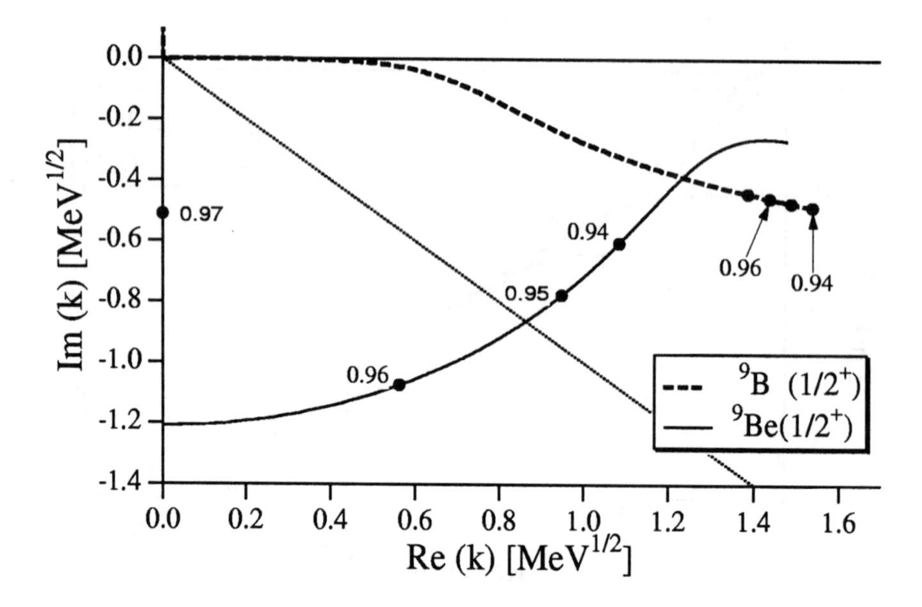

Figure 2. The trajectories of the $1/2^+$ states in ^{9}Be and ^{9}B as a function of the u parameter in the fourth quadrant of the complex k-plane. Here k is identified with $E^{1/2}$. The dotted line marks Im k=−Re k.

E=42 MeV and width Γ=197 MeV for ^{5}Li. It is obvious that these objects cannot be considered resonances.

The pole trajectory for ^{9}Be in Fig 2 is like that of a two-body s-wave case with no Coulomb barrier. The ^{9}Be resonance thus behaves like a ^{8}Be+n system. The pole trajectory of ^{9}B shows a confinement effect as it should, and is compatible with either a two-body resonance inside a (Coulomb) barrier or with a three-body resonance. Thus both a ^{8}Be+p and an $\alpha+\alpha+$p structure appear to be tenable. The physical value of u is 0.94. Since, however, in the case of ^{9}Be the pole moves along the trajectory at a pretty high rate, a slightly stronger force could make the state virtual.

Thus not only can we localize the unbound states by means of analytic continuation of bound-state calculations, but we can also say something about their nature within the uncertainties of the model.

5 Conclusion

The correlated-Gaussian approach is accurate for few-cluster systems. Our calculations give insight into the binding mechanism of exotic nuclei and reveal the nature of unbound states. As an example for these applications, we corroborated the core+n+n picture for the case of ^{6}He, but warned against oversimplified treatments of the core. With another example, we have shown that no observable $1/2^+$ resonance can exist for the five-nucleon systems, and that the $1/2^+$ resonance of ^{9}Be is a ^{8}Be+N-type structure. We also pointed out that the nature of the $1/2^+$ resonance of ^{9}B hangs in the balance owing to an anomalous sensitivity of the result to the actual strength of the force acting between the subsystems.

Acknowledgments

With this contribution the speaker (RGL) wished to express his appreciation to D. S. Delion, L. Gr. Ixaru, D. N. Poenaru, M. Rizea, N. Săndulescu and Gh. Stratan, members of the Horia Holubei National Institute, Bucharest, with whom he has been in friendly contact for many years and with some of whom he and the Debrecen nuclear theory group have had extremely fruitful cooperation.

This work was supported by the OTKA Grant No. T029003 (Hungary) and by a Grant-in-Aid for Scientific Research (No. 10640255) of the Ministry of Education, Science, Sport and Culture (Japan).

References

1. K. Varga and Y. Suzuki, *Phys. Rev.* C **52**, 2885 (1995).
2. Y. Suzuki and K. Varga, *Stochastic Variational Approach to Quantum Mechanical Few-Body Problems* (Springer-Verlag, Berlin, 1998).
3. K. Varga, Y. Suzuki and R.G. Lovas, *Nucl. Phys.* A **571**, 447 (1994).
4. K. Arai, Y. Suzuki and R.G. Lovas, *Phys. Rev.* C **59**, 1432 (1999).
5. N. Tanaka, Y. Suzuki, K. Varga and R.G. Lovas, *Phys. Rev.* C **59**, 1391 (1999).
6. V. I. Kukulin and V. M. Krasnopol'sky, *J. Phys.* G **3**, 795 (1977).
7. A. Csótó, *Phys. Rev.* C **48**, 165 (1993).
8. D. R. Thompson, M. LeMere and Y. C. Tang, *Nucl. Phys.* A **286**, 53 (1977);
   I. Reichstein and Y. C. Tang, *Nucl. Phys.* A **158**, 529 (1970).
9. V. I. Kukulin and V. M. Krasnopol'sky, *J. Phys.* A **10**, 33 (1977).

EVOLUTION OF NUCLEAR COLLECTIVITY: EMPIRICAL PHENOMENOLOGY AND MODEL INTERPRETATION

N.V.ZAMFIR

WNSL, Yale University, New Haven, Connecticut, USA
Clark University, Worchester, Massachusetts, USA
IFIN-HH, Bucharest, Romania

The energies of low-spin yrast states, $E(2_1^+)$ and $E(4_1^+)$, and the transition probabilities, $B(E2;2_1^+ \rightarrow 0_1^+)$, that are among the most revealing and the easiest to measure observables of collectivity, show global correlations that help our understanding of nuclear structure and nuclear phase transitions and provide new signatures to identify particular structures. These correlations combined with the behavior of two neutron separation energies show that the nuclear system presents a shape transition in its evolution from spherical vibrator to well deformed rotor. This behavior is supported by model calculations, as is the concept of phase/shape coexistence in the $N \sim 90$ region.

1 Introduction

The accumulation of vast quantities of data on nuclear observables offers the the opportunity to investigate the evolution of structure in an "horizontal" perspective over large mass ranges and to correlate the behavior of simple but critical observables [1]. It is well - known that the energy of the 2_1^+ state, the energy ratio $R_{4/2} = E(4_1^+)/E(2_1^+)$, and the transition probability $B(E2;2_1^+ \rightarrow 0_1^+)$ are among the most important and the easiest to measure characteristics of collectivity in even-even nuclei. Figure 1 shows typical stages in the evolution of nuclear structure with the addition of valence particles. The energy of the 2_1^+ state descends continuously from values $\sim$ 0.5 MeV near closed shells to $\sim$ 0.1 MeV for mid-shell well-deformed rotational nuclei. The energy ratio $R_{4/2}$ increases from $\sim$ 2.0 for vibrational nuclei to the limiting value of 3.33 in midshell regions corresponding to axially symmetric deformed nuclei. In this path, the $B(E2;2_1^+ \rightarrow 0_1^+)$ value is increasing from $\sim$ 10 W.u. to $\sim$ 200 W.u.. Despite these well-established trends, the detailed evolution of these basic observables over extended sequences of nuclei is not yet phenomenologically established. Moreover, the empirical distinction between the three principal classes of nuclear shapes: Spherical vibrator, γ - soft nuclei, and symmetric rotor generally entails the difficult measurements of extensive energy and transition rate data on higher - lying low-spin states over sequences of nuclei spanning the shape change.

We will discuss simple correlations involving these observables, in order

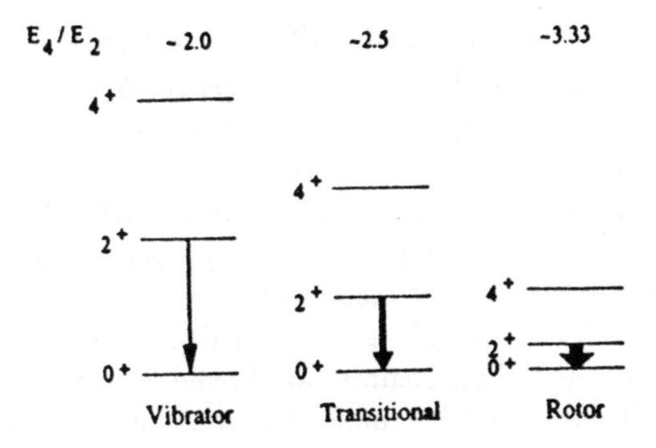

Figure 1. Typical stages in the evolution of nuclear structure from spherical vibrators near closed shell to well deformed midshell rotors

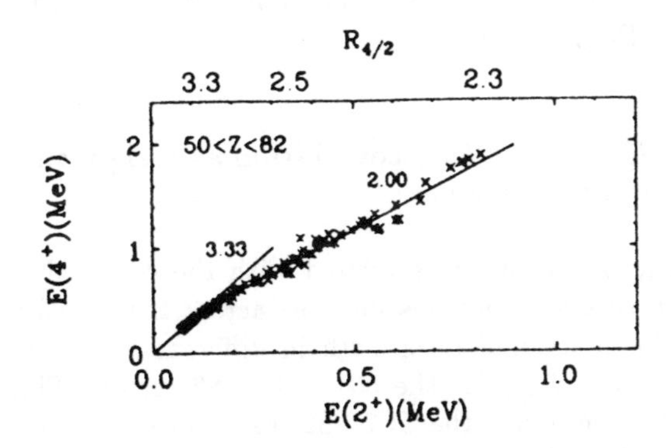

Figure 2. The correlation of $E(4_1^+)$ with $E(2_1^+)$ for all collective nuclei ($R_{4/2}$ >2.05 in the Z=50-82 shell. The rotor (3.33) and anharmonic vibrator (2.00) limits are shown.

to establish their role in classifying nuclear transition regions. The discussion will be related to the data in Z=50-82 even-even nuclei, since in this region a wealthy of data exist and is known to contain a large variety of structures.

2 Correlation of yrast energies

We first investigate the 2_1^+ and 4_1^+ energies. Figure 2 presents their correlation in all collective nuclei ($R_{4/2} \geq 2.05$) in the Z=50-82 shell.

The data are highly correlated and could be separated in two distinct regions, similarly to the classification for a larger range of nuclei [2]. For $E(2_1^+) \gtrsim$ 0.13 MeV the data fall along a compact smooth straight line with a slope of 2.0. This behavior is that of an anharmonic vibrator. With the decrease in $E(2_1^+)$ the ratio $R_{4/2}$ varies from $\sim$ 2.0 to nearly 3.15. With the decrease of $E(2_1^+)$ below $\sim$ 0.13 MeV the slope suddenly changes from the anharmonic value of 2.0 to the rotor value of 3.33. The entire evolution of collective nuclei in the Z=50-82 shell can, therefore, be described in terms of two regimes, the vibrator and rotor, and these regimes are linked by a rapid shape transition.

Although all the collective non-rotational nuclei behave in respect to $E(4_1^+)$ vs. $E(2_1^+)$ in the same way, as anharmonic vibrators, their structure is certainly not constant: vibrator to rotor transitional nuclei and γ-soft nuclei are included. In order to see if other observables related to these low-lying levels reflect the different shapes we will study the correlations between their energies and the B(E2) values.

3 The $B(E2; 2_1^+ \rightarrow 0_1^+)$ - $R_{4/2}$ correlation as a signature of nuclear shape transitions

The second stage of our study is related with the correlation $B(E2; 2_1^+ \rightarrow 0_1^+)$ - $R_{4/2}$. These two quantities increase across the transition region but they sample collectivity and its growth in different ways. Figure 3 shows $B(E2; 2_1^+ \rightarrow 0_1^+)$ vs. $R_{4/2}$ for the $50 < Z < 82$ region. Since there is no unique correlation we divide the Z=50-82 region in three subsets depending on the neutron half-shell:

a) Z=50-66 (proton particles), N=66-82 (neutron holes)
b) Z=50-66 (proton particles), N=82-104 (neutron particles)
c) Z=66-82 (proton holes), N=104-126 (neutron holes)

The correlation presents a clear bifurcation for $R_{4/2} \geq 2.5$ into two well-defined tracks which rejoin only near the rotor limit, $R_{4/2}$ $\sim$3.33. These two paths are empirically occupied by nuclei that undergo distinct types of transitional behavior: Nuclei along the upper track include those in the spherical vibrator $\rightarrow$ symmetric rotor regions (N=66-82 and N=82-104) while the lower track consists of nuclei with N=104-126 that have axial asymmetric shapes. The empirical behavior was shown [3] to be a natural outcome of simple models of nuclear shape transition regions, thereby providing a theoretical basis for the interpretation of this two-track correlation as a simple new signature for distinguishing spherical-vibrator $\rightarrow$ symmetric rotor from axially asymmetric $\rightarrow$ symmetric rotor transition regions.

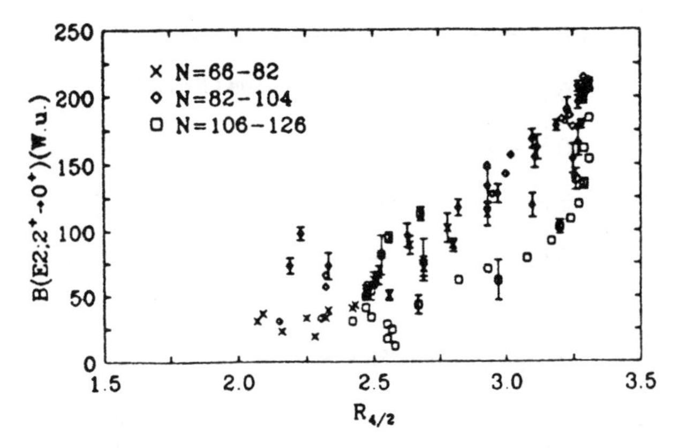

Figure 3. Plot of $B(E2;2_1^+ \rightarrow 0_1^+)$ values against $R_{4/2}$ for all even-even nuclei in the $50 < Z < 82$ region. Error bars are not shown if they are smaller than the symbols

4 Two neutron separation energies as a signature for nuclear shape transition

Other observables, such as two neutron separation energies, S_{2n}, reflect, of course, the structural changes which occur in the evolution of nuclear collectivity. In figure 4 are shown the empirical values of S_{2n} for the all collective ($R_{4/2} > 2.05$) Z=50-82 nuclei. We separate the data in the same three subsets as above according to the character, particles or holes, of the valence nucleons. For N=82-104, the behavior of S_{2n} is compact, with small fluctuations and a sharp break in trajectory in the region where the other observables (e.g., $E(4_1^+)$) show also a sudden shape transition. For the other neutron half-shells (N=66-82 and N=106-126) there are greater fluctuations and a gradual structural change but no evidence for a sharp phase transition.

5 Phase transitions and phase coexistence in phenomenological nuclear models

The two principal phenomenological approaches to collective behavior in nuclei are the Interacting Boson Model (IBA) of Iachello and Arima [5] and the geometric model of Bohr and Mottelson with its practical version the Geometric Collective Model of Gneuss and Greiner [6]. Both models suggest the existence of a phase/shape transition in the evolution of the nuclear structure from spherical vibrator to a deformed rotor.

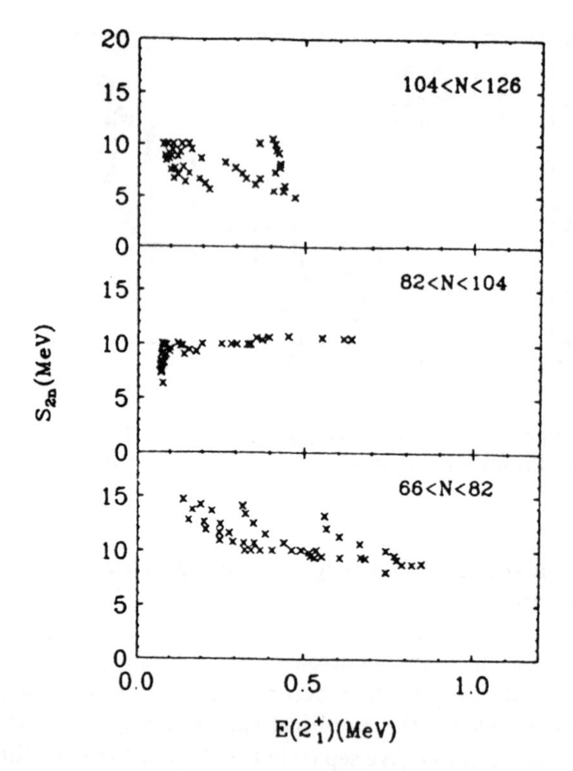

Figure 4. Separation energies S_{2n} (data from ref.[4]) as a function of $E(2_1^+)$. The S_{2n} values have a well known shift in values for each successive Z. To compare values for different elements S_{2n} for each Z is shifted by a constant amount to give equal S_{2n} values at N=76 for N=66-82, at N=88 for N=82-104, and at N=104 for N=106-126.

5.1 Interacting Boson Model

We consider the IBA Hamiltonian:

$$H = (1-\xi)n_d - \xi Q \cdot Q \tag{1}$$

where $Q = (s^\dagger \tilde{d} + d^\dagger s) - \frac{\sqrt{7}}{2} d^\dagger \tilde{d})^{(2)}$, The parameter ξ maps the transition from U(5) (vibrator) (ξ=0) to SU(3) (symmetric rotor) (ξ=1). In Figure 5(left) is shown the IBA energy surface [7] corresponding to the classical limit[5] of eq. (1). With the increasing ξ (evolution from vibrator to rotor), the location of the minimum in energy changes suddenly from β_{min}=0 to a large

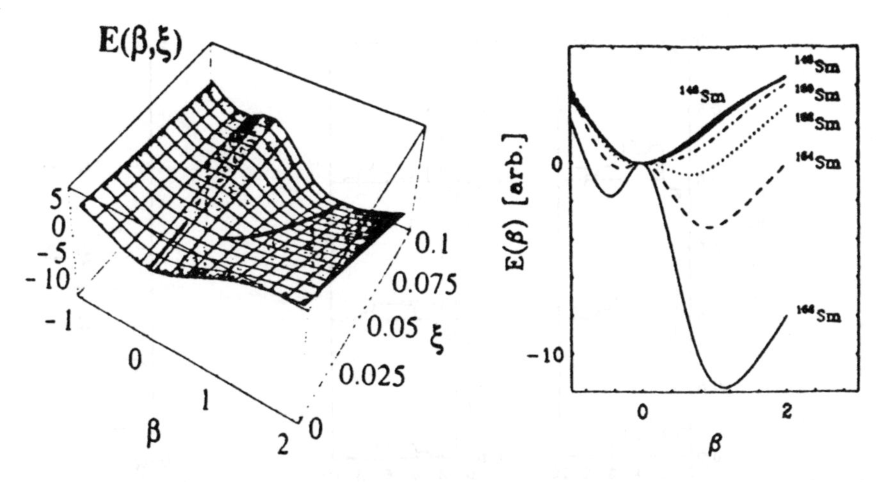

Figure 5. Classical limit analysis of IBA calculations. (left) Energy surface as a function of β (β in IBA is larger than β in the Geometric Model) and ξ for 10 bosons. (right) the IBA energy, $E(\beta)$, for the ξ values and boson numbers corresponding to the Sm isotopes [8]

value. The IBA indicates that nuclei change abruptly from near spherical to deformed at a critical value ξ_{crit}.

Figure 5(right) shows energy surfaces as a function of β for IBA parameters applicable to the Sm isotopes[8]. These surfaces range from near spherical shapes for ^{146,148}Sm to softer in ^{150}Sm, to the coexistence nucleus ^{152}Sm where two shallow minima occur, to the prolate deformed nuclei ^{154,156}Sm. In the phase transition region the single IBA Hamiltonian of eq. (1) generates, as can be seen in Fig. 6 for ^{152}Sm, a coexistence of two phases with some states (yrast) having wave functions appropriate to rotational [SU(3)] phase and others (non-yrast) having wave functions appropriate to spherical vibrator [U(5)] phase [9]. The new experimental data [10] in ^{152}Sm support this picture.

5.2 Geometric Collective Model

The coexistence picture emerges also in the GCM. In Fig. 7 is shown the potential energy surface corresponding to the calculation [11] reproducing the experimental characteristics of ^{152}Sm. The ground state and low energy yrast levels are trapped within the deformed minimum and the higher lying levels have significantly smaller expectation values of β.

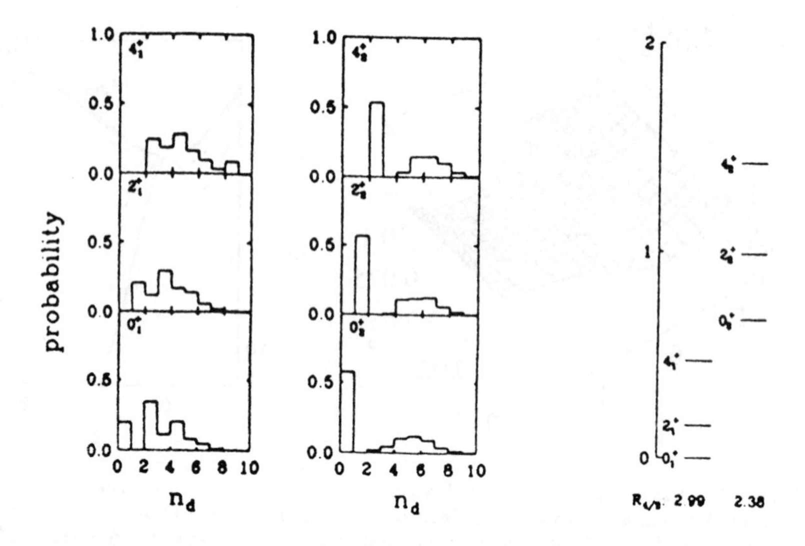

Figure 6. Distribution of IBA squared wave function amplitudes for yrast and yrare states in ^{152}Sm

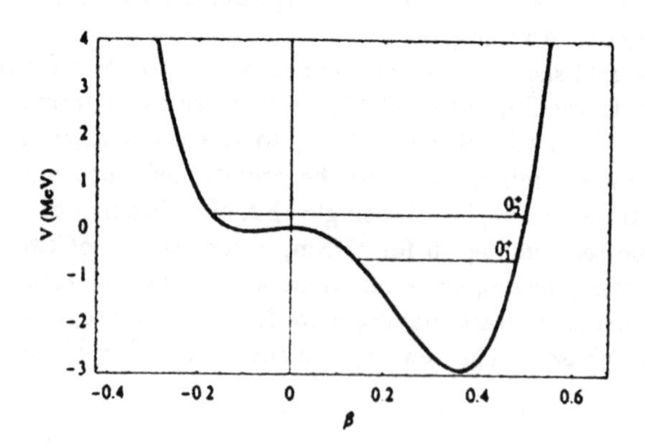

Figure 7. The potential in the GCM calculations.

6 Conclusion

A phenomenological study of the 4_1^+ and 2_1^+ energies, the $B(E2; 2_1^+ \rightarrow 0_1^+)$ and S_{2n} values shows that, these quantities sample the collectivity in different ways. Their correlations reveal a sharp shape transition in the evolution from spherical vibrator to symmetric rotor. The model calculations indicate that this behavior can be described as a phase transition and a phase coexistence at the critical point.

Acknowledgments

I am very grateful to R.F. Casten for collaboration in the research summarized here. I would like to thank F. Iachello and D. Kusnezov for many useful discussions related to phase transition and phase coexistence in nuclear systems. The collaboration with J.Y.Zhang and M. Caprio is gratefully acknowledged. Work has been supported under contracts DE-AC02-76CH00016 and DE-FG02-88ER40417 with the United States Department of Energy.

References

1. R.F. Casten and N.V. Zamfir, J.Phys. G22, 1521 (1996).
2. R.F. Casten, N.V. Zamfir, and D.S. Brenner, Phys. Rev. Lett. 71, 227 (1993)
3. N.V. Zamfir and R.F. Casten, Phys. Lett. B305,317 (1993)
4. G. Audi and A.H. Wapstra, Nucl. Phys. A565,66 (1993)
5. F. Iachello and A. Arima, The Interacting Boson Model (Cambridge University Press, Cambridge, England,1987)
6. G.Gneuss and W. Greiner, Nucl.Phys. A171, 440 (1971)
7. R. F. Casten, D. Kusnezov, and N.V. Zamfir, Phys. Rev. Lett. 82, 5000 (1999)
8. O. Scholten, F. Iachello, and A. Arima, Ann. Phys. (N.Y.) 115, 325 (1978)
9. F. Iachello, N.V. Zamfir, and R.F. Casten, Phys. Rev. Lett. 81, 1191 (1998)
10. N.V. Zamfir et al., Phys. Rev. C 60, 054312 (1999)
11. J.Y. Zhang, M.Caprio, N.V. Zamfir, R.F. Casten, Phys. Rev. C60, 061304 (1999)

IBM2 STUDY OF BAND CROSSING

A. GELBERG*,**, N. YOSHIDA***, T. OTSUKA**,****, A. ARIMA*****,
A. GADE*, A. DEWALD* AND P. VON BRENTANO*

*Institut für Kernphysik der Universität zu Köln, 50937 Köln, Germany
**Radiation Laboratory, Riken, 2-1 Hirosawa, Wako-shi, 351-0198 Japan
***Faculty of Informatics, Kansai University, Takatsuki-shi, 569-1095 Japan
****Department of Physics, University of Tokyo, 7-3-1 Hongo, Bunkyo-ku
Tokyo 113-0033 Japan
*****223 Sangiin-Giin-kaikan, 2-1-1 Nagata-cho, Chiyoda-ku
Tokyo 100-8962 Japan

The band crossing in the Ba region has been studied in the extended interacting boson model in which two neutrons or two ptorons are coupled to a boson core. Energy levels, E2 ratios and g-factors are analyzed. An interesting feature in ^{128}Ba is the mixing of three bands, which is reasonably explained by the mixing of the pure-boson band, the boson+neutron pair band, and the boson+proton pair band. In the analysis of ^{132}Ba, the isomeric 10^+ state is well described. The consistency of the boson-fermion interaction is investigated over even-even and odd-A isotopes in the region. For example, it turns out that the same quadrupole interaction between the odd neutron hole and bosons can account for the negative-parity states in ^{131}Ba and the high-spin states in ^{132}Ba.

1 Introduction

The interacting boson model (IBM)[1] has been extended so as to describe the high-spin states of even-even nuclei[2,3,4,5] in which one of the bosons is allowed to change into a pair of fermions in the high-j orbital.

Recently new data have been obtained around the band crossing spin in the Xe-Ba region including the crossing of three bands[6,7,8].

In the present report, we show the preliminary results of analysis of the energies and the electromagnetic transitions in the framework of the IBM+2qp model. We show the attempt of a consistent description with neighbouring odd-A nuclei.

2 The model

To describe the low-lying states in even-even nuclei, we use the IBM2, i.e., the proton-neutron interacting boson model which distinguishes proton bosons from neutron bosons. The model space for high-spin states consists of

I. Pure core states with boson numbers $N_\pi = N_{\pi_0}, N_\nu = N_{\nu_0}$ and no fermions.

$$|I> = |L>$$

II. States in which one proton boson has been converted into one proton pair, with $N_\pi = N_{\pi_0} - 1, N_\nu = N_{\nu_0}, n_\pi = 2$.

$$|I> = |L,\ j_\pi^2(J_\pi); I> \qquad (J_\pi = 4, 6, ..., 2j_\pi - 1)$$

III. States in which one neutron boson has been converted into a neutron pair, with $N_\pi = N_{\pi_0}, N_\nu = N_{\nu_0} - 1, n_\nu = 2$.

$$|I> = |L,\ j_\nu^2(J_\nu); I> \qquad (J_\nu = 4, 6, ..., 2j_\nu - 1)$$

The nucleon pairs coupled to $J = 0$ and $J = 2$ are excluded because they are already included in s- and d-bosons. Note that the total number of bosons and fermion pairs is conserved.

The hamiltonian is written as

$$H = H^{\mathrm{B}} + H^{\mathrm{F}} + V^{\mathrm{BF}}$$

where H^{B} is the boson hamiltonian:

$$H^{\mathrm{B}} = \epsilon_d n_d + \kappa Q_\pi^{\mathrm{B}} \cdot Q_\nu^{\mathrm{B}} + \frac{1}{2} \sum_{L=0,2,4;\rho=\pi,\nu} c_\rho^L([d_\rho^\dagger d_\rho^\dagger]^{(L)} \cdot [\tilde{d}_\rho \tilde{d}_\rho]^{(L)}) + M_{\pi\nu}$$

where n_d is the total number of d-bosons,

$$Q_\rho^{\mathrm{B}} = d_\rho^\dagger s_\rho + s_\rho^\dagger \tilde{d}_\rho + \chi_\rho [d_\rho^\dagger \tilde{d}_\rho]^{(2)} \qquad (\text{for } \rho = \pi,\ \nu),$$

$$M_{\pi\nu} = \frac{1}{2}\xi_2 \left((s_\nu^\dagger d_\pi^\dagger - d_\nu^\dagger s_\pi^\dagger) \cdot (s_\nu \tilde{d}_\pi - \tilde{d}_\nu s_\pi) \right) + \sum_{k=1,3} \xi_k \left([d_\nu^\dagger d_\pi^\dagger]^{(k)} \cdot [\tilde{d}_\pi \tilde{d}_\nu]^{(k)} \right).$$

The fermion hamiltonian and the boson-fermion interaction are

$$H^{\mathrm{F}} = E_\pi n_\pi + E_\nu n_\nu + V_\pi^{\mathrm{F}} + V_\nu^{\mathrm{F}},$$

$$V^{\mathrm{BF}} = \kappa Q_\pi \cdot Q_\nu - \kappa \left(Q_\pi^{\mathrm{B}} \cdot Q_\nu^{\mathrm{B}} \right) - \sum_{\rho=\pi,\nu} \frac{\Lambda_\rho}{\sqrt{2j_\rho + 1}} : \left[[a_{j_\rho}^\dagger \tilde{d}_\rho]^{(j_\rho)} [\tilde{a}_{j_\rho} d_\rho^\dagger]^{(j_\rho)} \right]_0^{(0)} :,$$

where E_π (E_ν) is the single particle energy of the proton (neutron), n_π (n_ν) is the number of protons (neutrons), and V_π^{F} (V_ν^{F}) is the two-body interaction between the protons (neutron). In V^{BF}, the total quadrupole operator is defined as

$$Q_\rho = Q_\rho^{\mathrm{B}} + \alpha_\rho [a_\rho^\dagger \tilde{a}_\rho]^{(2)} + \beta_\rho \left[[a_\rho^\dagger a_\rho^\dagger]^{(4)} \tilde{d}_\rho \right]^{(2)} - \beta_\rho \left[d_\rho^\dagger [\tilde{a}_\rho \tilde{a}_\rho]^{(4)} \right]^{(2)}$$

with $\rho = \pi, \nu$.

In calculating electromagnetic matrix elements, the transition operators are defined as

$$T(\mathrm{E2}) = e_\pi Q_\pi + e_\nu Q_\nu,$$

$$T(\mathrm{M1}) = \sqrt{\frac{3}{4\pi}}(g_\pi^{\mathrm{B}} L_\pi + g_\nu^{\mathrm{B}} L_\nu + g_\pi^{\mathrm{F}} j_\pi + g_\nu^{\mathrm{F}} j_\nu)$$

for E2 and M1 transitions, respectively. In $T(\mathrm{E2})$, the parameter $e_{\pi(\nu)}$ is proton (neutron) boson effective charge. The fermion part has been ignored because its contribution is negligibly small. In $T(\mathrm{M1})$, the boson and fermion g-factors are written as g_ρ^{B} and g_ρ^{F} respectively. The operators $L_{\pi(\nu)}$ and $j_{\pi(\nu)}$ represent the boson and the fermion angular momenta, respectively.

3 Calculations

3.1 ^{128}Ba

The isotope $^{128}_{56}\mathrm{Ba}_{72}$ has $N_\pi = (56-50)/2 = 3$ proton bosons and $N_\nu = (82-72)/2 = 5$ neutron bosons in the core. Note that the neutron bosons are holes because the nearest closed shell is $N = 82$. To describe high-spin states, the fermion high-j orbit $j_\pi = j_\nu = 0h_{11/2}$ is introduced both for protons and neutrons. The parameters for the core hamiltonian H^{B} are obtained by fitting the observed low-lying energy levels: $\epsilon_d = 0.8704$ MeV, $\kappa = -0.156$ MeV, $\chi_\nu = 0.2$, $\chi_\pi = -0.9$, $c = -0.205$ MeV, $\xi_2 = 0.24$ MeV, $\xi_1 = \xi_3 = -0.18$ MeV. In the fermion hamiltonian, the delta interaction with the strength $G = 0.36$ MeV has been taken to determine the two-body interaction. The fermion pair energies E_π and E_ν have been taken as adjustable parameters, determined from the experimental energies of the high-spin states. The obtained values are: $E_\pi = 1.6195$ MeV, $E_\nu = 1.499$ MeV. As the boson-fermion interaction, the quadrupole-quadrupole interaction and the exchange interaction are taken. The values obtained from fitting are: $\alpha_\pi = 3.00$, $\beta_\pi = 1.6$, $\alpha_\nu = -2.86$, $\beta_\nu = 0.8$; $\Lambda_\nu = 0.60$ MeV. After coupling to the fermion part, the boson core parameters have been multiplied by a factor 1.2 to adjust overall scale. The values given above have been actually the multiplied values.

Figure 1 shows the energy levels. The agreement is generally good up to the first 10^+ state. The calculation produces the second 10^+ state too close to the first 10^+. For $I \geq 12$, the calculated levels spacings are generally larger than observed spacings. This is a common character of models without the mechanism of the variable moment of inertia (VMI). As for the transitions, the boson effective charges: $e_\pi = e_\nu = 0.133$ eb have been determined so as to reproduce the experimental $B(\mathrm{E2}; 2_1^+ \to 0_g^+)$. Figure 2 shows the comparison

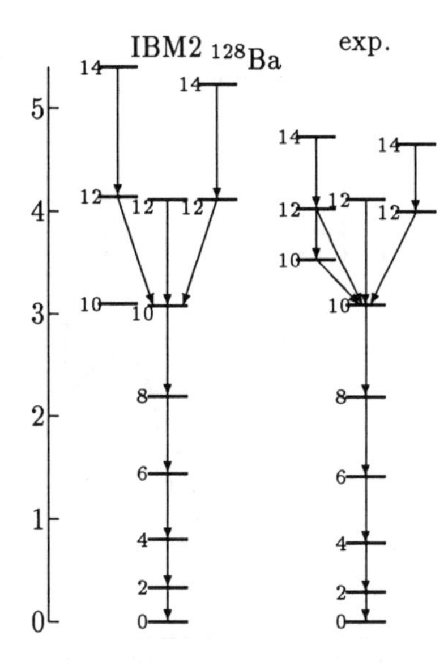

Figure 1. Energy levels in ^{128}Ba.

of calculated and experimental $B(\mathrm{E2})$ ratios. The calculation reproduces the general feature: the constant ratios around $I \approx 6$; the sudden drop that is explained as a result of the band-crossing; the second increase after $I = 14$.

3.2 ^{132}*Ba and* ^{131}*Ba*

In $^{132}_{56}\mathrm{Ba}_{76}$, it has been assumed[7] that the lowest two 8^+ states belong to the ground band and the quasigamma band. The lifetime of the isomeric 10^+_1 state has been measured. Together with the high-spin states, we attempt a consistent description of a neighbouring odd-A isotope $^{131}_{56}\mathrm{Ba}_{75}$.

First we determine the interaction for the odd-A nucleus. The nucleus ^{131}Ba is treated as a system where one neutron hole is coupled to a boson core of ^{132}Ba. The parameters for the core are determined from the energy levels of ^{132}Ba. The negative-parity states in ^{131}Ba are described with a neutron hole in $h_{11/2}$. The boson-fermion interaction conventionally taken for odd-A nuclei has the form:

$$V^{\mathrm{BF}} = \Gamma Q^{\mathrm{B}}_{\pi} \cdot [a^{\dagger}_{j} \tilde{a}_{j}]^{(2)} - \frac{\Lambda}{\sqrt{2j+1}} : [[a^{\dagger}_{j} \tilde{d}_{\nu}]^{(j)} [\tilde{a}_{j} d^{\dagger}_{\nu}]^{(j)}]^{(0)}_{0} : + C\, \hat{j} \cdot (\hat{L}^{\mathrm{B}}_{\nu} + \hat{L}^{\mathrm{B}}_{\pi}) + \ldots .$$

The parameters are fitted for the energy levels in ^{131}Ba. The obtained values

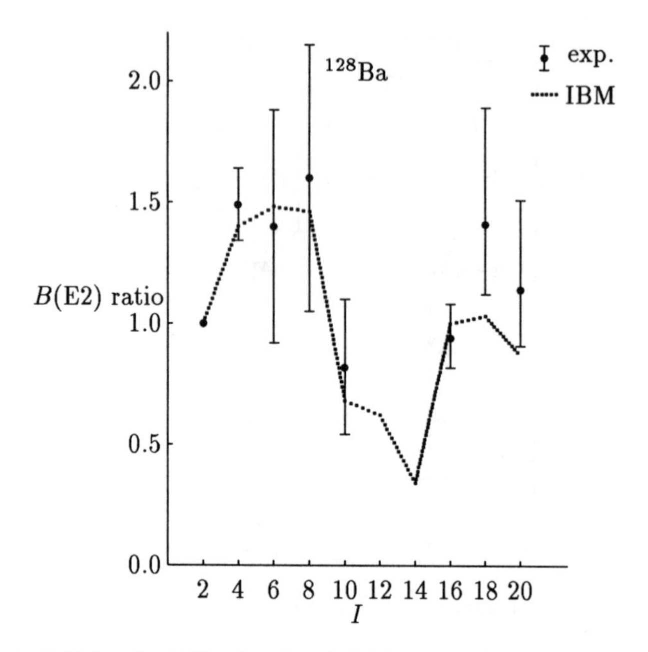

Figure 2. B(E2) ratios $B(\mathrm{E2}; I \to I-2)$ / $B(\mathrm{E2}; 2 \to 0)$ of the yrast states in ^{128}Ba.

are: $\Gamma = 0.46$ MeV, $\Lambda = 35.3$ MeV, $C = -0.016$ MeV. The obtained enengy levels are shown in Fig. 3. One sees that the observed general trends are well reproduced by the IBFM2. In calculating the transitions, the boson effective charges $e_\pi = e_\nu = 0.126\ e^2b^2$ have been determined from $B(\mathrm{E2}; 2_1^+ \to 0_g^+)$. The boson g-factors are: $g_\pi^{\mathrm{B}} = 0.85$, $g_\nu^{\mathrm{B}} = -0.23$. For the odd neutron, the Schmidt value with spin g-factor quenched by 0.7 is taken: $g_\nu^{\mathrm{F}} = -0.19$. The branching ratios are compared in Table 1. The experimental data are taken from ref.[9]. One sees that the branching ratios and the E2/M1 mixing ratios in the transitions from the states with $I = 13/2, 15/2$, are very well reproduced.

Then we study the high-spin states in ^{132}Ba. In order to have a consistent V^{BF}, we assume $\kappa\alpha_\nu = \Gamma$. ¿From the parameters of the core: $\kappa = -0.25$ MeV, $\Gamma = 0.46$ MeV, we fix $\alpha_\nu = \Gamma/\kappa = -1.85$. Other parameters are free. Figure 4 shows the comparison of energy levels in the form of a backbending plot. The drop of the rotational frequency ω from $I = 8$ to $I = 10$ reflects the crossing of the pure-boson and the neutron bands. Table 2 shows the comparison of B(E2) values and g-factors including the band-crossing region. The subtle feature of the E2 transitions from 10_1^+ is well described by the IBM2 calculation. As for the gyromagnetic factors, both $g(2_1^+)$ and $g(10_1^+)$ have been

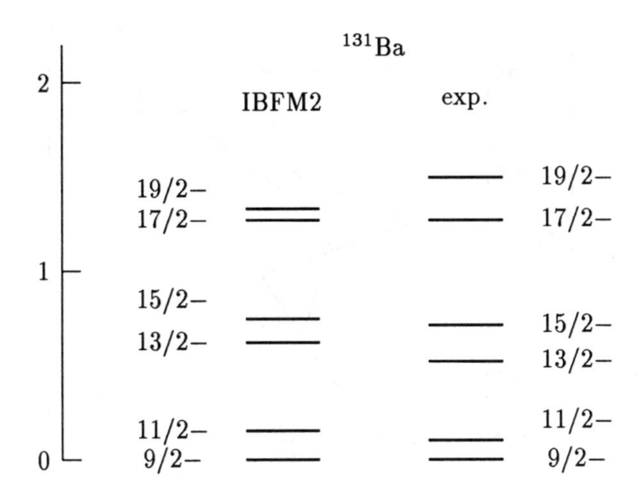

Figure 3. Energy levels of negative-parity states in ^{131}Ba.

Table 1. Branching ratios in ^{131}Ba.

initial	final	ratio (cal)	ratio (exp)	δ(E2/M1) (cal)	δ(E2/M1) (exp)
11/2(1)	9/2(1)			−1.41	−0.01(4)
13/2(1)	11/2(1)	100	100	−0.34	−0.32(7)
	9/2(1)	83	10		
15/2(1)	13/2(1)	1	15	−0.55	−0.24(12)
	11/2(1)	100	100		
17/2(1)	15/2(1)	24	100	−0.52	−0.42(9)
	13/2(1)	100	< 69		
19/2(1)	17/2(1)	1	6	−0.26	−0.19(10)
	15/2(1)	100	100		

reproduced by the calculation by taking boson g-factors of reasonable values.

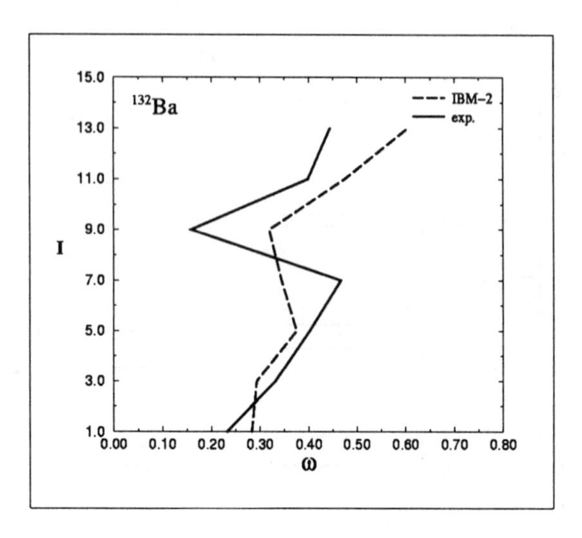

Figure 4. Backbending plot of ^{132}Ba.

Table 2. B(E2) values and and Gyromagnetic factors in ^{132}Ba.

B(E2) values (e^2fm^4)

Observable	exp.	IBM-2
B(E2;$2_1 \rightarrow 0_1$)	1720(172)	1711
B(E2;$10_1 \rightarrow 8_1$)	20.5(6)	25
B(E2;$10_1 \rightarrow 8_2$)	2.4(5)	0.9

Gyromagnetic factors

$g(I)$	exp.	IBM-2
$g(2_1)$	0.34(3)	0.33
$g(10_1)$	−0.156(11)	−0.164

4 Discussion

Looking at the wave functions in ^{128}Ba, the first and the third 10^+ states are both the mixture of the pure boson and the proton-pair states while the second 10^+ is an almost pure neutron-pair state. As for the $I = 12^+$ levels, the first state is a mixed state of the pure-boson band and the two-neutron band. It is interesting to know the real characters of these three bands. In ^{132}Ba, the wave functions of the yrast states are almost pure boson states up to spin $I = 8$. ¿From $I = 10$, the yrast states are s-band states dominated by the two-neutron components.

In the calculation for ^{132}Ba, we have taken into account the consistency with a neighbouring odd-A nucleus. The high-spin states in ^{132}Ba are described by coupling two neutron holes in $h_{11/2}$ to the boson core of ^{134}Ba. The negative-parity states in ^{133}Ba can be described by coupling a neutron hole hole in $h_{11/2}$ to the same core of ^{134}Ba. But only a few negative-parity levels have been observed in ^{133}Ba. So instead we took ^{131}Ba.

5 Conclusions

Calculation has been performed on ^{128}Ba involving the mixing of three bands in the backbending spin regions. Reasonable agreement has been obtained including the reduction of B(E2) after $I = 6$. The mixing of various components makes the interpretation complicated in ^{128}Ba. For ^{132}Ba and ^{131}Ba, a consistent quadrupole interaction has been taken.

More systematic calculations are being carried out for neighbouring nuclei.

References

1. F. Iachello and A. Arima, The Interacting Boson Model (Cambridge Univ. Press, Cambridge, 1987).
2. A. Gelberg ans A. Zemel, Phys. Rev. C **22**, 937 (1980).
3. I. Morrison, A. Faessler and C. Lima, Nucl. Phys. A **372**, 13 (1981).
4. N. Yoshida, A. Arima and T. Otsuka, Phys. Lett. B **114**, 86 (1982).
5. A. D. Efimov and V. M. Mikhajlov, Proc. Int. Symp. Perspectives for the Interacting Boson Model, 13-17 June, 1994 (World Scientific, 1994) p. 714.
6. A. Dewald et al., Phys. Rev. C **37**, 289 (1988).
7. E. S. Paul et al., Phys. Rev. C **40**, 1255 (1989).
8. S. Harissopulos et al., Phys. Rev. C **52**, 1796 (1995).
9. Evaluated Nuclear Structure Data File (ENSDF), National Nuclear Data Center, Brookhaven National Laboratory (http://www.nndc.bnl.gov/nndc/ensdf/).

SUPER- AND HYPERDEFORMED STATES IN THE ACTINIDE REGION

A. KRASZNAHORKAY[1], D. HABS[2], M. HUNYADI[1], D. GASSMANN[2], M. CSATLÓS[1], Y. EISERMANN[2], T. FAESTERMANN[3], G. GRAW[2], J. GULYÁS[1], R. HERTENBERGER[2], H.J. MAIER[2], Z. MÁTÉ[1], A. METZ[2], J. OTT[2], P. THIROLF[2] AND S.Y. VAN DER WERF[4]

[1] *Inst. of Nucl. Res. of the Hung. Acad. of Sci., H-4001 Debrecen, P.O. Box 51, Hungary*

[2] *Sektion Physik, Universität München, Garching, Germany*

[3] *Technische Universität München, Garching, Germany*

[4] *Kernfysisch Versneller Instituut, 9747 AA Groningen, The Netherlands*

The ^{233}U(d,pf)^{234}U, ^{235}U(d,pf)^{236}U and the ^{239}Pu(d,pf)^{240}Pu reactions have been studied with high energy resolution. The observed fission resonances were described as members of rotational bands with rotational parameters characteristic to super (^{240}Pu) and hyperdeformed nuclear shapes. Information on the K values of the bands and for the J values of the band members have been obtained from fission fragment angular distribution measurements. The level density of the most strongly excited states has been compared to the prediction of the back-shifted Fermi-gas formula and the energy of the ground state in the second (^{240}Pu) and third (^{234}U) minimum has been estimated. The fission fragment mass distribution of the hyperdeformed states in ^{236}U have also been measured. The width of the mass distribution, coincident with the hyperdeformed bands, is significantly smaller than the ones obtained in coincidence with background regions below and above the resonances.

1 Introduction

The investigation of the superdeformed (SD, ratio of long to short axis 2:1) and hyperdeformed (HD, axis ratio 3:1) states started with the discovery of fission isomers and studying their properties [1] and continued by the search for SD and HD states in high-spin γ-spectroscopy [2].

Recently, very effective, high resolution 4π gamma-ray spectrometers like EUROBALL and GAMMASPHERE have been developed for studying these highly deformed states, which is one of the most vital fields in modern nuclear structure physics. Using these arrays many SD bands have already been studied but no firm evidence could be obtained for any HD bands [3,4,5,6].

The theoretical description of fission isomerism is based on the introduction of shell corrections [7] to the smooth liquid drop potential energy surface, resulting in a double humped fission barrier with a second minimum containing SD states. Moreover, in the actinide region a third minimum in the potential energy (which contains HD states) was predicted already more than twenty years ago by Möller *et al.* [8]. According to recent calculations, in these nuclei the so-called third minimum of the potential barrier appears with deformation parameters $\beta_2 \approx 0.90$ and $\beta_3 \approx 0.35$ [7,9] and the depth is predicted to be much larger ($\Delta E \approx 3$ MeV [10]) than believed earlier [11].

The γ-spectroscopic studies of the SD states turned out to be very difficult in the actinide region because of the low cross sections and the high background produced by the fission fragments [2]. However, at higher excitation energy sub-barrier transmission resonances appear in the fission probability at the position of quasi bound states in the second and third minimum of the potential barrier. By measuring the fission probability as a function of the excitation energy one can map these SD and HD states. Early attempts of studying these transmission resonances critically suffered from either a limited energy resolution [12] or statistical significance [13].

The aim of the present experiments was to resolve the micro-structure of the vibrational resonances in the second and third minimum, to determine the J and K values as well as the fission decay properties of the SD and HD excited states.

2 Experimental setup

The experiments were carried out at the Debrecen 103-cm isochronous cyclotron at $E_d = 9.73$ MeV and in Munich with deuterons of 12.5 MeV from the Munich Tandem accelerator. Enriched (97.6 % - 99.89 %) 30 - 250 μg/cm^2 thick targets were used.

The energy of the outgoing protons was analyzed in Debrecen by a split-pole magnetic spectrograph, which had a solid angle of 2 msr and which was set at $\Theta_L = 140^o$ with respect to the incoming beam direction [14]. The position and energy of the protons were analyzed by two Si solid state position sensitive focal plane detectors.

In Munich the energy of the outgoing protons was analyzed by a Q3D magnetic spectrograph [15], which was set at $\Theta_L = 130^o$ relative to the incoming beam and the solid angle was 10 msr [16]. The position of the analyzed particles in the focal plane was measured with a light-ion focal-plane detector of 1.8 m active length [17]. Using that detector with the spectrograph a line-width of $\leq$ 3 keV has been observed for elastic scattering of 20 MeV deuterons.

The fission fragments were detected by two position sensitive avalanche detectors (PSAD) [18] having two wire planes (with delay-line read-out) corresponding to horizontal and vertical directions. The solid angle of the detectors varied between 10 % and 30 % of 4π in the different experiments [19].

The mass distribution of the fission fragments have been determined by using the time difference method. The fission fragments were detected with two PSAD's having active areas of 16 x 16 cm^2 and distances of 23 cm from the target resulting in a relatively large solid angle of 4% of 4π. The angle of the detectors was 55° and 125° with respect to the beam direction. The time resolution of the detectors were $\approx$ 1 ns resulting a mass resolution of about 10 amu.

3 Results

In program aiming at studying the super- and hyperdeformed states in the actinides we have already studied the SD states in ^{240}Pu [20] and the HD ones in ^{231}Th, ^{234}U [16] and in ^{236}U [14]. We are presenting now one example for the SD (^{240}Pu) and two examples for the HD ^{234,236}U states.

3.1 Superdeformed states in ^{240}Pu

Fig. 1 displays the energy spectrum of protons in coincidence with fission fragments measured in the ^{239}Pu(d,pf) ^{240}Pu reaction. The excitation energy of the ^{240}Pu pre-fission nucleus can directly be deduced from the kinetic energy of the protons. The fine structure of two enhanced structures of highly damped vibrational resonances around E* = 4.6 MeV and 5.1 MeV could successfully be resolved, each being a direct consequence of pure multi-phonon beta-vibrations with $K^\pi = 0^+$. We expect rotational sequences of $J^\pi = 0^+$, 2^+, 4^+ ... as a fine structure of the above broad resonances. For the lower resonance group around E* = 4.6 MeV rotational band members with a constant intensity ratio can clearly be identified for the first time. The described conception on the vibrational assignments and rotational bands was tested by a fit of the energy region first between 4.4 MeV and 4.8 MeV (included in Fig.1). The fitting procedure assuming $K^\pi = 0^+$ band heads led to a rotational parameter of $\hbar^2/2\Theta = 3.22 \pm 0.2$ keV (typical for SD bands), in good agreement with conversion electron measurements [21]. For the upper resonance group at E* = 5.1 MeV a sufficient fit can be reached by using the same intensity ratios and moment of inertia obtained from the lower group. The $K^\pi = 0^+$ assignments are consistently supported by the angular correlation coefficients extracted from the fission fragment angular correlations (see Fig 1b) and 1c).

The possibility of a complete spectroscopy of K = 0^+-states in the superdeformed second well opened up by the described technique enables the determination of the excitation energy of the fission isomer, i.e. the depth of the second well, exploiting the level density information. The experimental

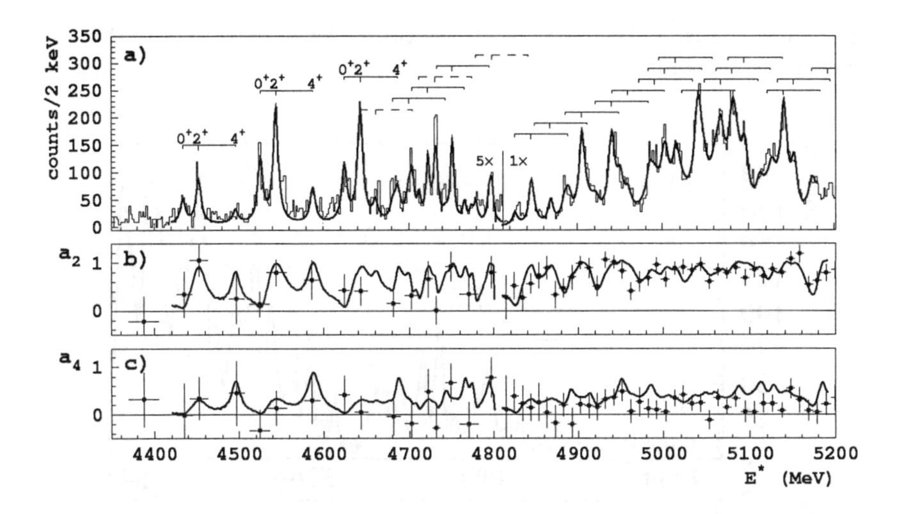

Figure 1. (a) Result of the fit to the observed resonance structures with $K^\pi=0^+$ rotational bands. (Dashed markers indicate the ambiguous presence of some additional bands). (b) and (c) Experimental and calculated a_2 and a_4 coefficients as a function of the excitation energy. The calculated values resulted from combining theoretical coefficients by the obtained fit of the spectrum.

distances of the identified $K = 0^+$ band heads have been fitted using the calculated average distances from the back-shifted Fermi gas model [22,23], assuming that the parameterization of the level density formula can also be applied in the SD well. As a result, the calculated function had to be shifted up in order to reproduce the experimental distances by 2.25 ± 0.2 MeV, corresponding to the isomer excitation energy E^*_{II}, confirming other measurements of E_{II} via excitation functions [1]. This establishes the described procedure as a reliable new method to determine the depth of the second well and justifies its use in the case of the third well in ^{234}U, where no other experimental information is yet available.

3.2 Hyperdeformed states in ^{234}U

In order to investigate the HD bands the excitation energy was chosen between the energy of the inner and outer barriers of the second well, i.e. between 4.5 and 5.2 MeV [16]. In this energy range the widths of the SD resonances in the second well should be much broader than those of the HD states due to the strong coupling to the normal deformed states. The widths of the HD states due to the higher outer barriers of the third well remain below the actual experimental resolution of ~5 keV.

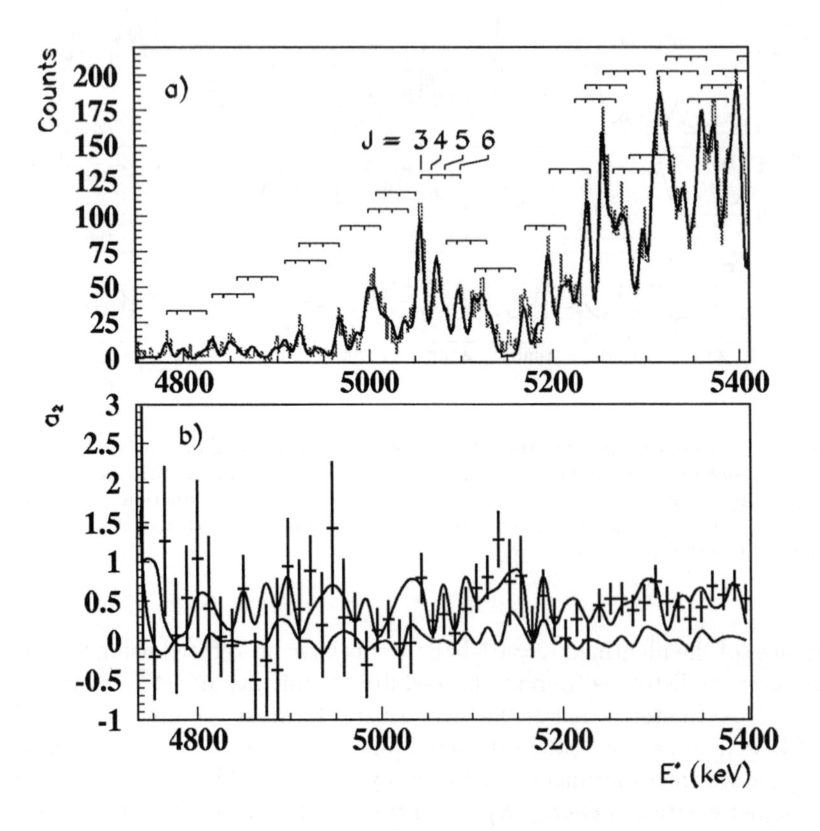

Figure 2. (a) Part of the measured proton energy spectrum fitted with 24 rotational bands with a common rotational parameter. The spectrum was divided into two parts at $E = 5150$ keV for the fitting; (b) Experimental fission-fragment angular-distribution coefficients as a function of excitation energy compared to the calculated ones using $K = 1$ (upper curve) and $K = 3$ (lower curve) for all of the bands. The $K = 0$ curve is very close to the $K = 1$ one while the $K = 2$ is in between the $K = 1$ and $K = 3$ ones. They are not shown.

Part of the proton spectrum measured in coincidence with the fission fragments is shown in Fig 2a). Assuming overlapping rotational bands with the same moment of inertia, inversion parameter and intensity ratio for the members in a band, we fit our spectrum using simple Gaussians for describing the different band members in the same way as we did it in our previous work [14]. The result of the fit is also shown in Fig. 2a). The details of the fitting procedure is described in our previous work [16].

Fission-fragment angular distributions were generated as a function of the excitation energy, normalized to the known (d,f) angular distribution and fitted with even Legendre polynomials (LP) up to fourth order. The a_2 angular distribution coefficient is shown in Fig. 2b) as a function of the excitation energy. In order to get information on the spins and K values of the observed rotational bands, or to check our assumptions made for fitting the energy spectrum, the angular distribution coefficients of the fission fragments have been calculated and compared to the experimental ones. For more details see ref. [16].

The density of $J = 3$ states has been determined from our experimental data and calculated as a function of the excitation energy using the back-shifted Fermi-gas description with parameters determined by Rauscher *et al.* [22]. In order to estimate the depth of the third well we compared the experimentally obtained and calculated values. We assumed that the same parameterization of the level density formula is valid in the third well, as was determined by Rauscher *et al.* [22] by fitting the level densities in the first well of the potential barrier and which we have already checked also in the second well in case of ^{240}Pu (for more details see ref. [16]). From the comparison we find a value of 3.1 MeV for the energy of the ground state in the third well.

In order to get some estimate for the precision of the level distance analysis described above we repeated the calculation of level distances by using the rotational parameter deduced in the present work. We also used two other formulas to estimate the level distances, which were parameterized by von Egidy *et al.* [23]. From the uncertainties of the calculated and measured level distances the error of the energy determination is estimated to be 0.4 MeV.

3.3 Determination of the mass distribution of the fission fragments in coincidence with the HD states

In our previous work we have found three HD bands in ^{236}U by studying the transmission resonances in the ^{235}U(d,pf)^{236}U reaction. Recently, we have repeated the experiment in Debrecen and measured the mass distribution in coincidence with the HD bands found previously at 5.28, 5.37, and 5.47 MeV. The mass distribution has been determined for different energy regions marked in Fig. 4. The result were fitted with two Gaussians having equal heights and widths. The full width of half maximum (FWHM) of the mass distributions at the bottom part of Fig. 4 for the different energy regions.

The mass distribution in coincidence with the HD bands has been found

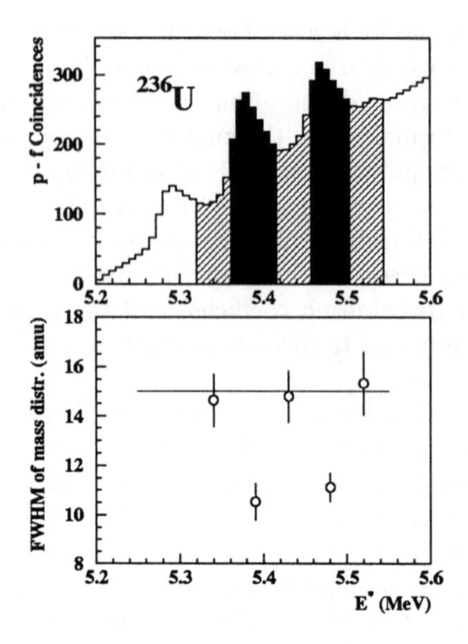

Figure 3. The observed HD resonances of ^{236}U. The marked regions selected the resonant and non-resonant parts of the spectrum, whose mass distributions were fitted by two Gaussians. The resulted widths of the mass distributions are shown in the lower figure.

significantly sharper compared to the other distributions measured above and below the peaks. The width of the mass distributions in coincidence with the HD bands is mostly caused by our mass resolution. Better mass resolution is required to determine the real widths of the mass distributions.

The HD states lying in the third well of the fission barrier may play a role of a doorway like state before fission, from which states the fission can only occur through a limited number of fission paths resulting in a sharper mass distribution.

In addition to this first observation of the sharpening of the mass distribution we have performed experiments very recently in Munich for ^{231}Th and obtained consistent results, but even more experimental data and theoretical calculations are needed to understand the nature of this phenomenon.

This work has been supported by DFG under IIC4-Gr 894/2 and The Hungarian Academy of Sciences under HA 1101/6-1, the Hungarian OTKA Foundation No. T23163, and the Nederlandse Organisatie voor Wetenschapelijk Onderzoek (NWO).

References

1. S. Bjornholm and J.E. Lynn, *Rev. Mod. Phys.* **52**, 725 (1980) and references therein.
2. D. Habs, *Nucl. Phys.* A **502**, 105c (1989).
3. A. Galindo-Uribarri *et al.*, *Phys. Rev. Lett.* **71**, 231 (1983).
4. G. Viesti *et al.*, *Phys. Rev.* C **51**, 2385 (1995).
5. D.R. LaFosse *et al.*, *Phys. Rev. Lett.* **74**, 5186 (1995).
6. D.R. LaFosse *et al.*, *Phys. Rev.* C **54**, 1585 (1996).
7. V.M. Strutinsky, *Nucl. Phys.* A **95**, 420 (1967).
8. P. Möller, S.G. Nilsson and R.K. Sheline, *Phys. Lett.* B **40**, 329 (1972).
9. V.V. Pashkevich, JournalNPA1692751971; P. Möller and J.R. Nix, (in Physics and Chemistry of Fission (IAEA, Vienna) **1** (1973) 103); J.F. Berger, M. Girod and D. Gogny, *Nucl. Phys.* A **502**, 85c (1989).
10. S. Ćwiok *et al.*, *Phys. Lett.* B **322**, 304 (1994).
11. W.M. Howard and P. Möller, *At. Data Nucl. Data Tables* **25**, 219 (1980), and references therein.
12. J. Blons *et al.*, *Nucl. Phys.* A **477**, 231 (1988).
13. P. Glässel, H. Röser and H.J. Specht, *Nucl. Phys.* A **256**, 220 (1976).
14. A. Krasznahorkay *et al.*, *Phys. Rev. Lett.* **80**, 2073 (1998).
15. H.A. Enge and S.B. Kowalsky, (in Proc. 3rd Int. Conf. on magnet technology, Hamburg (1970)).
16. A. Krasznahorkay *et al.*, *Phys. Lett.* B **461**, 15 (1999).
17. E. Zanotti *et al.*, *Nucl. Instrum. Methods* A **310**, 706 (1991).
18. P.C.N. Crouzen, (PhD Thesis, Rijksuniversiteit Groningen, (1988) Unpublished).
19. M. Hunyadi, (PhD Thesis, Lajos Kossuth University Debrecen, (1999) Unpublished).
20. M. Hunyadi et al., to be published.
21. H.J. Specht *et al.*, *Phys. Lett.* B **41**, 43 (1972).
22. T. Rauscher, F.K. Thielemann, K.L. Kratz, *Phys. Rev.* C **56**, 1613 (1997).
23. T. von Egidy, H.H. Smidt, A.N. Behkami, *Nucl. Phys.* A **481**, 189 (1988).

LIFETIME MEASUREMENTS IN THE PICOSECOND RANGE: ACHIEVEMENTS AND PERSPECTIVES

R. KRÜCKEN

A.W. Wright Nuclear Structure Laboratory, Physics Department, Yale University, New Haven, Connecticut 06520, U.S.A.

Recent developments in the measurement of lifetimes in the picosecond range using the recoil distance method (RDM) are reviewed. Results from recent RDM experiments on superdeformed bands in the mass-190 region, shears bands in the neutron deficient lead isotopes, and ground state bands in the mass-130 region are presented. New experimental devices for lifetime experiments at Yale, such as the New Yale Plunger Device (N.Y.P.D.), the SPEctrometer for Doppler-shit Experiments at Yale (SPEEDY) and the plans for the gas-filled recoil separator SASSYER are presented. Perspectives for the use of the RDM technique in the study of exotic nuclei and its potential use with radioactive beams are discussed.

1 Introduction

The knowledge of matrix elements for transitions between excited nuclear levels adds important insight into the structure of nuclei. While relative transition matrix elements can be determined by measuring gamma-ray branching ratios, absolute transition matrix elements are only accessible by measuring the lifetimes or the Coulomb excitation cross sections of the nuclear levels. The reduced transition probability B(E2) for the $2_1^+ \rightarrow 0_1^+$ ground state transition, for example, provides a measure of the charge quadrupole moment and thus the ground state deformation of an even-even nucleus. Other examples for the importance of transition matrix elements include the possibility to identify collective excitation modes on the basis of the magnitude and spin dependence of B(E2) values, or their sensitivity to the mixing of different configurations, in particular in the case of shape coexistence.

The precision achievable in lifetime experiments has dramatically increased with the availability of large, highly efficient γ-ray multi-detector arrays. These instruments have made exotic, weakly populated nuclear excitations accessible for lifetime measurements, such as superdeformed (SD) bands. In particular for such exotic nuclear excitations lifetime measurements have proven to be essential for a sound understanding of the excitation mechanism.

This contribution will review some recent achievements that have been made in measuring nuclear level lifetimes in the picosecond range, which is a critical lifetime range for low-lying collective excitations, as well as some levels in more exotic nuclear excitations. The most common method used is

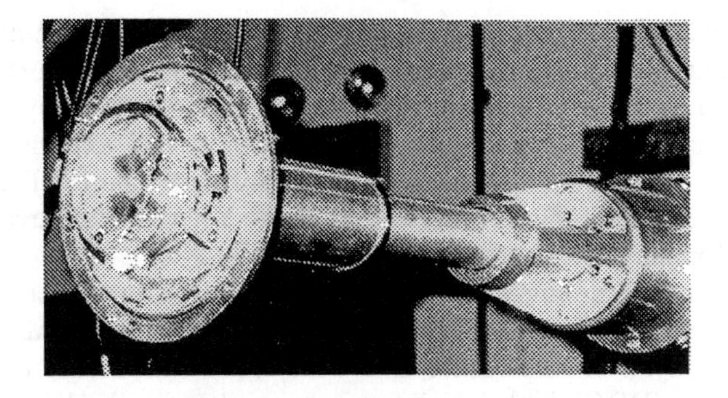

Figure 1. Picture of the New Yale Plunger Device [1].

the so-called recoil distance method (RDM), also called the recoil distance Doppler-shift (RDDS) technique. Both terms are used interchangeably in the course of this article.

2 The recoil distance method (RDM)

The RDM is the standard method to measure lifetimes of excited nuclear states in the picosecond range. The method uses a target chamber, called a plunger device, that contains a stretched target- and stopper-foil which are mounted parallel to each other at a variable distance. Excited states in the nuclei of interest are populated[a] in the target foil. The nuclei then recoil with a velocity of a few percent the speed of light in the direction of the stopper foil where they are stopped. The distances are chosen such that the flight time is on the order of the effective lifetime of the levels of interest and vary typically from about 10 μm to several millimeters. In some cases the second foil is chosen to slow the recoiling nuclei down without stopping it but the basic principle of the method stays the same. The lifetime of a level of interest is determined from the changing intensities of fully Doppler-shifted and stopped (or slowed) γ-ray components detected by the surrounding γ-ray detectors when varying the target-to-stopper distance.

In recent years new plunger devices were developed for the use with large multi-detector gamma-ray arrays. The New Yale Plunger Device (N.Y.P.D.) (Fig. 1) is one such device, which follows the design of the most recent Cologne plunger. The N.Y.P.D. fits into the center of arrays such as Gammasphere

[a]Typically a fusion evaporation reaction is used but also Coulomb excitation, transfer or the fission process can be used.

and the Yale Rochester Array for SpecTroscopy (YRAST Ball) [1] and can take advantage of the high efficiency of these arrays. More details on the N.Y.P.D. can be found in refs. [1,2]. Currently a new spectrometer is being constructed at Yale, specifically designed for Doppler-shift lifetime experiments. This SPEtrometer for Experiments with Doppler-shifts at Yale (SPEEDY) [1] concentrates all the high efficiency detectors of the YRAST Ball array at angles useful for Doppler-shift experiments.

Besides the developments in the design of plunger devices there has been a significant improvement of the reliability of lifetimes from RDM experiments due to the analysis of coincidence data with the so called Differential Decay-Curve Method (DDCM) [3,4], which extracts a value for the level lifetime for each target-to-stopper distance directly from observables. The DDCM with coincidence gates from above the level of interest eliminates various systematic uncertainties such as feeding and side-feeding times as well as the deorientation effect. At the same time the method provides a consistency check of the analysis.

3 Recent results from RDM measurements

The resolving power of modern γ-ray multi-detector arrays has enabled the application of the RDM to excited levels that are populated only with a few percent of the total fusion cross section. At the same time lifetimes for states along the yrast line can be measured with unprecedented relative accuracy of a few percent. In this section a few recent examples are presented where RDM experiments have had considerable impact on the understanding of the underlying structure of the nuclei under investigation.

3.1 The decay out of superdeformed bands in the mass-190 region

The sudden disappearance of the intensity at the bottom of SD bands has been a puzzling and intensely investigated problem. Only recently, major breakthroughs have been accomplished with the first observations of discrete linking transitions between the SD bands in ^{194}Hg [5,6] and ^{194}Pb [7,8] and the respective near yrast normal deformed (ND) levels. These observations have for the first time enabled the determination of the excitation energies, spins and parity of superdeformed states in the mass-190 region. It was also possible to measure lifetimes at the bottom of some SD bands in the mass-190 region using the RDDS technique [9,10,11].

The mechanism leading to the decay out of the SD bands can be understood in a simple mixing picture [11,12], where the lowest observed SD states mix with their nearest neighbouring normal deformed (ND) states with the same spin and parity. The transition quadrupole moments at the bottom of the SD bands did not show a significant reduction compared to the SD quadrupole moments at higher spin. This simple finding was the first experimental indication that the mixing between SD and ND states has to be very weak. A further analysis using a statistical model to estimate the level density and transition probabilities of the highly excited ND states revealed that the squared mixing amplitude of ND states mixed into the SD states is less then 4% for all SD states in the mass-190 region with a significant branch to ND states at lower excitation energy [9,10,11].

In the case of ^{194}Pb many direct transitions between the lowest members of the SD band and the near yrast states were observed [8] and multipolarities of pure E1 and mixed E2/M1 character were determined. B(E1) values of about 8×10^{-6} W.u. and upper limits for B(E2) and B(M1) values of 5×10^{-2} W.u. and 5×10^{-4} W.u., respectively, were determined for the pure ND states at the excitation energy of the SD states, which are consistent with a statistical decay [10]. This implies that the ND states mixing with the SD states are highly mixed and contain only small amounts of wave-functions that are very similar to the structure of the SD states.

3.2 Lifetimes in the ground state bands of mass-130 nuclei

The lifetime of the first excited 2^+ level of an even-even nucleus is a measure for the deformation of the nucleus in its ground state. Thus the ratio of the transition quadrupole moments for transitions between states in the ground band and that of the $2^+ \to 0^+$ ground state transition is an excellent measure for possible changes in the deformation within the ground band. However, small changes in the deformation can only be detected if the lifetimes are measured with very high precision. Such accurate measurements have recently been performed for the first time using large γ-ray detector arrays.

Table 1 shows the relative transition quadrupole moments within the ground state bands of several Xe, Ba and Nd nuclei [13]. The quadrupole moments are divided by the quadrupole moment for the $2^+ \to 0^+$ ground state transition. The lifetimes were measured with the GASP spectrometer of the INFN Laboratory Nazionali di Legnaro, Italy using the Cologne plunger and the DDCM for the analysis. The results in ^{126}Ba are most impressive since, for example, the lifetime of the 4^+ level was measured to be 8.59(18) ps [14] which represents an uncertainty of only 2% including systematic uncertainties. The ratios of the Q_t values are remarkably constant for most levels in the ground state bands below the crossing of the $\pi(h_{11/2})$ or $\nu(h_{11/2})$ intruder bands. An exception is the band in ^{126}Ba that shows clear deviations for the 6^+ and 8^+ levels. This dynamic shape effect is attributed to the underlying shell structure [14].

The example of ^{126}Ba demonstrates that high precision lifetime measurements can help to investigate nuclear structure beyond the standard approach of simple collective models. More such measurements need to be performed in order to carry out critical tests of these models.

3.3 Shears bands in the Pb region

Very regular rotational level sequences connected by strong magnetic dipole (M1) transitions in the neutron deficient Pb isotopes [15] have been observed in recent years. The band head of these M1 bands is generated by the perpendicular coupling of $h_{9/2}$ and $i_{13/2}$ protons to $i_{13/2}$ neutron holes. This leads to a large perpendicular component $\vec{\mu_\perp}$ of the magnetic moment with respect to the total angular momentum vector $\vec{J}$, which in turn leads to enhanced M1 transitions between the rotational levels.

Spin is generated in these M1 bands by the gradual alignment of the particle and hole angular momenta $\vec{j_\pi}$ and $\vec{j_\nu}$ with the total angular momentum vector $\vec{J}$. This is reminiscent to the closing of the blades of a pair of shears, leading to the name 'shears mechanism'. This behaviour arises naturally in the framework of the tilted-axis-cranking (TAC) model [16]. The closing of the particle and hole spin-vectors with respect to the total angular momentum vector leads to a specific large drop of B(M1) values with increasing spin, since $\mid \vec{\mu_\perp} \mid$ is decreasing with decreasing opening angle and B(M1)$\propto\mid \vec{\mu_\perp} \mid^2$.

Many of the features of the shears bands have been described by the shears mechanism as well as models that did not involve this new concept. Lifetime experiments were carried out at Gammasphere using the DSAM technique and RDDS technique in $^{193-199}$Pb [17,18] and ^{198}Pb [19], respectively. The life-

Table 1. Relative transition quadrupole moments Q_t for several even-even A≈130 nuclei.

	$R_4 = \frac{Q_t(4^+\to2^+)}{Q_t(2^+\to0^+)}$	$R_6 = \frac{Q_t(6^+\to4^+)}{Q_t(2^+\to0^+)}$	$R_8 = \frac{Q_t(8^+\to6^+)}{Q_t(2^+\to0^+)}$	$R_{10} = \frac{Q_t(10^+\to8^+)}{Q_t(2^+\to0^+)}$
^{122}Xe	0.96 (2)	0.99 (4)		
^{126}Xe	1.00 (3)	0.93 (6)	0.8 (1)	
^{124}Ba	1.04 (2)	1.00 (3)	0.99 (6)	
^{126}Ba	0.99 (2)	1.11 (3)	1.20 (5)	1.13 (7)
^{128}Ba	1.00 (3)	0.98 (6)	0.9 (1)	0.7 (4)
^{132}Nd	0.99 (2)	1.01 (2)		

time experiments have established the characteristic decrease of the B(M1) values with increasing spin. These key observations have provided the essential support for the concept of magnetic rotation as the underlying mechanism for the shears bands. This is a very impressive example for the importance of lifetime measurements.

4 Perspectives

The examples given in the previous section have highlighted that important insights can be gained by the study of absolute nuclear transition matrix elements. Since major thrusts of current and future nuclear structure research are aiming at the study of nuclei far from stability, it is useful to take a look at the prospective areas in which the measurement of nuclear level lifetimes can make an impact in accordance with these goals.

4.1 Combination of plunger with channel selection devices

Nuclei near the N=Z line and the proton dripline are currently intensely investigated. However, spectroscopic experiments are only possible by using auxiliary channel selection devices in combination with large gamma-ray detector arrays. Near the N=Z line it is essential to detect light charged particles as well as neutrons in order to pick out individual weak reaction channels of many open reaction channels. For heavier nuclei near the proton dripline the weak fusion reaction channels have to compete with a large fission background. Here it is essential to detect the fusion products directly by means of recoil spectrometers, such as the Argonne FMA or the Oak Ridge RMS, or using gas-filled recoil separators such as RITU in Jyväskylä or the BGS in Berkeley.

The combination of a plunger device with such channel selection devices provides a powerful tool for the study of exotic nuclei. First experiments have been performed in neutron deficient Sn isotopes using the Cologne plunger, the GASP spectrometer and its silicon ball ISIS. Future experiments employing the N.Y.P.D. and Gammasphere together with parts of the Microball and the Gammasphere Neutron wall can also be envisioned. In a first combination of a plunger device with a recoil detector the N.Y.P.D. has recently been combined with Gammasphere and the FMA at Argonne to measure lifetimes in ^{188}Pb in order to study the phenomenon of shape coexistence in this nucleus. The FMA will be used to detect evaporation residues and thus suppress the large fission background. Future experiments of this type will be possible at Yale. During the year 2000 the gas-filled separator SASSY 2 will move from

Berkeley to Yale where it will be combined with the gamma-ray detectors of YRAST Ball. It will then be called Small Angle Separator System at Yale for Evaporation Residues (SASSYER). The N.Y.P.D. will be available for experiments in conjunction with SASSYER for future experiments in regions of the nuclear chart, where the competing fission background is very large.

4.2 Neutron-rich nuclei in the mass-100 region

A currently much utilized technique to study neutron-rich nuclei is the spectroscopy of prompt or β-delayed γ-rays of fission fragments. Lifetimes of low lying states have in the past mostly been measured using β-delayed γ-rays and electronic time techniques but in a few cases the RDDS technique has also been employed [20].

The neutron-rich nuclei in the mass region around A=100-110 exhibit a variety of structural phenomena, which include shape coexistence, strong octupole correlations, the existence of low-lying intruder states, signs of triaxiality and γ-softness as well as vibrational excitations. However, very little lifetime information, which would help to classify the in structure in these nuclei is available.

RDM experiments in this region are possible with the N.Y.P.D. in conjunction with the YRAST Ball array or Gammasphere and an array of solar-cell detectors at backward angles. This technique was already used once by Mamane et al. [20] using only one Germanium detector and a thin ^{252}Cf fission source. The use of a multi-detector array will help significantly in expanding the knowledge of nuclear level lifetimes in this region[b].

4.3 RDM in inverse kinematics Coulex of radioactive ion beams

With the construction of second generation radioactive beam facilities such as the U.S. Rare Isotope Accelerator beams of exotic neutron rich isotopes will become available. One way to study these exotic nuclei is by Coulomb exciting them on a stable fixed target and performing spectroscopy using the emitted gamma radiation. From this type of data one can extract information about the level scheme as well as the transition and intrinsic matrix elements from the excitation cross sections. The reliability of the matrix elements is, however, sometimes dependent on the knowledge of all transitions going to and from a certain level as well as assumptions about the signs of the matrix elements involved. This can particularly become a problem when

---

[b]The Manchester group lead by A.G. Smith has recently performed a EUROBALL RDM experiment using a ^{252}Cf source.

multi-step Coulomb excitation plays a role. Here lifetime measurements with the RDDS technique can provide an alternative and model independent way to determine absolute transition matrix elements. The technique works by Coulomb exciting the beam in a thin, low-Z target foil. The Coulomb excited beam nuclei are then slowed in a retardation foil. Here the beam nuclei can also be Coulomb excited. One can distinguish between the excitations in the two foils since γ-rays emitted during the flight between the foils will be detected with a larger Doppler-shift than γ-rays emitted after the retardation foil. Thus spectra in coincidence with fully shifted higher lying transitions will contain only γ-rays from nuclei that were excited in the first foil. Such measurements will greatly benefit from the use of a highly efficient array of Ge-detectors such as Gammasphere.

Acknowledgements

Important contributions by R.M. Clark, A. Dewald and P. von Brentano are gratefully acknowledged. For the support of and help with the New Yale Plunger Device A. Dewald, J.R. Cooper, H. Tiesler, R. Peusquens and P. von Brentano are gratefully acknowledged. This work is in part supported by the U.S. Department of Energy under Grant No. DE–FG02–91ER–40609.

References

1. C.W. Beausang *et al.*, submitted to Nucl. Instr. Meth.
2. R. Krücken, Jour. Res. Natl, Inst. Stand. Technol., in press.
3. A. Dewald *et al.*, Z. Phys. A **334**, 163 (1989).
4. G. Böhm *et al.*, Nucl. Instr. Meth. **A 329**, 248 (1993).
5. T.L. Khoo *et al.*, Phys. Rev. Lett **76**, 1583 (1996).
6. G. Hackman *et al.*, Phys. Rev. Lett **79**, 4100 (1997).
7. A. Lopez-Martens *et al.*, Phys. Lett **B 380**, 18 (1996).
8. K. Hauschild *et al.*, Phys. Rev. C**55**, 2819 (1997).
9. R. Kühn *et al.*, Phys. Rev. C**55**, R1002 (1997).
10. R. Krücken *et al.*, Phys. Rev. C**55**, R1625 (1997) and references therein.
11. R. Krücken *et al.*, Phys. Rev. **C54**, 1182 (1996) and references therein.
12. E. Vigezzi *et al.*, Phys. Lett. **B 249**, 163 (1990).
13. P. von Brentano *et al.*, Abstract to the Symposium "New Spectroscopy and Nuclear Strucutre 1997", Copenhagen 1997.
14. A. Dewald *et al.*, Phys. Rev. C**54**, R2119 (1996).
15. R.M. Clark *et al.*, Nucl. Phys. **A562**, 121 (1993), and references therein.
16. S. Frauendorf, Nucl. Phys. **A 557**, 259c (1993).
17. R.M. Clark *et al.*, Phys. Rev. Lett. **78**, 1868 (1997) .
18. R.M. Clark *et al.*, Phys. Lett. B 440, 251 (1998).
19. R. Krücken *et al.*, Phys. Rev. C58, R1876 (1998).
20. G. Mamane *et al.*, Nucl. Phys. **A 454**, 213 (1986).

GAMMA-RAY SPECTROSCOPY STUDIES AT THE NIPNE TANDEM ACCELERATOR

D. BUCURESCU, I. CĂTA-DANIL, G. CĂTA-DANIL, M. IVAŞCU,
N. MĂRGINEAN, L. STROE AND C.A. UR

Horia Hulubei National Institute of Physics and Nuclear Engineering, Bucharest, Romania

1 Introduction

Nuclear structure studies by in-beam γ-ray spectroscopy constituted a continuous component of the work of our group since the first delivery of a heavy-ion beam by the tandem Van de Graaff accelerator of our Institute. After reviewing shortly some of the earlier results, present results and preoccupations will be shown, including some related topics from our collaboration at the GASP array (Legnaro).

2 Early results

2.1 A≈80 nuclei

We have studied a series of nuclei in the A $\sim$ 80-90 region, on the neutron deficient side (Sr, Y, Zr) as well as higher mass Tc, Ru and Rh. The Sr to Zr isotopes have been reached by reactions of ^{14}N, ^{16}O and ^{19}F beams on Ge and Se isotopically enriched targets, and medium-high spin states in these nuclei were studied by in-beam γ-ray spectroscopy methods. The results for nuclei such as ^{83}Sr [1], ^{85}Y [2], ^{86}Zr [3], indicated that the collectivity of the yrast states increases when N decreases towards the middle of the shell. Thus, it appeared that N (or Z) = 38 or 40 does not always behave as a quasi-magic number. To understand these results we performed Nilsson-Strutinsky calculations for nuclei in the A $\sim 80 - 100$ mass region [4,5]. These calculations reproduced well the sudden onset of deformation known in the $A \sim 100$ region (Sr, Zr and Mo isotopes), and also, with carefully chosen Nilsson model parameters, the nuclei in the middle of the 28 to 50 shell, and especially those with $N \approx Z =$ 38 or 40 were predicted strongly deformed, due to large deformed shell gaps of the single particle (s.p.) energy levels. Also, due to the relatively low density of s.p. levels the nuclear shape changes quite rapidly with the particle number, therefore it is a small 'island' of deformation. Interacting boson model for the neutron-deficient Sr and Zr nuclei showed similar characteristics [6].

These features of the $A \sim 80$ region made it very attractive for further experimental studies. It was indeed discovered that the $N = Z$ nuclei ^{76}Sr [7] and ^{80}Zr [8] are among the nuclei with the largest known quadrupole deformation ($\beta_2 \approx 0.40$). We shall present later some of our results for such nuclei studied with the GASP array.

3 Recent results

3.1 High spin state spectroscopy with heavy-ion induced reactions

Our present experimental setup includes four intrinsic Ge detectors (20 - 25% efficiency), several NE213 neutron detectors and one or two Silicon $\Delta E - E$ telescopes. This allows a good determination of the reaction channels and reasonable statistics for γ-γ coincidences measured in beamtimes of a few days.

Most of our recent work has been centered on odd-odd $A \sim 90$ nuclei which are not so well known. It is interesting to study such nuclei with only a few nucleons outside the closed shell $Z(N)$=50. The maximum spin that can be constructed by aligning the individual spins of the valence nucleons is not so large, so that one can easily excite higher spin states which should result from nucleons excited across the closed shell.

^{96}Tc. This nucleus has been populated in the reaction ^{82}Se(^{19}F,5nγ) at energies around 68 MeV. Gamma-rays have been assigned to this channel by coincidences with neutrons (showing a multiplicity of 5), non-coincidence with charged particles, as well as coincidence with γ-rays known from the (α,nγ) study [9]. Analysis of γ-γ coincidence matrices led to the level scheme shown in Fig. 1, where two γ-ray cascades have been added at higher spins, above the states 2601 keV, 13^+ and 2399 keV, 11^+ [9]. These cascades could be seen in our experiment up to the states assigned as $J^\pi=20^+$. Multipolarities and spin values have been deduced from the analysis of γ-ray angular distributions measured in coincidence with the neutrons.

The irregular pattern of the two quasiband structures shown in Fig. 1 is a sign of angular momentum built up by progressive alignment of the spins of individual nucleons. While lower spins can be understood, e.g., as coming from the $\pi g_{9/2} \otimes \nu d_{5/2}$ configuration, it is interesting to learn how are the higher spins constructed (excitations of other nucleons from the $g_{9/2}$ orbital across the closed shell may contribute). For this, shell model calculations are in progress [10].

^{94}Nb. This nucleus has been populated in the same reaction with ^{96}Tc, as the α3n channel. Its γ-rays have been assigned by coincidence with α-

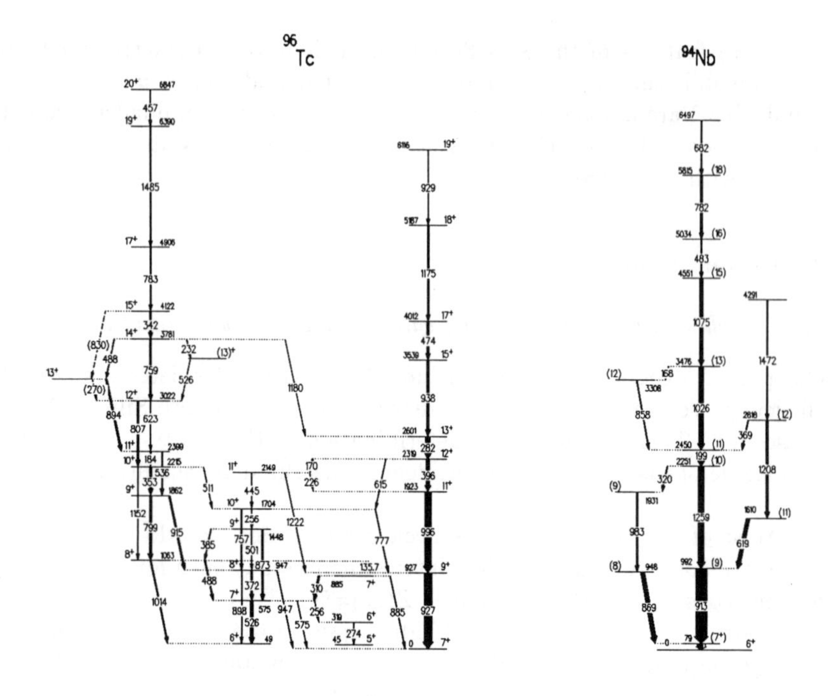

Figure 1. Level schemes of ^{96}Tc and ^{94}Nb as deduced from our experiments (see text)

particles, and with neutrons (multiplicity of 3). The γ-rays thus assigned were found in coincidence with the known transition of 78.7 keV [?] between a low-lying (7^+) state and the 6^+ ground state, so that they were placed above this state. The observed structure (Fig. 1) is, very likely, of positive parity.

^{86}Y. We have previously studied this nucleus, by the crossed reactions ^{74}Ge(^{16}O,p3nγ) and ^{76}Ge(^{14}N,4nγ), but due to the low efficiency of the γ-detectors only its yrast line above the known E_x=218.3 keV, 8^+ isomer has been proposed [11]. We have recently concluded new measurements which led to a new, much richer level scheme, which is presented in Fig. 2. The yrast line itself has been modified and extended in the upper part, and the observation of the weak transitions of 1101, 1202 and 1356 keV allowed a clear definition of both signatures of this quasiband up to J^π=15^+ (the highest

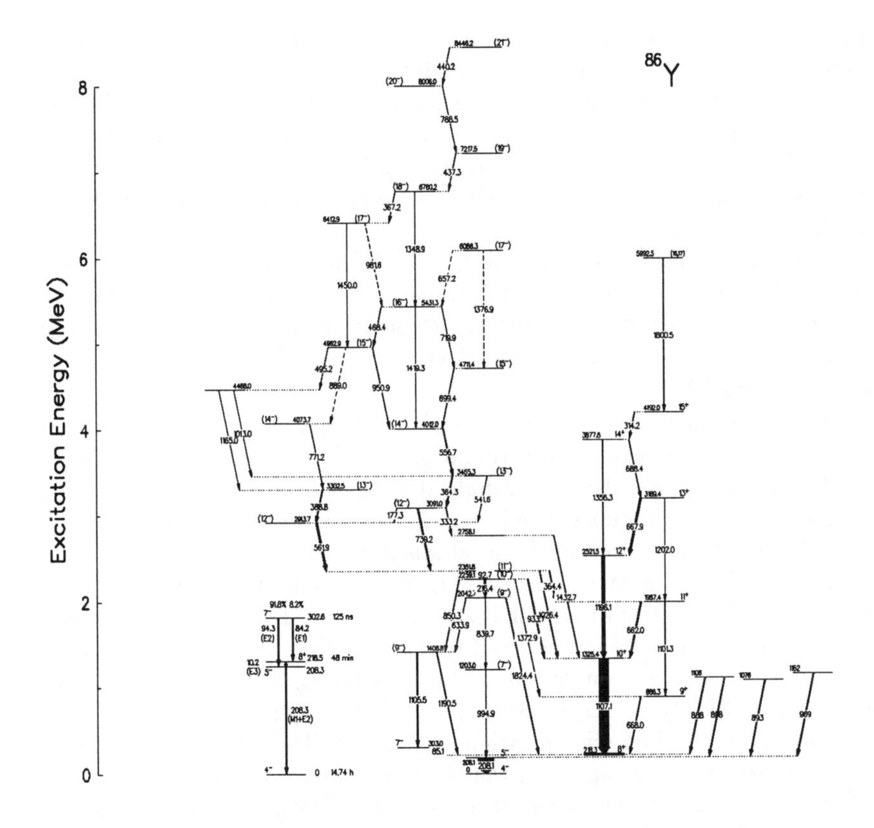

Figure 2. Level scheme of ^{86}Y as deduced from our experiments (see text)

spin that can be obtained by aligning the spins of the nucleons from the $g_{9/2}$ orbital). In addition, we have assigned a rich negative parity level structure. A new 7^- isomeric level at E_x=302.6 keV has been populated and studied with the (p,nγ) reaction [12], and it has been weakly seen also in the heavy ion reaction. The negative parity structure develops a quasi-regular ΔJ=1 structure above the E_x=3091 keV, (12^-) state, which may be an indication of magnetic rotation.

3.2 Studies of $A \approx 80$ nuclei at GASP

Due to our special interest in this mass region, we concentrated, during the last years, on the study of nuclei on the $N = Z$ line and close to it. In addition to the points of interest stated above (large deformation and rapid transition towards (probably) spherical shape), another one is the hope that the structure of these nuclei might show effects (even at low spins) due to the neutron-proton T=0 pairing. These nuclei are difficult to reach and study; with the presently available reactions they are populated with very small cross-sections, simultaneously with a huge background of other channels. Therefore we have made these studies at GASP [13], operated in coincidence with the ISIS Silicon ball.

^{84}Mo. By using GASP and the ^{58}Ni+^{28}Si(90 MeV) reaction, we have identified the second yrast $4_1^+ \rightarrow 2_1^+$ transition [15], in addition to the first one known [14]. This shows that after the strongly deformed region is already over at the $N = Z$ nucleus ^{84}Mo, which is transitional nucleus: $R_{4/2} = E(4_1^+)/E(2_1^+)$=2.52.

^{84}Nb. This nucleus was observed in the same reaction as above. Several band structures with regular, rotational aspect, have been observed up to spins of about $22\hbar$, indicating a strongly deformed nucleus [16]. These bands have been built on a complex low-spin structure, including a previously known isomeric level but not placed in the level scheme [17].

^{81}Zr. No information existed on this T_z=1/2 nucleus before our work, except for the $3/2^-$ assignment for its ground state. This nucleus has been studied as the αn channel in the reaction above. Two band structures with rotational aspect have been found, [18] and assigned the Nilsson structure $[422]5/2^+$ and $[301]3/2^-$, respectively, for wich Bogolyubov-Strutinsky TRS calculations predict a large deformation of $\beta_2 \approx 0.37$. These bands fit well into the systematics of similar bands in the N=41 isotones ^{77}Kr and ^{79}Sr.

^{80}Y. This nucleus has been studied in the ^{24}Mg + ^{58}Ni (180 MeV) reaction. Similarly to ^{84}Nb, several rotational bands have been assigned to this nucleus [19].

To conclude this section, we would like to emphasize what kind of knowledge we have accumulated by studying systematically a wide range of nuclei in the A $\sim$ 80-90 mass region. We take as an example the Yttrium isotopic chain. We can now follow the evolution of the structure of the odd-odd isotopes, from ^{80}Y to ^{86}Y. For example, the systematic of the positive parity yrast band in these nuclei, assigned to the $\pi g_{9/2} \otimes \nu g_{9/2}$ configuration shows quite regular, rotational features in the light isotopes and a more irregular one in ^{86}Y. We performed calculations with the Interacting Boson-Fermion-Fermion (IBFFM) model which account well for the features of this band in the lighter, well deformed isotopes (correct excitation energies and signature splitting), but cannot describe the signature splitting observed in the ^{86}Y isotope. This is likely due to the non-collective nature of this structure in ^{86}Y.

3.3 Lifetime measurements for nuclear excited states

We have systematically measured lifetimes of excited states in different nuclei where such data were very scarce. During the last few years we have used the Doppler shift attenuation method (DSAM) for excited states populated with the (p,nγ) reaction. Although the recoil velocity in such a reaction is very small, which implies the measurement of very small γ-ray energy shifts with the detection angle, there are nevertheless some advantages in using such a reaction. One is that the measurements can be performed reasonably close to the threshold of the state(s) of interest, which means recoiling nuclei in the forward direction and avoiding of feedings from the higher lying levels. Another is the rather non-selective population of practically all the low spin states. We have thus measured lifetimes of low-lying, low-spin states in nuclei such as ^{115,117}Sb [20], ^{111}In [21], ^{73}As [22], and ^{71}Ge [23]. In all these cases, by using all the spectroscopic data available, quite extensive checks could be made for the Interacting Boson-Fermion model (IBFM).

Here we shall illustrate these measurements with the ^{71}Ge case [23]. In this nucleus no lifetimes were known with the exception of two low-lying isomers. Previous IBFM-1 analyses of this nucleus proposed that two states, 831 keV, $3/2^-$ and 1212.5 keV, $5/2^-$, respectively, are likely "intruder" states, resulting from couplig of the valence nucleon to intruder configurations of the ^{70}Ge core, so that it was interesting to investigate better the properties of the low-lying states. The ^{71}Ge levels were populated via the ^{71}Ge(p,nγ) reaction at 3.0 and 3.5 MeV incident energies, on a thick target sticked on a cooled Ta backing. A very good stability of the spectroscopic chain was needed since it was found that all levels up to $\sim$2 MeV in this nucleus have lifetimes above 0.5 ps and therefore very small centroid shifts had to be measured (maximum $F(\tau)$ value found was 15%). Lifetimes (or upper limits) could be measured for 12 excited states of low spins up to 2.0 MeV excitation, and some δ(E2/M1) values could be determined or confirmed. The derived B-values, together with other known quantities such as static moments, γ-ray branching ratios, δ(E2/M1) mixing ratios and one-neutron transfer spectroscopic factors, could be used to make a rather detailed check of multishell IBFM-1 calculations. It was found that the calculations account very well for the properties of *all* the experimental levels up to 1.5 MeV excitation [23]. Further lifetime measurements using both the DSAM and the recoil-distance method (in the heavy-ion reactions) are planned for other medium mass nuclei.

3.4 Study of excited 0+ in even-even nuclei

The phenomenology of the excited 0^+ states in nuclei is known to be very complex, while their properties are not known too well in many nuclei, including the rare earths region. Their γ-ray decay modes, when known in detail, may give quite precise signatures concerning the nature of such states. An excellent example is that of the 2_3^+ state in ^{152}Sm whose decay gave the decisive argument for the phase coexistence phenomenon in nuclei [24]. We plan to make precise measurements of the decays of such states, both with in-beam and activation measurements. Preliminary measurements have been made with the ^{169}Tm(p,2n) reaction at 12.0 MeV, to determine the properties of three known excited 0^+ states in ^{168}Yb.

4 Conclusions

After a short review of earlier results, we have presented some of our recent γ-ray spectroscopy results in medium mass nuclei, in three main directions: γ-ray spectroscopy of high-spin states with heavy ion fusion-evaporation reactions, determinations of lifetimes of excited nuclear states, and precise determinations of γ-ray decay schemes of low-spin states (0^+, 2^+) in even-even nuclei. We think that even with modest experimental means one can still achieve meaningful results in such experiments performed at our accelerator. Moreover, such experimental studies can be well correlated with others, performed with state-of-art γ-ray spectroscopy instruments, (such as GASP, EUROBALL) and thus be extended into new nuclear regions and at higher spins.

References

1. D.Bucurescu, G.Constantinescu, M.Ivascu, N.V.Zamfir, M.Avrigeanu, J. Phys. **G7**,399(1981)
2. D.Bucurescu, G.Constantinescu, M.Ivascu, N.V.Zamfir, M.Avrigeanu, D.Cutoiu, J. Phys. **G7**,667(1981)
3. M.Avrigeanu, V.Avrigeanu, D.Bucurescu, G.Constantinescu, M.Ivaşcu, D.Pantelica, M.Tanase, N.V.Zamfir, J. Phys. **G4**,261(1978)
4. D.Bucurescu, G.Constantinescu, M.Ivascu, Rev. Roum. Phys. **24**,971(1979)
5. D.Galeriu, D.Bucurescu, M.Ivascu, J. Phys. **G12**,329(1986)
6. D.Bucurescu, G.Cata, D.Cutoiu, G.Constantinescu, M.Ivascu, N.V.Zamfir, Nucl. Phys. **A401**,22(1983)
7. C.J.Lister et al., Phys. Rev. **C42**,R1191(1990)
8. C.J.Lister et al., Phys. Rev. Lett.**59**,1270(1987)

9. H.A.Mach, M.W.Johns, J.V.Thompson, Can. J. Phys. **58**,174(1980)
10. K.Ogawa, private communication (1999)
11. D.Bucurescu, G.Constantinescu, D.Cutoiu, M.Ivascu, N.V.Zamfir, A.Abdel-Haliem, J. Phys. **G10**,1189(1985)
12. D.Bucurescu et al., IPNE, Ann. Rep., Dep. Heavy Ion Physics (1984-1985), p.21; M.Ionescu-Bujor et al., ibidem (1992-1994), p. 7; M.Ionescu-Bujor and A.Iordăchescu, NIPNE Ann. Rep. (1998)
13. D.Bazzacco, in *Proceedings of the International Conference on Nuclear Structure at High Angular Momentum*, Ottawa, 1992, Report No. AECL 10613, Vol. II, p.376
14. W.Gelletly et al., Phys. Lett. **B253**,287(1991)
15. D.Bucurescu, C.Rossi-Alvarez, C.A.Ur, N.Mărginean, P.Spolaore, D.Bazzacco, S.Lunardi, D.R.Napoli, M.Ionescu-Bujor, A.Iordachescu, C.M.Petrache, G.de Angelis, A.Gadea, D.Foltescu, F.Brandolini, G.Falconi, E.Farnea, S.M.Lenzi, N.H.Medina, Zs.Podolyak, M.De Poli, M.N.Rao, R.Venturelli, Phys. Rev. **C56**,2497(1997)
16. N.Mărginean, D.Bucurescu, C.A.Ur, D.Bazzacco, S.M.Lenzi, S.Lunardi, C.Rossi Alvarez, M.Ionescu-Bujor, A.Iordăchescu, G.de Angelis, M.De Poli, E.Farnea, A.Gadea, D.R.Napoli, P.Spolaore, A.Buscemi, Eur. Phys. Journal **A4**,311(1999)
17. P.H.Regan et al., Acta Phys. Pol. **28**,431(1997)
18. N.Mărginean, D.Bucurescu, C.A.Ur, D.Bazzacco, S.Lunardi, S.M.Lenzi, C.Rossi Alvarez, G.de Angelis, A.Gadea, D.R.Napoli, M.De Poli, P.Spolaore, Phys. Rev. **C**, in press (1999)
19. D.Bucurescu, C.A.Ur, D.Bazzacco, C.Rossi-Alvarez, P.Spolaore, C.M.Petrache, M.Ionescu-Bujor, S.Lunardi, N.H.Medina, D.R.Napoli, M.De Poli, G.de Angelis, F.Brandolini, A.Gadea, P.Pavan, G.F.Segato, Zeit. Phys. **A352**,361(1995)
20. D.Bucurescu, I.Cata-Danil, G.Ilas, M.Ivascu, L.Stroe, C.A.Ur Phys. Rev. **C52**,616(1995)
21. D.Bucurescu, I.Cata-Danil, G.Ilas, M.Ivascu, N.Marginean, L.Stroe, C.A.Ur, Phys. Rev. **C54**,2313(1996)
22. D.Bucurescu, I.Căta-Danil, M.Ivaşcu, N.Mărginean, L.Stroe, C.A.Ur, N.Dinu, Intern. J. of Modern Physics **E8**,17(1999)
23. M.Ivaşcu, N.Mărginean, D.Bucurescu, I.Căta-Danil, C.A.Ur, Phys. Rev. **C60**,024302(1999)
24. R.F.Casten et al, Phys. Rev. **C57**,R1553(1998); F.Iachello, N.V.Zamfir, R.F.Casten, Phys. Rev. Lett. **81**,1191(1998)

SHAPE COEXISTENCE PHENOMENA IN MEDIUM MASS NUCLEI

A. PETROVICI
National Institute for Physics and Nuclear Engineering - Horia Hulubei, R-76900 Bucharest, Romania
E-mail: spetro@ifin.nipne.ro

The theoretical description of some coexistence phenomena identified at low and high spin states in even mass $A \simeq 70$ nuclei within the *complex* versions of the Excited Vampir and Fed Vampir variational approaches is presented. The coexistence and mixing of oblate and prolate deformed configurations at low spin could explain the appearance of the irregular sequences of states and the isomeric decays found at low excitation energy in even-even nuclei. At higher spins the coexistence of states with different deformations causes complicated multiple band structures and strong fragmentation of the B(E2) strengths. It also influences the occurence of almost identical bands at normal deformation. Furthermore, the shape coexistence suggests a possible mechanism to produce high spin isomeric states in odd-odd nuclei. Our investigation indicates that particular neutron-proton matrix elements of the effective interaction play an essential role for the shape mixing.

1 Introduction

The medium mass nuclei in the $A \simeq 70$ mass region manifest a rich variety of shapes and rapid changes in structure with particle number, angular momentum and excitation energy. These effects are generally believed to be caused by the occurence of large gaps in the single-particle spectra at different deformations which lead to a strong competition between the different many-body configurations based on the corresponding deformations. Particular strong effects are expected in the self-conjugate N=Z nuclei. Here, in addition, a strong competition between or even simultaneous occurence of neutron and proton alignment are to be expected. One also expects that neutron-proton pairing correlations play a significant role in N$\simeq$Z nuclei, but very little is known about their nature.
For a unified description of the complex behaviour of these nuclei at low as well as at high angular momenta one needs a model which can account for the delicate interplay between collective and single particle degrees of freedom and does not rely on educated guesses of the various underlying structures. Furthermore, the like-nucleon and neutron-proton pairing correlations should be treated on the same footing and the handling of realistic model spaces as well as general two-body forces should be numerically feasible. These requirements are fulfilled by the *complex* Excited Vampir and Excited Fed Vampir

approaches. These models are based on chains of variational calculations using symmetry projected Hartree-Fock-Bogoliubov (HFB) vacua which account for neutron-proton pairing and unnatural-parity corelations. Using essentially *complex* HFB transformations these models allow to account for all kinds of nucleon-nucleon correlations. Thus the HFB vacua are constructed from like as well as unlike two nucleon pairs coupled to arbitrary spin and parity quantum numbers even though time reversal and axial symmetry are still imposed on the underlying HFB transformations. In each nucleus the lowest few states of a given spin and parity are approximated within the *complex* EXCITED (FED) VAMPIR approach using a rather large model space. The variational calculations for all the investigated states are independent from each other and thus arbitrary drastic changes in structure with increasing angular momentum (e.g., the shape) and excitation energy (e.g., shape coexistence and via the final diagonalization also shape mixing) can be accounted for.
Since the Vampir approaches allow the use of rather large single particle basis systems and of general two-body interactions, *large-scale* nuclear structure studies going far beyond the abilities of the conventional shell-model configuration-mixing approach have become possible. Furthermore, since these approaches provide detailed spectroscopic information for the lowest few states for each spin and parity, they also have considerable advantages with respect to the shell-model Monte Carlo method in which only expectation values of operators in the statistical ensemble or the ground state can be calculated.
In the last years these approaches have been used extensively for the study of both even-even as well as odd-odd nuclei in the $A \simeq 70$ mass region and in many cases an almost quantitative description of the changes in structure with angular momentum and excitation energy could be achieved. Some of these results will be reviewed in the following.
Though many of the theoretical results have been confirmed experimentally, one has to stress that the theoretical predictions do strongly depend on the choice of the effective interaction and thus contain a substantial amount of uncertainty. So, e.g., the onset of deformation does depend sensitively on the strength of the isoscalar neutron-proton interaction and the relative importance of like-nucleon and neutron-proton pairing can be influenced strongly by manipulating the strength of the interaction in the corresponding particle-particle channels. This, however, is not only a drawback but also a chance to learn something valuable on the effective interaction in this mass region.
In the next section some recent results on shape coexistence and shape transition at low and high spins in even-even and odd-odd nuclei will be discussed. Conclusions are drawn in the last section.

2 Results and discussion

For all the results presented here a ^{40}Ca core was assumed and the model space restricted to the $1p_{1/2}$, $1p_{3/2}$, $0f_{5/2}$, $0f_{7/2}$, $1d_{5/2}$ and $0g_{9/2}$ oscillator orbits for both protons and neutrons. As effective two body interaction a renormalized nuclear matter G-matrix derived from the Bonn one boson exchange potential (Bonn A) is used.
As already mentioned in the introduction, due to the occurence of large gaps in the deformed single particle spectra one expects rapid changes in structure with the nucleon number. This effect should be strongest in N=Z nuclei, since there the gaps occur simultaneously for protons and neutrons and even may be enhanced by the strong residual neutron-proton interaction.

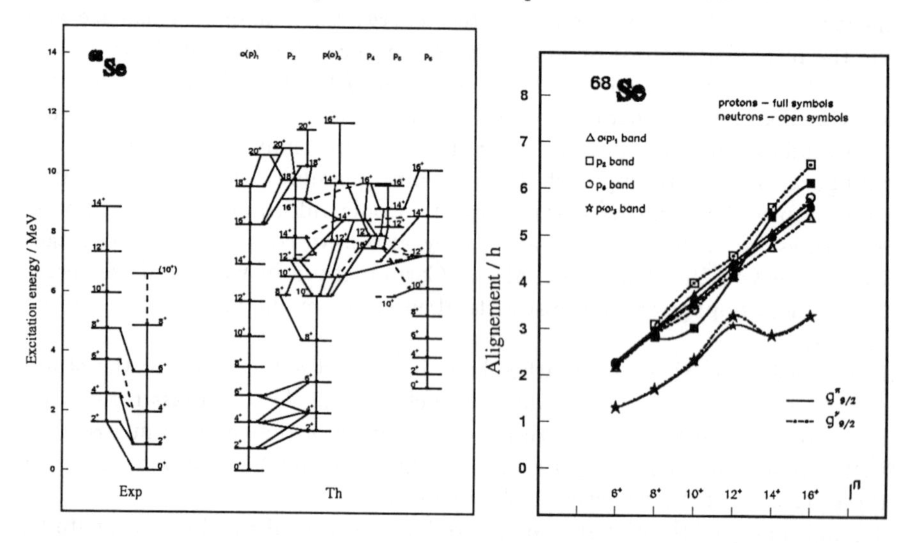

Figure 1. The theoretical spectrum of ^{68}Se for even spin positive parity states calculated within the *complex* EXCITED VAMPIR approximation is compared to the experimental results[2]. The labels o_i and p_i are for states based on intrinsically oblate and prolate deformed configurations, respectively. Significant $M1, \Delta I = 0$ transitions are indicated by dashed lines. The alignment plot is given for selected theoretical bands.

We investigated the shape transition in the N=Z chain of even-even nuclei from selenium to molybdenum[1]. In ^{68}Se and ^{72}Kr nuclei the first minima for spins 0^+ and 2^+ are oblate deformed in the intrinsic system and the first excited minimum is prolate deformed as can be observed from Figs. 1 and 2. Between N=Z=36 and N=Z=38 one obtains a shape transition from oblate to prolate deformation in the ground-bands. While in ^{68}Se and ^{72}Kr two almost degenerate and strongly mixed oblate and prolate bands are coexisting, these bands change their relative position in ^{76}Sr and the mixing almost vanishes.

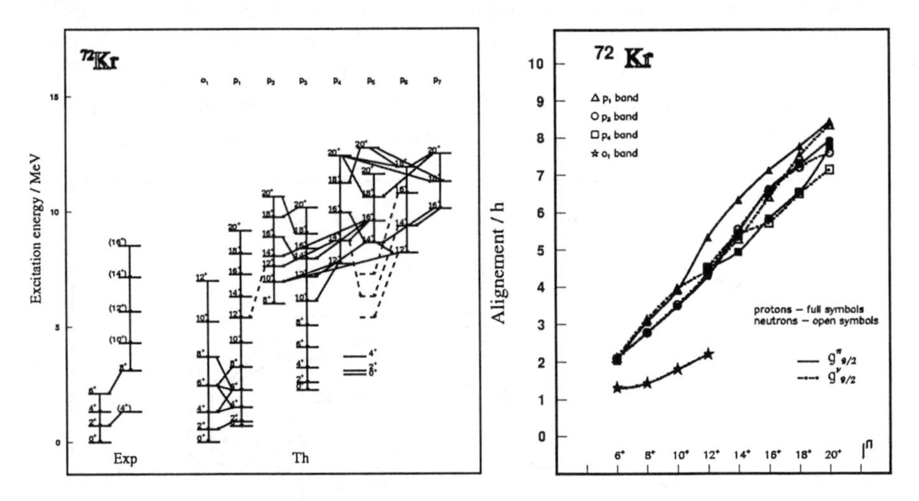

Figure 2. The same as in Fig. 1, but for ^{72}Kr nucleus. The experimental data are taken from ref.[3].

Even in ^{72}Kr the yrast band becomes prolate deformed above angular momentum 8^+. Strongly deformed prolate states have been found in the ground bands of ^{76}Sr and ^{80}Zr nuclei, with side bands very high in energy and decreased prolate quadrupole deformation of the ground state band for the nucleus ^{84}Mo. The maximum deformation in the region was found for the systems with $N \simeq Z = 38(40)$. The hexadecapole deformation is not too large, but is changing moving from one system to the other in the investigated chain of nuclei. Almost no octupole deformation was found.
As it is suggested by the theoretical spectra of ^{68}Se and ^{72}Kr presented in Figs. 1,2 at intermediate spins a high density of states was obtained after diagonalizing the residual interaction between the Excited Vampir solutions for each spin. The strong mixing and the high level density for certain angular momenta yield a rather complicated decay pattern. Many significant E2 decaying branches have been obtained for particular intermediate and high spin states. The decay pattern of the high spin states becomes even more complex due to the presence of strong $\Delta I = 0$, M1 transitions which are indicated in Figs. 1,2 by dashed lines. Some of the calculated states even cannot be attached to any particular band structure and at intermediate spins some of the excited bands fade away into other structures.

Recently was reported the identification of an isomeric decay in ^{74}Kr[4], interpreted as the E0 decay from an excited 0^+ state, confirming our prediction on high deformation prolate-oblate shape coexistence and mixing at low spins[5,6]. A different scenario based on shape mixing was found as a possible mechanism to populate isomeric states at intermediate spins in the odd-odd nucleus ^{68}As. In this case the different behaviour of the high spin negative parity states with respect to the low spin ones could cause the appearance of an isomeric state[7].

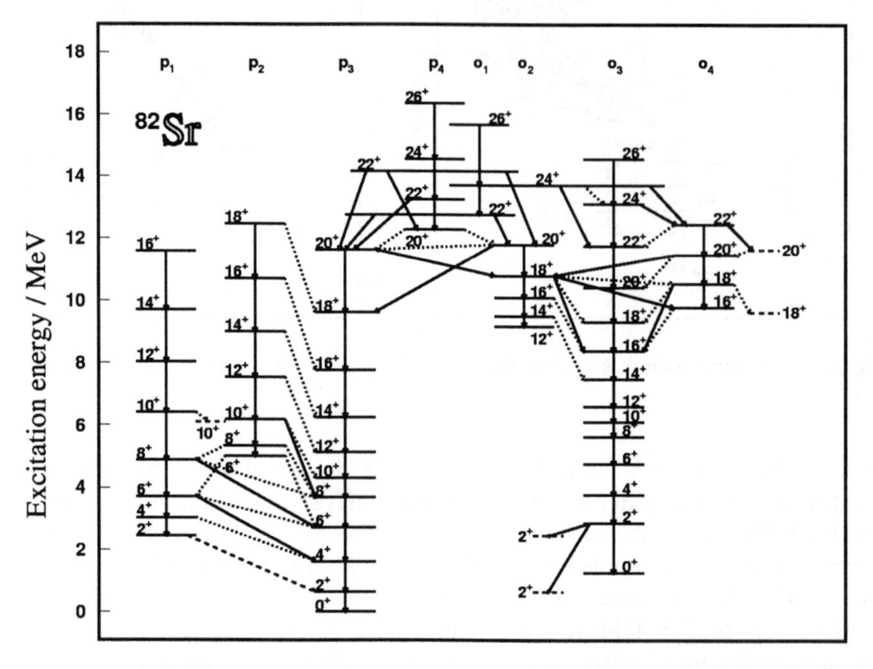

Figure 3. The theoretical spectrum of ^{82}Sr for even spin parity states. Significant M1, $\Delta I = 0$ transitions are indicated by dotted lines.

In the nucleus ^{82}Sr the prolate-oblate coexistence seems to persist up to intermediate spins, while for the high spin states in normally deformed bands the oblate configurations seem to be energetically favoured with respect to the prolate ones. The transition quadrupole moments of some normally deformed bands show a decrease at the highest populated spins. This fact could be interpreted as band termination. Our microscopic investigations offer an alternative explanation. The Excited Vampir results[8] indicate at high angular momenta strong mixing especially between the various oblate bands and a rather high level density (see Fig. 3). This leads to a strongly fragmented stretched B(E2) decay and may explain the trends observed experimentally.

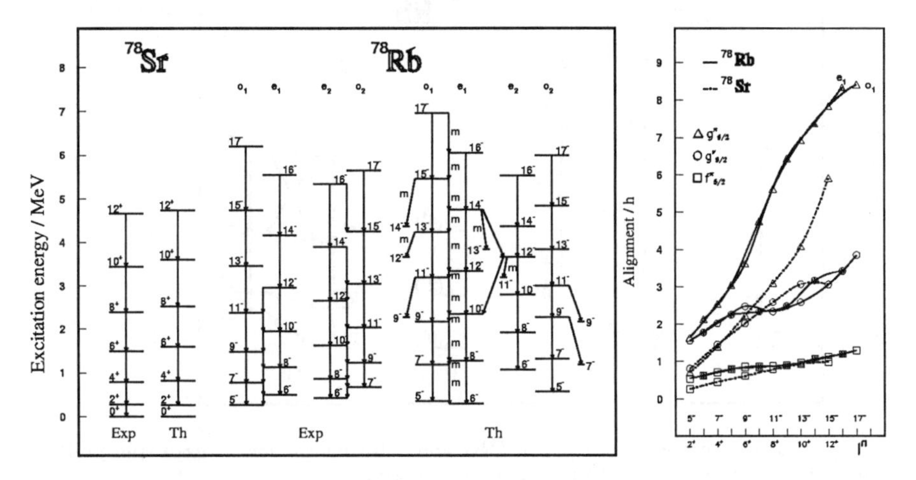

Figure 4. The theoretical spectrum of ^{78}Sr for even spin positive parity states and the lowest odd- and even-spin negative parity bands in ^{78}Rb are compared with the experimental results. The alignment plot for the ground-state band in ^{78}Sr is compared with those for the odd spin sequence o_1 (open symbols) and the corresponding even spin sequence e_1 (full symbols) in ^{78}Rb.

Many nuclei in the $A \simeq 70$ mass region exibit shape coexistence phenomena and changes in spin manifest themselves in quite dramatic changes in shape sometimes, too. In this context the occurence of identical bands in ^{78}Sr and ^{78}Rb[9] as well as the nearly identical bands in ^{74}Kr and ^{74}Rb[10] becomes a challenge for the theoretical models.

We found[1] that the ground-state band in ^{78}Sr is strongly prolate deformed, the side bands are very high in energy with respect to the yrast one and no shape mixing effects are present. A completely different picture is obtained for the negative parity bands in ^{78}Rb[5,11]. The Excited Vampir states are bunched in a small excitation energy interval for each spin and strongly mixed for particular values of angular momentum. Strong similarities appear between the calculated ground state band in ^{78}Sr and the most deformed negative parity band in ^{78}Rb (the odd spin sequence o_1 and its even spin analogue e_1 in Fig. 4). However, for higher angular momenta this similarity is deteriorating because of the configuration mixing obtained for the Rb bands.

A particular picture is obtained for ^{74}Kr and ^{74}Rb nuclei: the oblate and prolate minima are almost degenerate and strongly mixed up to spin 8^+ for both nuclei[5,6,12]. The oblate-prolate mixing disappears above spin 8^+, but the

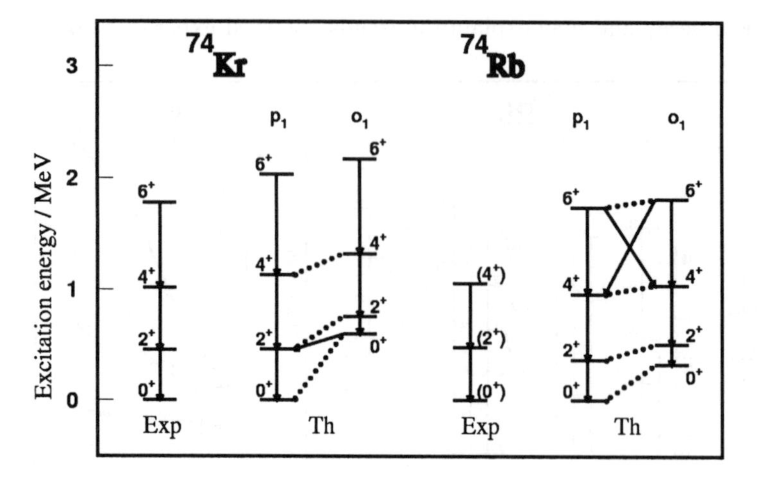

Figure 5. The theoretical spectrum of ^{74}Kr and ^{74}Rb for the lowest even spin positive parity states are compared with the experimental results.

presence of other prolate minima close in energy with the first one deteriorates the similarity of the investigated states for the two nuclei with increasing spin. It is worthwhile to mention that particular monopole shifts in the neutron-proton T=0 matrix elements have a strong influence on the scenario for feeding the yrast line at low spins in nuclei dominated by oblate-prolate coexistence. On the other hand neutron-proton pairing correlations play an essential role in the structure of $N \simeq Z$ nuclei. Recent results[13,14] indicate the dominant role played by the isovector neutron-proton pairing correlations in the N=Z nuclei and the reduction of their importance with increasing neutron excess. Furthermore, the neutron-proton correlations play an essential role for the alignment of particular bands in odd-odd nuclei[14].

3 Conclusions

We presented recent microscopic results concerning the shape coexistence phenomena in medium mass nuclei. Over all, the theoretical interpretation of the complex spectroscopy of the nuclei in the $A \simeq 70$ mass region within the *complex* versions of the Excited Vampir and Excited Fed Vampir approaches has been rather successfull. Many of the rapid changes in structure with angular momentum, excitation energy and nucleon numbers observed in these nuclei could be described rather well within these theoretical approaches. The

dominant feature in this mass region seems to be the coexistence of configurations with different shapes at comparable excitation energy. This coexistence is responsible for all kinds of shape mixing effects and causes rather complicated fragmented decay patterns.
In order to achieve a deeper understanding of the fascinating phenomena encountered in this mass region there still remains a tremendeous amount of work to be done, both theoretically as well as experimentally. However, it should not be forgotten that the theoretical methods used here are present "state of the art". We are convinced that their results which have been reviewed here provide some valuable information on the way towards a better understanding of the complex nuclear structure phenomena encountered in the medium mass nuclei of the $A \simeq 70$ mass region.

References

1. A. Petrovici, K.W. Schmid, A. Faessler, *Nucl. Phys.* A **605**, 290 (1996).
2. C.J. Lister *private communication.*
3. G. de Angelis *et al*, *Phys. Lett.*. B **415**, 217 (1997).
4. C. Chandler *et al*, *Phys. Rev.* C **56**, R2924 (1997).
5. A. Petrovici, K.W. Schmid, A. Faessler, *Progr. Part. Nucl. Phys.* **38**, 161 (1997).
6. A. Petrovici, K.W. Schmid, A. Faessler, *Nucl. Phys.* A *accepted for publ.*.
7. A. Petrovici *et al*, *Phys. Rev.* C **53**, 2134 (1996).
8. C.-H. Yu *et al*, *Phys. Rev.* C **57**, 113 (1998).
9. C. Gross *et al*, *Phys. Rev.* C **49**, R580 (1994).
10. D. Rudolph *et al*, *Phys. Rev. Lett.* **76**, 376 (1996).
11. A. Petrovici, K.W. Schmid, A. Faessler, *Z. Phys.* A **359**, 19 (1997).
12. A. Petrovici *et al*, *Progr. Part. Nucl. Phys.* **43**, 485 (1999).
13. A. Petrovici, *J. Phys* G: Nucl.Part.Phys. **25**, 803 (1999).
14. A. Petrovici, K.W. Schmid, A. Faessler, *Nucl. Phys.* A **647**, 197 (1999).

NEUTRON PRE-EMISSION AT THE FUSION OF ^{11}LI HALO NUCLEI WITH LIGHT TARGETS

M.PETRASCU

H.Hulubei National Institute of Physics and Nuclear Eng. POB MG6, Bucharest, Romania

E-mail: mpetr@ifin.nipne.ro

The fusion of ^{9,11}Li with Si and C targets has been experimentally investigated. The neutron spectra in the 0.5-22.5 MeV energy range, corresponding to ^{9}Li and ^{11}Li projectiles, have been measured by the aid of the RIKEN "long neutron wall". In the case of ^{9}Li the neutron spectrum is in good agreement with the Monte Carlo calculated fusion-evaporation spectrum. In the case of ^{11}Li the neutron spectrum is showing that an important number of fusions, are preceded by the breakup of one or two halo neutrons. A model for the neutron pre-emission probability estimation has been worked out. Good qualitative agreement with the experimentally measured value has been obtained. The recent results showing that the neutrons within the observed "narrow component" of the position spectra are emitted in pairs, will also be presented.

1 Introduction

The neutron halo nuclei characterized by very large matter radii, small separation energy and small internal momentum of the valence neutrons, were discovered by Tanihata[1] and co-workers. The designation of "halo" for the low density matter around the core was first introduced in ref.[2]. Until now, the halo nuclei were investigated mostly by elastic, inelastic scattering and breakup processes (e.g., see[3,4]). As concerns fusion with halo nuclei, one has to mention two experiments that were performed up to now, both using ^{11}Be as projectile, and heavy targets[5,6], ^{238}U and ^{209}Bi. A recent review on fusion with unstable nuclei is presented in ref.[7]. As far as it is known, no other experiment on fusion of a halo nucleus with a light target, except the one to be presented here, has been accomplished until now.

It was recently predicted[8], that due to the very large dimension of ^{11}Li, one may expect, that in a fusion experiment on a light target, the valence neutrons will not be absorbed together with the ^{9}Li core, but will be emitted in the early stage of the reaction process. The experiment aiming to check this expectation, was performed at RIKEN-RIPS facility.

In the following, the results obtained in the neutron pre-emission probability determination will be outlined. Some recent results on neutron pair pre-emission will be presented afterwards.

2 The neutron pre-emission probability determination

The experimental arrangement is described in ref.[9-14] This experimental approach turns out to be especially suitable for the case of low intensity radioactive beams.

The neutrons resulting from the Si(^{9,11}Li, fusion) were measured by the time of flight technique. The position on the "wall" of the detected neutrons, could be also determined.

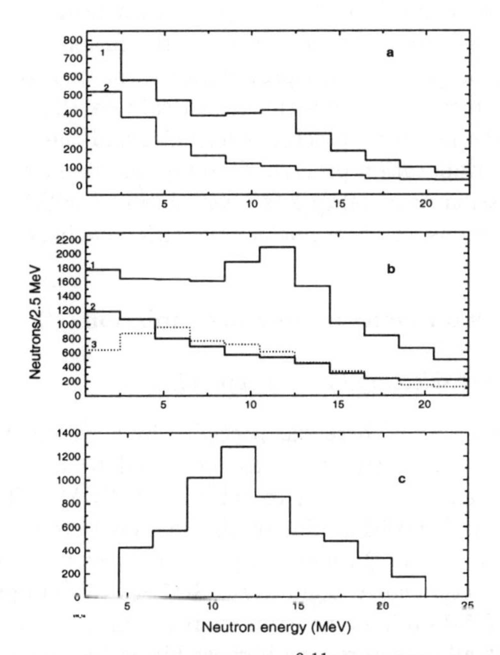

Figure 1. The neutron energy spectra from the ^{9,11}Li+Si fusion reaction for the 11.2-15.2 projectile energy range. (a): The histograms 1 and 2 represent the measured neutron energy spectra corresponding to Si(^{11}Li, fusion) and to Si(^{9}Li, fusion) respectively. (b): The histograms 1 and 2 represent the spectra from (a) corrected by the neutron detection efficiency. Histogram 3 represents the Monte Carlo calculated neutron evaporation spectrum corresponding to Si(^{9}Li, fusion). (c): The differences between spectra 1 and 2 (Fig.1b), after normalisation of both spectra in the 0.5-6.5 MeV energy range.

The measured neutron spectra for ^{11}Li and ^{9}Li, normalised to the incident beam, are shown in Fig.1a. One can see in Fig.1a, that there is a marked difference between the two spectra. This difference is better visible, when the energy dependent efficiencies are introduced, as it is shown in Fig.1b. In Fig.1b, the Monte Carlo calculated evaporation neutron spectrum, showing good agreement with the spectrum corresponding to ^{9}Li, is also represented. Therefore the difference in shape of the spectrum in the case of ^{11}Li projectiles is to be understood by the contribution of a large amount of pre-emission processes.

In Fig.1c, the differences between the neutron spectra due to ^{11}Li and ^{9}Li are shown. The sum of all counts in Fig.1c, corrected by a 10% background due to the cross-talk between the neutron detectors, divided by the total number of fusion events produced by ^{11}Li projectiles, leads to the contribution of the pre-emission of one or two neutrons in the fusion process. According to ref.[12], the neutron pre-emission probability turns out to be $\sim(40\pm12)\%$. The assigned error includes besides the statistical one, the uncertainty in the determination of the neutron detector efficiencies and the uncertainty in the subtraction of the evaporation process. Good qualitative agreement with the measured pre-emission probability has been obtained within a sharp cut-off model[12–14]. Predictions for other nuclear systems have been made in ref.[15,16]

3 Recent results on neutron pair pre-emission

3.1 The criteria for selecting the n-n coincidences

The analysis to be described here was preceded by a systematic investigation of the neutron position distribution, as measured by all 15 front neutron detectors of the "wall". According to ref.[10,11] (FWHM$\approx$ 20 chan. for the narrow neutron group), position bins of 20 channels were considered for this investigation. In Fig.2 , the 15 neutron position spectra for Si(^{11}Li, fusion), are shown. These spectra correspond to the 8-15 AMeV ^{11}Li projectile energy range, and to the 6.5-14.5 MeV neutron energy range. In this figure one can see that for almost all detectors, the highest bin is the central one, situated at 0 position. This central bin is highest for detectors #7 and #8 of the wall. Next are the symmetric detectors #6 and #9, being somewhat lower than detectors #7 and #8. Due to the fact that the total width corresponding to these 4 detectors is comparable to the width of the narrow neutron group, they were all included in the group of central detectors for which the coincidence effect has been studied. The position distribution spectrum corresponding to these four central detectors is shown in Fig 3 . In building this spectrum, the detection efficiencies were taken into account. More specific one may say, that spectrum of Fig.3 is the spectrum of incoming neutrons on the detector, while the spectra of Fig.2 are the spectra of detected neutrons.

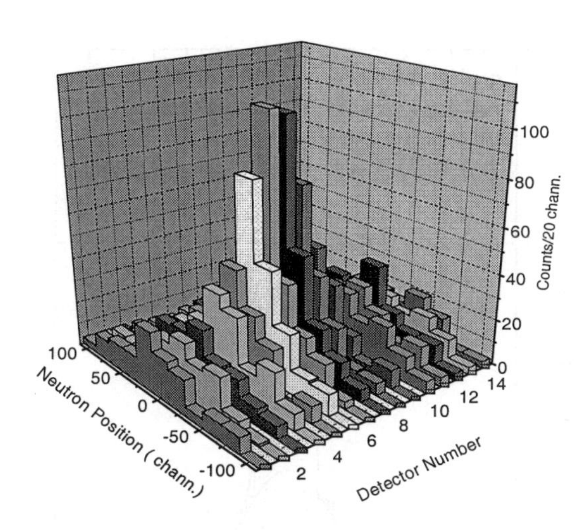

Figure 2. The neutron position distribution spectra in bins of 20 channels, measured by the 15 front detectors of the "RIKEN Long Neutron Wall". Reaction: Si(^{11}Li, fusion). Beam energy range: 8-15 AMeV. Neutron energy range: 6.5-14.5 MeV.

One can see that the ratio R of the narrow group area to the total area, in Fig.3 is R=0.35. This value is important for subsequent discussion. Detectors #1-#4 and #12-#15 were considered for the cross-talk determination. The problem was to check if for the group of central detectors, there are present coincidences beyond the cross-talk background.

Two kind of coincidences between the detectors of any of the groups were considered: a) coincidences between adjacent detectors named in the following as "first order coincidences" and b) coincidences between detectors separated by one detector, named in the following as "second order coincidences".

The following criteria were adopted for selecting the coincidences:

1. The arrival time of both detected neutrons to be within the same gate generated by the trigger of the experiment.
2. The energy of both detected neutrons to be within the 6.5-14.5 MeV

energy range.
3. The position of both detected neutrons to be within the n_i-n_i+20 channels range.

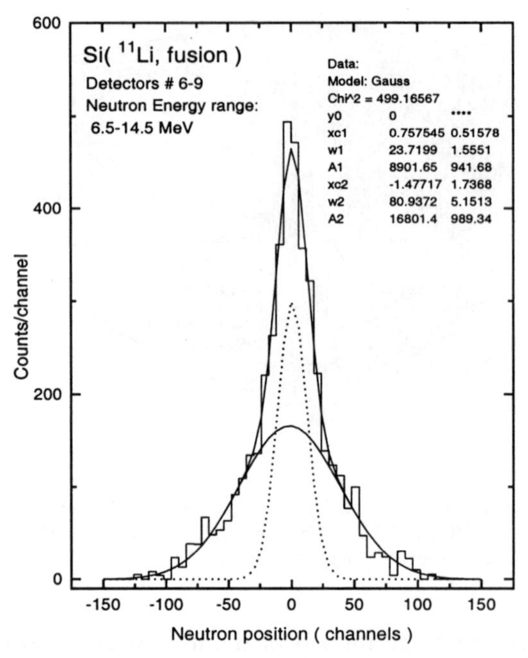

Figure 3. The neutron position spectrum corresponding to the central #6-#9detectors (1 chan.=1.1 cm). Reaction: Si(^{11}Li, fusion). Beam energy range: 8-15 AMeV. Neutron energy range: 6.5-14.5 MeV. The neutron detection efficiencies were applied.

3.2 The position spectrum of coincidences for the central #6-#9 detectors

In Fig.4a, the position spectrum for all neutrons recorded by the detectors #6-#9, in the 6.5-14.5 MeV energy range, is represented.

In Fig. 4b, the position spectrum of the total number of detected coincidences (first and second order) is represented by solid line. By dotted line, the cross-talk background (see ref.[13–14]) is represented.

In Fig.4c the second order coincidences for the central detector group (#6-#9) (solid line) together with the calculated cross-talk background (dotted line) are shown.

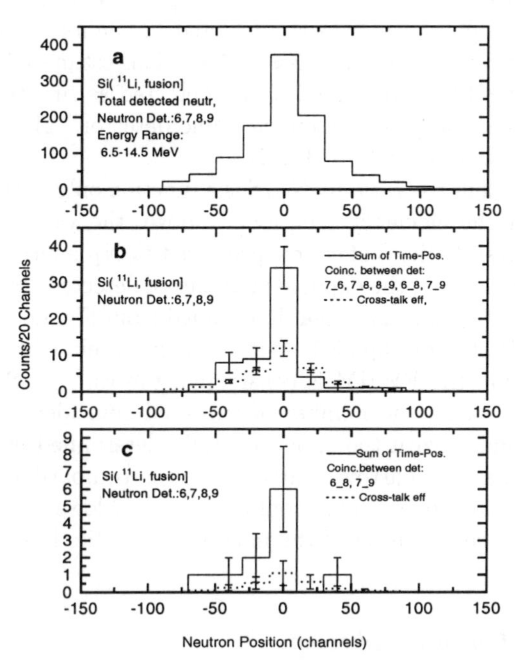

Figure 4. a: The position spectrum of all neutrons detected by detectors #6-#9, in the 6.5-14.5 MeV energy range. b: The position spectrum of the total number of detected coincidences (first and second order). The dotted line represents the cross-talk background. c: The second order coincidences detected by the cental #6-#9 detectors. The cross-talk background is represented by the dotted line.

One can see in Fig.4b, that the number of coincidences (first and second order) in the bin at zero position is far beyond the cross-talk background. After subtraction of the background the net number of counts is 21±6.2. This number is significant against 0 at a 99% confidence level.

From Fig.4c, after subtraction of background it follows that the net number of second order coincidences is significant against 0 at ~ 95% confidence level.

3.3 Estimation of expected number of coincidences

Let consider a number N_P of neutron pairs falling on a neutron detection system. The number of detected single neutrons is: $N_S= 2\epsilon(1-\epsilon)N_P$, in which ϵ is the neutron detection efficiency. The mean value of the detection efficiency is $\epsilon = 0.21$, in the case of the "long neutron wall" detector, and for the energy range considered here. The number of detected pairs is given by: $N_C=N_P.\epsilon^2$ In the case all these pairs are properly detected (2 counts for one pair), the total statistics S of counts will be $S=N_S +2*N_C$. It is estimated for the "long neutron wall" that ~1/2 of coincidences are detected as singles. This occurs always when the neutron pair falls on the same detector. Therefore the real statistics of counts will be $S=N_S +3/2*N_C$.

One can see that the ratio $p=N_C/S$ is a function only of the ϵ efficiency. With the value of the efficiency mentioned above, the value of p is 0.055. Consequently, for a statistics S of counts produced by N_P neutron pairs falling on the "long neutron wall", the expected number of detected coincidences will be 0.055*S. The statistics S has been determined from Fig.4a, by taking into account the ratio R of paragraph 3.1. With correction related to the fact that only statistics within the FWHM is considered, the value of S turns out to be 258. With this value, the number of expected coincidences is ~15. By comparing this number with the experimentally determined one, 22.1± 6.2, one could assert that the narrow neutron group is compatible with a 100% composition of neutron pairs. One could establish a lower limit of at least ~70% neutron pair composition of the narrow group, the level of confidence being ~95%.

3.4 Discussion of the obtained results

The coincidence results of Fig.4b are compatible with a 100% neutron pair composition of the narrow neutron group. Since this group cannot be explained by the aid of the Cluster Orbital Shell Model Approximation (COSMA)[19] , one may emphasize that the narrow distribution could be an effect of quantum interference[20] between the highly coherent two neutrons pre-emitted in the fusion process. In other break-up processes, the coherence of the two neutrons removed from ^{11}Li is lost, because the main reaction mechanism is the two step process via the intermediate resonances[21] in ^{10}Li. It appears that fusion experiments could be a way to prepare coherent neutron pairs, since due to the very short time in which the 2 neutrons are separated from the colliding ^{9}Li core-target, any resonances in ^{10}Li are avoided. An interesting point that has to be underlined is that the quantum interference experiments could be an alternative way to determine the size of the region[20] from which the halo neutrons are expelled.

The next experiment[22] on n-n coincidences will be performed at by the aid of an array[17] detector. This detector will allow to investigate ~5 times more coincidences in comparison with the "long neutron wall" (2 times more, due to the geometry of the array detector and 2.5 times more, due to the higher efficiency for coincidence detection). By using also a Si target 2 times thicker, one is expecting to detect 10 times more coincidences per day than in the previous experiment.

This new experiment, will be performed by using also a considerably improved in resolution MUSIC chamber[18], for the separation from the background of the evaporation residues inclusive spectra.

A further possibility is related to the fact that a reaction to produce a higher yield of neutrons in the narrow group is C(^{11}Li, fusion). Preliminary experimental results on this reaction[23] are indicating that about two times more neutrons in the narrow group than in the case of Si(^{11}Li,fusion) are present. These results are supported by a sharp cut-off calculation.[24]

4 Summary

A novel effect of neutron pre-emission in the fusion of ^{11}Li halo nuclei with Si targets, confirming a prediction earlier made, has been observed. The precise measurements of neutron pre-emission probabilities, compared with results of calculations, are an alternative way to get information upon the size of the neutron halos.

In the position distribution of the pre-emitted neutrons, a very narrow group has been found. An investigation of neutron-neutron coincidences revealed that the narrow neutron group consists predominantly of neutron pairs.

It follows that a posibility to explain the narow group, is by the quantum interference of the pre-emitted coherent neutron pairs.

Exciting questions are expected to be answered in future experiments.

Acknowledgements

The results presented here are based on a common work with:
I. Tanihata, T. Kobayashi, H. Petrascu, A. Korsheninnikov, Zs. Fülöp, H. Kumagai, K. Morimoto, A. Ozawa, T. Suzuki, F. Tokanai, K. Yoshida, H. Wang, (RIKEN),
A. Isbasescu, C. Bordeanu, M. Giurgiu, I. Lazar, I. Mihai, G. Vaman, (Bucharest),

E. Nikolski (Moskow).
The author wishes to express his gratitude to Prof. A. Arima, and to Prof. S. Kobayashi, for the hospitality and support during the measurements performed at the RIKEN Ring Cyclotron, and to Prof. H. Morinaga, for many valuable discussions.

References

1. I.Tanihata *et al*, *Phys. Rev. Lett.* **55**, 2676 (1985).
2. P.G.Hansen and B.Jonson, *Europhys.Lett.* **4**, 409 (1987).
3. I.Tanihata, *Prog.Part.Nucl.Phys.* **35**, 505 (1995).
4. K.Riisager, *Rev.Mod.Phys.* **66**, 1105 (1994).
5. V.Fekou-Youmbi *et al*, *Nucl.Phys.* **A583**, 811c (1994).
6. A.Yoshida *et al*, in *Perspectives in Heavy Ion Phys.*, 2nd Japan-Italy Joint Symposium, May 1995 RIKEN, Japan, M.Yshihara, T.Fukuda,C.Signorini eds. (World Sci. Singapore 1996).
7. C.Signorini, *Nucl.Phys.* **A616**, 262c (1997).
8. M.Petrascu *et al*, *Balkan Phys.Lett.* **3(4)**, 214 (1995).
9. M.Petrascu, in *Collective Motion and Nuclear Dynamics*, Proc. Int. Summer School, Predeal, Romania, Aug. 28-Sept. 9,1995, A.A.Raduta, D.S. Delion, I.I.Ursu eds. (World Sci. Singapore 1996).
10. M.Petrascu *et al*, *RIKEN Rep.* AF-NP-237, Oct. 1996.
11. M.Petrascu *et al*, *Phys.Rev.Lett.* **B405**, 224 (1997).
12. M.Petrascu *et al*, *Rom.J.Phys.* **43**, 307 (1998).
13. M.Petrascu *et al*, *Rom.J.Phys* **44** n1-2 suppl. 83 (1999).
14. M.Petrascu, in *Structure and Stability of Nucleon and Nuclear Systems*, Proc. Int. Summer School, Predeal, Romania, Aug. 24-Sept. 5,1998, A.A.Raduta, S.Stoica, I.I.Ursu eds. (World Sci. Singapore 1999).
15. M.Petrascu *et al*, *J.Phys.G* **25** 799 (1999).
16. M.Petrascu *et al*, *Rom.J.Phys* **44** n1-2 suppl. 71 (1999).
17. M.Petrascu *et al*, *Rom.J.Phys* **44** n1-2 suppl. 115 (1999).
18. H.Petrascu *et al*, *Rom.J.Phys* **44** n1-2 suppl. 105 (1999).
19. V.Zhukov *et al*, *Phys. Rep.* **231**, 151 (1993).
20. M.I.Podgoretsky, *Particles & Nuclei* **20**, 628 (1989).
21. T.Kobayashi in *Collective Motion and Nuclear Dynamics*, Proc. Int. Summer School, Predeal, Romania, Aug. 28-Sept. 9,1995, A.A.Raduta, D.S. Delion, I.I.Ursu eds. (World Sci. Singapore 1996).
22. M. Petrascu, RIKEN PAC-Seminar Jan. 1998.
23. M.Petrascu *et al*, *inpc/98 abstracts*, p 626.
24. M.Petrascu *et al*, *to be published.*

CORRELATED OVERLAP FUNCTIONS AND NUCLEON REMOVAL REACTIONS FROM ^{16}O

M. K. GAIDAROV, M. V. IVANOV, A. N. ANTONOV, K. A. PAVLOVA, S. S. DIMITROVA AND M. V. STOITSOV

Institute of Nuclear Research and Nuclear Energy, Bulgarian Academy of Sciences, Sofia 1784, Bulgaria

C. GIUSTI

Dipartimento di Fisica Nucleare e Teorica, Università di Pavia, Istituto Nazionale di Fisica Nucleare, Sezione di Pavia, Pavia, Italy

D. VAN NECK

Laboratory for Theoretical Physics, Proeftuinstraat 86, B-9000 Gent, Belgium

H. MÜTHER

Institut für Theoretische Physik, Universität Tübingen, Auf der Morgenstelle 14, D-72076 Tübingen, Germany

Using the relationship between the one-nucleon overlap functions related to bound states of the $(A-1)$-particle system and the one-body density matrix for the ground state of the A-particle system the overlap functions and spectroscopic factors are calculated within different correlation methods, such as: i) the Jastrow correlation method; ii) the Correlated Basis Function Theory; iii) the Green Function Method and, iv) the Generator Coordinate Method. The resulting bound-state overlap functions are used to calculate the cross sections of the (p,d), $(e,e'n)$, $(e,e'p)$ and (γ,p) reactions on ^{16}O. The nucleon-nucleon correlation effects are studied and various correlation methods are tested in the comparison with the experimental data.

1 Introduction

The main aim of this work is to present a test for the extent to which different types of nucleon–nucleon (NN) correlations are included in various theoretical methods. This is done by the use of the overlap functions obtained in these methods for calculations of cross sections of electron-, photon- and proton-induced one–nucleon removal reactions.

It has been shown recently that absolute spectroscopic factors and overlap functions (OF) for one-nucleon removal reactions can be extracted from the one-body density matrix (ODM) of the target nucleus [1]. This procedure (discussed in Sect.2) avoids the complicated task of calculating the total nuclear spectral function. The method for extracting bound-state OF has been applied [2,3,4,5,6] to ODM obtained within various correlation methods such as the Jastrow correlation method (JCM) [7], the Correlated Basis Function (CBF) theory [3,8] and the Green function method (GFM) [9]. It has been shown that these functions are of particular importance, since they contain nucleon correlations which are accounted for in the various theoretical methods considered. The applicability of the theoretically calculated OF has been tested in the description of the ^{16}O(p,d) pickup reaction [4,5,6] and of the ^{40}Ca(p,d) reaction (within the JCM) [4]. It has been found a good overall agreement between the calculated and the empirical cross sections. It has been pointed out also that acceptable spectroscopic factors can be obtained with the method proposed. Considering the role of the short–range (SRC) and tensor correlations, it has been concluded that the long-range correlations (LRC) corresponding to collective degrees of freedom have to be taken also into account in order to achieve a better agreement with the (p,d) data.

In this work, we discuss in Sect.3 some of the results on the ^{16}O(p,d) reaction cross sections calculated in the above mentioned correlation methods. Secondly, in Sect.4 we give the results of calculations on ^{16}O$(e,e'p)$ knockout reactions obtained in [10] and in the present work using the OF from the same correlation methods as well as from the approach [11,12,13] within the Generator Coordinate Method (GCM). These calculations are based on the same nonrelativistic DWIA treatment [14] already used for the analysis of many experimental data. Third, the s.p. OF are used to calculate the cross section of the ^{16}O(γ,p) reaction (Sect.5). For the photon-induced reaction we have adopted the theoretical treatment of ref. [15], where the contributions of the direct knockout mechanism (DKO) and of meson-exchange currents (MEC) are evaluated consistently. The comparison of calculations, with consistent theoretical ingredients and constrained parameters, for the $(e,e'p)$ and (γ,p) cross sections can enable us to check, in comparison with data, the consistency of the theoretical description of the two reactions.

2 Single–particle overlap functions and spectroscopic factors

The single–particle OF in quantum–mechanical many–body systems are defined by the overlap integrals between eigenstates of the A–particle and the $(A-1)$–particle systems:

$$\phi_\alpha(\mathbf{r}) = \langle \Psi_\alpha^{(A-1)} | a(\mathbf{r}) | \Psi^{(A)} \rangle, \tag{1}$$

where $a(\mathbf{r})$ is the annihilation operator for a nucleon with spatial coordinate $\mathbf{r}$ (spin and isospin operators are implied). In the mean–field approximation $\Psi^{(A)}$ and $\Psi_\alpha^{(A-1)}$ are single Slater determinants and the overlap functions are

Table 1. Spectroscopic factors for the $p_{1/2}$ and $p_{3/2}$ quasihole states in ^{16}O: column I–deduced from the calculations with different ODM of ^{16}O; column II–additional reduction factors determined through a comparison between the $(e, e'p)$ data of [17] and the reduced cross sections calculated in DWIA with different overlap functions; column III–total spectroscopic factors obtained from the product of the factors in columns I and II.

ODM	$1p_{1/2}$			$1p_{3/2}$		
	I	II	III	I	II	III
HF	1.000	0.750	0.750	1.000	0.550	0.550
JCM [2]	0.953	0.825	0.786	0.953	0.600	0.572
CBF [3]	0.912	0.850	0.775	0.909	0.780	0.709
CBF [8]	0.981	0.900	0.883	0.981	0.600	0.589
CBF [16]	0.983	0.880	0.865	0.983	0.630	0.619
GFM [9]	0.905	0.800	0.724	0.915	0.625	0.572
GCM [13]	0.988	0.700	0.692	0.988	0.500	0.494

identical with the mean–field single–particle wave functions. Of course, this is not the case at the presence of correlations where both, $\Psi^{(A)}$ and $\Psi_{\alpha}^{(A-1)}$, are complicated superpositions of Slater determinants. The norm of OF defines the spectroscopic factor $S_\alpha = \langle \phi_\alpha | \phi_\alpha \rangle$.

The lowest $(n_0 lj)$ bound-state overlap function (for neutrons) is determined by the asymptotic behaviour of the corresponding partial radial contribution of the one-body density matrix $\rho_{lj}(r, r')$ $(r' = a \to \infty)$ [1]:

$$\phi_{n_0 lj}(r) = \frac{\rho_{lj}(r, a)}{C_{n_0 lj}\ \exp(-k_{n_0 lj}\ a)/a}\ , \tag{2}$$

where the constants $C_{n_0 lj}$ and $k_{n_0 lj}$ are completely determined by $\rho_{lj}(a, a)$. In this way the separation energy

$$\epsilon_{n_0 lj} \equiv E^{A-1}_{n_0 lj} \ - \ E^A_0 = \frac{\hbar^2\ k^2_{n_0 lj}}{2m_{\rm n}} \tag{3}$$

and the spectroscopic factor can be determined as well. As shown in [1], the procedure also yields in principle all bound-state overlap functions with the same multipolarity if they exist.

The overlap functions from different correlation methods mentioned in the Introduction (JCM [2], CBF theory [3,6], GFM [6] and GCM [10]) have been used to calculate cross sections of one–nucleon removal reactions on the ^{16}O nucleus. The calculated spectroscopic factors are listed in Table 1.

3 (p,d) reactions

The differential cross section for the $^{16}O(p,d)$ reaction can be written in the form:

$$\frac{d\sigma_{pd}^{lsj}(\theta)}{d\Omega} = \frac{3}{2}\frac{S_{lsj}}{2j+1}\frac{D_0^2}{10^4}\sigma_{DW}^{lsj}(\theta), \tag{4}$$

where S_{lsj} is the spectroscopic amplitude, j is the total angular momentum of the final state, $D_0^2 \approx 1.5 \times 10^4\ MeV.fm^3$ is the $p-n$ interaction strength in the zero–range approximation and σ_{DW}^{lsj} is the cross section calculated by the DWUCK4 code [18]. For our purposes the standard Distorted Wave Born Approximation (DWBA) form factor has been replaced by the s.p. overlap function derived from the ODM calculations. In this case no extra spectroscopic factor S_{lsj} in Eq.(4) is needed, since our overlap functions already include the associated spectroscopic factors (see column I of Table 1). The results for the differential cross sections for the transitions to the ground $1/2^-$ state in ^{15}O nucleus at different incident proton energies E_p=31.82, 45.34 and 65 MeV are given in ref.[6] compared with the experimental data from [19]. Considering the role of the short-range and tensor correlations, we can conclude that the LRC corresponding to collective degrees of freedom have to be taken also into account in order to achieve a better agreement with the (p,d) data.

4 $(e,e'N)$ reactions

The cross sections of the $^{16}O(e,e'p)$ and $^{16}O(e,e'n)$ reactions have been calculated with the code DWEEPY [14], which gave a good description of the $(e,e'p)$ experimental momentum distributions, for transitions to different final states, in a wide range of nuclei and in different kinematics (see, e.g., [20] and, more specifically for the analysis of the $^{16}O(e,e'p)$ reaction, [17]). Here we would like to emphasize that the use of the OF derived from different calculations of the ODM is the correct theoretical procedure which in principle has to be applied for an accurate description of $(e,e'N)$ knockout reactions.

Our analysis is made for the transitions to the $1/2^-$ ground state and to the first $3/2^-$ excited state of the residual nucleus (at excitation energy $E_x = 6.18$ MeV for ^{15}O in the case of the $(e,e'n)$ reaction and at $E_x = 6.3$ MeV for ^{15}N in the case of the $(e,e'p)$ reaction), representing a knockout from the valence $1p$ shell of ^{16}O. The reduced cross sections for the $^{16}O(e,e'p)$ reaction are shown in Fig.1 in comparison with the data taken at NIKHEF [17] in the parallel kinematics. It can be seen that they are sensitive to the shape of the various overlap functions used. In order to reproduce the size of the

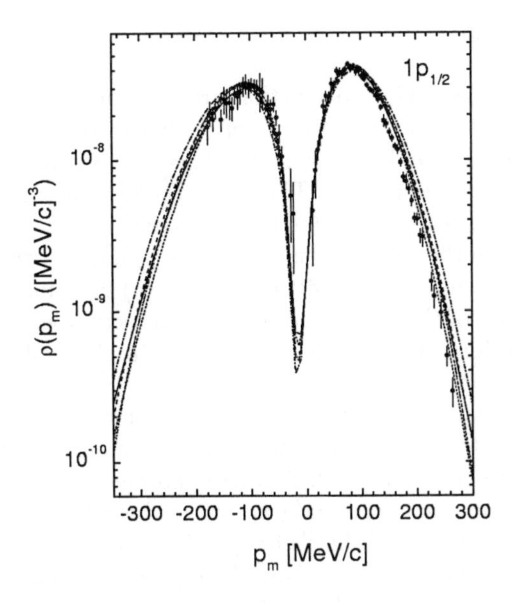

Figure 1. Reduced cross section of the $^{16}O(e,e'p)$ reaction with incident electron energy of 520.6 MeV and an outgoing proton energy of 90 MeV. Overlap functions derived from the ODM of GFM [9] (solid line), CBF [3] (dashed line), CBF [8] (dot-dashed line), JCM [2] (double dot-dashed line) and GCM [13] (short-dashed line). The dotted line is calculated with the HF wave function. The experimental data are taken from [17]. The theoretical results have been multiplied by the reduction factor given in column II of Table 1.

experimental cross section a reduction factor has been applied to the theoretical results. These factors, which have been obtained by a fit of the calculated reduced cross sections to the data over the whole missing-momentum range considered in the experiment, are also listed in Table 1 (column II). In general, a good agreement with the shape of the experimental distribution is achieved. The best agreement with the data, for both transitions, is obtained with the overlap functions [6] obtained using the ODM calculated within the GFM [9] and with OF from the GCM [13]. In Table 1 we give in addition, in column III, the factor obtained by the product of the two factors in columns I and II. This factor can be considered as a total spectroscopic factor and can be attributed to the combined effect of SRC and LRC.

5 (γ, p) reactions

Calculations for the $^{16}O(\gamma, p)$ reaction cross sections have been performed within the theoretical framework of ref.[15], where one–body and two–body currents are included and both contributions of DKO mechanism and MEC can be evaluated consistently. The same theoretical ingredients, i.e. s.p. overlap functions, spectroscopic factors and consistent optical potentials, have been adopted as in the calculations of the $(e, e'p)$ cross section. Moreover, the reduction factor determined by fitting the $(e, e'p)$ data has been applied also in the comparison of the calculated (γ, p) cross section with data. This made it possible to perform a consistent study of the $(e, e'p)$ and (γ, p) reactions on the ^{16}O nucleus.

Our calculations are for energies $E_\gamma = 60$ and 72 MeV, where $^{16}\mathrm{O}(\gamma, p)$ data are available for the transition to the $1/2^-$ ground state [21,22,23] and to the $3/2^-$ excited state at 6.3 MeV [23]. At these photon–energy values it is possible to sample in comparison with data p_{m} values between 250 and 400 MeV/c. The angular distribution of the $^{16}\mathrm{O}(\gamma, p)^{15}\mathrm{N}_{\mathrm{g.s.}}$ reaction at $E_\gamma = 72$ MeV is displayed in Fig.2. It can be seen from the Figure that much better agreement with (γ, p) data is obtained when MEC are added to the DWIA result. The contribution of the two–body current is large and significantly affects both size and shape of the calculated cross section. The results, however, are very sensitive to the overlap function used. The HF wave function and the OF obtained from the ODM [2] are able to reproduce the size of the experimental cross section, but only at low values of the outgoing proton angle. Much better agreement with the shape of the experimental distribution is given by the other correlated overlap functions. Only the OF from the ODM [8] largely overshoots the data. A fair agreement with the (γ, p) data is obtained using the OF from the ODM of ref.[9] and from [13] and, to a lesser extent, also by that from [3]. These overlap functions also give the best agreement with $(e, e'p)$ data for the $1/2^-$ state in Fig.1.

6 Conclusions

We can summarize the results of the present work as follows:
i) The single-particle overlap functions obtained on the basis of ODM for the ground state of ^{16}O within different correlation methods have been used to calculate the cross sections of the (p, d), $(e, e'n)$, $(e, e'p)$ and (γ, p) reactions on ^{16}O for the transitions to the $1/2^-$ ground state and the first $3/2^-$ excited state of the residual nucleus. The theoretical results for the reaction cross sections show that they are sensitive to the shape of the different overlap

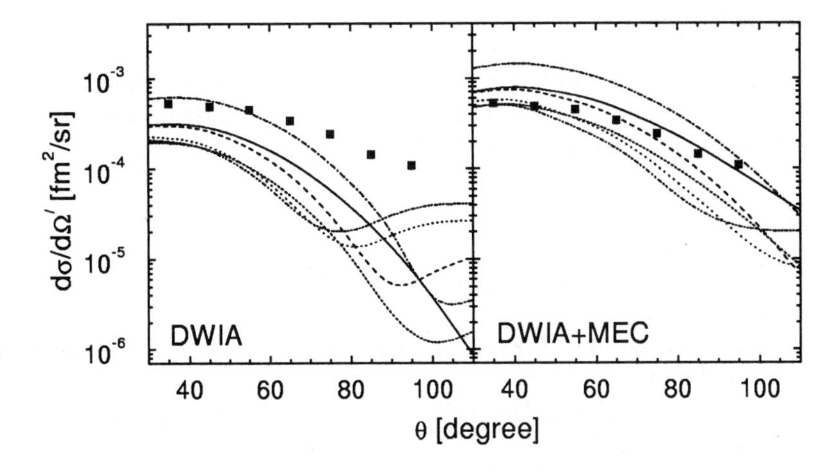

Figure 2. Angular distribution of the cross section of the $^{16}O(\gamma, p)$ reaction for the transition to the $1/2^-$ ground state of ^{15}N at $E_\gamma = 72$ MeV. The separate contribution given by the one-body current (DWIA) and the final result given by the sum of the one-body and the two-body seagull current (DWIA+MEC) are shown. Line convention is as in Fig.1. The optical potential is from [24]. The experimental data are taken from Ref.[23]. The theoretical results have been multiplied by the reduction factors listed in column II of Table 1, consistent with the analysis of $(e, e'p)$ data.

functions and are generally able to reproduce the shape of the experimental cross sections.
ii) In order to describe correctly the size of the experimental data a reduction factor must be applied to the calculated $(e, e'p)$ reduced cross sections. This factor, that is extracted from a fit to the data, can be considered as a further spectroscopic factor to be mostly ascribed to LRC, which also cause a depletion of the quasihole states. The spectroscopic factors accounting for SRC and LRC obtained in the present analysis are in reasonable agreement with those given by previous theoretical investigations (e.g. for $1/2^-$ state the total spectroscopic factors are in agreement with the spectroscopic factor (0.76) calculated in [25], where both SRC and LRC are consistently included).
iii) For the transition to the ground state of ^{15}N the best agreement with the (γ, p) data is achieved using the overlap functions from the Green function method [9] and from the Generator Coordinate Method [13] which give also the best description of the $(e, e'p)$ data. This is a strong indication in favour of a consistent analysis of both $(e, e'p)$ and (γ, p) reactions.

Acknowledgments

The authors are grateful to Dr. G. Co' for providing us the results for ODM from [8,16]. This work was partly supported by the Bulgarian National Science Foundation under the Contract Φ–809.

References

1. D. Van Neck, M. Waroquier, and K. Heyde, *Phys. Lett.* B **314**, 255 (1993).
2. M.V. Stoitsov, S.S. Dimitrova, and A.N. Antonov, *Phys. Rev.* C **53**, 1254 (1996).
3. D. Van Neck, L. Van Daele, Y. Dewulf, and M. Waroquier, *Phys. Rev.* C **56**, 1398 (1997).
4. S.S. Dimitrova, M.K. Gaidarov, A.N. Antonov, M.V. Stoitsov, P.E. Hodgson, V.K. Lukyanov, E.V. Zemlyanaya, and G.Z. Krumova, *J. Phys.* G **23**, 1685 (1997).
5. M.K. Gaidarov, K.A. Pavlova, S.S. Dimitrova, M.V. Stoitsov, and A.N. Antonov, in: *Proceedings of XIV Intern. Seminar on High Energy Physics Problems*, Dubna (Russia), 1998, in print; A.N. Antonov, in: *Structure and Stability of Nucleon and Nuclear Systems*, edited by A.A. Raduta et al., (World Scientific, Singapore, 1999), p.225.
6. M.K. Gaidarov, K.A. Pavlova, S.S. Dimitrova, M.V. Stoitsov, A.N. Antonov, D. Van Neck, and H. Müther, *Phys. Rev.* C. **60**, 024312 (1999).
7. M.V. Stoitsov, A.N. Antonov, and S.S. Dimitrova, *Phys. Rev.* C **47**, R455 (1993); *Phys. Rev.* C **48**, 74 (1993); Z. Phys. A **345**, 359 (1993).
8. F. Arias de Saavedra, G. Co', A. Fabrocini, and S. Fantoni, *Nucl. Phys.* **A605**, 359 (1996).
9. A. Polls, H. Müther, and W.H. Dickhoff, in: *Proceedings of Conference on Perspectives in Nuclear Physics at Intermediate Energies*, Trieste, 1995, edited by S. Boffi, C. Ciofi degli Atti, and M.M. Giannini, (World Scientific, Singapore, 1996), p.308.
10. M.K. Gaidarov, K.A. Pavlova, A.N. Antonov, M.V. Stoitsov, S.S. Dimitrova, M.V. Ivanov, C. Giusti, *Phys. Rev.* C. **61**, 014306 (2000).
11. A.N. Antonov, P.E. Hodgson and I.Zh. Petkov, *Nucleon Momentum and Density Distributions in Nuclei* (Clarendon Press, Oxford, 1988).
12. A.N. Antonov, P.E. Hodgson and I.Zh. Petkov, *Nucleon Correlations in Nuclei* (Springer–Verlag, Berlin-Heidelberg-New York, 1993).
13. A.N. Antonov, Chr.V. Christov, and I.Zh. Petkov, *Nuovo Cim.* **91A**, 119 (1986); A.N. Antonov, I.S. Bonev, Chr.V. Christov, and I.Zh. Petkov,

Nuovo Cim. **100A**, 779 (1988); A.N. Antonov, I.S. Bonev, and I.Zh. Petkov, Bulg. J. Phys. **18**, 169 (1991); A.N. Antonov, I.S. Bonev, Chr.V. Christov, and I.Zh. Petkov, *Nuovo Cim.* **103A**, 1287 (1990).

14. C. Giusti and F.D. Pacati, *Nucl. Phys.* **A473**, 717 (1987); *Nucl. Phys.* **A485**, 461 (1988).
15. G. Benenti, C. Giusti, and F.D. Pacati, *Nucl. Phys.* **A574**, 716 (1994).
16. F. Arias de Saavedra, G. Co', and M.M. Renis, *Phys. Rev.* C **55**, 673 (1997).
17. M. Leuschner, J.R. Calarco, F.W. Hersman, E. Jans, G.J. Kramer, L. Lapikás, G. van der Steenhoven, P.K.A. de Witt Huberts, H.P. Blok, N. Kalantar-Nayestanaki, and J. Friedrich, *Phys. Rev.* C **49**, 955 (1994).
18. K. Langanke, J.A. Maruhn, and S.E. Koonin, *Computational Nuclear Physics 2: Nuclear Reactions*, (Springer-Verlag, Berlin-Heidelberg-New York), 88 (1993).
19. B.M. Preedom, J.L. Snelgrove, and E. Kashy, *Phys. Rev.* C **1**, 1132 (1970).
20. S. Boffi, C. Giusti, F.D. Pacati, and M. Radici, *Electromagnetic Response of Atomic Nuclei, Oxford Studies in Nuclear Physics* (Clarendon Press, Oxford, 1996).
21. D.J.S. Findlay and R.O. Owens, *Nucl. Phys.* A **279**, 389 (1977).
22. F. de Smet, H. Ferdinande, R. Van de Vyver, L. Van Hoorebeke, D. Ryckbosch, C. Van den Abeele, J. Dias, and J. Ryckebusch, *Phys. Rev.* C **47**, 652 (1993).
23. G.J. Miller, J.C. McGeorge, J,R.M. Annand, G.I. Crawford, V. Holliday, I.J.D. MacGregor, R.O. Owens, J. Ryckebusch, J.-O. Adker, B.-E. Andersson, L. Isaksson, and B. Schröder, *Nucl. Phys.* A **586**, 125 (1995).
24. P. Schwandt, H.O. Meyer, W.W. Jacobs, A.D. Bacher, S.E. Vigdor, M.D. Kaitchuck, and T.R. Donoghue, *Phys. Rev.* C **26**, 55 (1982).
25. W.J.W. Geurts, K. Allaart, W.H. Dickhoff, and H. Müther, *Phys. Rev.* C **53**, 2207 (1996).

COLLISIONS OF COLD ELECTRONS WITH COOLED IONS IN CRYRING

R. SCHUCH, H. DANARED*, N. EKLÖW, M. FOGLE+, P. GLANS,
E. JUSTINIANO+, E. LINDROTH, S. MADZUNKOV, M. TAREK, W. ZONG

Department of Atomic Physics, Stockholm University, S-104 05 Stockholm, Sweden
**Manne Siegbahn Laboratory, Stockholm University, S-10405 Stockholm*
+Permanent address:Department of Physics, East Carolina University, Greenville, NC 27858

We have studied collisions of electrons with ions at very low energies and obtained recombination rate coefficients as well as high precision spectroscopic data. The low collision energies were reached with the upgraded electron cooler in the storage ring CRYRING at Manne Siegbahn Laboratory in Stockholm. With the cooled beams, the mean relative energy between electrons and ions could be varied from 10^{-6} eV to $\sim$ 100 eV. Experimental data are reported for bare ions and compared with calculations for radiative recombination, which is expected to be the dominant recombination process. Here, the measured rate coefficients are found systematically higher than the theoretical predictions. Examples of high resolution measurements of dielectronic recombination resonances are presented here for the Li-like ions F^{6+} and Ar^{15+} and compared to many-body theories of the electron-ion interaction.

1 Introduction

An important process for diagnostics and modeling of plasmas [1] and astrophysics [2,3] is electron-ion recombination. Astrophysical objects are studied through analysis of their radiation spectra [3], thus requiring accurate knowledge of electron-ion recombination. Plasma modeling and plasma diagnostics are based on the knowledge of cross section for recombination of ions of nuclear charge Z. Recombination rates are important also to accelerator physics, as they lead to particle losses during electron cooling. On the other hand, these processes may provide an efficient mechanism for antihydrogen production in a trap filled with antiprotons and positrons [4,5]. At the end of the 80's new experimental techniques for studying electron-ion collisions were invented with heavy-ion cooler storage rings[6,7] and electron beam ion traps[8]. These devices allow investigations of reactions between electrons and ions in almost any charge state with high resolution, signal-to-background ratio, and luminosity. Primarily recombination processes are studied, but work on electron impact excitation and ionization is done as well. The fundamental processes of recombination are: $Z^{q+} + ne \rightarrow Z^{(q-1)+} + (n-1)e + h\nu$. Where for radiative recombination (RR) the photon is emitted directly ($n = 1$). In dielectronic recombination (DR) ($q < Z$, n=1) and the emission of photons is from an intermediate doubly-excited state in the ion of nuclear charge Z. In three-body recombination (TBR) a neighboring electron carries away recombination energy ($n > 1$).

Recent studies of recombination of free electrons with multi-charged heavy ions revealed an unexpected deviation from standard radiative recombination theory[9,10]. The measured rates show a strong increase and enhancement over the calculated ones for electron energies below the electron temperature [10,11,12]. At zero energy, the measured enhanced rates increase approximately as Z^3 with the ion charge Z [12]. First, no clear enhancement was observed for D^+ [12]. However, with further improvement of the cooler with an electron beam having a factor 10 lower temperature, an enhancement is clearly seen even for D^+. One possible cause of the enhancement could be three-body recombination (TBR) [12], although calculations indicated that it might not be efficient enough [16]. Three-body recombination should show a characteristic quadratic dependence of the rate on the electron density. This dependence has been investigated [11] with Ne^{10+} over a factor of 7 in density. However, the experimental rate coefficient α_{exp} was found constant as a function of electron density, although being enhanced by a factor of 3 at $E_{rel} \approx 0$. Thus TBR seems to be excluded. It has also been suggested that the weak guiding magnetic field (0.03T) causes the enhancement. To find an answer this question magnetic field dependencies have been investigated, using D^+ [13], Au^{25+} [14] and F^{6+} [15]. In those experiments an increasing rate coefficient with the B-field was found.

With non-bare ions dielectronic recombination can occur as well. In this process an intermediate doubly excited state (n,n') is formed which has a very large probability to autoionize and loose the electron again. However, it may instead emit a photon and end up in a singly excited state. This last step completes dielectronic recombination. Due to the resonance character one can obtain accurate data on energies, autoionization and radiative rates of singly resolved terms in few-electron atomic systems. Thus, besides of the interest in the cross sections for plasma energy balances and diagnostics, these resonances can serve for accurate tests atomic structure and quantum electrodynamic (QED) tests[17,18,19,20,21]. We report here some examples for recent measurements of dielectronic recombination done at CRYRING with Li-like ions, F^{6+} and Ar^{15+}. The resonance energies are determined with 1 to 20 meV resolution and in absolute scale with about 1 to 10 meV accuracy, dependent on their position above the threshold. The absolute rates are determined to better than 20%.

Here we compare the experimental data to the result of a fully relativistic calculation which also accounts for correlation to high orders. The contributions to the energy positions from QED corrections, from the Breit interaction, and correlations is different in the F and Ar case. For Ar the contributions from QED corrections, as well as from the Breit interaction are 100-200 meV. Correlation contributions from the inner electrons ($1s^2 2\ell$) are of the same size and for the low energy DR resonances even the detailed interaction between the outer electron (with $n \geq 6$) and the inner electrons reaches the size of nearly 100 meV in some cases. For F the Breit and the QED contributions are one order of magnitude smaller than in Ar, while the correlation between the inner electrons still contributes with 100-200 meV. The correlation between the 2p electron and the outer electron contribute with as much as 300 meV in some cases.

2 Experiment

The experiments were performed in the ion storage ring CRYRING at the Manne Siegbahn Laboratory. The ions, produced from an electron-beam ion source (EBIS), were injected into the ring at 300 keV/amu via an RFQ, and accelerated to 11 MeV/amu prior to storage. During electron cooling, the ions were merged over an effective interaction length of $l = 0.8m$ with a velocity matched electron beam, confined by a solenoidal magnetic field to a diameter of normally 40 mm. The expanded electron beam of the cooler [22] was used. By the expansion ratio (ε) of the solenoidal magnetic field (B_i) to the gun magnetic field (B_g), a transverse temperature component ($T_\perp$) between 1 meV/k_B and 100 meV/k_B could be adjusted. The longitudinal temperature component ($T_\parallel$) should be around 0.1 meV/k_B determined mainly by the electron density of about $n_e = 10^7/cm^3$. From fitting DR resonances and from electron-cooling measurements [22] with different ions we find electron beam temperature components close to the adjusted values.

In the measurements of the recombination rates we detected the charge changed projectiles at the exit of the first bending magnet after the e^--cooler by a surface barrier detector with unity detection efficiency. The motional electric field in this bending magnet determines the upper limit of quantum states $n_{max} = \left(6.2 \times 10^8 \, Z^3/\mathcal{E}\right)^{1/4}$ with the electric field strength $\mathcal{E}$ in units of V/cm. In these experiments, $n_{max} \approx 6, 23$, and 46 for D^+, F^{6+}, and Ar^{15+}, respectively. After cooling the beam for typically 1-3 sec., the relative velocity between the ion and electron beam was tuned through the energy region of interest. This is done by sweeping the cathode voltage of the cooler up and down from cooling voltage [23]. The rate coefficients are determined by: $\alpha_{exp} = \gamma^2 R_{det} L / n_e N_i l$, where γ is the Lorentz factor, R_{det} is the background corrected counting rate of recombined particles, N_i is the number of stored ions, and $L = 51.63$ m is the ring circumference. The data was corrected for the electron capture background of up to a few percent of R_{det} at the pressure of 10^{-11} Torr. Most of the uncertainties in the rate values are from the beam current measurement, which is estimated to vary from 10% to 30%. Details of the experiment are described elsewhere[23].

3 Recombination rates for bare ions

Figure 1 shows the measured recombination rate coefficients of D^+ as function of the relative energy E, together with corresponding theoretical RR predictions. One should notice that E represents the energy resulting from the mean longitudinal velocity ($v_{\parallel}$) component difference between electrons and ions. The measured rate coefficients in the figure are absolute. The experimental uncertainty is about 15%. Mainly determined by the uncertainty in the effective interaction length and the knowledge of the number of stored ions.

In the calculation of RR, the semiclassical Kramers cross section corrected by the Gaunt factor [24] is used, and convoluted with a flattened electron velocity distribution characterized by longitudinal and transverse temperatures. Data and theory for three nominal transverse temperatures are compared in Fig. 1: $T_{\perp}$=100, 10 and 1 meV/k at a longitudinal temperature of $T_{\parallel} = 0.1$ meV/k. The values of transverse temperatures are expected from initial electron temperature and beam expansion factor. They are approximately confirmed by cooling force measurements [22]. The longitudinal temperature is estimated from [25], together with about 40% contribution by transverse-longitudinal relaxation calculated from [26,27]. A rather accurate determination of both temperature components is obtained by fits to isolated DR resonances (e.g. He^{+} [28], Ar^{13+} [23], Ar^{15+} [21], C^{3+} [29,30], F^{6+} [31]). However, a somewhat larger transverse temperature is often found from these fits. This can possibly be caused by overlapping resonances or beam misalignment. In our predictions we use the nominal transverse temperature which is the lower limit and gives thus the upper limit for the RR theory Fig. 1.

The measured D^+ rates agree well with the calculated RR curve for the cases of $T_{\perp}$=100 and 10 meV/k [10,32]. Further, there is a rather good agreement with the measured rates for $T_{\perp} = 1$ meV/k above $E \sim 10$ meV. Which indicates that the differences at low energies for $T_{\perp} = 1$ meV/k cannot arise from background corrections or normalization. Under the static conditions of cooling the mean longitudinal velocities of electrons and ions are equal, and

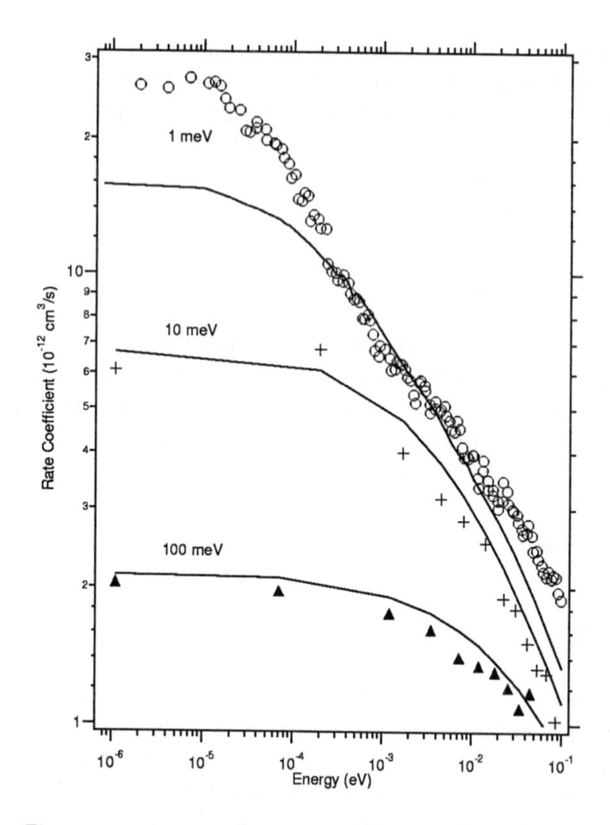

Figure 1. Comparison of rates α_{exp} as function of "relative" energy. The curves are calculated RR rates based on electron temperatures of $T_{\perp} = 100, 10$, and 1 meV/k and $T_{\|} = 0.1$ meV/k. The meaning of the "relative" energy is described in the text.

the values of the rates are found to correspond to those at E below 10^{-4} eV.

In the regime of energies reaching down to below 10^{-5} eV, the average center of mass energies of the electron-ion systems should be constant and close to $kT_{\perp}$. It is therefore striking that the recombination rates change strongly for variations of E much below $kT_{\perp}$. This finding may be explained by assuming a modification of the electron velocity distribution, such that the transverse motion of the electrons is "frozen" regarding the process that causes this enhancement. An argument for the latter may be found by considering the guiding solenoidal field of B_{sol}=0.03 T in the cooler which forces the electrons onto cyclotron orbits with a mean radius of about 8 μm, less than the average distance between electrons of 23 μm. The transverse motion could thus be confined in these orbits and the electrons collide with each other mainly via the longitudinal direction, resulting in the observed characteristics of the enhanced rates at such small E. In order to find further indications for this effect we studied the variation of the rate with B_{sol}.

Results of total recombination rates for D^+ as function of the nominal transverse temperature and different values of B_i are shown in Fig. 3. The measurements were made at cooling conditions for the following settings: B_i=0.03T ε=100, 50, 25 $n_e = 7$ $[10^6 cm^{-3}]$; B_i=0.06T ε=50, 25 $n_e = 8$ $[10^6 cm^{-3}]$; B_i=0.09T ε=33, 17 $n_e = 10, 14$ $[10^6 cm^{-3}]$; and B_i=0.12T ε=25, 12.5 $n_e = 8, 16$ $[10^6 cm^{-3}]$. The systematic error is 11% and the statistical error is about 6%. It was not possible to check the values of the temperatures for the different adjustments of the magnetic field for all these cases independently. Thus, a variation of the temperatures with magnetic field can not be excluded.

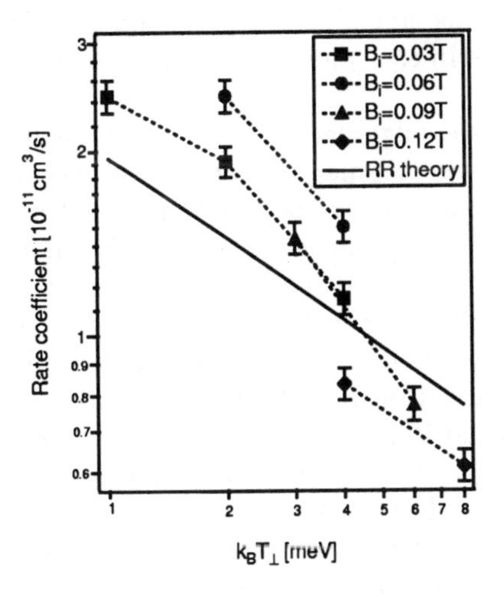

Figure 2. Recombination rate coefficient α for different guiding magnetic fields B_i and different expansions ε plotted as function of $k_B T_\perp$ in logarithmic scales. Full line shows RR-theory, obtained by Kramers formula with the Gaunt correction.

It has been shown that the recombination cross section folded with the electron velocity distribution scales with $\frac{1}{\sqrt{k_B T_\perp}}$ [16]. This means that the logarithm of α should show a linear dependence on the logarithm of $k_B T_\perp$. Thus, we can test this scaling law with the above data, as shown in Fig. 2:

The slopes of the measured α for fixed B_i agree fairly well with the slope of the theoretically calculated curve. The values of the RR rate coefficients are obtained by Kramers formula with the Gaunt correction.

Our measured values indicate a dependency of α_{exp} on B_i. The largest enhancement factor of about 1.7 is found for the data taken with $B_i = 0.06$T. From the data it seems that a maximum in the enhancement exists in the region 0.03T-0.09T. This result is partly in agreement with what was found in the experiment at TSR on F^{6+} [15]. There the rate coefficient was found to increase rapidly for $B_i \sim 0.02$T-0.04T and then more slowly at higher fields. However, in that experiment the maximum field was limited to 0.07T and it is not clear whether higher fields would lead to a decreasing rate coefficient as our results indicate. The single-pass experiment at GSI [14] on Au^{25+}, on the other hand, seem to contradict our result of a maximum enhancement in the 0.03T-0.09T region. In fact, their data showed a rate coefficient which increased monotonically with the field in the region between 0.27T and 0.33T, however, the experimental conditions in this single-pass experiment is very quite different from those of the ring experiments and, hence, the magnetic field dependence might not be the same.

This measurement is very sensitive to the alignment of the electron beam with the ion beam. Any angle between the two beams will increase the actual transverse temperature and thus lower the measured recombination rate. Due to the space charge potential built up by the electrons in the electron beam it is crucial that the about 1mm in diameter ion beam is aligned to the center of the 1cm to 4cm electron beam. Going off center will increase the electron velocity spread due to the radially rising gradient of the space charge potential. A larger electron density increases the gradient of the potential and makes it sensitive to a optimum centering. It is hard to estimate the effect of the alignment on α_{exp}. But for one of the points where we went back to the same setting of the magnetic fields, but at half the electron density, α_{exp} was nearly doubled. It can therefore not be excluded that the values reported here are not the optimum ones and that the lower rates at higher fields could be due to a misalignment at the higher electron densities used. However, it should be noted that for α_{exp} always the maximum values are used. An explanation for the enhanced rates measured with bare ions is still lacking.

4 Recombination rates for few-electron ions

The resolution of dielectronic recombination resonances is determined by the energy spread (temperature) of the electron beam. As the ions are cooled, and their mass is orders of magnitude larger then the one of electrons, their velocity spread is negligible. At small resonance energies $E_{nl} \leq k_B T_\perp$ the transverse energy spread will usually dominate the resolution since it remains unaffected by the transformation to the centre-of-mass system. At larger collision energies the longitudinal energy spread $E_{nl} < k_B T_\perp$ of the electron beam starts to become important.

Measurements of DR with lithium-like ions provide excellent tests for calculations of doubly excited states since the spectral features are still simple enough to be handled accurately in extended structure calculations, and they contain rich information on correlation, relativistic, and quantum electrodynamical effects. As measured quantities one obtains are the energy positions E_{nl}, width Γ, and strengths S of the resonances. The value of S is determined by the transition rate into the doubly excited state, i. e. the autoionization rate of the doubly excited state in the time reversed process; and the radiative branching ratio from this state to all states below the ionization threshold. S is thus governed by the rate of the weakest of these two channels; autoionization or radiative decay. For all the resolved resonances in the experimental data we found that S is determined by the radiative decay rate.

Recently, DR with Li-like F (F^{6+}) was measured at TSR (Heidelberg storage ring)[15] and at CRYRING [31]. Calculations, which include correlation to all orders in a relativistic treatment and first order QED effects, were made by Lindroth. Complex rotation is employed in order to deal with autoionizing states and the calculation gives directly the widths together with the energy positions. The theory predicts that two series, $1s^2 2s_{1/2}$ to $1s^2 2p_{1/2}(nl)$ and $1s^2 2p_{3/2}(nl)$ with $n \geq 6$, are seen. It is particularly interesting here that two of the 2p6l resonances are very close to the first ionization threshold ($\leq$ 10 meV) and the states are spread out to both below and above the first ionization threshold. One of these resonances is narrow $\Gamma \leq k_B T_\perp$ and one is very wide, $\Gamma \gg k_B T_\perp$. The narrow one with its vicinity to the threshold is very suitable for determining $T_\perp$.

The results from the measurements at CRYRING are shown in Fig. 3.

Essential details of the transformations into CM energies and calibration by the Schottky frequency method are described in ref.[21]. The energy scale is logarithmic up to 0.1 eV and from there on it is linear. With the new super-conducting cooler, the value of $T_\perp$ should be 1 meV. A fit to the first resonance gave a somewhat larger value of 3 meV. The figure contains also theoretical predictions from the calculation mentioned above, folded with this temperature in absolute scale, both in height and energy. RR is included and at very low electron energies the enhancement effect is clearly seen. The results from the DR resonances with Ar^{15+} after transformation into CM energies are summarized in Fig. 4. The spectra show the Rydberg series of doubly

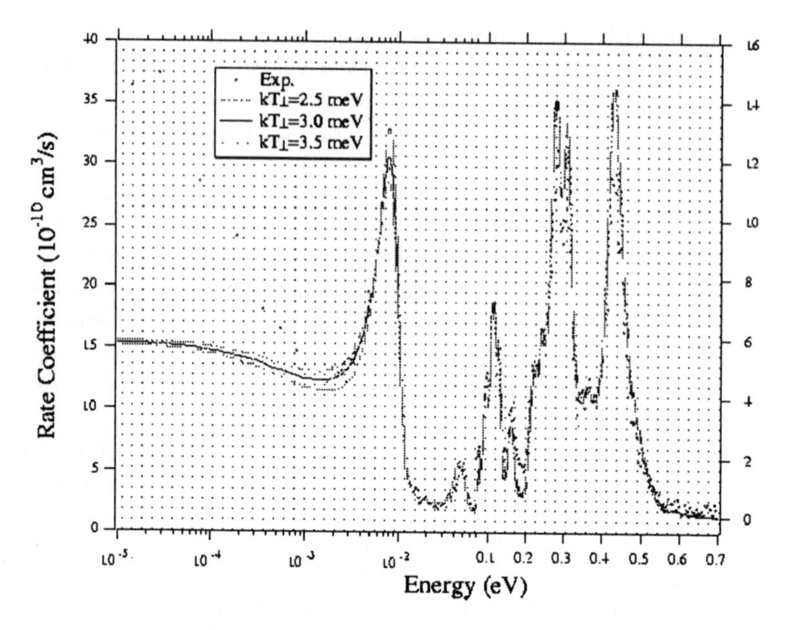

Figure 3. Measured (dots) and calculated recombination rate coefficients for F^{6+} vs. relative energy. A fit to the first resonance with different values of the transverse electron temperature.

excited states formed by $\Delta n = 0$ DR transitions in Ar^{14+}. The theory predicts that two series, $1s^2 2s_{1/2}$ to $1s^2 2p_{1/2}(nl)$ and $1s^2 2p_{3/2}(nl)$ with $n \geq 10$, are seen. A labeling of some of the main resonances is given in Fig. 4. For the CM energy approaching zero, the experimental rate rises sharply. This rise, which is not predicted by the DR calculation, could be interpreted to be due to radiative recombination. However, the experimental rate under cooling is $4.5 \cdot 10^{-9} cm^3 s^{-1}$. The RR prediction peaks at $1.47 \cdot 10^{-9} cm^3 s^{-1}$ which is around a factor of 3 below the experimental rate (see discussion above). Thus, also here the major part of the rate coefficient at zero relative energy seems not to be described by standard RR theory. A detailed comparison of the data with calculated resonances is shown in Fig. 5. only for the $1s^2 2p_{1/2}(10\ell)$ resonance group. The theoretical cross sections are folded with the electron beam temperatures (in this case with 20 meV/k_B transverse temperature and 0.13 meV/k_B longitudinal temperature) In the data one sees clearly the $1s^2 2p_{1/2}(10s)$ and $(10p)$ resonances resolved and separated from the higher ℓ.

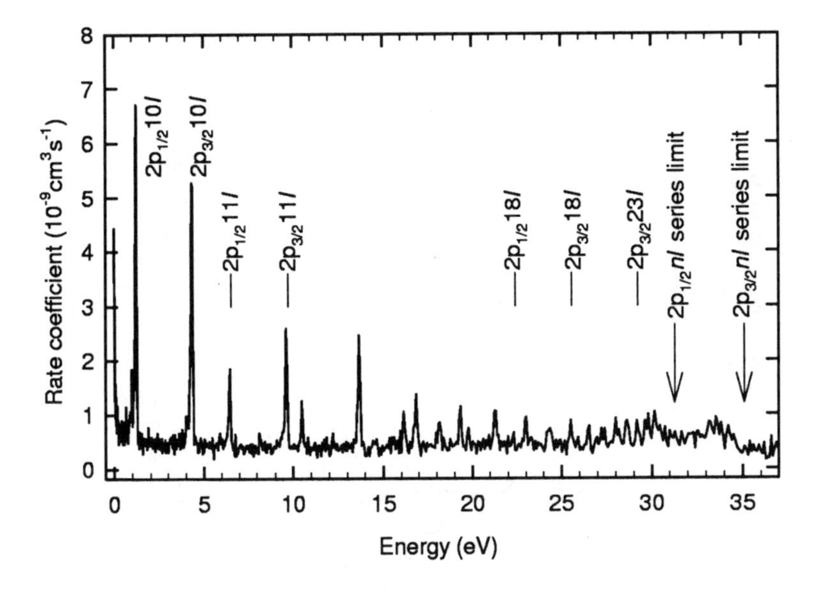

Figure 4. Recombination rate coefficient for Ar^{15+} vs. relative energy The main single electron quantum numbers of the active electrons are indicated.

In the experimental data, the 10s resonance is at 0.65 eV, the 10p resonance group is at around 0.95 eV, and the 10d resonance group starts at around 1.1 eV. The rest of the series converges in a large peak at around 1.2 eV Fig.5.

The calculation of DR resonance strengths and energies is done with a method which combines relativistic many-body perturbation theory in an all-order formulation, described for three and four electron ions in Ref. [33], with complex rotation. For the inner electrons ($n = 1$ and 2) correlation is included due to the Coulomb as well as due to the Breit interaction. Important radiative effects, self-energy and vacuum polarization, have been calculated for lithium-like systems with high accuracy by Blundell[35] and were added to our calculation. The outer electron is first described as moving in the Dirac-Fock- Breit potential from the $1s^2$-core and a spherical symmetric potential accounting for the main screening effects from the inner $2p$ electron and then the detailed Coulomb interaction between the outer electron and the $2p$ electron is treated as a perturbation. The contributions from this perturbation is quite different for states coupled to different total J. The agreement in the absolute rate coefficient between theory and experiment is within 15 per-

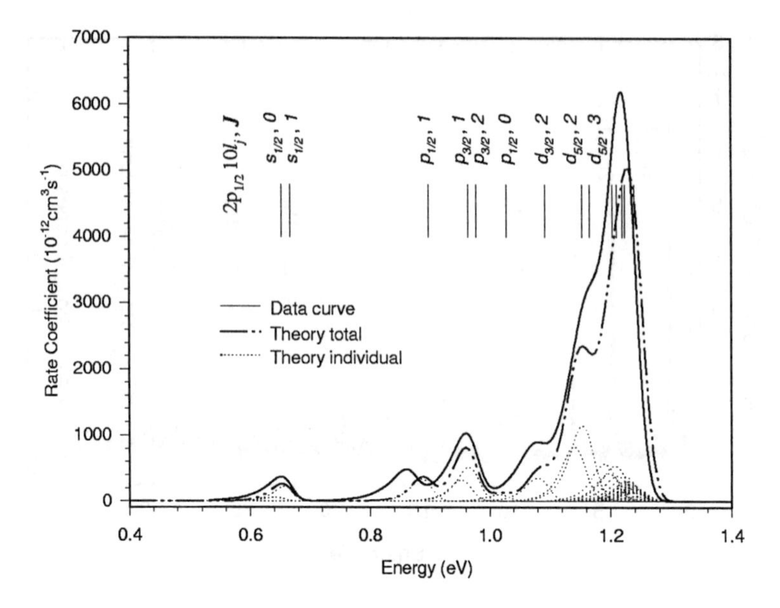

Figure 5. Recombination rate coefficient for Ar^{15+} vs. relative energy for the $2p_{1/2}(10\ell)$ resonances only. The calculated spectrum (full line) is folded with 20 meV$/k_B$ transverse temperature and 0.13 meV$/k_B$ longitudinal temperature) The angular momentum values of the highest electron are indicated.

cent and thus within the experimental error bars. On the energy scale the agreement is in the meV range. It should be noticed that the inclusion of QED effects and a fully relativistic treatment of correlation to a high order is necessary to reach such a good agreement.

5 Conclusion

For the very low transverse temperature of the highly expanded electron beam of the CRYRING cooler an enhancement of recombination rates is observed even for singly charged ions. The enhanced rates may reveal a striking dependence on the external magnetic field. However, it should be emphasized that no currently available theoretical model properly describes these effects. It was shown that dielectronic recombination resonances can be measured with an accuracy in the order of 1 meV for the examples of Li-like F and Ar. Energy positions and sizes of the cross sections are found in reasonable good agreement with the values calculated in many-body perturbation theory when QED effects, relativistic effects, and correlation is taken into account. Sensitive tests of QED in many particle systems and strong fields can be done with such experiments.

Acknowledgments

The authors thank the staff of CRYRING for their assistance. This work was supported by the Swedish Natural Science Research Council and the Knut and Alice Wallenberg Foundation.

References

1. H.P. Summers and W.J. Dickson *Applications of Recombination*, NATO ASI Series B, Physics, Plenum Press, New York (1992).
2. W.H. Tucker, *Radiation Processes in Astrophysics*, MIT University Press, Cambridge, MA (1975).
3. K.P. Kirby, Physica Scripta. T **59**, 59 (1995).
4. G. Gabrielse, S.L. Rolston, L. Haarsma, and W. Kells W, Hyp. Interact. **44**, 287 (1988).
5. M. Holzscheiter *et al.* AIP Conference Proceedings 457, 1998, p.65
6. R. Schuch in "Review of Fundamental Processes and Applications of Atoms and Ions" ed. by C.D. Lin, World Scientific Publ., Singapore (1993).
7. F. Bosch in Physics of Electronic and Atomic Collisions ed. T. Andersen and B. Fastrup, publ. by Adam Hilger IOP Publishing Ltd (1993).
8. R.E. Marrs, P. Beiersdorfer, D. Schneider , Phys. Today, Oct. 94, p.27.
9. A. Wolf, J. Berger, M. Bock, D. Habs, B. Hochadel, G. Kilgus, G. Neureither, U. Schramm, D. Schwalm, E. Szmola, A. Müller, M. Wagner, and R. Schuch, Z.Phys. D **21**, S69 (1991).
10. H. Gao, D.R. DeWitt, R. Schuch, W. Zong, S. Asp, and M. Pajek, Phys. Rev. Lett. **75**, 4381 (1995).
11. H. Gao, S. Asp, C. Biedermann, D.R. DeWitt, R. Schuch, W. Zong and H. Danared, Hyp. Interact. **99**, 301 (1996).
12. H. Gao, R. Schuch, W. Zong, E. Justiniano, D.R. DeWitt, H. Lebius, and W. Spies, J.Phys.B **30**, L499 (1997).
13. N. Eklöw, P. Glans, W. Zong, R. Schuch, and H. Danared Hyp. Interact. to be publ.
14. A. Hoffknecht, O. Uwira, S. Schennach, A. Frank, J. Haselbauer, W. Spies, N. Angert, P.H. Mokler, R. Becker, M. Kleinod, S. Schippers, A.

Müller, J. Phys. B 31 (1998) 2415.
15. A. Hoffknecht, T. Bartsch, S. Schippers, A. Müller, N. Eklöw, P. Glans, M. Beutelspacher, M. Grieser, G. Gwinner, A. A. Saghiri and A. Wolf, Physica Scripta. Vol. T80 (1999) 298.
16. M. Pajek and R. Schuch, Hyp. Interact. **108**, 185 (1997).
17. D .R. DeWitt, et al. *et al.*, Phys. Rev. **A50**, 1257 (1994); D .R. DeWitt, *et al.*, Journal of Phys. **B28**, L147 (1995); D.R. DeWitt, et al. Phys. Rev. **A53**, 2327, (1996).
18. W. Spies, et al. Phys. Rev. Lett. **69** (1992) 2768.
19. H. T. Schmidt, et al. Phys. Rev. Lett. **72**, p.1616 (1994).
20. S. Mannervik, et al. Phys. Rev. **A55**, 1810 (1997)
21. W. Zong, R. Schuch, E. Lindroth, H. Gao, D.R. DeWitt, S. Asp, and H. Danared, Phys. Rev. A **56**, 386, (1997).
22. H. Danared, G. Andler, L. Bagge, C.J. Herrlander, J. Hilke, J. Jeansson, A. Källberg, A. Nilsson, A. Paál , K.-G. Rensfelt, U. Rosengård, J. Starker, and M. af Ugglas, Phys. Rev. Lett. **72** 3775 (1994).
23. D.R. DeWitt, R. Schuch, H. Gao, W. Zong, S. Asp, C. Biedermann, M.H. Chen, and, N.R. Badnell, Phys. Rev. A **53**, 2327 (1996).
24. L.H. Andersen, J. Bolko, and P. Kvistgaard, Phys. Rev. Lett. **64**, 729 (1990) , L.H. Andersen and J. Bolko, Phys. Rev. A **42**, 1184 (1990).
25. A.V. Aleksandrov, *Proceedings of the Workshop on Electron Cooling and New Cooling Techniques*, ed R Calabrese and L Tecchio, p. 279, World Scientific, Singapore (1991).
26. S. Ichimaru and M.N. Rosenbluth, Phys. Fluids **13**, 2778 (1970).
27. D. Montgomery, G. Joyce, and L. Turner, Phys. Fluids **17**, 2201 (1974).
28. D.R. DeWitt, R. Schuch, T. Quinteros, H. Gao, W. Zong, S. Asp, H. Danared, M. Pajek and N. Badnell, Phys. Rev. A **50**, 1257 (1994).
29. S. Mannervik, D.R. DeWitt, E. Lindroth, J. Lidberg, R. Schuch, Phys. Rev. Lett. **81**, 313 (1998), and S. Mannervik, S. Asp, L. Broström, D.R. DeWitt, J. Lidberg, R. Schuch, and K.T. Chung, Phys. Rev. A **55**, 1810 (1997).
30. R. Schuch, W. Zong, P. Glans, W. Spies, and H. Danared, Hyp. Interact. **115**, 123 (1998).
31. P. Glans, N. Eklow, W. Gwinner, W. Zong, and R. Schuch, Nucl. Inst. Meth. **B154**, 997 (1999).
32. T. Quinteros, H. Gao, D.R. DeWitt, R. Schuch, M. Pajek, S. Asp, and Dz. Belkic, Phys. Rev. A **51**, 1110 (1995).
33. E. Lindroth and J. Hvarfner, Phys. Rev. **A45**, 2771 (1991).
34. E. Lindroth, Phys. Rev. **A49**, 4473 (1994).
35. S. A. Blundell, Phys. Rev. **A47**, 1790 (1993).

ELECTRON TRANSFER PROCESSES IN COLLISIONS OF RELATIVISTIC IONS WITH ATOMS

ZIEMOWID SUJKOWSKI
A. Sołtan Institute for Nuclear Studies 05-400 Świerk, Poland, e-mail: sujkowsk@ipj.gov.pl

The main processes occuring to a relativistic ion colliding with an atom are the radiative and the non-radiative electron capture into the vacant states of the ion and the stripping (ionization) of electrons from the ion. A brief review of these processes is given separately for the heavy and for the light ions. The results of recent experiments at the RCNP Osaka on the electron capture to He^{++} projectiles and the stripping of electron from He^{+} ions are presented. The astrophysical relevance of this information is discussed.

1 Introduction

The last decade has witnessed a rapid development of the techniques of nuclear physics, notably those of stripping the atoms to any desired degree of ionization, of accelerating the highly stripped ions to high energies and of storing and cooling the ion beams. This has brought about an increased interest in a new, separate subfield of physics: the *High Energy Atomic Physics*. Among the problems addressed here we can mention:

i) the physics of very strong (extreme?) electromagnetic fields,

ii) the testing of QED predictions at the level of high order corrections,

iii) the testing of approximations used in atomic physics to describe **many body systems** (though the forces are known, various approximations are still needed),

iv) the dynamics of atomic interactions (collisions ion-ion, ion-atom, particle-atom, ...),

v) the interplay of atomic and nuclear phenomena (including searches for "new physics"),

vi) applications to plasma physics and astrophysics.

The present paper concentrates on item (iv), i.e. on the dynamics of atomic interactions at high energies. Understanding of the relevant processes can be considered as the necessary prerequisite for the work on the other items.

The dominant processes occuring to a fast ion colliding with an isolated atom or traversing a solid are the emission of the bremsstrahlung radiation and the capture of electrons from the atom to the vacant states in the ion and the stripping of the bound electrons from the ion. The essential features of the latter two processes are described below. The recent experiments on the electron transfer processes involving a very light, relativistic projectile, the $^3He^{++}$ ion, are outlined in some detail, both because of their potential astrophysical relevance and as an example of pushing the atomic physics measurements to the extremes thanks to the use of modern nuclear physics spectrometric tools.

2 Heavy relativistic projectiles: electron capture and stripping

The capture of electrons by a passing ion proceeds via two distinctly different processes: the Radiative Electron Capture, REC, and the Non-Radiative Electron Capture, NREC. The REC process can be considered as the time reversed photoelectric effect on the projectile atom in an ionized state (Fig.1):

$$Z_p^{q+} + \hbar\nu \Rightarrow Z^{q+1^+} + e^-$$

The implicit assumption is that the captured electron is free and at rest. If the capture involves an electron bound in the target atom, the kinematics of the photon spectrum will change accordingly and the REC photon line will be broadened. The NREC process occurs exclusively for the bound atomic electrons: the third body, the atomic nucleus, is necessary to satisfy the momentum and energy conservation conditions. The process has a very sharp dependence on the atomic number of the target, Z_t. It occurs mainly at the velocity matching condition $v_p \approx v_e$, where v_p and v_e are the velocities of the projectile and of the bound target electron, respectively. For $v_p > v_e$ the cross-section, σ_{NREC}, roughly changes as:

$$\sigma_{NREC} \sim Z_p Z_t^5 E_p^{-6}$$

where E_p is the energy of the projectile. A review of the electron capture processes for relativistic heavy ions has recently been given by Stoehlker [1].

Fig. 2 shows the energy dependence of the capture cross-section for bare uranium ions impinging on gaseous nitrogen target [2]. It is seen that at low energies, below about 100 MeV/amu, the dominating contribution to the total cross-section is due to the NREC process, while already at 400 MeV/amu that contribution is marginal compared to that due to REC. The rapid decrease

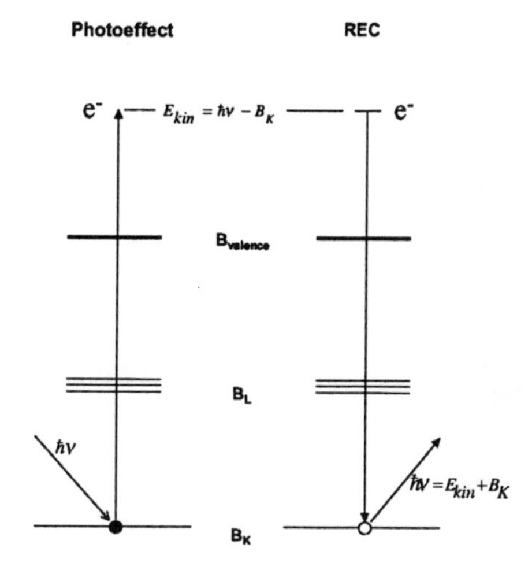

Figure 1. Schematic presentation of the REC process as the time reversed photoelectric effect

of the cross-section with energy is also worth noting. Fig. 3 shows the same cross-section as a function of the target atomic number for a selected uranium ion energy of about 220 MeV/amu. Again, the relative contribution of REC and NREC change very sharply, with REC prevailing for low Z_t and NREC raising quickly for heavy elements [2].

The theoretical description of the REC process using the electric dipole approximation [3] seems to be satisfactory for low bombarding energies, at least as the total cross-sections are concerned. Recently, deviations from these predictions have been observed [4] for differential cross-section at the forward photon emission angles for the few hundred MeV/amu uranium projectiles. This is interpreted as evidence for a magnetic dipole contribution (the so-called transverse component).

While the angular distributions of photons are presumably the most sensitive observables for this effect, the magnetic component can also be identified in another process - namely that of the total electron stripping cross-section,

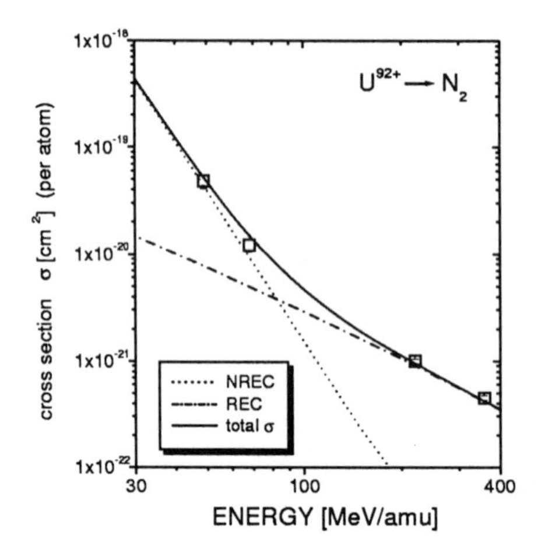

Figure 2. Electron capture cross-section for bare uranium ions as function of bombarding energy

σ_{ion}, albeit at much higher energies. This is illustrated in Fig 4 where the total ionization (stripping) cross-section for hydrogen-like gold ion, ${}_{79}Au^{78+}$, is plotted as a function of energy. The low energy data (up to 1 GeV/amu) are from [5], the 12 GeV/amu point is from [6]. While the electric dipole approximation predicts a maximum of σ_{ion} at about 200 MeV/amu and a slow decrease above that value, the data show a rather quick increase above ~ 1 GeV/amu. This can be attributed to the magnetic (transvertial) component, correctly reproduced in the complete relativistic calculation [7].

3 Light relativistic projectiles: electron capture and stripping

The very sharp Z_p dependence of the capture cross-section makes experiments with fast light ions very difficult. Very special techniques are required. The technique developed at the RCNP in Osaka to study the (^3He, t) charge exchange reactions was recently applied to measure the REC and NREC cross-section for semi-relativistic $^3He^{++}$ ions ($E_p = 150\,\text{MeV/amu}$, $v = 0.51c$). The technique consists in recording the triton spectra in a large magnetic

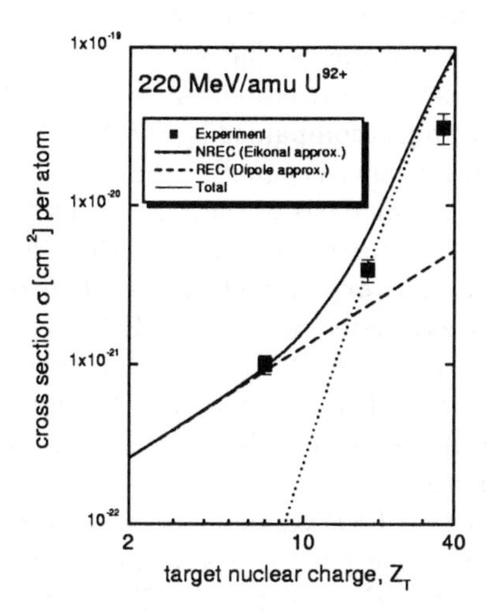

Figure 3. Electron capture cros-section for 220 MeV/amu bare U ions as function of Z_t

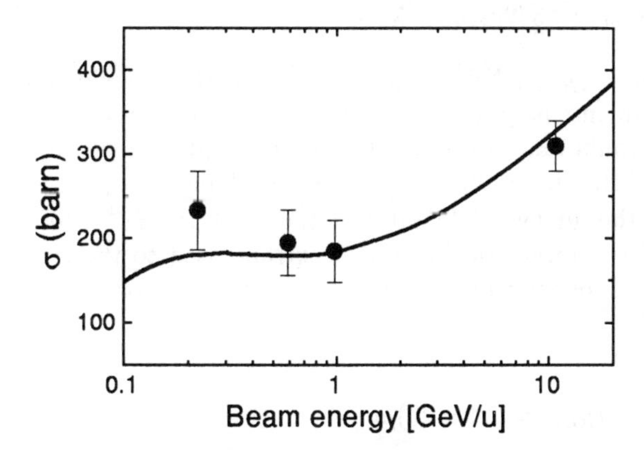

Figure 4. Ionization cross-section for hydrogen - like gold ions as function of energy

spectrograph set at $\theta = 0^o$ with respect to the beam. The magnetic rigidity of the $^3He^+$ ions resulting from the atomic charge exchange process (the electron capture) is approximately the same as that of the triton ejectiles , $^3H^+$, following the nuclear charge exchange reaction. As a result, the triton spectra recorded at $\theta = 0^o$ are invariably accompanied by the $^3He^+$ peak. The yield ratio of the He^+ to the beam intensity is equal to the ratio of the total capture to stripping cross-sections. An analysis of these ratios as function of Z_t is presented in [8]. Recently, an experiment dedicated to the REC process has been carried out at the RNCP. The experiment consisted in measuring coincidences between the photons associated with the REC effect and the $^3He^+$ ions corresponding to those initial He^{++} ions which have captured an electron. The $^3He^+$ ions were recorded in the Grand Raiden magnetic spectrometer [9] set at $\theta = 0^o$, i.e. in the set-up identical to that used in the (^3He, t) charge exchange reaction studies [10].

The photons were registered with a LEP Ge detector positioned sequentially at $\theta = 80^o$ and at $\theta = 130^o$ with respect to the beam. Two carbon targets were used, $46\,\mu g/cm^2$ and $9.6\,\mu g/cm^2$ thick, respectively. The photon spectra are practically due only the Radiative Electron Capture effect, with negligible background. The REC peaks observed at the two angles appear with the correct Doppler shifts with respect to the center-of-mass energy $E_{REC} = 84\,keV$. The coincidence yields were analysed according to the formula

$$\frac{\sigma_{REC}}{\sigma_{tot}} = \frac{N^{REC}}{3/2\varepsilon sin^2\Theta} \frac{1}{N^{He^{++}}}$$

where $\sigma_{tot} = \sigma_{REC} + \sigma_{NREC}$, N^{REC} and $N^{He^{++}}$ are the coincidence photon yield and the beam intensity, respectively, and ε is the photon detection efficiency. The formula takes into account the REC photon angular distribution according to [11]. The result is $\frac{\sigma_{REC}}{\sigma_{tot}} = 0.58 \pm 0.08$.

This, combined with the singles yield ratio of the He^+ to He^{++} ions measured as a function of the target thickness and extrapolated to the zero thickness value permits to independently determine the three cross-section values:

$$\sigma_{REC} = (54 \pm 20)\mu b$$

$$\sigma_{NREC} = (40 \pm 20)\mu b$$

$$\sigma_{ion} = (375 \pm 80)kb$$

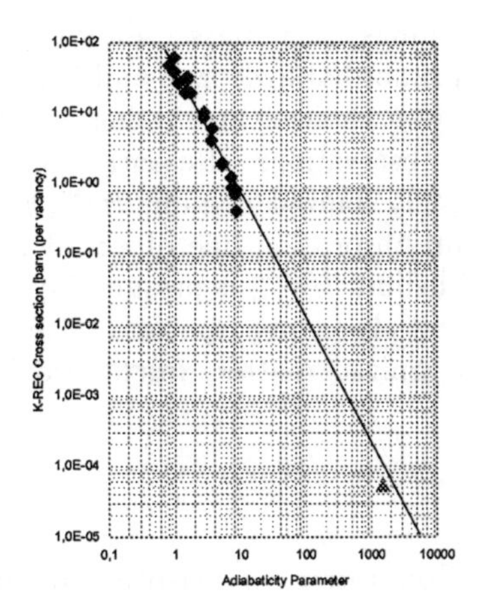

Figure 5. Systematics of REC cross-sections as function of adiabaticity parameter

The errors are statistical and do not include the calibration of the Faraday cup measuring the He^{++} intensity. The rather large error values are due mainly to the extrapolation procedure (only three targets were used - the two thin ones mentioned above and one 2000 $\mu g/cm^2$ thick; the saturation of the yield ratio sets in at about 100 $\mu g/cm^2$).

The measured σ_{REC} value is significantly smaller than that calculated according to the state-of-the-art theory [12], equal to 115 μb. The theoretical values of σ_{NREC} and σ_{ion} are 20 μb and 435 kb, respectively.

Fig. 5 shows the available systematics [1] of σ_{REC} values versus the adiabaticity parameter, $\eta = (v_p/v_e)^2$. The previously available data end at $\eta \simeq 10$, with the corresponding $\sigma_{REC} \approx 1b$. The present σ_{REC} value at $\eta = 1500$ extends this systematics by nearly five orders of magnitude down. This demonstrates the power a modern nuclear physics spectrometric facility such as the Grand Raiden in Osaka, in measuring an atomic cross-section under extreme conditions.

An interaction of fast He ions with free or quasi-free electrons is of primary astrophysical interest. Helium and hydrogen are the two main primordial elements. Any observable pertaining to these elements may be relevant for deducing, among other things, the mass balance in the Universe. In particular, searching for the primordial hydrogen and helium in intergalactic space is hampered by the fact that these elements there are fully ionized and thus escape the usual optical observations. Recently, evidence has been found for the presence of the singly ionized helium, He^{+}, in the space. This was deduced from the 30.4 nm absorption line observed in the quasar spectra [13], [14]. The fully ionized helium He^{++} is presumably much more abundant but it gives no signal in the absorption spectra. The REC process, which may occur for the totally ionized matter, might provide a useful observable in this search.

ACKNOWLEDGEMENT

This work was supported by the Polish Committee for Scientific Research KBN Grant No 2P03B11615.

References

1. Th. Stöhlker, habilitation thesis, GSI Darmstadt, 1999 and to be published.
2. Th. Stöhlker, T. Ludziejewski *et al.*, *Phys. Rev.* **A58**, (1998), 2043
3. J. Eichler, *Phys. Rep.* **193**, (1993), 165
4. Th.Stöhlker, T. Ludziejewski *et al.*, *Phys. Rev. Lett.* **82**, (1999), 3232
5. Th. Stöhlker *et al.*, *Phys. Rev. Lett.* **79**, (1997), 3270
6. N. Claytor *et al.*, *Phys. Rev. A* **55**, (1997), R842
7. J. Eichler *et al.*, *Phys. Rev. A* **51**, (1995), 3027
8. D. Chmielewska *et al.*, *Proc. of the 8th International Conference on Nuclear Reaction Mechanisms*, Varenna 1997, Ricerca Scientifica ed Educazione Permanente, p. 703
9. T. Noro, *Proc. of the 14th RCNP OSAKA International Symposium: Nuclear reaction Dynamics of Nucleon-Hadron Many Body System*, World Scientific (1996), 86
10. M. Fujiwara *et al.*, *Acta Phys. Pol.* B **29**, (1998), 179
11. E. Spindler *et al.*, *Phys. Rev. Lett.* **42**, (1979), 832
12. J. Eichler , priv. communication
13. P. Jakobsen *et al.*, *Nature* **370**, (1994), 35
14. A. Songalia *et al.*, *Nature* **375**, (1995), 124

DATA ANALYSIS IN POSITRON SPECTROSCOPY

GH. ADAM AND S. ADAM
Institute of Physics and Nuclear Engineering, IFIN-HH, Department of Theoretical Physics, P.O. Box MG-6, R76900 Bucharest-Măgurele, ROMANIA
E-mail: adamg@ifin.nipne.ro, adams@ifin.nipne.ro

The identification of the specific Fermi surface jumps in n-axis-projected positron spectroscopy histograms, measured by the technique of two-dimensional angular correlation of the electron-positron annihilation radiation (2D-ACAR) in high-T_c superconducting single crystals, is reviewed. Four topics are highlighted: (i) the main features of a 2D-ACAR histogram; (ii) the data reformulation from the laboratory frame to the crystal frame; (iii) statistical noise smoothing and background removal; (iv) analysis illustration on a c-axis projected histogram, collected on the archetypal $YBa_2Cu_3O_{7-\delta}$ single crystal at the University of Geneva.

1 Introduction

At the time of the discovery of the high-T_c superconductivity by Bednorz and Müller in 1987, the positron spectroscopy was a well established method for the investigation of the Fermi surface in solids[1]. It came therefore as no surprise that the technique of two-dimensional angular correlation of the electron-positron annihilation radiation (2D-ACAR) was considered by its practitioners as one of the most promising tools to probe the occurrence of the Fermi surface sheets in high-T_c superconductors. Until the first convincing result was obtained[2], however, all the positron spectroscopy topics had to be scrutinized again: the structural characterization of the high-T_c superconductors, the positron-electron interaction in such samples, the methods of growing high-T_c single crystals, the accuracy of the crystal-detector tuning, and, last but not least, the off-line analysis of the measured histograms.

In this paper, we consider the identification of the Fermi surface jumps inside a single crystal of orthorhombic symmetry, from a 2D-ACAR histogram which records the projection, along a principal crystallographic axis **n**, of the momentum density coming from zero-spin positron-electron pairs (an *n-axis-projected histogram*). We discuss an approach to the 2D-ACAR signal analysis which was able to solve[3,4,5] the electron Fermi surface sheets in high-T_c superconductors under tight control of the critical parameters which could spoil the resolution, or modify the shape, of the expected Fermi surface sheets.

The organization of the paper follows the main steps which secure a consistent signal analysis. Sec. 2 summarizes the main features of the 2D-ACAR histograms. Sec. 3 discusses the histogram redefinition from the laboratory frame to the crystal frame. Sec. 4 considers the statistical noise smoothing and background removal. The paper ends with an illustration, in Sec. 5, of the analysis of a 2D-ACAR histogram measured, at room temperature on the archetypal $YBa_2Cu_3O_{7-\delta}$ single crystal, at the University of Geneva.

2 The main features of the 2D-ACAR histograms

1. The initial stages of the positron interaction with the lattice of a high-T_c single crystal proceed similarly to those in normal metals. Before annihilation, the positrons reach thermal (or nearly thermal) equilibrium with the lattice and get spatially distributed along a characteristic "implantation profile", which is sufficiently deep such that the greatest fraction of the thermalized positrons annihilates in the bulk of the sample under study. Because of the strong Coulomb interaction of the positrons with electrons and ions, and as a consequence of the lamellar structure of the crystalline high-T_c superconductors, two features of the 2D-ACAR technique are to be noted: (a) the "positron energy band" formed by the low-energy positron states induces in the measured histograms specific *positron wavefunction effects*, which have to be dropped off from the output of the signal analysis; (b) the usefulness of the method is limited to the resolution of the Fermi surface sheets coming from those crystal regions which attract the positrons.

2. The transverse momentum distribution inside the 2D-ACAR histogram plane can be split into *Umklapp components*, which are related to the n-axis-projection of the first Brillouin zone (1BZ) of the sample in the *extended zone representation*, with an amplitude of the useful signal that rapidly decays towards the higher Umklapp components. Let

$$\{D_M | M = 0, 1, ..., M_{max}\}. \tag{1}$$

denote the areas centered at Γ_{00}, the *zero momentum projection of the center* Γ *of* 1BZ *onto the histogram plane*, and covering the first M-Umklapp components inside the histogram plane. The edges of D_0 are then given respectively by appropriate projections, $\{p_x, p_y\}$, of the primitive translation vectors of the reciprocal crystal lattice onto the histogram plane. The edges of D_M ($M \geq 1$) are given respectively by $\{(2M+1)p_x, (2M+1)p_y\}$. Thus, D_M consists of a collection of regions of areas D_0, centered respectively at the points Γ_{mn} of coordinates

$$p_x^{(m)} = mp_x; \; p_y^{(n)} = np_y; \; m, n \in \{-M, \cdots, M\}. \tag{2}$$

The number M_{max} of the Umklapp components available within a particular **2D-ACAR** histogram is defined by the geometry of the experiment and the angular aperture of the setup. Usually, $M_{max} = 2$ or 3.

3. The measured **2D-ACAR** signal originates in positron annihilations with electrons within bands crossing the Fermi level, bound electrons, electrons at crystal inhomogeneities, and in artifacts.

   The fraction which originates in the conduction electrons will show a quasi-periodic distribution over the histogram plane, with symmetries of the transverse momentum spectrum around the points Γ_{mn}, Eq. (2), as emerging from the subduction to the histogram plane of the symmetry properties of the electron energy bands of the crystal in the extended zone representation. In high-T_c superconductors, the magnitude of this *useful* fraction of the momentum distribution does not exceed a few percent of the accumulated statistics[6] and this is the main reason why giga-count statistics is needed to get well-resolved Fermi surface sheets[2].

   The fraction which originates in the bound electrons will show a unique symmetry center, Γ_{00}, with an overwhelming radially symmetric component and a localized component defined by the symmetry of the energy band crystal field splittings.

   The fraction which originates in electrons at crystal imperfections is responsible for the enhanced spectral intensities of the experimental data at low momenta in comparison with theoretical data calculated under the assumption of perfect crystalline periodicity [7,8,9]. Similar to the previous fraction, there is a unique symmetry center of it, Γ_{00}, with an overwhelming radially symmetric component[9].

   While the systematic instrumental artifacts are ruled out by specific procedures[3], there still remain residual artifacts the distribution and magnitude of which are unknown. The occurrence of *zero mean randomly distributed residual artifacts* represents the most favorable hypothesis on the quality of the **2D-ACAR** histogram under analysis. If, however, the artifacts have their own center, different from Γ_{00}, then the computed center of the **2D-ACAR** distribution will be pulled away from Γ_{00} [10]. As a consequence, the off-line data processing has to include criteria proving that the magnitude and distribution of the artifacts will have a negligibly small effect on the inferences of the data analysis.

4. Within a **2D-ACAR** experiment, the data is *discretized* in square bins of side a_D defined by the detectors and the geometry of the setup. In what

follows, if not specified otherwise, the in-plane coordinates and lengths are given in discretization step units a_D. The bins are then labeled by pairs of integers, (i, j), which correspond to the bin centers inside *the detector defined histogram plane*, $Op_x^D p_y^D$. As shown in detail elsewhere[3], this discretization defines a *piecewise constant continuous* momentum distribution over the histogram plane.

5. Within each histogram bin, the counts obey a Poisson statistics.

3 Data redefinition from laboratory frame to crystal frame

Essential for the success of the signal analysis is the operation of *histogram redefinition from the laboratory frame* (LF), $Op_x^D p_y^D p_z^D$, (defined by the setup and within which the data acquisition is performed) *to the crystal frame* (CF) $\Gamma_{00}p_x p_y p_z$ (identified with the projection of the canonical reference frame of the 1BZ along the n-axis and within which the various steps of the off-line analysis of the spectrum are legitimate). An n-axis-projected 2D-ACAR spectrum is secured by tuning the crystal axes to the detector axes. Under perfect tuning, the LF axis Op_z^D, normal at O to the histogram plane $Op_x^D p_y^D$, and the CF axis $\Gamma_{00}p_z$, are parallel to each other. Within data analysis, this condition is assumed to be fulfilled provided the Euler angle θ_0, Eq. (4), relating the CF to LF is negligibly small.

Thus, a consistent off-line analysis has to begin with the definition, if at all possible, of the CF and then to continue with the reformulation and processing of the 2D-ACAR histogram in this frame, under close scrutiny of the magnitude of the departures of the histogram features from the ideal ones.

Since the positron annihilation technique is non resonant, the derivation of the crystal frame from the LF data is to be done by a mathématical procedure able to maximize the *characteristic symmetry pattern* induced in a 2D-ACAR histogram by the occurrence of the Fermi surface, while minimizing the other contributions listed in Sec. 2, item #3. The concept of *signature of crystal symmetry* (SCS) in a 2D-ACAR histogram[4,11], fulfills this requirement.

The SCS of interest for signal analysis is defined both for the raw histogram $H = (h_{pq})$ and for the histogram $S = (s_{pq})$ obtained from H by a procedure of statistical noise smoothing *which does not involve symmetrization* (Sec. 4). It consists of a set of chi-square *densities per bin* which measure the goodness of the assumption that some suitable (κ, λ) sites inside the histogram plane $Op_x^D p_y^D$ can be taken for Γ_{00}. For single crystals of orthorhombic symmetry, admissible (κ, λ) sites are *bin centers, bin corners* and *centers of bin sides*, inside a convenient neighborhood of the center O of H.

For the histogram H, the SCS elements are chi-square sums

$$\chi_0^2(\kappa, \lambda; \widetilde{D}_M; H) = \frac{1}{N_M} \sum_{(i,j)\in\widetilde{D}_M} \sum_{(k,l)\in(i,j)^\star} (h_{ij} - h_{kl})^2/\sigma_{ij}^2. \quad (3)$$

Here, the summation region $\widetilde{D}_M$, centered at (κ, λ), denotes the closest *entire bin approximation* of the M-Umklapp area (1). The notation $(i, j)^\star$ denotes the manifold of histogram bins which can be obtained from the (i, j)-th bin by point group symmetry operations. The quantity σ_{ij} is estimated taking into account Sec. 2, item #5. Finally, the normalization factor N_M equates the total number of symmetry defined pairs of distinct bins inside $\widetilde{D}_M$.

For a 2D-ACAR histogram characterized by zero mean randomly distributed residual artifacts, the $M_{max} + 1$ obtained SCS sets would result in a *same set of* CF *parameters*

$$\{\widetilde{\gamma}_x = \kappa_0 + \xi_0, \quad \widetilde{\gamma}_y = \lambda_0 + \eta_0, \quad \phi_0, \quad \theta_0, \quad \psi_0\}, \quad (4)$$

where the first two quantities denote the components (entire and fractionary respectively) of the *translation vector* from O to Γ_{00}, while the last three denote the Euler angles relating the CF axes to the LF axes.

Unfortunately, the solutions (4) show a strong D_M dependence[4,11]. We are therefore obliged to devise alternative validation criteria allowing us to infer whether among the $M_{max} + 1$ areas D_M there is one showing small enough residual artifacts. Our analysis[4,11] showed that a dozen of validation criteria have to be used to pick the most reliable CF solution. In particular, the alternative definition of Γ_{00} as *center of symmetry* of $\widetilde{D}_M$ and as *center of gravity* of $\widetilde{D}_M$ singled out the area D_0 as being the least affected by artifacts.

4 Statistical noise smoothing and background removal

4.1 Statistical noise smoothing

The occurrence of Umklapp components, with a rapid decay of the useful signal towards the histogram borders (Sec. 2, item #2) suggests an off-line analysis over the various $M_{max} + 1$ areas D_M, Eq. (1). Thus, at fixed D_M, a local *window least squares* (WLS) statistical noise smoothing will be done at bins inside D_M only (the *central bins of the histogram*), while it will be skipped at the bins outside D_M (the *border bins of the histogram*).

The most convenient shape of the *smoothing window*, $C_r(K, L)$, around the (K, L)-th central bin satisfies the following requirements[3]: $C_r(K, L)$ is centered at (K, L) site; it consists of an *integer number of bins* (see Sec. 2, item #4); in the limit of an in-plane *continuous* point distribution, $C_r(K, L)$ is a *circular surface* around the point of coordinates (K, L).

As a consequence, to perform statistical noise smoothing, we draw around each *(K, L)-th central bin* a *smoothing window* $C_r(K, L)$ of *quasi-circular shape* which includes inside it all the bins the centers (κ, λ) of which satisfy

$$(\kappa - K)^2 \; + \; (\lambda - L)^2 \; \leq \; (2r + 1)^2/4, \tag{5}$$

where the *free parameter* r is the *window radial parameter* (WRP).

To accommodate both the data discretization into bins and the possibility to predict noise-free values *inside the bins*, the approximating space of noise-free data is spanned by a basis set of polynomials of continuous variables, $P_m(x, y)$, orthogonal over $C_r(K, L)$.

Within each smoothing window (5), *constant bin weights*, $\sigma_{K+\kappa,L+\lambda} = \sigma_{KL}$, may be assumed[3]. This results in a *constant weight* WLS *smoothing formula* of radial parameter r (CW-WLS(r)). At the fractionary coordinates (ξ, η), inside the (K, L)-th bin, the CW-WLS(r) yields a value[3,5]

$$s_{K+\xi,L+\eta} \; = \sum_{K+k,L+l \in C_r(K,L)} h_{K+k,L+l} \; G(\xi, \eta; k, l), \tag{6}$$

which defines the (K, L)-th element of the smoothed 2D-ACAR histogram S.

Since the values of ξ and η may be *arbitrary* within the range $[-0.5, 0.5]$, the discretization of the S histogram may be *different* from that of the raw histogram H. Discretization steps a_D^x and a_D^y may be tailored which are *rational fractions* of the 1BZ edges p_x and p_y respectively, allowing straightforward exploitation of the space group properties of the crystal lattice[4,5].

There is a unique value of the WRP r in Eq. (5), which selects the *best function within the admissible class* spanned by the basis polynomials. In[3,4,12], this parameter was only roughly optimized: it was kept constant inside a given M-Umklapp manifold, but different values were used at various M-Umklapp components. Later on it was shown[5] that the *study of the distribution of the signs of the residuals* of the smoothed data allows the derivation of *optimal* WRP *values at each individual bin*.

4.2 Background removal

Except for the tiny fraction of conduction electrons, the contributions to a 2D-ACAR histogram measured in a high-T_c superconductor (Sec. 2, item #3), form a useless background, removed so far by three kinds of methods:

(i) Subtraction from the smoothed histogram S of the same histogram rotated by 90 degrees. This procedure is very easy to use. It was able to resolve 2- and 3-Umklapp components of the Fermi surface ridge[2,3,12].

(ii) Computation of the spectrum component which is radially symmetric with respect to $\Gamma_{00}p_z$, as an inner envelopatrix of S, and its subtraction to get the anisotropic part of the spectrum. This approach, which is much more difficult to implement, allows signal resolution inside the central region D_0, characterized by the highest signal to noise ratio[4].

(iii) In an ideal single crystal, the `2D-ACAR` resolved Fermi surface jumps yield specific jumps in the distribution of the optimally adapted to the data `WRP`s, Eq. (5). In a real crystal, the smearing of the derived `WRP` optimal values, which originates from positron annihilations with electrons at crystal imperfections, is ruled out by median smoothing of the obtained distribution, over symmetry defined stars of bins. We are thus left with a subtractionless procedure of resolving the Fermi surface jumps[5].

5 Illustration on a typical example

The *c*-axis-projection of the `2D-ACAR` spectra on the archetypal $YBa_2Cu_3O_{7-\delta}$ single crystal was extensively investigated. It was found by various authors[6,9] that there is no investigation method able to resolve *all* the existing Fermi surface sheets. As it concerns `2D-ACAR`, this was able to resolve both electronic Fermi surface sheets predicted by the energy band theory.

The one-dimensional ridge, which crosses the whole `1BZ` along ΓX, shows the strongest signature inside a `2D-ACAR` histogram. Its specific 0-Umklapp component jumps were inferred without doubt within the `SCS` pattern inside D_0 as well as from the periodicity of `SCS` with respect to the S histogram rotation around the normal at Γ_{00} to the histogram plane. Its graphic representation[4] showed a remarkable similarity with the theoretically predicted representation. The 2-Umklapp component of the ridge, the first to be evidenced[2], was also solved within either window[3] and slice[12] representations. Finally, the first resolution of the 3-Umklapp component[3] was later confirmed theoretically[12] as well as by alternative off-line processing[13].

The pill-box around the point S of `1BZ` shows a very weak signature in the data. At room temperature, it was nevertheless evidenced by two alternative methods of analysis: by Lock-Crisp-West folding (i.e., signal enhancement by translation of the four corners of D_0 to a single place[4]), as well as by the recently described subtractionless method[5].

Acknowledgments

We are indebted to Professor Martin Peter for introducing us to the topics of positron spectroscopy, for help and encouragement, and for permission to use data measured at the University of Geneva. We are also indebted to Prof. A.A. Manuel, Dr. L. Hoffmann, and Dr. B. Barbiellini for advice.

The project was developed under financial support from the Ministry of Science and Technology of Romania, a Go West fellowship from the European Communities, and computing support from the Abdus Salam ICTP, Trieste.

References

1. S.Berko in *Positron Solid State Physics, Proc. of the Internat'l School "E. Fermi", Course 83*, eds. W. Brandt and A. Dupasquier (North Holland, New York, 1983) pp. 64–145. Reprinted in *Positron Studies of Solids, Surfaces and Atoms*, eds. A.P. Mills, Jr., W.S. Crane and K.F. Canter (World Scientific, Singapore, 1986) pp. 246–327.
2. H. Haghighi, J.H. Kaiser, S. Rayner, R.N. West, J.Z. Liu, R. Shelton, R.H. Howell, F. Solal and M.J. Fluss, *Phys. Rev. Lett.* **67**, 382 (1991).
3. Gh. Adam, S. Adam, B. Barbiellini, L. Hoffmann, A.A. Manuel and M. Peter, *Nucl. Instr. and Meth.* **A 337**, 188 (1993).
4. Gh. Adam and S. Adam, *Int. J. Modern Phys.* **B 9**, 3667 (1995).
5. Gh. Adam and S. Adam, *Computer Phys. Commun.* **120**, 215 (1999).
6. L.P. Chan, K.G. Lynn and D.R. Harshman, *Mod. Phys. Lett.* **B 6**, 617 (1992).
7. L. Hoffmann, W. Sadowski, A. Shukla, Gh. Adam, B. Barbiellini, and M. Peter, *J. Phys. Chem. Sol.* **52**, 1551 (1991).
8. L.C. Smedskjaer, A. Bansil, U. Welp, Y. Fang and K.G. Bailey, *Physica* **C 192**, 259 (1992).
9. R. Pankaluoto, A. Bansil, L.C. Smedskjaer and P.E. Mijnarends, *Phys. Rev.* **B 50**, 6408 (1994).
10. L.C. Smedskjaer and D.G. Legnini, *Nucl. Instr. and Meth.* **A 292**, 487 (1990).
11. Gh. Adam and S. Adam, *Romanian J. Phys.* **42**, 689 (1997), *ibid.* **43**, 75 (1998), *ibid.* **43**, 89 (1998).
12. Gh. Adam, S. Adam, B. Barbiellini, L. Hoffmann, A.A. Manuel, M. Peter and S. Massidda, *Solid State Commun.* **88**, 739 (1993).
13. L. Hoffmann, A.A. Manuel, M. Peter, E. Walker, M. Gauthier, A. Shukla, B. Barbiellini, S. Massidda, Gh. Adam, S. Adam, W.N. Hardy, Ruixing Liang, *Phys. Rev. Lett.* **71**, 4047 (1993).

DIRAC EXPERIMENT AND TEST OF LOW-ENERGY QCD

M. PENTIA

ON BEHALF OF
THE DIRAC COLLABORATION

NIPNE-HH, P.O.Box MG-6, 76900, Bucharest-Magurele, ROMANIA
E-mail: pentia@ifin.nipne.ro

B. ADEVA[O], L. AFANASEV[L], M. BENAYOUN[D], V. BREKHOVSKIKH[N], G. CARAGHEORGHEOPOL[M],T. CECHAK[B], M. CHIBA[J], S. CONSTANTINESCU[M], A. DOUDAREV[L], D. DREOSSI[F], D. DRIJARD[A], M. FERRO-LUZZI[A], T. GALLAS TORREIRA[A,O], J. GERNDT[B], R. GIACOMICH[F], P. GIANOTTI[E], F. GOMEZ[O], A. GORIN[N], O. GORTCHAKOV[L], C. GUARALDO[E], M. HANSROUL[A], R. HOSEK[B], M. ILIESCU[E,M], N. KALININA[L], V. KARPOUKHINE[L], J. KLUSON[B], M. KOBAYASHI[G], P. KOKKAS[P], V. KOMAROV[L], A. KOULIKOV[L], A. KOUPTSOV[L], V. KROUGLOV[L], L. KROUGLOVA[L], K.-I. KURODA[K], A. LANARO[A,E], V. LAPSHINE[N], R. LEDNICKY[C], P. LERUSTE[D], P. LEVISANDRI[E], A. LOPEZ AGUERA[O], V. LUCHERINI[E], T. MAKI[I], I. MANUILOV[N], L. MONTANET[A], J.-L. NARJOUX[D], L. NEMENOV[A,L], M. NIKITIN[L], T. NUNEZ PARDO[O], K. OKADA[H], V. OLCHEVSKII[L], A. PAZOS[O], M. PENTIA[M], A. PENZO[F], J.-M. PERREAU[A], C. PETRASCU[E,M], M. PLO[O], T. PONTA[M], D. POP[M], A. RIAZANTSEV[N], J.M. RODRIGUEZ[O], A. RODRIGUEZ FERNANDEZ[O], V. RYKALINE[N], C. SANTAMARINA[O], J. SCHACHER[Q], A. SIDOROV[N], J. SMOLIK[C], F. TAKEUTCHI[H], A. TARASOV[L], L. TAUSCHER[P], S. TROUSOV[L], P. VAZQUEZ[O], S. VLACHOS[P], V. YAZKOV[L], Y. YOSHIMURA[G], P. ZRELOV[L]

[a] *CERN, Geneva, Switzerland ;* [b] *Czech Technical University, Prague, Czech Republic ;* [c] *Prague University, Czech Republic ;* [d] *LPNHE des Universites Paris VI/VII, IN2P3-CNRS, France ;* [e] *INFN - Laboratori Nazionali di Frascati, Frascati, Italy ;* [f] *Trieste University and INFN-Trieste, Italy ;* [g] *KEK, Tsukuba, Japan ;* [h] *Kyoto Sangyou University, Japan ;* [i] *UOEH-Kyushu, Japan ;* [j] *Tokyo Metropolitan University, Japan ;* [k] *Waseda University, Japan ;* [l] *JINR Dubna, Russia ;* [m] *National Institute for Physics and Nuclear Engineering NIPNE-HH, Bucharest, Romania ;* [n] *IHEP Protvino, Russia ;* [o] *Santiago de Compostela University, Spain ;* [p] *Basel University, Switzerland ;* [q] *Bern University, Switzerland*

The low-energy QCD predictions to be tested by the DIRAC experiment are revised. The experimental method, the setup characteristics and capabilities, along with first experimental results are reported. Preliminary analysis shows good detector performance: alignment error via Λ mass measurement $m_\Lambda = 1115.6\ MeV/c^2$ with $\sigma = 0.92\ MeV/c^2$, $p\pi^-$ relative momentum resolution $\sigma_Q \approx 2.7\ MeV/c$, and evidence for $\pi^+\pi^-$ low momentum Coulomb correlation.

1 Introduction

Quantum Chromodynamics (QCD), responsible for the strong interaction sector of the Standard Model (SM) has successfully been tested only in the perturbative region of high momentum transfer ($Q > 1\ GeV$) or at short relative distance $\Delta r \sim \hbar/Q$ ($\Delta r < 0.2 fm$). Here the constituent quarks behave as weakly interacting, nearly massless particles. The QCD in the perturbative region, as any gauge theory with massless fermions, presents chiral symmetry. In the nonperturbative region of low momentum transfer (low-energy), say $Q < 100\ MeV$, or equivalently at large distance ($\Delta r > 2fm$), asymptotic freedom is absent, and quark confinement takes place. In the low energy region the chiral symmetry of QCD must be spontaneously broken.

The *Chiral Perturbation Theory* (ChPT)[1,2] seems to be the candidate theory for low energy processes. It exploits the mechanism of spontaneous breakdown of chiral symmetry (SBChS), or, in other words, the existence of a quark condensate. In order to test the existence of the quark condensate, the particularly significant symmetry breaking effect refers to the *S-wave $\pi\pi$ scattering lengths*. From the theoretical point of view, $\pi\pi$ scattering is a deeply studied problem. Within standard ChPT, Gasser and Leutwyler[1,2,3] and also Bijnens and collaborators[4] as well as within the *Generalized Chiral Perturbation Theory* (GChPT), Stern and collaborators[5,6], have obtained expressions for the $\pi\pi$ scattering amplitude in the chiral expansion.

The leading order expansion of the scattering amplitude is[6]

$$A(s;t,u) = \alpha \frac{M_\pi^2}{3F_\pi^2} + \frac{\beta}{F_\pi^2}\left(s - \frac{4}{3}M_\pi^2\right) + \mathcal{O}(p^4) \tag{1}$$

where α and β encode the information on the strength of the quark condensate. In the limit of a strong quark condensate, one has $\alpha \approx 1$, $\beta \approx 1$. A substantial departure of α from unity signals a much smaller value of the condensate.

The values predicted for the isospin $I = 0$ and $I = 2$ S-wave scattering lengths a_0^0 and a_0^2 can be confronted with the future experimental values of the DIRAC experiment. The available experimental data[7] for the scattering length is $a_0^0 = 0.26 \pm 0.05$. Based on these data there is no possibility to estimate the strength of the quark condensate and so to measure the extent of chiral symmetry breaking.

The DIRAC experiment[8] aims to determine the difference of the scattering lengths $\Delta = |a_0^0 - a_0^2|$ with 5% accuracy, by measuring the lifetime of pionium ($\pi^+\pi^-$ bound state). For the first time experimental evidence in favour of or against the existence of a strong quark condensate in the QCD vacuum could be within reach.

2 Pionium lifetime

Pionium is a metastable bound state, produced by π^+ and π^- electromagnetic interaction and decaying into $\pi^0\pi^0$ due to strong interaction. To obtain pion scattering lengths in a model independent way, a measurement of the lifetime of pionium has been proposed many years ago by Nemenov[9]. The measurement of the pionium lifetime (τ) will allow to determine the difference $|a_0^0 - a_0^2|$ of the strong S-wave $\pi\pi$-scattering lengths for isospin $I = 0$ and $I = 2$.

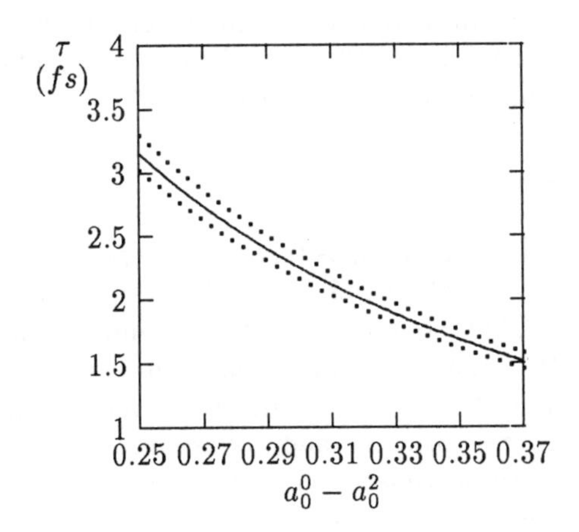

Figure 1. The pionium lifetime, in units of 10^{-15} s, as a function of the combination $(a_0^0 - a_0^2)$ of the S-wave scattering lengths. The band delineated by the dotted lines takes account of the uncertainties, coming from theoretical evaluations, low energy constants and a_1^1. (Thanks to H.Sazdjian hep-ph/9911520).

The general expression for the pionium decay width (Γ), according to ChPT[10] at leading and next-to-leading order in isospin breaking, is

$$\frac{1}{\tau} = \Gamma = \frac{2}{9} \cdot \alpha^3 \cdot p^* \cdot \left(a_0^0 - a_0^2 + \epsilon\right)^2 (1 + K), \tag{2}$$

where α is the fine structure constant ;

$p^* = \left(M_{\pi^+}^2 - M_{\pi^0}^2 - \alpha^2 M_{\pi^+}^2/4\right)^{1/2}$ is the CM momentum of π^0 ;
$\epsilon = (0.58 \pm 0.16) \cdot 10^{-2}$; $K = 1.07 \cdot 10^{-2}$.

Using Eq.(2), $a_0^0 = 0.206$ and $a_0^2 = -0.0443$, Gasser et al.[10] have evaluated the pionium lifetime in the ground state $\tau = 3.25 \cdot 10^{-15} s$. In the isospin symmetry limit $\epsilon = 0$; $K = 0$, Eq.(2) becomes the Deser type formula[11]. Eq.(2) allows to get a relative error of the scattering lengths difference $\frac{\Delta(a_0^0 - a_0^2)}{(a_0^0 - a_0^2)} \approx 5\%$, if DIRAC measures the lifetime with a $\frac{\Delta\tau}{\tau} \approx 10\%$ error.

According to GChPT[13] the τ vs. $(a_0^0 - a_0^2)$ dependence is presented in Fig.1. Hence the experimental determination of the pionium lifetime could be interpreted in quark condensation terms for three different cases: First, if the central value of τ is close to 3×10^{-15} s, lying above 2.9×10^{-15} s, then the strong condensate assumption of ChPT is firmly confirmed, since its predictions of $(a_0^0 - a_0^2)$ lie between 0.250 and 0.258. Second, if the central value of τ lies below 2.4×10^{-15} s, it is the scheme of GChPT, which is confirmed, since the corresponding central value of α would lie above 2. The third possibility is the most difficult to interpret. If the central value of τ lies in the interval $2.4 \div 2.9 \times 10^{-15}$ s, then, because of the uncertainties, ambiguities in the interpretation may arise.

3 Experimental method

In an abundent production of oppositely charged pions the Coulomb interaction can form atomic $\pi^+\pi^-$ bound states. If the pions have a small relative momentum in their CM system ($Q \sim 1\ MeV/c$) and are much closer than the Bohr radius (387 fm), then pionium atoms are produced with a high production probability due to the large wave function overlap. Such pions originate from short-lived sources (ρ, ω, Δ), and not from long-lived ones (η, K_s^0), because in the latter case the separation of the two pions is in most cases larger than the Bohr radius.

The pionium production cross section is proportional to the double inclusive cross section $d\sigma_s^0/(d\vec{p}_1 d\vec{p}_2)$ for $\pi^+\pi^-$ pairs from short-lived sources, without Coulomb interaction in the final state[9], and to the squared atomic wave function of nS-states at the origin $|\psi_n(0)|^2$

$$\frac{d\sigma_n^A}{d\vec{p}_A} = (2\pi)^3 \frac{E_A}{M_A} |\psi_n(0)|^2 \left.\frac{d\sigma_s^0}{d\vec{p}_1 d\vec{p}_2}\right|_{\vec{p}_1 = \vec{p}_2 = \frac{\vec{p}_A}{2}} \tag{3}$$

where $\vec{p}_A, E_A$ and M_A are momentum, energy and mass of the pionium atom in the Lab system respectively; $\vec{p}_1$ and $\vec{p}_2$ are the π^+ and π^- momenta in the Lab system, and they must satisfy the relation $\vec{p}_1 = \vec{p}_2 = \vec{p}_A/2$ to form the atomic bound state. Atoms formed in this way are in a S-state.

After production in hadron-nucleus interaction relativistic pionium atoms ($2\ GeV/c < p_A < 6\ GeV/c$) are moving in the target. They can decay or, due to electromagnetic interaction with the target material, get excited or broken-up (ionized). Using atomic interaction cross sections, for a given target material and thickness, one can calculate the break-up probability for arbitrary values of pionium momentum and lifetime. In Fig.2 there are presented these dependencies for the pionium momentum $p = 4.7\ GeV/c$.

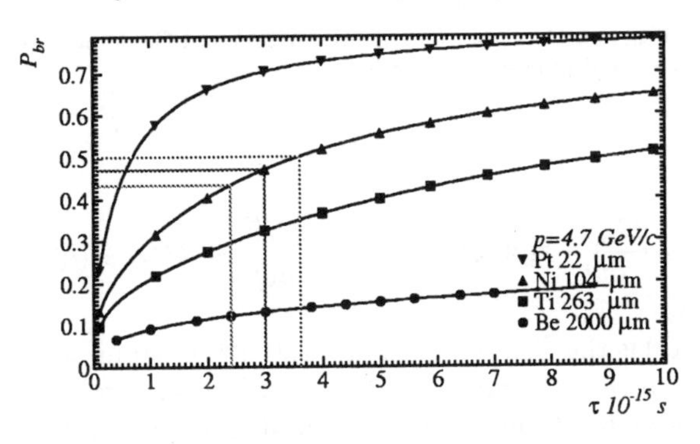

Figure 2. Probability of pionium break-up in the target.

Comparison of the measured break-up probability $P_{br} = n_A/N_A$ (ratio of broken-up - n_A and produced - N_A pionium atoms) with the calculated dependence of P_{br} on τ (see Fig.2) gives a value of the lifetime.

The break-up process gives characteristic $\pi^+\pi^-$ pairs, called *atomic pairs*. They have a small relative momentum in their CM system ($Q < 3\ MeV/c$), a small opening angle ($\theta_\pm \approx 6/\gamma \approx 0.35\ mrad$ for $p_A = 4.7\ GeV/c$) and nearly identical energies in the Lab system ($E_+ = E_-$ at 0.3 % level).

3.1 Determination of broken-up pionium atoms (n_A)

The measurement of broken-up n_A pionium atoms is realized through the analysis of the experimental distribution in Q of $\pi^+\pi^-$ pairs.

The free pion pair distribution can be written as the sum of the non-Coulomb (nC) (no final state interaction) and the Coulomb (C) pair distribution (Fig.3):

$$\frac{dN^{free}}{dQ} = \frac{dN^{nC}}{dQ} + \frac{dN^{C}}{dQ}$$

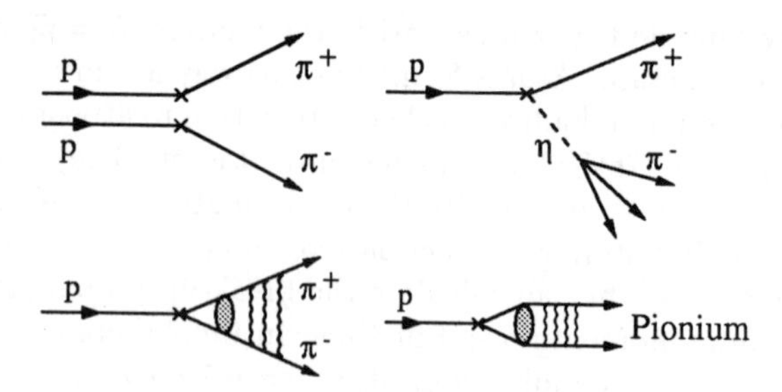

Figure 3. Accidental, non-Coulomb (long-lived sources) and Coulomb (short-lived sources) pion pair production and pionium production

$$\frac{dN^{nC}}{dQ} \sim \frac{dN^{exp}_{acc}}{dQ} \equiv \Phi(Q) \qquad ; \qquad \frac{dN^{C}}{dQ} \sim \Phi(Q)A_c(Q)(1+aQ)$$

where $A_c(Q)$ is the Coulomb and $(1+aQ)$ is the strong correlation factors. Here we assumed that the non-Coulomb distribution of $\pi^+\pi^-$ pairs (without FSI) can be extracted from the experimental distribution of accidental pairs $\Phi(Q)$. The free pion pair distribution is then given by

$$\frac{dN^{free}}{dQ} = \frac{dN^{nC}}{dQ} + \frac{dN^{C}}{dQ} = N_0\Phi(Q)\left[f + A_c(Q)(1+aQ)\right], \tag{4}$$

N_0, f, a - free parameters. In the region $Q > 3\ MeV/c$, there are mainly free pairs. This part of the distribution is fitted with the function (4) containing the experimental distribution $\Phi(Q)$ of $\pi^+\pi^-$ accidental pairs. The extrapolation of the approximation function to the region $Q < 2\ MeV/c$ yields the number of free pairs in this region. Hence the value n_A of atomic pairs is

$$n_A = \int\limits_{Q<2} \left(\frac{dN^{exp}}{dQ} - \frac{dN^{free}}{dQ}\right) dQ. \tag{5}$$

From the measured break-up probability $P_{br} = n_A/N_A$, where N_A is the calculated total number (according to (3)) of produced pionium atoms, and from the dependence of P_{br} on the lifetime τ, one can derive a pionium ground state lifetime and hence a value for $\Delta = \left|a_0^0 - a_0^2\right|$.

4 Experimental setup

The experimental setup[12] (Fig.4) has been designed to detect pion pairs and to select atomic pairs at low relative momentum with a resolution better than 1 MeV/c. It was installed and commissioned in 1998 at the ZT8 beam area of the PS East Hall at CERN. After a calibration run in 1998, DIRAC has been collecting data since summer 1999.

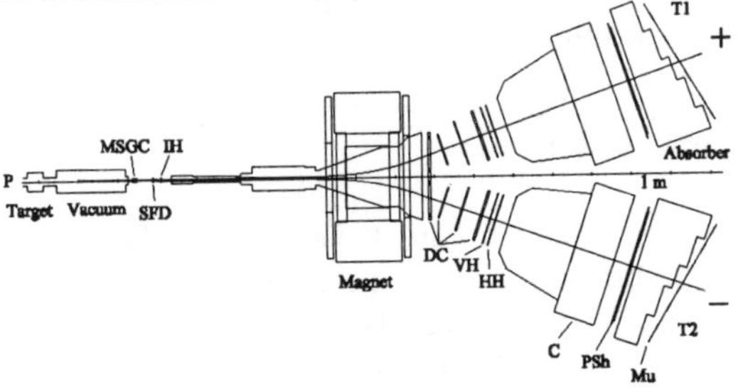

Figure 4. Experimental setup

The 24 GeV/c proton beam extracted from PS is focused on the target. The secondary particle channel, with an aperture of 1.2 msr, has the reaction plane tilted upwards at 5.7° relative to the horizontal plane. It consists of the following components: 4 planes of Micro Strip Gas Chambers (MSGC) with 4×512 channels; 2 planes Scintillation Fiber Detector (SciFi) with 2×240 channels; 2 planes Ionization Hodoscope (IH) with 2×16 channels; 1 Spectrometer Magnet of 2.3 Tm bending power. Downstream to the magnet the setup splits into two arms placed at $\pm 19°$, relative to the central axis. Each arm is equipped with a set of identical detectors: 4 Drift Chambers (DC), the first one common to both arms and with 6 planes and 800 channels, the other DC's have altogether per arm 8 planes and 608 channels; 1 Vertical scintillation Hodoscope (VH) plane with 18 channels; 1 Horizontal scintillation Hodoscope (HH) plane with 16 channels; 1 Cherenkov detectors (Ch) with 10 channels; 1 Preshower scintillation detector (PSh) plane with 8 channels; 1 Muon counter (Mu) plane with (28+8) channels.

For suppressing the large background rate a multilevel trigger was designed to select atomic pion pairs. The trigger levels are defined as follows: $T_0 = (VH \cdot PSh)_1 \cdot (VH \cdot PSh)_2 \cdot IH$, fast zero level trigger; $T_1 = (VH \cdot HH \cdot \overline{Ch} \cdot PSh)_1 \cdot (VH \cdot HH \cdot \overline{Ch} \cdot PSh)_2$, first level trigger from the downstream detectors; $T_2 = T_0 \cdot (IH \cdot SciFi)$, second level trigger from the upstream detectors, which selects particle pairs with small relative distance; T_3 is a logical trigger which applies a cut to the relative momentum of particle pairs. It handles the patterns of VH and IH detectors. T_3 did not so far trigger the DAQ system, but its decisions were recorded.

An incoming flux of $\sim 10^{11}$ protons/s would produce a rate of secondaries of about 3×10^6/s in the upstream detectors and 1.5×10^6/s in the downstream detectors. At the trigger level this rate is reduced to about 2×10^3/s, with an average event size of about 0.75 Kbytes.

With the 95 μm thin Ni target, the expected average pionium yield within the setup acceptance is $\sim 0.7\times10^{-3}$/s, equivalent to a total number of $\sim 10^{13}$ protons on target to produce one pionium atom.

5 First experimental results

The data taking has been done mainly with $\pi^+\pi^-$ and $p\pi^-$ pairs and also e^+e^- pairs for detector calibration. For the first data analysis only the most simple events were selected and processed, those with a single track in each arm, with signals in DC, VH and HH. The tracks in the DC's were extrapolated to the target plane crossing point of the proton beam. A cut was applied along X and Y distances between the extrapolated track and the hit fiber of the SciFi planes (18mm divided by the particle momentum in GeV/c, to take into account the multiple scattering effect). Finally, these events were interpreted as $\pi^+\pi^-$ or $p\pi^-$ pairs produced in the target. The difference in the time-of-flight Δt between the positive particle (left arm) and negative particle (right arm) of the pair at the level of VH is presented in Fig.5.

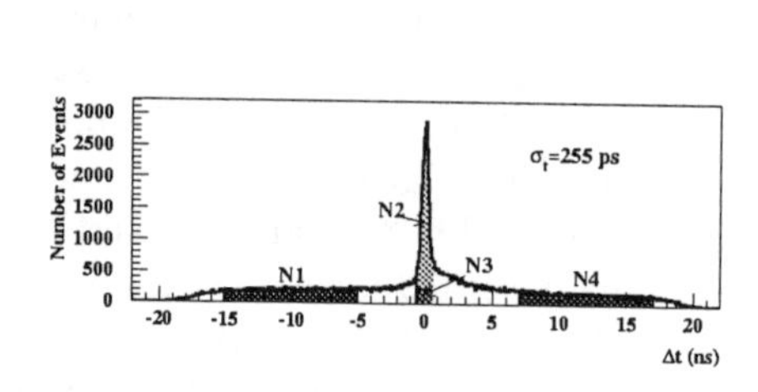

Figure 5. VH time-of-flight difference distribution for pair events

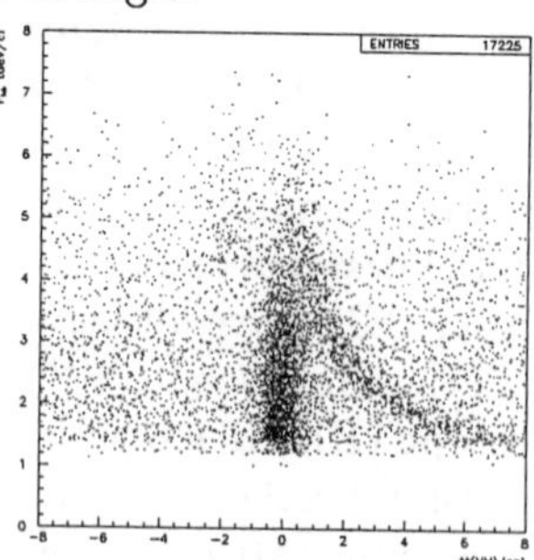

Figure 6. Positive particle momentum versus VH time-of-flight difference for particle pairs

The first interval $-20 < \Delta t < -0.5$ ns corresponds to accidental hadron pairs (mainly $\pi^+\pi^-$). In the second interval $-0.5 < \Delta t < 0.5$ ns one observes the peak of coincidence hits associated to correlated hadron pairs over the background of accidental pairs. The width of the correlated pair peak yields the time resolution of the VH ($\sigma_t \approx 250\ ps$). The asymmetry on the right side of the peak is due to admixture of protons in the π^+ sample, that are $p\pi^-$ events. Hence the third interval $0.5 < \Delta t < 20\ ns$ contains both accidental pairs and $p\pi^-$ events.

This time-of-flight discrimination between $\pi^+\pi^-$ and $p\pi^-$ events is effective for momenta of positive particles below 4.5 GeV/c. This is demonstrated in Fig.6, where the scatter plot of positive particle momentum versus difference in time-of-flight Δt in VH is shown. The single particle momentum interval accepted by spectrometer is $1.3 \div 7.0$ GeV/c.

For correlated $\pi^+\pi^-$ pairs Coulomb interaction in the final state has to be considerated, because it increases noticeably the yield of $\pi^+\pi^-$ pairs with low relative momentum in CM ($Q < 5\ MeV/c$). For accidental pairs this enhancement is absent.

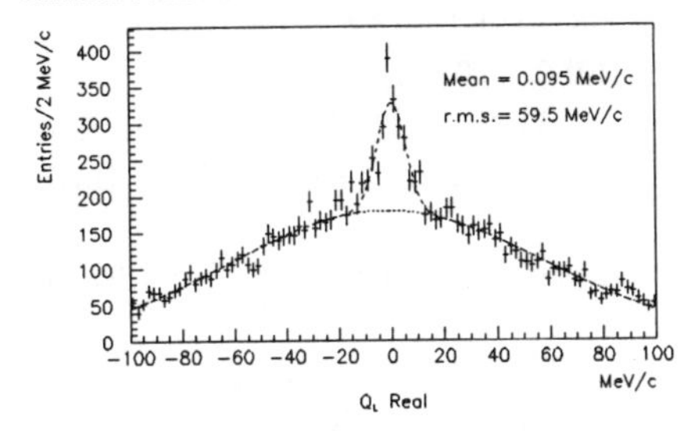

Figure 7. Correlated $\pi^+\pi^-$ pairs with positive particle momenta $p_{lab} <$ 4.5 GeV/c and $Q_T <$ 4 MeV/c.

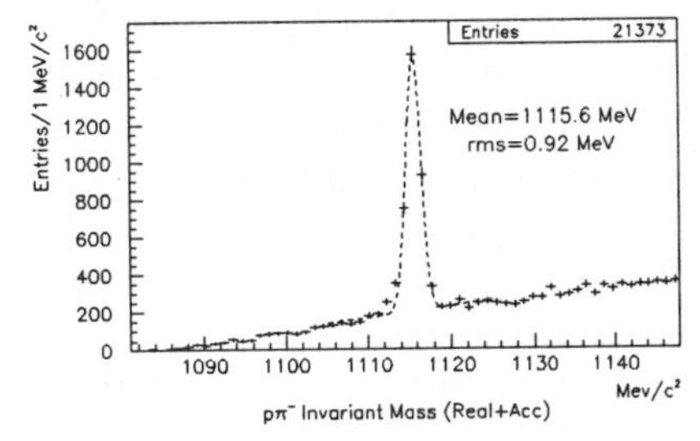

Figure 8. $p\pi^-$ invariant mass for proton momenta $p_{lab} >$ 3 GeV/c.

Fig.7 shows the distribution of the longitudinal component Q_L (the projection of Q along the total momentum of the pair) for correlated pairs. There are plotted pair events with positive particle momentum $p_{lab} < 4.5\ GeV/c$, occuring within the "correlated" Δt peak and with transversal component $Q_T < 4\ MeV/c$ to increase the fraction of low relative momentum pairs. In the region $|Q_L| \leq 10\ MeV/c$ there is a noticeable enhancement of correlated $\pi^+\pi^-$ pairs due to Coulomb attraction in the final state. The most important parameter for data analysis is the resolution in Q_L and Q_T. This has been measured by the reconstruction of the invariant mass of $p\pi^-$ pairs. The distribution of $p\pi^-$ invariant mass is presented in Fig.8. Positive particles are restricted to momenta larger than 3 GeV/c, and the time-of-flight must lie in $0.5 < \Delta t < 18\ ns$. A clear peak at the Λ mass $m_\Lambda = 1115.6\ MeV/c^2$ with a standard deviation $\sigma = 0.92\ MeV/c^2$ can be seen. These mass parameter values show a good detector calibration and coordinate detector alignment, with an accuracy in momentum reconstruction better than 0.5 % in the kinematic range of Λ decay products. This gives for the relative momentum resolution $\sigma_Q \sim 2.7\ MeV/c$. For $\pi^+\pi^-$ pairs a better resolution can be obtained, due to the different kinematics.

6 Conclusion

The DIRAC setup test and calibration have been done, and the DIRAC experiment began data taking. To achieve the goal - measurement of the pionium lifetime with 10% precision - we have to consider a number of at least 20000 recorded "atomic pairs". Improvements in hardware and software will continue this year. These will result in a better data quality.

References

1. J.Gasser and H.Leutwyler, Ann.Phys.158, (1984), 142.
2. J.Gasser and H.Leutwyler, Nucl.Phys.B250, (1985), 465.
3. J.Gasser and H.Leutwyler, Phys.Lett. B125, (1983), 325; J.Bijnens, G.Colangelo, G.Ecker. J.Gasser and M.E.Sainio, Phys.Lett. B374, (1996), 210.
4. J.Bijnens, G.Colangelo, G.Ecker, J.Gasser and M.Sainio, Nucl.Phys., B508, (1997), 263.
5. J.Stern, H.Sazdjian and N.H.Fuchs, Phys.Rev.D47, (1993), 3814.
6. M.Knecht, B.Moussallam, J.Stern and N.H.Fuchs, Nucl.Phys. B457, (1995), 513.
7. L.Rosselet et al., Phys.Rev. D15, (1977), 574; M.M.Nagels et al., Nucl.Phys. B147, (1979), 189.
8. B.Adeva et al., "Lifetime Measurement of $\pi^+\pi^-$ Atoms to Test Low Energy QCD Predictions", Proposal to the SPSLC, CERN/SPSLC 95-1, SPSLC/P 284, (1994).
9. L.L.Nemenov, Sov.J.Nucl.Phys. 41, (1985), 629.
10. A.Gall, J.Gasser, V.E.Lyubovitskij and A.Rusetsky, Phys.Lett. B462, (1999), 335.; J.Gasser, V.E.Lyubovitskij and A.Rusetsky, e-print hep-ph/9910438, (1999).
11. S.Deser, M.L.Goldberger, K.Baumann and W.Thiring, Phys.Rev. 96, (1954), 774.; J.L.Uretsky and T.R.Palfrey, Jr., Phys.Rev. 121, (1961), 1798.; S.M.Bilenky, Nguyen Van Hieu, L.L.Nemenov and F.G.Tkebuchava, Sov.J.Nucl.Phys. 10, (1969), 469.
12. A.Lanaro, e-print hep-ex/9912029, (1999).
13. H.Sazdjian, e-print hep-ph/9911520, (1999).
14. H.Jallouli and H.Sazdjian, Phys.Rev. D58, (1998), 014011.

RODOS - THE DOMESTIC COUNTERPART

DAN V. VAMANU

Horia Hulubei National Institute for Nuclear Physics and Engineering, IFIN-HH Bucharest

E-mail: vamanu@fx.ro

One term of reference for validly teaming up with Project RODOS (*Real-Time On-Line Decision Support System for Off-Site Nuclear Emergencies in Europe*) is to contribute research initiatives geared towards emulating the RODOS functions on novel conceptual pathways; expand system's coverage of issues that are relevant in the nuclear crisis management; and bring in a domestic perspective on how the system may best respond to local needs and constraints in each and every national - *i.e.* legal, managerial, logistic, and cultural, environment. On this line, one reviews the approach taken in the IFIN-HH RODOS Group to address the challenge, in the post-Chernobyl trail of trends and events.

1 Introduction

In the month of May, 1986, Romania has started to assimilate the Nuclear Safety culture, the hard way. True enough, the public health impact of the Chernobyl accident was minimal, with no deterministic effects ever detected and stochastic effects still debatable. The economic effects, mainly temporary food export bans and short-term withdrawal of food and feedstock were quickly absorbed by the centrally-planned system of the time. System routines also favored a play-down of the public anxiety and concerns. Of consequence in the context was, however, the institutional response to the Ukrainian crisis, that may at best be characterized as improvised. A few lessons were learned. Among these:

a) The recognition of the fact that 40 years of sound domestic intellectual and practical experience in Nuclear Physics and Engineering is, while a necessary, not a sufficient condition to get oneself equipped for responding to a real-life radiological emergency. The body of knowledge, it was discovered, has to be streamlined into operation-oriented subsets, and packaged into experimentally-verified, internationally-accepted, practically-proved as applicable, *rules* that can easily be communicated as make-sense recommendations for action to the laymen environment, including the decision makers, the operatives, and the public at large.

b) With the first nuclear power unit going critical on its territory (April 1996), Romania joins the club of nations that share in a special responsibility

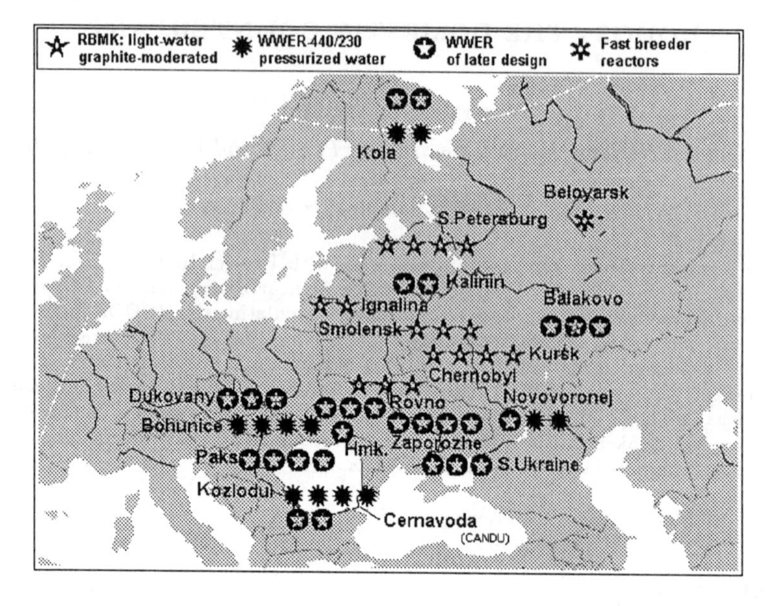

Figure 1. Nuclear power in Eastern Europe

for the environmental integrity and human health across borders. While the assets of the CANDU technology adopted for the Romanian power reactors, mainly relating to the amplitude of the defense-in-depth barrier, the relative independence of the process- and safety systems and the redundancy of the latter, would bring probabilistic risk estimates down to reassuring levels (an expected frequency of a severe core damage of 5.0E-6/ year, and an expected frequency of a power excursion initiator plus failure to shut down of 3.0E-8/year), the Chernobyl event has been revealing as to the bearing that a 'low-probability-high-consequence accident' may have, as well as, and particularly, to the weakest link in the operation of the nuclear power: the human factor – a primary concern for the young domestic industry of the kind. And,

c) The fact that, whether or not going nuclear, Romania finds itself well into the risk-riddled perimeter of one the most sensitive and controversial nuclear power parks of the Northern Hemisphere (Fig.1), originating in the technology, and revolving around the territory of the former Soviet Union. Apart from the allegedly positive-void coefficient RBMKs (Chernobyl-like), operating in the area are a number of elder-generation WWER-440 power reactors, many of which within a meteorologically-relevant distance from the Romanian territory. A 15-month evaluation by experts in 21 countries and the

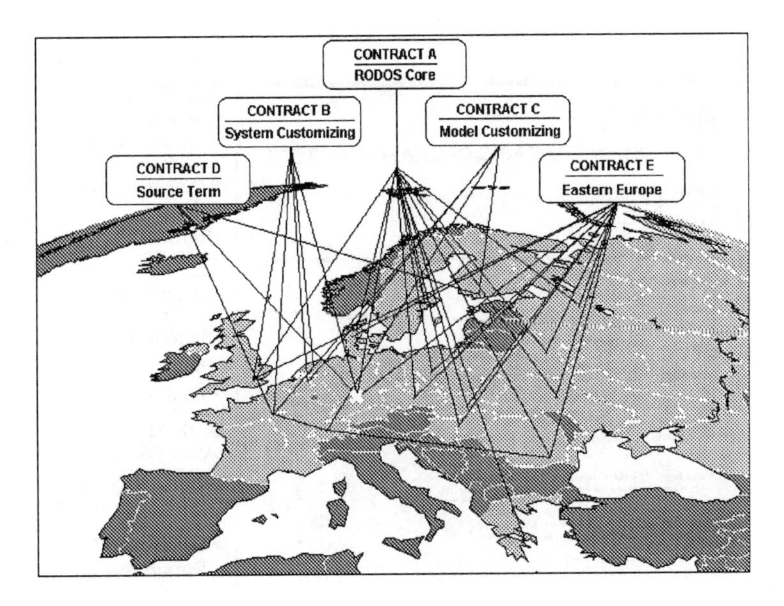

Figure 2. RODOS – coverage and themes.

IAEA, of 10 WWERs in Bulgaria, the former Czechoslovakia and Soviet Union has evidenced ca. 100 safety irregularities, of which 60 per cent of 'a high safety concern'. This is bound to remain a standing justification for securing a credible domestic capability of response to possible nuclear emergencies.

2 Overview

The nuclear emergency response system in Romania consists of an association of varied institutions and governmental departments including the National Commission for Nuclear Activity Control, hosting the requisite expertise and logistics, and a focal point represented by the National Command Unit for Nuclear Accidents and Fall of Extra-Atmospheric Objects, integrated with the Civil Defense and the military. This framework – the identity of which follows from the Romanian Law – is supposed to behave like a coherent structure (a) on drills, and (b) in the event of an actual crisis. On the other hand, at all times the self-awareness and business-as-usual part of the emergency preparedness effort rests largely with a supportive ring of several research institutes including the national Meteorology, the Environmental Engineering, the Public Health, the Food and Agriculture, the Physics (IFIN-HH) and, not

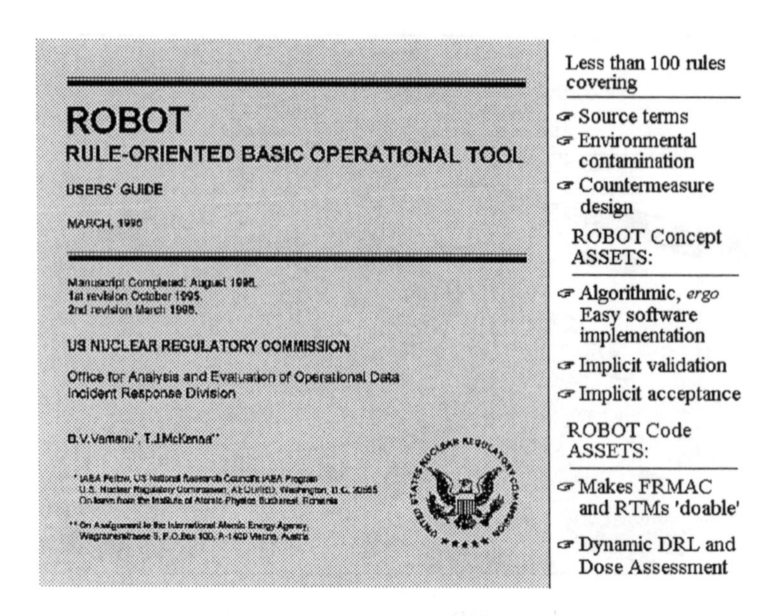

Figure 3. ROBOT – software support for the U.S.NRC/EPA/DOE emergency procedures.

the least, the research branch of the Nuclear Utility itself. Emphasizing the R & D supportive ring of this assorted collection is not a parochial vanity: in fact, and on a day-by-day basis, it is largely the research that ensures the interface with the international community, without which the domestic preparedness effort may lack relevance, acceptance, and resources.

Back in 1993 one link in the ring – IFIN-HH – was elected to become an associate RODOS contractor. RODOS is a post-Chernobyl international project initiated by the European Atomic Energy Commission on behalf of the European Communities, having as objective the development and implementation of an uniformly-agreed **R**eal-Time, **O**n-Line **D**ecision Support System for **O**ff-**S**ite Nuclear Emergencies in Europe. Comprehensive as far as topical profile and also observant of the geopolitical features of the targeted territory (Fig.2), RODOS endeavors to offer the national authorities a shared platform of knowledge and effective tools enabling an expeditious and reliable assessment of a severe nuclear accident, the collection and processing of verifiable data as well as the issuing of appropriate recommendations for response action, together with action prioritizing and an evaluation of their effectiveness. In pursuing this goal, the RODOS Management Group has actively promoted an early awareness on the importance of a comprehensive commitment of the

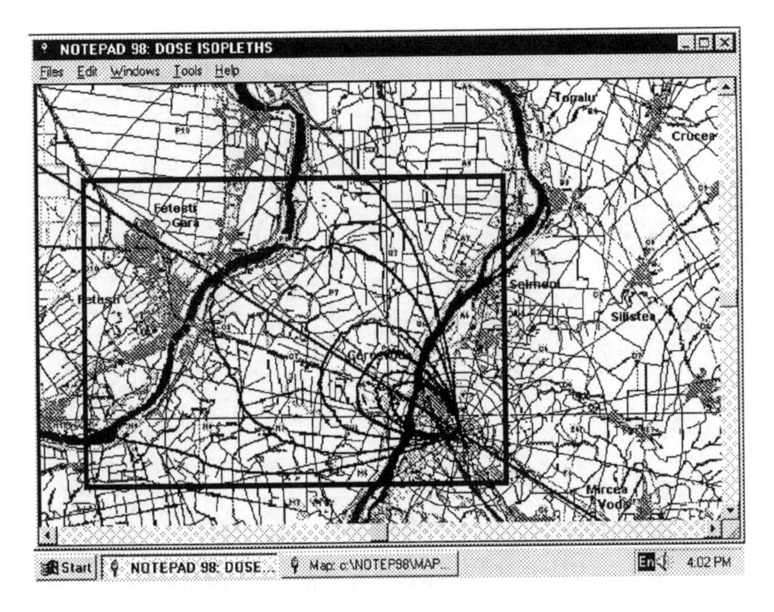

Figure 4. DSS outputs – isodose tagging of the plant near-field.

constituents in the very development of the system and in the customization of the tools it offers, to fit national variations as far as data availability, legislation, practices, and culture.

It is on this working assumption that IFIN-HH has deployed a proportionate effort to develop and maintain a domestic counterpart to its RODOS-related activities. It took the form of a hands-on self-training in developing a variety of versions of decision support systems (DSS) drawing upon the current background and experience, and looking at what was progressively thought to be the most relevant pieces of knowledge, methodology, algorithms, and code design styles available in the open literature. One purpose was to offer the pertinent authorities, at each phase into the project, interim yet convincing demonstrations on the value and feasibility of having available practical tools to guide, assist, and satisfactorily document response decisions in the event of a drill-, or no-drill emergency.

In retrospect, three work-phases can be identified:

(i) **A phase-one, pre-RODOS**, period (1990-1994), involving topical field orientation, feasibility demonstrations and expertise acquisition in environmental radiology modeling and DSS code development, the resulting products – codes in the 'APUD' and 'ACA-IFA' family, including custom versions

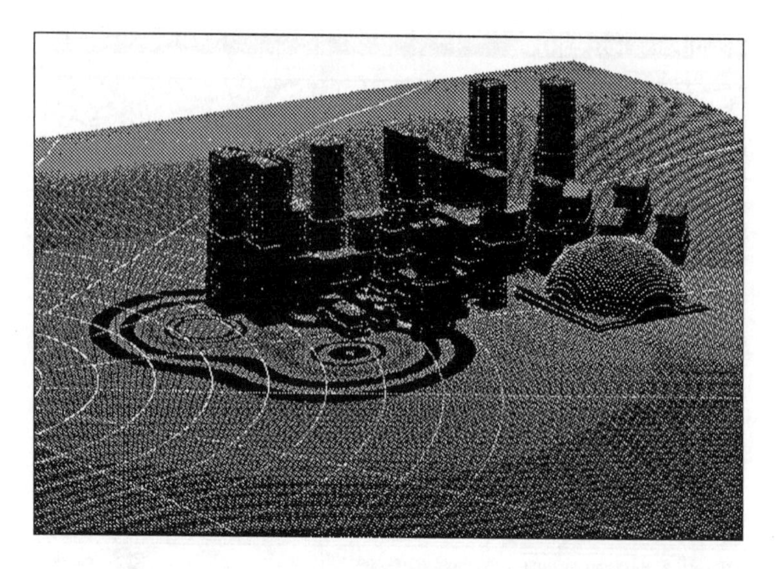

Figure 5. Contamination footprint in a complex terrain.

developed by appointment of the IAEA Safety Section and AECL-Research, Ontario, Canada – being favorably reviewed by the NEA/OECD Data Bank.

(ii) **A phase-two, upgrading and consolidation** period (1994-1996), assisted by the IAEA fellowship system and the U.S.A. Government, through the Nuclear Regulatory Commission in Washington, D.C., having on record an effective participation in the development of the US NRC, EPA, and DOE's *FRMAC Assessment Manual – The Federal Manual for Assessing Environmental Data during a Radiological Emergency*, and the *RTM-95 International Technical Response Manual*; and the research initiative to develop a convenient software support for manuals implementation (Fig.3), the code package ROBOT(*Rule-Oriented Basic Operational Tool*) winning the appreciation of several factors that are directly instrumental in carrying out the US Federal policy of emergency response.[1]

(iii) **A phase-three period** (1996-), evolving within the RODOS framework and featuring, *inter alia*:

(a) The application of the ROBOT concept and methodology in the development of a domestic, standalone software for decision support in field operations, code-named NOTEPAD, designed as an open-ended modular structure running from small IBM compatibles including laptops, to be effectively used in nuclear accident drills, expert services ordered by authorities on issues such as siting, health and environmental impact assessment, and evaluation of operational radiological incidents, and mainly targeting the near-field radiological situation at the site of an abnormal event (Fig.4); and likewise – an experiment in the summary assessment of the far-field consequences of an atmospheric radioactive release (the codes 'QuickFIX').

(b) A research initiative dealing with the atmospheric dispersion in terrains with a complex topography (Fig.5), using adapted, scaled and linearized equations of the mesoscale meteorology and heuristic concepts, developed and successfully benchmarked in Switzerland as a customized application. [2 7]

(c) The development, by appointment of the Swiss Federal Institute of Technology ETH-Zurich and under the joint guidance of *Institut de protection et de surete nucleaire* CEA-IPSN and *Kernforschungszentrum Karlsruhe* - the RODOS Coordinator, of the technical specifications, the demonstration, and the working-prototype, versions of the 'CONTAINMENT' module of the software package STEPS (*Source Term Estimation based on Plant Status*), meant to offer a reference in the European source term analysis and to interface with RODOS (Contract RODOS D, to be continued under the 5th EU Framework Programme).[3]

3 Conclusion

The approach described, combining RODOS-relevant domestic initiatives with scientific service provided by appointment to extra-national centers of excellence has proven a valid and profitable endeavor. It favored a better understanding of the RODOS philosophy and substance; has resulted in several working concepts and end-products that were subject to independent peer-reviews and testing by the appointing parties;[4] served well, on an interim basis until the final RODOS version is released, the local DSS needs in both domestic and international nuclear alert exercises (*viz. e.g.* the INEX Series jointly supported by the Nuclear Energy Agency – NEA of the OECD and the International Atomic Energy Agency in Vienna); and encouraged a decent margin of local creativity within an otherwise firmly planned and timed international project. Upon these, it is envisaged that the work continue, mainly on the following lines: (i) *Model Acquisition and Development*: new reference models will be adapted and assimilated into the local software platforms, with Surface- and Ground Water Contamination as items on the forthcoming agenda. (ii) *Improved Operability and Maintenance*: database and GIS platforms will be upgraded to better cover the national and regional scales; the user interface will be amended, streamlined and brought in style with the trends prevailing in the trade; Internet capabilities are contemplated.

(iii) *Consolidation*: independent testing, benchmarking, and participation in drills will continue, together with the resulting feedback integration; selective assistance will be provided to the local RODOS test bench as appropriate, based on the positive experience gained *e.g.* in designing trial source terms for CANDU reactors. (iv) *Customer-Oriented Applications and Spin-Offs*: based on the experience gained with the evaluation of minor, no-drill radiological incidents on record, software modules may be derived to serve specific needs; the continued communication and co-operation with third parties will further track topical and interdisciplinary matters of election, and primarily the Chemical Accident Consequence Assessment; the Risks in the Transportation of Dangerous Goods;[5] and issues in the Multicriterial Risk Analysis.[6] In particular, the agenda of the Disaster Risk Management Institutes, contemplated by the Swiss Federal Institutes of Technology (SFIT) and the World Bank will be monitored and targeted.[7]

References

1. D. Vamanu, T.J. McKenna, *ROBOT: Rule-Oriented Basic Operational Tool*, U.S. Nuclear Regulatory Commission, Washington, D.C., March 1996.
2. D. Vamanu, A. Gheorghe, Int. J. Environment and Pollution, Special Issue, 6 4-6 462-490 1996. Also: Int. J. Environment and Pollution 9 4 352-370 1998.
3. D. Vamanu, *STEPS - Source Term Estimation based on Plant Status: CONTAINMENT Module Technical Specifications*, EAEC Project F14P-CT96-0048. ETH-Zurich & CEA-IPSN, Fontenay-aux-Roses, May 1997.
4. A. Gheorghe, D. Vamanu, *Emergency Planning Knowledge*, VDF Verlag der Fachvereine Zurich, 1996, ISBN 3 7281 2201 7.
5. A. Gheorghe, D. Vamanu *et al*, *Decision Support System for Transportation of Dangerous Goods. Optimal Routing and Emergency Preparedness Modeling*, Proc. Int. Conf. Probabilistic Safety Assessment & Management, PSAM 4, New York, 13-18 September 1998.
6. A. Gheorghe, D. Kogelschatz, K. Fenner, M. Engel, A. Harder, W. Kroeger, D. Vamanu, *Integrated Risk Assessment in the Transportation of Dangerous Goods. Introducing the 'Hot Spots' Concept as a Solution*, Proc. European Safety and Reliability Conference ESREL'99, Munich-Garching, 13-17 September 1999.
7. * * *, *DRM – World Institutes for Disaster Risk Management*, SFIT & the World Bank, published by ETH Center Zurich, 1999.

PROFESSOR HORIA HULUBEI, THE FATHER FOUNDER OF THE INSTITUTE OF ATOMIC PHYSICS

G. STRATAN

Department of Theoretical Physics, Institute of Atomic Physics, 76900-Bucharest, Romania, E-mail: stratan@theor1.theory.nipne.ro

A hero of WW I, Horia Hulubei (b. November 15, 1896, d. November 22, 1972), was one of the most prominent Romanian scientists of all time, leader and teacher of several generations of Romanian scientists during more than four decades. Graduated from Jassy University, he took his Ph.D. in Paris with Marie Curie and Jean Perrin in 1933. A few years later, Horia Hulubei was nominated Directeur de Recherches at the French National Centre of Scientific Research and elected Corresponding Member of Paris Academy of Sciences. Back in Romania, Hulubei was nominated professor and Rector of Bucharest University (1941). Professor Hulubei had a broad field of interests, from Classical to Atomic and Nuclear Physics, but his main achievements are connected with the Physics of X-rays (the first spectra of noble gases, the multiple Compton effect, the search for elements 87 and 85, etc.). The Institute of Atomic Physics (IPA) in Bucharest (1949), was the third research institution founded and directed by him. Following Hulubei's initial design, IPA was, and, in spite of the past and actual difficulties, remains the flagship of Romanian scientific research. Along the years, IPA influenced beneficially the development of the post-war Romania and established many collaborations abroad.

PACS numbers: 01.65.+j, 01.75.+m, 33.20.Rm, 25.10.+s

1 The young years

Born in Jassy, in North-Eastern Romania, on November 15, 1896 to an intellectual family, Horia Hulubei was a brilliant student of the Scientific branch of the college, graduating in 1915 as the first of his class. In the same year, he attended the Jassy University, as a student in Physics and Chemistry of the Faculty of Sciences.

His studies were interrupted in 1916, when Romania entered the First World War. The young Hulubei enlisted and, after the participation to the first battles in Southern Moldavia, was selected by General Berthelot, the chief of French military mission in Romania, to join the Air Force. Hulubei left for France, became a fighter pilot and, being severely wounded in action, was awarded with the well known "Legion d'Honneur". It is interesting to mention that Hulubei didn't like to speak about his feats of arms, in spite of being very open to his collaborators and friends in other matters.

The experience accumulated as a pilot helped Horia Hulubei in the post-

war period, when financial shortcomings prevented him from resuming the studies. Between 1920 and 1921, he worked as a head of the new established Aircraft Division of the Romanian Ministry of Communications, contributing to the development of the national aviation and to the extension of the international flights connecting Romania with other countries. It was a heroic period of the beginnings of civil aviation and the fact that Hulubei knew personally the famous French pilots of war time, being recognized by them as one of the family, played a benefic role in the success of his mission.

Hulubei returned to the Jassy University in 1922. He graduated in 1926 with "Magna cum Laude" and became Assistant Professor at the Department of Physical Chemistry, led by Professor Petru Bogdan, whom Hulubei recognized as his first master. From Jassy period dates Hulubei's first scientific paper dealing with the internal pressure of liquids.

The next year, Horia Hulubei went in Paris, at the Physical Chemistry Laboratory of Sorbonne, to work under Jean Perrin's supervision[a].

2 Studying in Paris (1927-1933)

Starting with the mid-19th century, the Capital of France played a special role in the formation of Romanian intellectuals[b]. For example, contemporary with Hulubei in Paris were active famous Romanians like the physicist Alexandru Proca[c], the engineer and inventor Henri Coanda[d], the sculptor Constantin Brancusi[e], the composer, violinist and conductor George Enesco[f]. Hulubei's relations with them were of a reciprocal esteem and friendship.

We must imagin Horia Hulubei looking quite different from the usual Ph. D. students, both from France and from abroad. France and Paris were not unknown for him, and his French was excellent. In 1927, he was 31, and his experience during the war and the post-war periods gave him the confidence

[a]Jean-Baptiste Perrin (1870-1942), received the Nobel Prize in 1926 for his work on Brownian motion.

[b]This is true mainly for two historical provinces of Romania – Moldavia (where Hulubei came from) and Walachia. The third one, Transylvania, was under the influence of German Culture. However, this division was nor strict, nor exclusive.

[c]Alexandru Proca (1897-1955) predicted (independently from Yukawa) the existence of mesons. Proca's ecuation plays a very important role in actual field theories.

[d]Henry Coanda (1885-1972), was the founder of modern fluidic science, the inventor and builder of the first jet aeroplane and the discoverer of the fluid jet deviation due to curved surfaces (the so-called Coanda Effect).

[e]Constantin Brancusi (1876-1955), Rodin's pupil, Modigliani's master and the founder of modern abstract style.

[f]George Enesco (1881-1955), author of Romanian Rapsodies, Oedip (opera), Yehudi Menuhin and David Oistrach's master.

of a self made man. Professor Jean Perrin was conquered by the newcomer and Hulubei received the keys of the laboratory where he was supposed to start the research. As Hulubei liked to recount, when he unlocked the door to have a first glimpse on the X-rays laboratory, he couldn't believe his eyes: the room was large, but empty... Fortunately, the disappointment didn't last. Hulubei received money to plan and organize the future laboratory together with Mademoiselle Yvette Cauchois and other French collaborators. In a short time, the X-rays laboratory of Sorbonne became one of the best, allowing advanced experiments. Before the completion of X-ray laboratory – and, after that, in parallel – Hulubei performed experiments in the field of combined diffusion of visible and ultraviolet light. From this early period of activity, his outstanding qualities revealed a scientist of great expectations. Improving the experimental settings and developing new methods[1], Hulubei was able to study the Raman spectra[2] for different substances under the variation of external factors[3], establishing also a standard spectroscopic methodology recognized and quoted in the literature.

The main field of Hulubei's research in Paris was certainly the study of atomic structure using the X-ray spectra. The complexity of X-lines, the multitude of faint lines, the absence of a systematic classification of spectra, the deviations from Moseley's law, etc. made from the X-ray spectroscopy a challenge field for many laboratories. To understand the systematics of lines emitted by different elements, a very important but difficult task was to obtain the X-ray spectra of noble gases. Together with Yvette Cauchois, Horia Hulubei invented a new technique of excitation using a fascicle of electrons introduced in a chamber containing the studied gas. The first X-rays spectra of matter in gaseous state (namely for Krypton[4] and Xenon[4], including the attribution of faint lines[6]) obtained by Hulubei in collaboration with Mademoiselle Yvette Cauchois, were met with great interest in laboratories around the world.

Hulubei was interested to clarify the origin of faint lines present in different spectra of diffusion of X-rays and one of his hypothesis was that some of them are a result of multiple quantum diffusion. Until then, only the single diffusion (the Compton effect) was known. Hulubei calculated and put into evidence the multiple Compton effect[7]. The multiple Compton effect, (see also[8]), was Hulubei's favourite intellectual child[g]. Hulubei's Ph. D. thesis (ref. [8]) sustained in July, 1933, included this subject, together with the

[g]Hulubei was married to Alice, but the Hulubei couple didn't have children. Later, when Hulubei was professor at Bucharest University, the students used a trick to pass the Atomic Physics examination: they deviated the discussion to the multiple Compton effect. The effect was really multiple: it saved many students ...

study of Raman spectra modifications under the change of different physical conditions. The Ph. D. Commission, presided by Marie Curie, included Professor Jean Perrin, and Professor C. Mauguin[h] Hulubei received his Ph. D. with the mention "très honorable". Very honourable was also the proposal from Professor Perrin to remain in his laboratory to continue the study of X-ray spectra, gladly accepted by Hulubei.

3 A brilliant career in Paris and in Romania (1933-1940)

After the Ph. D., Hulubei divided evenly his time between France and Romania. In 1932, he was appointed associate professor at the Faculty of Sciences of Jassy University, where he started to teach a course on radioactivity and structure of matter. Hulubei's nomination as a full professor in 1938 was an aknowledgement of his outstanding academic career in France.

Indeed, Hulubei received the professional degree of "Maître de Recherches" and, soon after that, of "Directeur de Recherches" at the French National Centre of Scientific Research, being the only foreigner detaining such responsabilities in that time. He succesfully continued his research in the field of X-rays, this time devoted mainly to the search for the "missing" elements from the Mendeleev Table.

In the mid-thirties, the completion of the inner part of the Periodical Table was almost finished. Only three places remained empty, having the atomic numbers Z=87, 85 and 43 and a rush started for the discovery of the corresponding elements. The stake was high: the discoverer had the right to name the new element and Hulubei's dream was to find one by means of X-ray identification. He was well placed to try such an exploit due to the improvements done by him to the methods of spectroscopical measurements, which enabled the determination of an element with a relative abundance of 10^{-7}. The name was also ready: the 87th element should be called Moldavium, after the Hulubei's native place.

The first paper, published in 1936, didn't present conclusive results[9], so, Hulubei tried again the next year, with a refined technique[10] and looking for other lines in the X-ray spectra of different minerals containing alkalines. His main supposition was that the element 87 was stable or long lived. In 1937, Perrier and Segré discovered ^{43}Tc, an element obtained artificially by bombarding Molybden with deuterons. After this feat, the search for the two remained "inner elements" was quickened, as well as the search for the

[h]Professor of Mineralogy at Faculty of Sciences. Many of Hulubei's studies involved the use of cristals.

Figure 1. Horia Hulubei in 1939

transuranian ones. Hulubei developed both directions of research, reporting about some X-ray lines attributable to the 85th element[11], or to the 93th one[12]

Unfortunately for Hulubei, his main hypothesis about a long lived Moldavium proved to be wrong. In 1937, M-lle Marguerite Perey, from the Curie Institute, discovered by rapid chemical identification the element 87 known as Francium (with a lifetime of 21 minutes). M-lle Perey found Francium in minerals containing actinides. It was a beginning of the radiochemistry and a marvellous achievement, but, in the perspective of future developments of the search for the transuranian elements, the X-ray analysis will prove to be better placed for the identification of the very short lived elements.

Horia Hulubei published between 1932 and 1947 a number of 43 articles in the field of X-ray physics concerning spectra of more than 25 elements. Many of his results were included in classical tables like Landolt-Börnstein. The French Academy gave him two awards: "Henry de Jouvenel Prize" (1937) and "Henry Wilde Prize" (1938) and elected Hulubei Correspondent Member (1940).

4 Building Science in Romania (1940-1972)

Once again the war interrupted Hulubei's activity: he left France, returned to Romania and was appointed to plan and lead a modern research institute in Cluj, (today, Cluj-Napoca) but the Vienna Dictate prevented him to accomplish this mission. He went in Bucharest as professor at the Department of Physical Chemistry and Structure of Matter, and had to start once again from scratch: in spite of its name, the department hadn't any laboratory of Structure of Matter. In 1941, Hulubei was nominated Rector of Bucharest University.

During the difficult years of war, Hulubei equipped the laboratory and edited a journal entitled "*Disquisitiones Mathematicae et Physicae*" where the French physicists published their papers, a feat impossible in their country occupied by the Germans.

In the post-war period, Hulubei was prosecuted by the Communist authorities under the unfair accusation of being nominated Rector by Marshal Antonescu, the Romanian military dictator. The charges against Hulubei were retired under the insistence of the French scientists occupying important positions in the first Government after W. W. II. So, in 1948, Hulubei was nominated director of the Institute of Physics, which, the next year, was splitted and one branch became the Institute of Atomic Physics (IAP). It was already an established fact that Hulubei had to start again and again from zero. Romania was under Soviet occupation, many people were in prison on political charges, or in USSR as POW, the country was distroyed by the war and after two years of terrible drought. Nevertheless, Hulubei persuaded the rulers of Romania to give money for the development of Science. In a few years, the campus in Magurele hosted the new buildings and laboratories and started to acquire the facilities. The activity begun vith less than 25 trained people, but, appointing young specialists of different professions and graduate students and training them in the field of Nuclear and Atomic Physics, the first research activities were put forward.

In 1957-58, IAP was equipped with a 2MW research reactor and a cyclotron bought from Soviet Union. Around these facilities, an intense activity started in the field of fundamental and applied research.

Meanwhile, Hulubei pursued his research in the X-ray spectroscopy and approached the new field of Nuclear Physics using the recent facilities of IPA. Together with his Romanian collaborators, he published between 1948 and 1972 more than 50 scientific papers devoted to a large cathegory of subjects, from Elementary Particle Physics to Nuclear Reactions and applications. An enormous effort was displayed by Hulubei in planning the IPA development, in guiding the young generations of scientists and in supervising the general research activity.

IPA contributed essentially to the development of Romanian scientific research and, through the applied science, to various fields of economy, from Medicine, to Agriculture. The adoption of the National Program of Nuclear Energetics is due also to Hulubei's vision about the future of our country[i]. In Hulubei's conception, Institute of Atomic Physics had as a mission to develop a large front of both fundamental and applied research, ready to answer to the problems of Romanian society and in a permanent contact with the international science.

Horia Hulubei remained director of the Institute of Atomic Physics until 1968. It was the time of retirement, but Hulubei was nominated Presidential Adviser for Science[j]. It was merely a honorary position, used by Professor Hulubei to foster the position of the scientific research in Romania and to plan the extension of campus in Magurele, which was finished in 1974 in the actual form.

Along his career, Hulubei was elected member of many learned societies in France, Germany, USA, etc, as well as Vice-President of Board of Governors of the IAEA in Vienna and member of the JINR-Dubna Scientific Council.

In 1956, Hulubei was elected Member of Romanian Academy and in 1963,

[i]Using the argument of safety, Hulubei convinced Nicolae Ceausescu to adopt the Western type of reactors. Hulubei's decisive argument was the independence of Romania from the Soviet Union in the field of energy generation.

[j]In the same period, the Presidential Adviser for Technology was Hulubei's friend Henri Coanda.

President of Physical Section of Romanian Academy.

Hulubei was the founder of the Chair of Structure of Matter at the Faculty of Physics. His courses of Atomic Physics were attended by many students from different faculties of Bucharest University attracted by the Hulubei's fame. In fact, they weren't usual lessons, but a live presentation of his personal experience and a vivid description of great personalities of science met by him. Hulubei took care of each student who manifested interest for science and many physicists of mature generation owe him their career.

The people who knew Hulubei, remember him as a charming person, with a profound sense of humour, frank and ready to listen to other's opinion.

Horia Hulubei died on November 22, 1972, in Bucharest.

The celebration of the Centenary of his birth in 1996 and the comemoration of 25 years from his death in 1997 have shown that the Romanian scientific community has a vivid rememberance of Horia Hulubei, as a great scientist and organizer of Science, ardent patriot and outstanding model for younger generations.

Acknowledgments This contribution to the Symposium is largely based on the book **Horia Hulubei** *Selected papers* edited by the Central Institute of Physics, Bucharest, 1986, on the occasion of the celebration of ninety years from Hulubei's birth. In the mid-eighties, the appearison of such a book was very difficult: the homages were reserved for the Presidential couple only. This fact explains the modest graphical conditions of *Selected papers.* Nevertheless, this book is an invaluable source for the people interested in Horia Hulubei's life and achievements.

We owe sincere aknowledgements to the following editors: I. Ursu, M. Ivascu, A. Berinde, C. Besliu, A. Corciovei (deceased), O. Gherman, Th. V. Ionescu (deceased), M. T. Magda, N. Martalogu (deceased), V. Mercea (deceased), Al. Mihul, M. Peculea, M. Petrascu, I. Purica (deceased), V. Tutovan and technical editors: I. A. Dorobantu, M. Dumitriu, M. T. Magda.

References

1. H. Hulubei et Y. Cauchois, *Dispositif simple et lumineux pour l'étude de l'effet Raman*, C. R. Ac. Sci. Paris, **192**, 935 (1931).
2. Horia Hulubei et Yvette Cauchois, *Excitation monochromatique des spectres Raman dans l'ultraviolet. Applications.* C. R. Ac. Sci. Paris, **192**, 1640, (1931).
3. Horia Hulubei, *Contribution a l'étude du spectre Raman de l'eau*, C. R. Ac. Sci. Paris, **194**, 1474 (1932).
4. Y. Cauchois, Horia Hulubei, *Emission X caractéristique d'éléments à*

l'état gaseux. Spectre K du krypton., C. R. Ac. Sci. Paris, **196**, 1590 (1933).

5. Horia Hulubei, Y. Cauchois *Emission X caractéristique d'éléments à l'état gaseux. Spectre K du xénon.*, C. R. Ac. Sci. Paris, **197**, 644 (1933).
6. Y. Cauchois, Horia Hulubei, *Emission X caractéristique d'éléments à l'état gaseux. Raies faibles dans le spectre K du krypton.*, C. R. Ac. Sci. Paris, **197**, 681 (1933).
7. Horia Hulubei, *Mise en evidence de la diffusion Compton multiple*, C. R. Ac. Sci. Paris, **195**, 1249 (1932).
8. Horia Hulubei, THÈSES, *1. Contribution a l'étude de la diffusion quantique des rayons X, 2. Modification des spectres Raman sous l'action de difféerents agents physiques*, juillet, 1933, Paris, Masson et Co.
9. Horia Hulubei, *Recherches relatives à l'élément 87* C. R. Ac. Sci. Paris, **202**, 1927 (1936).
10. Horia Hulubei, *Nouvelles recherches sur l'élément 87 (Ml)* C. R. Ac. Sci. Paris, **205**, 854 (1937).
11. H.Hulubei, Y. Cauchois, C. R. Ac. Sci. Paris, **209**, 39 (1939); C. R. Ac. Sci. Paris, **210**, 696 (1940).
12. Horia Hulubei, Y. Cauchois, C. R. Ac. Sci. Paris, **206**, 181 (1938) and **209**, 476 (1939).